Blogging
the Moon

We acknowledge the financial support of the Government of Canada through the Book Publishing Industry Development Program for our publishing activities.

Published by Apogee Prime, A division of Griffin Media

Blogging the Moon /Paul D. Spudis

First Edition

© 2010 Apogee Prime/The Author   ISBN 978-1926837-17-8

All photographs are courtesy of NASA unless otherwise noted.

Printed and bound in Canada

Blogging the Moon

The Collected "Once and Future Moon" blog of Air and Space Magazine

Paul D. Spudis

Contents – Blogging the Moon

Preface		**9**
India Aims for the Moon	Oct 21, 2008	11
The Moon, Space and Other Things	Nov 9, 2008	15
Hitting a Bull's-eye on the Moon	Nov 15, 2008	17
Another "Roadmap"	Nov 18, 2008	19
A Decade of the International Space Station	Dec 1, 2008	23
The Vision for Space Exploration (VSE) and Project Constellation	Dec 12, 2008	26
Forty Years Ago, Three Men Left for the Moon	Dec 20, 2008	28
Moon Water – Again	Dec 23, 2008	30
Space Goals – One More Time	Jan 9, 2009	33
Radar Mapping the Moon	Jan 17, 2009	35
What Apollo Was …. and Wasn't	Jan 25, 2009	37
The Strange Story of Lunar Magnetism	Feb 8, 2009	41
Another Strategic Plan Misfires	Feb 20, 2009	43
Human Spaceflight: What Value to Science? (Part 1)	Feb 28, 2009	46
Human Spaceflight: What Value to Science? (Part 2)	Mar 1, 2009	48
Of Science and Cathedral-Building	Mar 15, 2009	50
Mini-SAR Nears Completion of its First Mapping Cycle	Mar 29, 2009	52
moon vs. Moon: A Study in Arrant Pedantry	Apr 2, 2009	54

Those Were the Days…. — Apr 10, 2009 — 56

The Deadly Dust of the Moon — Apr 10, 2009 — 59

Return to the Moon: Outpost or Sorties? — May 5, 2009 — 61

What the Augustine Committee Didn't Know in 1990 — May 15, 2009 — 65

Can We Be "Resourceful" on the Moon? (Part 1) — May 30, 2009 — 68

Lunar Resources: Changing Our Approach to Spaceflight (Part 2) — Jun 5, 2009 — 71

First, Nail Down the Mission — Jun 25, 2009 — 73

Would More Money Improve NASA? — Jul 8, 2009 — 75

Space Program vs. Space Commerce — Jul 16, 2009 — 81

Can You Legally Own a Piece of the Moon? — Jul 24, 2009 — 83

Next Step or No Step — Aug 3, 2009 — 87

Two Views of the Vision — Aug 11, 2009 — 96

Scientists vs. The Icy Commander — Aug 21, 2009 — 99

I Aim at the Stars…But Sometimes I Only Make Viewgraphs — Sep 9, 2009 — 102

Water, Water Everywhere…. — Sep 25, 2009 — 107

Space Exploration Sets Sail on Lunar Water — Oct 4, 2009 — 110

LCROSS: Mission to HYPErspace — Oct 12, 2009 — 115

Paradigms Lost — Oct 23, 2009 — 119

Caves on the Moon? — Oct 27, 2009 — 124

A Rainbow on the Moon — Nov 14, 2009 — 126

Contents

Thanksgiving on the Moon: A Lunar Feast — Nov 22, 2009 130

Another Moon-Forming Collision? — Dec 7, 2009 132

Arguing About Human Space Exploration — Dec 16, 2009 134

Cataclysmic Events on the Moon — Dec 16, 2009 145

Robotic Sample Return and Interpreting Lunar History:
The Importance of Getting it Right — Jan 11, 2010 147

Beyond LEO – Flexible Path Revisited — Jan 23, 2010 150

Have We Forgotten What Exploration Means? — Jan 25, 2010 155

Vision Impaired — Feb 3, 2010 161

Confusing the Means and the Ends — Feb 13, 2010 168

A Lunar Visionary — Feb 23, 2010 174

Talismanic Thinking — Feb 27, 2010 176

Ice at the North Pole of the Moon — Mar 1, 2010 180

Stuck in Transit – Unchaining Ourselves From the Rocket Equation — Mar 11, 2010 183

Value for Cost: The Determinate Path — Mar 24, 2010 195

NASA Lost Its Way — Apr 2, 2010 202

To Do The Heavy Lifting — Apr 14, 2010 215

"We've been there before. Buzz has been there." — Apr 16, 2010 226

The Four Flavors of Lunar Water — May 2, 2010 245

Using the Earth to Study the Moon — May 15, 2010 253

It's the Space Economy, Stupid! ————————————— May 21, 2010   256

American Heroes ————————————— May 28, 2010   271

A Wetter Moon Impacts Understanding of Lunar Origin ————— Jun 19, 2010   283

Malice, Mischief, and Misconceptions ————————— Jun 26, 2010   286

Searching for the Moon's mantle ————————— Jul 7, 2010   298

NASA's New Mission and the Cult of Management ————— Jul 10, 2010   301

The Moon, asteroids, and space resources ————— Jul 23, 2010   314

BONUS CHAPTER

The New Space Race ————————————— Feb 2, 2010   323

Preface

In October of 2008, I got a phone call from Tony Reichhardt, an editor at the Smithsonian Institution's *Air and Space* Magazine. I was getting ready to travel once again to India, this time for the launch of the Chandrayaan-1 mission to the Moon. I was the Principal Investigator of the Mini-SAR imaging radar experiment on that mission. Its goal was to map the dark areas near the poles of the Moon with radio waves. We were looking for evidence of water ice that we suspected might occur in these extremely cold regions.

Tony asked if I would be interested in "live blogging" the launch of Chandrayaan-1 for their magazine's website. I have published my scientific findings over the years and written popular articles for the press and public science magazines, but I had never done any blogging and was a novice in understanding what it entailed. I was pleased to be asked to contribute to a publication that I had long admired. I sent e-mail dispatches to Tony, which he posted on the magazine's web site (collected here as Chapter 1). I enjoyed my first steps onto the blogosphere stage; it made me feel connected, both to home in the USA and to the wider public interested in space exploration and the Moon.

Soon after I returned home, *Air and Space* asked if I would like a regular gig blogging about lunar science and exploration on their web site. I was already hooked and readily accepted. My first blog was an introductory few paragraphs on my background and why I titled it *The Once and Future Moon* (http://blogs.airspacemag.com/moon/). It appeared on November 9, 2008. Now, 65 essays later and at a time our country is wrestling with what direction and leadership role our country will play in future space exploration, I've put together this collection to draw as many people as possible into the debate about the importance of the Moon and why they should care about it.

Originally, I think that Tony envisioned a more frequent, less wordy running commentary on various things lunar, but *The Once and Future Moon* naturally gravitated toward a more quasi-periodic, op-ed column. I must admit that I didn't intend to do this; I think that, having admired and read such columns by others over the years (particularly the columns in *Natural History* by Stephen Jay Gould, which sparked within me an interest in the history of scientific ideas), I moved in that direction. Hopefully, one thousand words is long enough to develop a complex idea in some detail, yet not so exhaustive as to bore the reader excessively. In any event, this book holds the product of that experiment. If you want to see what the original blog concept might have looked like, check out *The Daily Planet* (http://blogs.airspacemag.com/daily-planet/); Tony gave up on me doing it properly, so he took on that task with his own lively, informative blog.

My short essays cover lunar science and exploration, as well as space policy over the last 22 months. This period has seen startling changes and developments, both in our understanding of how the Moon works and about our strategic direction in space. At this writing (summer 2010), it is still unclear what form our national space policy will take. We're discovering that the Moon is even more interesting and useful than we had ever dreamed. However, this exciting news has been met with a vigorous campaign to abandon the Moon as a goal.

In my work, I have tried to articulate the reasons why the Moon is an important destination and how our attempts to return there with people have faltered and why. By collecting this extended rationale into a single source, those who believe in the significance of the Moon, as I do, will have at their fingertips a compact sourcebook of useful information, from both a technical and policy perspective. I also wish to reach those who are just beginning to understand the scope and importance of space exploration and settlement and hope that they too will find these writings informative and helpful.

I have reproduced my blog entries in the order they appeared, as the historical dimension may be instructive; the reader of this book finds out about things just as I did. I use links in each post to source the authority of a quote or fact; these source links appear in this book as numbered URL web links after the text of each post. The blog has a comment function and my readers are not shy about telling me exactly what they think about what I have written or believe. I appreciate their interest and for taking the time to comment. I reproduce these comments with minimal changes, correcting only obvious spelling and formatting errors. I have omitted most simple "Pingback" comments (i.e., those that only note the appearance of a new post), but have retained any that have substantive commentary associated with them. Some excessively discursive and irrelevant posts are also omitted; any reader wishing to see all of the comments in their unedited glory has only to follow the appropriate hyperlink to the original web posting. *The Once and Future Moon* blog is still active and I invite all readers of this book to visit and join the ongoing discussion. The opening (1) and closing (65) chapters are not part of the blog *sensu stricto*, but contain information and ideas that I developed while doing the blog and are an important piece of whatever central theme or philosophy this book may be said to possess.

I am grateful for the many friends and colleagues through the years who have helped me develop the ideas on lunar exploration and space policy expressed in these pages. In particular, I thank Ben Bussey, Jeff Plescia, Steve Mackwell, Clive Neal, Steve Baloga, Tony Lavoie, John Gruener, Dean Eppler, Nancy Ann Budden, Stu Nozette, Don Pettit, Tom Rogers, Tom Cremins, Doug Cooke, Dennis Wingo, Bill Readdy, Jack Frassanito, and Klaus Heiss for extended discussion on many of these ideas and concepts.

I also thank my fellow space bloggers on the web (even though we may disagree on some policy issues) for continued information and inspiration: Keith Cowing, Rand Simberg, Clark Lindsey, Mark Whittington, Alan Boyle, Jim Oberg, Dana Mackinzie, Emily Lakdawalla, Ken Murphy, Marcel Williams, Joel Raup, and Warren Platt. Tony Reichhardt of the Smithsonian Institution invited me to do this weblog and I thank him for his continued support and counsel over the life of the feature.

Finally, and most importantly, I thank the best editor, critic and friend I have ever had or ever will have. This blog would not exist without my wonderful wife Anne, who reads every post that I write, corrects my mistakes, improves my expression, stops me from being excessively nasty, and sharpens my thinking. Whenever I despair of the direction things are going, she makes me want to keep at it.

Houston, Texas July 2010

India Aims for the Moon

A U.S. scientist reports from the scene of India's first lunar launch. airspacemag.com, October 21, 2008

Chandrayaan-1 lifts off for the Moon

Through the fog, into the fog (October 21, 8:00 a.m.)

I'm sitting in a hotel room in Chennai, India, attempting to recover from my jet lag. Houston to India is a 26-hour trip (one way) and although there's plenty of time to snooze, I never sleep well on planes. Through my brain-fog, I have CNN International on in the room. At the top of the hour, there's a detailed report on the Chandrayaan-1 mission to the Moon, now less than 24 hours away from launch. The report describes the mission as well as ISRO's (the Indian space agency's) four-year effort to build the spacecraft, and has interviews with key mission personnel, including Mylswamy Annadurai, the mission director, with whom I have formed a close friendship during the last few years. The news item is enthusiastic and thorough.

This matches my previous experience in India closely—the Indian people are genuinely excited about going to the Moon. I've talked to porters, room cleaners, taxi drivers, airport security people and many others during my seven trips here over the past four years. When they find out that I'm here to work on the Chandrayaan mission, they are not only very interested, but very excited and well informed—about space, the Moon, and India's first journey into the solar system.

Over the past 50 months, we've designed and built our instrument, the Mini-SAR, which will fly to the Moon tomorrow on an Indian PSLV rocket. Mini-SAR is an imaging radar designed to map the poles of the Moon. Because radar provides its own illumination, Mini-SAR will map the dark areas near the lunar poles and search for evidence of the presence of water ice. This has been a controversial subject for the last decade. Now, we're going to collect information on these deposits by mapping them from an instrument in lunar orbit, a first in the exploration of the Moon.

So now I sit here in Chennai, gazing out my hotel window into a gray, drizzly day. I hope the weather is better just up the coast, but the monsoon is with us and rain is a fact of life for tropical India for the next six months. ISRO is determined to get the mission on its way to the Moon (having been delayed several times), but they do have minimum launch conditions. I don't know what they are, but I hope to find out this afternoon, as we head up the coast to the ISRO launch site, SHAR, in Sriharikota, about 80 kilometers from Chennai. Less than 24 hours—and counting!

A long and tedious journey (October 21, 4:00 p.m.)

No, I'm not talking about the trip to the Moon. I'm talking about the three-hour, 100-km (65 mile) car trip I've just endured from Chennai to the Indian space launch center. Solid bumper-to-bumper traffic for two hours—and that was just to get out of Chennai! India has almost (but not quite) achieved total traffic gridlock in their cities, and the time getting out of the center city was most of the trip. After we reached the suburbs, our speed of progress increased substantially.

The Indians launch their missions from a space center known as the Satish Dhawan Space Centre, or SHAR (from its location, Sriharikota). It sits on a low-lying spit of land that borders the Indian Ocean. They launch from here for the same reason that the Americans launch from Cape Canaveral—to ensure that any falling debris from an exploded rocket falls harmlessly into the ocean.

SHAR has a lot of the same ambiance as the Cape. It's rather isolated (as was Cape Canaveral early in its history) and it's flat, humid and warm. Scrub palm and thorny brush cover the landscape. Sea birds dot the tidal and mud flats as we drive across what seems like an endless causeway connecting the mainland to the spit on which the launch pad lies. One interesting difference here is that you must always keep your eyes on the road—you're liable to run into goats, cows, chickens, pigs and an endless stream of stray dogs that run heedlessly across and along the road.

We're staying at the ISRO (Indian space agency) guesthouse, a large block building that has a college dorm atmosphere. The big influx of foreign visitors arrived today; I would guess that we have about 20 to 30 visitors here. The press is also here in force. I saw around 15 remote vans and cars outside the main gate of SHAR, all getting ready to provide live coverage of tomorrow's launch for Indian television.

As I was talking to Jitendra Goswami, the Chief Scientist for Chandrayaan-1, in the courtyard of the guesthouse, a reporter from Indian television saw us and ran over to get a talking head sound bite. Ben Bussey, a colleague from Johns Hopkins University's Ap-

plied Physics Laboratory, is here with me, so we both did our turn on camera. It's always interesting to see how these short interviews get edited; sometimes, they don't make you look particularly intelligent.

The weather is currently looking a lot better. It rained very heavily here yesterday, creating large, deep puddles in the parking lot to go with the high humidity and heat. The rain is not as much a concern for launch as the possibility of lightning. Goswami told me that the meteorologists are measuring continuously the electrical potential of the cloud cover. Pictures of the launch pad at SHAR show it to be surrounded by four very large red metal towers, all designed to serve as giant "lightning rods" to protect the vehicle.

We're waiting around now to hear the status of the mission. We'll probably be briefed at dinner tonight, which will be held in the dining hall nearby. I've found out that cameras are banned from SHAR, so I won't be taking any pictures of the launch. But it will be intensely photographed by ISRO personnel.

A warm, rainy morning (October 22, 7:30 a.m.)

I wake at 3 a.m. Might as well get up, as my alarm would be going off shortly anyway. It's pitch dark out here in the Indian boondocks. The small television in my room tells me that the countdown is proceeding smoothly. It is now about two and a half hours until launch.

Having heard light constant rain all night as I slept fitfully, I go outside with some anticipation about a weather delay in the launch of Chandrayaan. Outside, it's calm and beautiful; a last-quarter Moon smiles down on SHAR from directly overhead, and the brighter stars twinkle through some high clouds.

We may just get this thing off today! No time for breakfast as the VIP contingent boards several large buses in the dark. They are taking us out to a special launch viewing site set up especially for us. The drive takes about 20 minutes, even though it cannot be more than a few miles away. In the warm, close dark morning, we pass the occasional stone sign, like one for the "S-Band Precision Tracking Station." We finally arrive at an old, abandoned rocket assembly tower, a site that has been re-configured into a special viewing area for the launch.

As I wander about this site, I suddenly see the PSLV rocket on its pad, about three miles away. It is floodlit and surrounded by lightning arrestors. We have a clear view of the vehicle and it's only about an hour and a half until launch. ISRO has set up tents with large video screens, showing the activities of Mission Control. The countdown has gone so smoothly that it makes me slightly worried. Weather is no problem, as we have broken rain clouds at low altitude with hazy cirrus above. Our view stand should give us a spectacular view of the flight as the rocket curves over the Indian Ocean (which I cannot see from here; dunes block the view).

I strike up a conversation with Raj Chengappa, the managing editor of India Today, a news magazine. He wants to know all about our experiment, the Chandrayaan mission, and the value of the Moon. We have a great time in this discussion, as he is very well informed and we talk about the long term value of the Moon. I give him my lunar "stump speech"—that the Moon is a stepping stone to the rest of the solar system, a source of materials and energy to enable new spaceflight capabilities. Chandrayaan is a key pathfinder in our voyage back to the Moon.

The countdown continues, slowly ticking by until it's just two minutes to launch before I even realize it. I stop talking to my friends and the people around me. I want to immerse myself in what is about to come.

Fire, thunder and water (October 22, 8:15 a.m.)

As the voice over the loudspeaker counts below 20 seconds, I strain my eyes to look out over the coastal scrub between me and the gleaming white monument in the distance. As the count reaches below 5 seconds, I first see the bright orange glow of rocket ignition. It is surprising, even though I expected to see it. In the demi-light of early morning, it is startling. As the count reaches zero, I finally see the entire vehicle—until now I could only view the upper two-thirds. It's a beautiful white needle, with a huge ball of orange flame beneath it.

It first rises very slowly, but when it clears the launch tower, it is absolutely spectacular! The launch pad is surrounded by a thick plume of white smoke around the base of the tower. As it streaks through the sky, it is still dead silent—the rocket sound has not yet reached us at our viewing site. The rocket quickly disappears into the low morning rain clouds—it's moving astonishingly quickly. Then I hear the deep roar of the engines. The low frequencies of the engine noise beats on my chest. The crowd seems disappointed that the rocket vanished so quickly, but I suspect it will re-appear soon.

It does! A bright orange spotlight rises above the low clouds, arcing over the ocean in a magnificent streak. We have it in continuous sight only for a few tens of seconds, but from these glimpses, I can get a good feel for the trajectory, taking the rocket east-southeast over the Indian Ocean, toward orbit. When the rocket goes out of sight a second time, the crowd rushes into the nearby tents, which are set up with computer readouts and video of the Mission Control Center. We all sit in the plastic lawn chair seats provided inside a very pleasant, air-conditioned tent. A plot of time versus velocity and time versus speed is on the screen, showing the rocket as a bright spot over a curve of the planned trajectory. As near as I can tell, it is absolutely spot on the money. It's moving like a bat out of hell—after only five and a half minutes, the PSLV has already achieved orbital velocity.

As we all gather in the tent to watch Chandrayaan reach orbit, an enormous downpour occurs outside. The heavy monsoon rain pounds our tent roof. The space gods have smiled upon on us this day—the rain held off until after we had left Earth.

We all watch the trajectory information intently. Now, a mere 20 minutes after launch, Chandrayaan is on its way to the Moon. The crowd relaxes and applauds enthusiastically. It has been a memorable morning. This was my third launch; I attended the launch of Clementine to the Moon in 1994 (from Vandenberg AFB, on a surplus Titan II, the rocket that launched the Gemini astronauts). I also went to a space shuttle launch in 2001, a particularly memorable launch that arced over a full Moon, rising above the Atlantic. Both of those were striking experiences.

But I think this one actually exceeds the other two. The tension released after a launch is enormous. You work on an experiment for years, nursing it through financial and technical difficulties. You baby-sit it during testing and integration with the spacecraft. So much rides on something so dangerous. You have visions and nightmares of exploding rockets and time and effort wasted.

I do not have those thoughts this morning. This warm, rainy day in southern India, I feel wonderful. Our spacecraft got a superb ride this morning. It's on its way to the Moon. Now I think ahead—what new adventures await us on the remainder of this voyage of discovery?

On its way (October 23, 4:00 a.m.)

After the launch, we all come back to the guesthouse for a late breakfast and celebration. Team Chandrayaan, the dedicated group of ISRO scientists and engineers who have worked on this mission, are all excited and jocular. There was tremendous pressure on them to deliver this mission, and with a perfect launch they are well along the road to success. I run into Madhavan Nair, the head of the Indian Space Agency. He is clearly tired, but very happy to have a "perfect" launch under his belt. I offer my congratulations and express our team's gratitude for giving us a good start to the mission. For the first time, I also meet Chandra Dathan, the SHAR center director. He is all smiles and is clearly basking in the exultation of the moment. We know that the mission has only begun, but having a picture perfect launch has created a success vibe. Good karma for a space mission is always welcome.

After lunch, we make the long drive back from SHAR to Chennai. I am exhausted, having never really caught up on sleep since arriving two days earlier. But knowing that Chandrayaan is successfully on its way to the Moon is a great feeling. That evening, the Mini-SAR team celebrates the day's events with a few gin and tonics, the one undeniable contribution to western civilization by the departed British Raj. (They contain quinine—anti-malarial, don't you know.)

The print and electronic media are filled with stories on Chandrayaan. Few space missions get this level of attention in America. The launch and orbit inject was magnificent, and the stories cover the mission objectives, spacecraft instruments, and flight profile. While at SHAR, I was interviewed by two different Indian television networks. One of our team members, Bill Marinelli of NASA, tells me that one of those interviews just aired, although he caught only the end of it. The press coverage is overwhelming, positive and appears to be demand-driven, not an attempt to impose or simulate an excitement that doesn't really exist. Today, all of India is proud of its space program. It has the right to be.

Back at the hotel in Chennai, we prepare to leave India and fly back home late that evening. We have lots of work ahead of us before Chandrayaan gets to the Moon next month. Next week, we'll have our first opportunity to turn on our instrument, point the antenna at the Earth and calibrate it using the large radio-telescopes at Greenbank, West Virginia and the giant hole-in-the-ground dish at Arecibo in Puerto Rico. We will carefully map out the signal pattern of the Mini-SAR antenna and test its performance in space. We will then use these calibration tests to learn how to extract the maximum amount of knowledge from our data.

So far, so good. Now, we begin to look forward to the Moon.

The Moon, Space and Other Things

November 9, 2008

Your humble narrator

The editors of Air & Space Magazine have asked me to continue blogging on lunar exploration, the space program in general, and the relationship of both to broader society. I am happy to do so. This is my first post on the new blog, "The Once and Future Moon."

A brief word about that name. People who know me will recognize it as the title of a book I wrote over a decade ago (*The Once and Future Moon*[1], Smithsonian Institution University Press, 1996, 300 pp.). At that time, no return to the Moon was being contemplated by the American space program and international intentions were unclear. I believed then – as I believe now – that the Moon is our logical next destination in space, a natural space station where we can learn the skills and develop the technologies to live and work on another world. Thus, that choice of title reflected my conviction that the Moon has an important story to tell both about its history and the history of the early solar system, but also about its role as a critical asset to the future movement of humanity into space.

Since that book was published, much has changed. America's new strategic direction in space includes the Moon. Moreover, many other nations, particularly in Asia, have set their sights on the Moon. These ongoing lunar missions are producing information that will revolutionize our understanding of the history and processes of that body.

Why has the Moon become interesting again? What do we hope to accomplish there? How can the Moon become a useful object in enabling our exploration of space? Will we take advantage of these opportunities? In future posts, I hope to address these and other related questions.

A brief word about me by way of introduction. I am a geologist by training and have studied the Moon and lunar science for the past 30 years. I have worked at the U. S. Geological Survey in Flagstaff, Arizona, the Applied Physics Laboratory in Laurel Maryland and both previously and currently, the Lunar and Planetary Institute in Houston, Texas. My work focused initially on lunar geological history, but for the last 20 years, I also have worked for a return to the Moon, including serving on two White House groups assembled to examine technologies and architectures for lunar return. I am currently involved in characterizing the environment and processes of the polar regions of the Moon, with the aim of understanding whether they could be appropriate places to establish a human foothold on the Moon. The poles of the Moon are of extreme interest and exactly why they are so important is a topic I hope to develop in some detail in future entries in this column.

So, with that out of the way, what's new on the Moon?

Links and references

1. *The Once and Future Moon*, Smithsonian Institution Press, Washington DC, 1998, 308 pp. http://tinyurl.com/25y7dhr

Hitting a Bull's-Eye on the Moon

November 15, 2008

Close-up of the lunar surface from the Chandrayaan-1 Moon Impact Probe

I am in Bangalore, sitting awake in my hotel room at 4 am. Last night was a memorable and exciting experience. Chandrayaan-1[1], in lunar orbit since last Saturday, released its Moon Impact Probe (MIP)[2], designed to descend and hit the Moon at high velocity, sending images and other data as it went. This part of the mission had to go well in order to fully deploy the Chandrayaan antenna, required to send all the mapping data to Earth. Once the antenna is deployed tomorrow, Chandrayaan will begin the job of mapping the Moon from a polar orbit for the next two years.

What a great linked-together world we live in! As I sit here in India, in the middle of the night, on the opposite side of the Earth, my good friend Astronaut Don Pettit and six of his colleagues are getting ready to blast into orbit on the Space Shuttle Endeavour. I am watching the launch countdown on NASA TV over the internet on my laptop computer. In addition, I'm getting live launch status and weather reports via e-mail from my colleague Dr. Ben Bussey of the Applied Physics Laboratory, who is down at the Cape for the launch. Ben just sent me a Blackberry cell phone image he took of the Shuttle on the pad, lit up by flood-lights; I reciprocated by sending him an image sent down last night from the Moon by the MIP[3] just a few minutes before its impact. And just now, the NASA TV feed shows a nearly full Moon, slowly rising over the Atlantic and the Shuttle launch site. Three friends, separated by the globe and involved in two very different space missions, but looking over each others' shoulders. I think about it and find it profound and deeply satisfying.

Back to the MIP impact last night. It has the air of a Hollywood premier about it. The press is packed along the street outside the Mission Control Center, the Chairman of the Indian Space Research Organization (ISRO)[4] and the former President of India are in the viewing room, along with hundreds of ISRO engineers and scientists. The air is electric with anticipation. Although the mission of Chandrayaan has been nearly flawless so far, the release, descent and impact of the MIP is a critical event, not only for its data, but also for the pride India feels in this mission, all riding on a little subsatellite with the national flag painted on its side.

We have little to look at in the control center.

Large computer projection displays show technical data and graphs. We observe major mission events by noting a change in velocity here or a drop in voltage there. I am somewhat lost, but one of the investigators on the MIP mass spectrometer helps me interpret what I'm seeing. The MIP separates from Chandrayaan, begins to spin for stabilization, and then fires a solid rocket motor to brake itself out of lunar orbit. The descent to the surface takes about 25 minutes. We see none of the data in real time; the images and measurements are being relayed to the main spacecraft and will be sent down to Earth in 90 minutes, on the next orbital pass. We see that everything is working fine through the cryptic and delightful expedient of observing Chandrayaan's data recorder filling up – with numbers in hexadecimal, no less! – presumably with data from the probe. Suddenly, the graphs show a large spike or a drop to zero; the probe has hit the Moon! Applause and cheers!

There is much backslapping and good cheer in the room. President Abdul Kalam[5] makes a brief and moving statement, dedicating the Moon probe to India's children as a symbol of hope for the future. My colleague from APL and I decide that it's been a long day and it's best to leave the facility now; there'll be plenty of time to look at the scientific results tomorrow. As we emerge from our cocoon, I am amazed to see the press frenzy outside. Long lines of reporters are doing TV remotes and newspaper people pace the sidewalk. They have been kept out of the control center and only a few are being let in for interviews. Therefore, anybody coming outside is fair game and they pounce on the two Americans (not particularly inconspicuous over here) for scraps of information. Who are we? Are we part of the mission? Was the probe impact successful? (This last question surprises me – haven't they been told yet that all went well?) We try to answer their questions, but our car is waiting and the last thing we want is to get caught in the obviously approaching massive traffic jam. So we leap for the car doors and speed off into the night.

The MIP impact is a major public and media event. It is a source of great pride to the Indians and ISRO has done a marvelous job on the mission. But the real work is yet to come. In the next two years, Chandrayaan will map the Moon in unprecedented detail. Future milestones may be less splashy, but they will largely make up the ultimate value of the Chandrayaan-1 mission.

Topic: Lunar Exploration

Links and References

1. Chandrayaan-1 mission, http://www.isro.org/Chandrayaan/htmls/home.htm
2. Moon Impact Probe, http://www.isro.org/Chandrayaan/htmls/mip.htm
3. Data from MIP, http://www.isro.org/pslv-c11/photos/moon_images.htm
4. Indian Space Research Organization, http://www.isro.org/index.htm
5. India President Abdul Kalam, http://en.wikipedia.org/wiki/Abdul_Kalam

Comments

1. Dr. Spudis,

Thanks for sharing your thoughts in this post and previous ones. How long might it take for the Mini_RF project to come to a definitive conclusion about the presence of water in the polar regions?

Comment by Vish — November 15, 2008 @ 11:53 pm

2. We missed you at the Pettit send-off party, Paul. So good to celebrate the initial success of both missions. See you in Houston.

Comment by Micki Pettit — November 16, 2008 @ 1:41 pm

3. Vish,

Mini-SAR gets its first mapping cycle in mid-January; it lasts about a month. After that, we will have complete mosaics of both poles. Three months later, we get another month's worth of mapping. After those two cycles, we should have complete coverage of sufficient quality to see if any areas in the permanent dark regions are interesting. Stay tuned.

Comment by Dr. Paul D. Spudis — November 18, 2008 @ 9:07 am

4. Hi Micki,

I really wanted to be there; sorry I couldn't. Glad it was a such a great and successful launch. Looking forward to Don coming home safely.

Cheers from India!

Comment by Dr. Paul D. Spudis — November 18, 2008 @ 9:09 am

5. Please convey my heartiest congratulations to all concerned. Good Show!

Comment by Zoe Brain — November 19, 2008 @ 8:32 pm

Another "Roadmap"

November 18, 2008

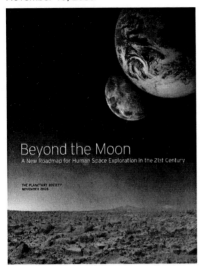

The Planetary Society speaks.

Considerable buzz was generated in space circles last week when The Planetary Society[1], the keepers of Carl Sagan's flame, released a report[2] that recommended a re-orientation of the Vision for Space Exploration[3]. This report was based in part on the results of an invitation-only workshop held at Stanford University last February. The object of that workshop was to examine U. S. national space policy with the specific aim of determining whether goals intermediate to a human mission to Mars other than the Moon were feasible and desirable.

Anyone who knows the history of Sagan and The Planetary Society (TPS) could probably guess what the conclusion of this workshop was going to be. Carl was famous for his opinion that our Moon was "boring" whereas Mars, as a possible cradle of early primitive life, was a fitting object for intensive scientific exploration. This premise in fact has guided most of the robotic exploration of the Solar System by NASA for the past 30 years. While lunar scientists couldn't convince NASA to send a polar orbiter to the Moon until the ultra-cheap Lunar Prospector[4] in 1998, Mars has seen a flotilla of spacecraft probe its secrets, measuring and mapping every quantity known to man both from orbit and the surface.

Wonder of wonders! The new report recommends that exploration of the Moon be "de-ferred." In favor of what? Human missions to the Lagrangian points[5] (areas in space that are fixed relative to large bodies, such as the Earth and Sun) and a near-Earth asteroid. So instead of the "boring Moon" – a little planet with a complex history closely tied to the origin and evolution of the Earth – they want human missions to empty points in space and to investigate the abundant and varied problems posed by large orbiting rocks.

A lot of the chatter about the Planetary Society report has focused on whether the report missed the point of going to the Moon – the "we-must-learn-to-crawl-before-we-can-walk" argument of using the Moon as a training ground for Mars. Some point out that we have very little experience in living and working on other planetary surfaces. We do not have the long-lived, ultra-reliable subsystems needed to fly missions of very long duration and limited abort capability. By going to the Moon, we can develop those technologies and learn those skills needed to live, work on, and explore any planet.

I submit that their counter-arguments miss the point. The basic assumption of the TPS report is that the purpose of the space program is scientific exploration and that academic scientists should determine where and how program money is spent. This worldview (called "Saganism" by space advocates) has long prevailed among space scientists, the Science sections of NASA, and portions of the space community. A corollary of Saganism is that the Quest for Life Elsewhere is and ought to be the principal mission of NASA. Former Administrator of the agency Dan Goldin, for one, subscribed to this view.

One of the most interesting (and little-known) aspects of the Vision is that its objective is much greater than exploration alone. A startling aspect of the original Vision was making the use of lunar and space resources a key objective of lunar return. Almost everyone ignored both the intent and purpose of such an inclusion. Fundamentally, it's about expanding the economic sphere of mankind from low Earth orbit to cislunar space and beyond. In other words, we hope to use the material and energy resources of space to create new markets, new capabilities and new opportunities. This aspect of the Vision was brilliantly illuminated and elaborated upon in an overlooked speech[6] by Presidential Science Advisor John Marburger at the Goddard Symposium in 2006.

Space is a big place and ripe with many possibilities. The Planetary Society wants to keep it a sanctuary for science, regulated and ruled by scientists for scientific purposes. The Vision is about expanding opportunities in space for many different and varied parties, including scientists. The Moon is included because it is the first place near the Earth that has the material and energy resources to allow us to understand if using space resources is possible and if so, how difficult that might be. This objective is not merely designed to lower the costs of future space missions, but to understand what it takes for humankind to live off-planet. If people are to have a future in space, we simply must learn how to extract and produce what they need from what's already there; we cannot drag everything we need with us from the deep gravity well of the Earth.

Science or Settlement? The Quest for Life Elsewhere or the Quest for Prosperity Here? Policy and direction set by an elite priesthood or by a free market? Those arguments are at the root of the policy debate initiated by the TPS report.

Topics: Lunar Exploration

Links and References

1. The Planetary Society, http://www.planetary.org/home/
2. Beyond the Moon: A New Roadmap for Human Space Exploration in the 21st Century, http://www.planetary.org/programs/projects/space_advocacy/20081113.html
3. A Renewed Spirit of Discovery, http://www.spaceref.com/news/viewpr.html?pid=13404
4. Lunar Prospector mission, http://lunar.arc.nasa.gov/
5. Lagrangian points, http://en.wikipedia.org/wiki/Lagrangian_points
6. Keynote Address, 44th Goddard Symposium, Greenbelt MD http://www.spaceref.com/news/viewsr.html?pid=19999

Comments

2. *"we cannot drag everything we need with us from the deep gravity well of the Earth."*

And then put it at the bottom of another one! Phobos First! (But only after L1, L2, Moon, ESL2, an assortment of NEOs...) With regard to your critique, Henry agrees and so do I. BTW Nice to hear of your involvement with Chandrayaan. Looking for independent confirmation that the impactor hit Shackleton! Can you oblige?

Comment by Vacuum Head — November 18, 2008 @ 3:17 pm

3. A fascinating response to an otherwise unpublicized event. I took away two key points here:

1. To overlook the Moon as a point of exploration is akin to saying that all scientific exploration of the Moon has been exhausted. And yet, we know very little about this planet, the closest neighbor to our own, and, as the article points out, one with a "complex history closely tied to the origin and evolution of the Earth."

4. Very well put, Paul, but I must take issue with your terminology re: priesthood.

True priests try to keep their minds open to newly revealed wisdom and always question, with thorough deliberation and consideration of potential new insights, if the precepts of their subscribed religion conform to a greater more accurate truth.

TPS has always struck me as more akin to a Mars-obsessed cult, self-appointed keepers of an exclusive vision based on a "superior" knowledge. Describing them as an elite priesthood is insulting to many priests.

The greatest irony here remains the fact that it was a TPS-sponsored study, co-chaired by the current NASA administrator, that begat the ESAS architecture (Big Stick & all) that is currently hobbling the original, grander purpose of the VSE that was articulated so well by both the President and Marburger. How ridiculous it is that they themselves would now seek to derail its foundational premise of lunar return as a stepping-stone to the rest of the solar system.

Comment by Bob Mahoney — November 18, 2008 @ 6:06 pm

5. *"they want human missions to empty points in space and to investigate the abundant and varied problems posed by large orbiting rocks."*

Hey! Why are you dissing asteroids? You are making both a scientific case for understanding the origins and evolution of the solar system, and for the use of space resources to expand our economic sphere. Surely you would agree that asteroids are essential targets for both of those areas. They provide information about the original building blocks of the solar system that has been obscured in the Moon's very complicated and possibly massively violent history, and they have abundant resources that are not trapped in a deep gravity well like the Moon's.

While the Moon is a very worthy and under-appreciated target both scientifically and economically, I think you do your case disservice by trying to cast aspersions on asteroids as targets simply because they are advocated by a report you didn't like. It seems to me that your stated interests in a human expansion into space and the understanding of the evolution of the solar system would lead you to strongly advocate both Lunar and asteroid missions.

Comment by Mark — November 18, 2008 @ 6:19 pm

6. 3 cheers for Paul Spudis, the Scientist for the rest of us! I've cancelled my membership with TPS and urged all my friends to do the same. Although some of their sponsored projects have merit (e.g. solar sail, SETI at home, asteroid search), their management has lost their way. You'd think that their tumbling membership numbers over the past few years would give them a clue. But apparently not. This time they have crossed the line. They need to get off their elitist academic haunches and realize that space exploration is not just for scientific weenies who worship Carl Sagan. They have a valid role to play – but not by burning the current precariously built bridges exemplified by VSE. They need to get in sync with NSS, SFF, Moon Society, and the industrial space advocacy partnerships. These are the groups that will advocate space exploration for the rest of us!

Comment by enb_48 — November 18, 2008 @ 6:30 pm

7. Mr. Spudis,

Thank you for the post. Well done. I was a member of TPS years ago... let my membership lapse. The lapse will continue.

Comment by Jim Rohrich — November 18, 2008 @ 8:59 pm

8. The question is not Moon Or Mars...it's the first step...getting into space. It the human launch issue the next NASA administrator must address...buy foreign or fly shuttle... issue addressed in my webpage...appreciate comments.

Comment by Don Nelson — November 19, 2008 @ 9:37 am

9. In support of Carl Sagan

Ah, as a famous politician once said: "There you go again" Dr. Spudis. Despite a recent NRC report supporting the scientific importance of returning to the Moon, the arguments for spending huge sums to send US robots and astronauts back to the Moon can not be justified upon close examination notwithstanding such an illustrious endorsement. (See "A Critique of the NRC Report: The Scientific Context for Exploration of the Moon-Final Report," Space Times, Sept/Oct 2007 issue By Donald Beattie) If placed in a priority list of what are the most important programs and questions that should be addressed with NASA's limited resources, studying the Moon would be low on the list. We are not in a race with other countries to land on the Moon despite what some proclaim. If they want to spend the huge sums needed to repeat what we accomplished 40 years ago, so be it.

Even NASA says it is not a top priority but would be done in conjunction with other objectives NASA has listed for returning, also of questionable importance. NASA space exploration should not be dominated by one objective, a priority that I doubt will be embraced by the next administration and congress. Consider the April GAO report that estimates to complete Constellation will require $230 billion, and that does not include Ares V or other requirements to put an outpost on the Moon. ALL NASA programs must be placed in priority, including aeronautical research, and then probable future budgets applied accordingly.

Comment by Don Beattie — November 19, 2008 @ 12:00 pm

10. Mark,

My comment in regard to asteroids was only in the context of Carl Sagan's famous opinion that the Moon is "boring." If he thought the Moon boring, how is it that asteroids could be "interesting" as they display fewer planetary processes than the Moon does.

I am not against asteroid missions, but think that we have work to do on the Moon first. Moreover, the materials processing experience we get on the Moon is directly applicable to asteroid mining.

Comment by Dr. Paul D. Spudis — November 20, 2008 @ 7:44 am

11. There are Jupiter-size potholes in The Planetary Society's Space Exploration Roadmap.

Quoting from the press release: "The Planetary Society today outlined a vigorous new approach to space exploration for the consideration of the new U.S. Administration and Congress." Right. It's so vigorous that it almost completely ignores the fundamental source of vigor that will be necessary if the Roadmap is to be achieved. Not to mention two other significant flaws.

While the Roadmap makes some good points about interplanetary flight, the need for international commitment and cooperation, and for letting capability drive the timeframe, it is still too mired in the ClassicSpace paradigm of national space programs. I simply do not see enough evidence to believe that any nation's government space program, or collective multi-national programs, can maintain the consistent political and funding support required to achieve the Roadmap's goals. The best solution is to use a wise combination of public-private efforts that take advantage of the competitive nature of the emerging NewSpace industry. Unfortunately, there is barely a breath about the emerging entrepreneurial space industry (NewSpace) in the Roadmap, and certainly not to any degree that reflects an understanding that NewSpace must be the fundamental underpinning rather than a mere supporting player.

The Planetary Society's solution for insufficient funding leads to another significant flaw. Again, quoting from the press release, the plan calls for "deferring humans landing on the Moon until the costs of the interplanetary transportation system and shuttle replacement are largely paid." What kind of vision is this? It makes no sense to bypass what will become humanity's first celestial economic subsidiary.

I can only assume that the proposed deferring of human lunar landings is a direct outcome of The Planetary Society's inability or unwillingness to move beyond the ClassicSpace focus on exploration. And therein lies the third flaw in the Roadmap; the complete exclusion of the word "settlement," which also occurred in the very useful June 2004 "Report of the President's Commission on Implementation of United States Space Exploration Policy" (aka, The Aldridge Report). Note that "Exploration" is in the title of both documents.

Whether it's President Bush's Vision for Space Exploration or The Planetary Society's Roadmap, neither offers the economic sustainability and New World excitement that settlement would provide, which will only occur with the private sector leading the way.

Comment by Jeff Krukin — November 21, 2008 @ 7:41 am

13. I think TPS is partly right. I just think they don't go far enough. NASA should get out of the transportation business altogether. So, fine. Concentrate on science. Concentrate on finding life elsewhere. If we're talking about letting the free market dictate policy, then let's let NASA move out of the way and allow private companies settle on the Lunar surface. Let private industry shoulder the burden of funding trips to, from, and stays on the Moon. It would be safer, cheaper, and faster than anything any state program could come up with. All they need to do is get out of the way. Let them have their science. Let them further not keep us from reaping the benefits of going ourselves.

Comment by guthrie — December 20, 2008 @ 2:22 pm

A Decade of the International Space Station

December 1, 2008

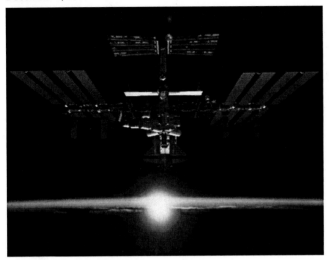

Building the ISS has taught us a lot.

The Space Shuttle Endeavour safely landed at Edwards yesterday, completing a highly successful 16-day mission to the International Space Station[1] (ISS), which celebrated a decade of continuous operation[2] last week. It's common in my business of planetary science to complain about the ISS, how it sucks up money that should be used for scientific exploration, and numerous other sins. Today, I write of its benefits and how it is relevant to our expansion into the Solar System.

The idea of a space station is very old; Wernher von Braun made it one of the first pieces of a broadly based transport system in space in his original Collier's articles[3]. In its original incarnation, a manned space station would do many jobs – Earth observation, including weather forecasting, communications relay, astronomy, a servicing port for trips beyond low Earth orbit. In fact, almost all of these tasks came to pass, but not at ISS. We watch and monitor the Earth with robotic spacecraft. An unmanned communications satellite (comsat) network 22,000 miles above the Earth acts as a relay network that ties the entire Earth together. The Hubble Space Telescope observes the heavens with crystal clarity. Many of the jobs von Braun envisioned for a space station are done today, but by robotic satellites, not humans at ISS.

What about the station's role as a jumping off point for voyages beyond Earth? There hangs an interesting tale. When NASA first began thinking about lunar bases in the early 1980s, Space Station Freedom (as it was then called) served exactly that role. Station would be a transportation hub between the Moon and other destinations in cislunar space[4], such as geosynchronous orbit[5] (where comsats reside). The Shuttle would deliver people and goods to the station in low Earth orbit and from there, they would journey to the Moon and beyond.

But Space Station changed. To maintain political support for the program, it morphed from Space Station Freedom into the International Space Station, with significant participation by Russia. For the Russians to be able to access station from their launch facilities in Kazakhstan, the inclination from the equator of the plane of its orbit changed from 28 degrees to 52 degrees. This new inclination enabled station to fly over 80% of the globe (good for Earth studies) but made it more difficult for Shuttle to reach (lowering the total

23

mass Shuttle could bring up) and making the station much less attractive as a staging node for voyages beyond LEO.

These consequences were known at the time and as no future human missions were planned beyond LEO for the foreseeable future, made operational sense of a sort. Now we find ourselves with a space station, but apparently one not optimally placed to support our movement into the solar system. This is one of the many criticisms of station – that it is essentially a dead end and cannot be used as a "jumping off point" for future missions.

But I contend that ISS is useful for future lunar and planetary exploration. For one thing, building and operating a million-pound spacecraft for over a decade has surely taught us something about space faring. One of the most remarkable facts about ISS is that it went from drawing board (more accurately, from computer-aided design bits) to working hardware in space, without numerous prototypes and precursors, and it worked the first time it was turned on. By any standard, that is a remarkable achievement. We have learned how to assemble and operate complex spacecraft in orbit, in many cases solving deployment problems and coaxing balky equipment into operation, as exemplified by the recent experience of Don Pettit and Mike Fincke with the renowned urine conversion machine. Assembling complex machines and making them work in space is a key skill of any space faring society. Building and operating ISS over the last decade has taught us much about that skill.

The station could be made even more important and relevant to future operations in space. A key requirement of routine operations in cislunar space is the ability to manage, handle and transfer rocket fuel, particularly the difficult to manage cryogenic liquid oxygen and hydrogen. We could begin to acquire real experience working with these materials at ISS – transfer a quantity of water, crack it into its component hydrogen and oxygen using solar-generated electricity on orbit, and experiment with different methods of handling, conversion and storage of these materials. None of this requires a new module, but some specialized equipment could allow us to experiment with cryogenic fuel in microgravity, mastering a skill of vital importance to future operations in space and on the Moon.

A decade of building the ISS has taught us much about real spaceflight and the ex-

perience gained will be vital to future, long duration human missions. We should take advantage of this asset to explore further the technologies and techniques we will need to create a true space faring infrastructure, one of von Braun's original goals of an orbiting space station.

Topics: Space Transportation

Links and References

1. International Space Station, http://www.nasa.gov/mission_pages/station/main/index.html
2. ISS, 10 years in the making http://www.usatoday.com/tech/science/space/2008-11-19-issassembly_N.htm
3. Collier's articles, http://home.flash.net/~aajiv/bd/colliers.html
4. cislunar space, http://en.wikipedia.org/wiki/Outer_space#Geospace
5. Geosynchronous orbit, http://en.wikipedia.org/wiki/Geosynchronous_orbit
6. cryogenic liquids, http://en.wikipedia.org/wiki/Cryogenics

Comments

2. Paul, While I'm all for getting some use out of the ISS, is it really the best place to be doing propellant depot related research?

I've been involved with some of the people doing research in the field, and there are some other routes for testing orbital Cryo Fluid Management (CFM) stuff that are worth mentioning. The Centaur guys at United Launch Alliance have a concept they've developed a bit called the Centaur Test Bed, which they've written some AIAA papers on in the past. The idea is to have a secondary payload on the Centaur that takes unused propellants from the Centaur after the primary payload is delivered. They've also got some other interesting related approaches in the works.

There's always the possibility of using suborbital RLVs to do shorter-duration experiments. There's actually some history in the past of using suborbital vehicles to test out Cryogenic Fluid Management techniques. David Chato of NASA Glenn had a good paper on the topic. I'm biased on that last approach, since it's one we've proposed in the past.

BTW, would you like me to dig up some links on the topic FWIW?

Comment by Jonathan Goff — December 1, 2008 @ 2:28 pm 4. For more illustrations from von Braun's Collier's articles see http://www.fabiofeminofantascience.org/COLLIERS/COLLIERS1.html

Comment by Paolo Amoroso — December 1, 2008 @ 3:43 pm

5. Hi Jon,

Thanks for the comment.

the ISS, is it really the best place to be doing propellant depot related research?

Perhaps not, but I'm not arguing that it is. I'm saying that there are exploration-related engineering research tasks that can be done on the ISS and station should be part of the effort to extend human reach beyond LEO. We have this superb instrument for microgravity research already in orbit; it should be used. And it can be useful.

The ISS currently has a lot of potentially hazardous materials and configurations (virtually all spacecraft do). I'm confident that procedures and protocols for safe handling of cryogenic liquids could be developed.

Please feel free to link to any studies and web pages you think are relevant.

Comment by Dr. Paul D. Spudis — December 2, 2008 @ 6:42 am

6. Paolo,

Thanks for the links — great scans of the original Bonestell figures from von Braun's articles!

Comment by Dr. Paul D. Spudis — December 2, 2008 @ 6:45 am

7. Here's the reference to the Centaur Test Bed article:

http://www.ulalaunch.com/docs/publications/Atlas/Centaur_In-Space/Centaur_Test_Bed_(CTB)_for_Cryogenic_Fluid_Management_2006-4603.pdf

And here's the article by David Chato on Cryo Fluid Management flight testing:

http://gltrs.grc.nasa.gov/reports/2006/TM-2006-214261.pdf http://gltrs.grc.nasa.gov/reports/2006/TM-2006-214262.pdf

Hope you find those interesting,

Comment by Jonathan Goff — December 2, 2008 @ 3:08 pm

The Vision for Space Exploration (VSE) and Project Constellation

December 12, 2008

Two Different things: Constellation and the Vision

There's a huge hubbub in the press revolving around alleged "obstructionism" at NASA toward the Presidential Transition team. As this rather overwrought piece at the Orlando Sentinel[1] has been posted and commented upon endlessly at several web sites, I do not propose to rehash it. Instead, I want to comment on a theme that I see running through many of the reader comments, viz., that the Vision for Space Exploration[2] (VSE) is dead, that it was a stupid idea to begin with, and the Constellation project[3] (NASA's Shuttle-replacement spacecraft system) will be and should be terminated.

As I have discussed previously[4], the VSE was an attempt to give a long-term strategic direction to our national space program after the tragic loss of Space Shuttle Columbia in 2003. It called for the return of Shuttle to flight, completion of the International Space Station, retirement of the Shuttle, development of a new manned spacecraft, a return to the Moon and finally, human missions to Mars and other destinations. Unlike President Kennedy's Apollo challenge to reach the Moon "before this decade is out", the motivation for the VSE was to create a long-term, continuing commitment to human spaceflight. Toward that end, it specified what we were to do beyond low Earth orbit – to understand and use the resources of space to create new space faring capability. Such an expansion of capability was the purpose for making the use of local resources a principal activity of lunar return in the original Presidential speech.

Many people have conflated the Vision with NASA's implementation of it, but they are two very different things. Project Constellation is the architecture that NASA has chosen to implement the VSE. In its essentials, Constellation is a launch system, a spacecraft, and a mission design. NASA chose to develop a new series of launch vehicles, the Ares I and V rockets[5], the Orion crew "capsule" (formerly called the CEV), and a craft designed to land on the Moon, the Altair lunar lander. The mission design is to launch the crew in the Orion capsule on an Ares I into low Earth orbit, launch the Altair lander and rocket departure stage separately on the Ares V, rendezvous and dock with the lander and depart from Earth orbit to the Moon. The crew would land and explore the Moon from the Altair spacecraft, return to the Orion in lunar orbit, and return to Earth in that vehicle.

Much of the criticism of NASA in recent

years is actually criticism of this architectural plan, not necessarily of the goals of the Vision (although some have questioned it[4]). But this architecture is an implementation of the VSE; it is not the VSE itself. The Vision specified long-range goals and objectives, not the means to attain them. To briefly review, we are going to the Moon to learn the skills and develop the technologies needed to live and work productively on other worlds. And there are many ways to skin that cat.

NASA spent many months and thousands of man-hours developing the architecture to implement the Vision. From the start, it was controversial, particularly the decision to develop a new launch vehicle (Ares I) using a single Shuttle solid rocket booster, whose sole purpose is to transport the crew vehicle to low Earth orbit. Note well: this vehicle cannot send people to the Moon. Ares I can only transport the crew in Orion to low Earth orbit. To go to the Moon, a second launch of a much larger rocket (Ares V) is required, carrying the lunar lander and an Earth departure stage. Critics allege that by developing Ares I, we are not moving towards the Moon, but rather creating an Earth to LEO system that is less capable than the existing Shuttle. The true objective of the Ares I program is not to build such a vehicle but rather to develop the pieces needed for the big rocket, the Ares V. These pieces include the 5-segment solid rocket motor, cryogenic upper departure stage, and flight avionics, all of which are needed to build the Ares V, which is capable of launching a lunar spacecraft.

One can criticize this architecture on a number of grounds, ranging from the technical to the programmatic, but it is important to distinguish the Constellation program from the Vision for Space Exploration. They are two separate and unequal things. Regardless of what happens to this architecture in the coming months of uncertainty, the VSE remains a logical and forward-looking set of strategic space goals for the nation.

Topics: Lunar Exploration, Lunar Resources, Space Politics, Space Transportation

Links and References

1. NASA has become a transition problem for Obama, http://blogs.orlandosentinel.com/news_space_ thewritestuff/2008/12/nasa-has-become.html
2. Vision for Space Exploration, http://www.spaceref.com/ news/viewpr.html?pid=13404
3. Constellation project, http://www.nasa.gov/mission_pages/ constellation/main/index.html
4. Another roadmap, http://blogs.airspacemag.com/ moon/2008/11/18/another%e2%80%9croadmap%e2%80 %9d/
5. Ares rockets, http://www.nasa.gov/mission_pages/ constellation/ares/index.html

Comment

1. [...] Spudis has related thoughts: Many people have conflated the Vision with NASA's implementation of it, but they are two very [...]

Pingback by Transterrestrial Musings » Blog Archive » A Big Sloppy Wet Kiss — December 12, 2008 @ 5:43 pm

Forty Years Ago, Three Men Left For the Moon

December 20, 2008

The view from the Apollo 8 CSM, Christmas 1968

Tomorrow marks the 40th anniversary of the launch of the Apollo 8 mission[1], America's first human mission to the Moon and by any measure, still a remarkable achievement. It's difficult from our position so many years later to appreciate what a bold, giant leap this mission was, in some ways even greater than the subsequent lunar landing of Apollo 11[2]. Before Apollo 8, no one had ever ventured more than a few hundred kilometers above the Earth. No one had ever seen, with their own eyes, the glowing disk of a full Earth nor the cratered dusty surface of the Moon up close. And no one had ever experienced the isolation of being on the far side of the Moon, cut off from all contact with the Earth and everything the human race has ever known.

It is not an exaggeration to say that Apollo 8 changed everything. It won the race to the Moon for the United States, although most didn't realize it at the time. Apollo 8 was supposed to test the lunar module (LM) in Earth orbit, but technical problems meant that delivery of the LM was going to be late. Planned originally as a repeat of the Apollo 7 mission (the Command Module in Earth orbit), the lunar orbital mission was substituted instead because NASA had intelligence that the Soviets were planning a human lunar flyby[3] before the end of the year. The successful flight of Apollo 8, coupled with the catastrophic explosions of their N-1 lunar rocket, convinced the Soviets that they had lost the Moon race.

The Apollo 8 mission was a positive, uplifting development for both the lunar program and the nation as a whole. The program had been resurrected from the ashes of the Apollo 1 fire by the successful flight of Apollo 7 in October 1968, but serious questions remained about the reliability and space worthiness of the system. Apollo 8 demonstrated that the hardware and architecture devised for lunar flight would work well, boosting confidence in continuing forward with the goal of a man on the Moon before the end of the decade. Moreover, after a 1968 full of nothing but bad news, the Apollo 8 mission, coming at Christmastime, charmed even the famously cynical American press corps, enough so that Time magazine changed their choice of "Man of the Year" (from "The Protester") to the crew of Apollo 8[4].

Although it took some time to develop, Apollo 8's most lasting legacy was a permanently changed human perspective. Much had been written about how the various scientific revolutions (Copernican, Darwinian) removed man from the center of the universe. This is true enough, but so often, the most lasting changes come from images. The famous Apollo 8 picture of Earthrise over the lunar horizon was a stunningly beautiful image. Even though it had been photographed earlier by the first robotic Lunar Orbiter mission[5], both the magnificent color of the Apollo 8 picture and the fact that a human had taken it, changed our view of the Earth and ourselves. Earth became "a grand oasis in the big vastness of space" to use Jim Lovell's memorable phrase. This single image did more to raise a "global consciousness" — for good and ill — than did the tons of books, protest marches and pamphlets produced by the environmental movement.

The flight of Apollo 8 changed the way we perceive our world and the cosmos. For the first time in human history, people had traveled beyond the gravitational sphere of Earth and looked back upon it. That event changed history in many ways, some of which we are still trying to comprehend. What similar change in perspective and history awaits when people actually live on the Moon, and routinely look into the black sky to watch a constantly changing Earth?

Topic: Lunar Exploration

Links and References

1. Apollo 8 mission, http://www.lpi.usra.edu/expmoon/Apollo8/Apollo8.html
2. Apollo 11 mission, http://www.lpi.usra.edu/lunar/missions/apollo/apollo_11/
3. Soviet lunar plans, http://www.fas.org/spp/eprint/lindroos_moon1.htm
4. Time Men of the Year, 1968, http://www.sixties60s.com/1968.htm
5. Lunar Orbiter mission images, http://www.dailymail.co.uk/sciencetech/article-1087106/Restored-posterity-The-historic-moment-Earth-pictured-space-time.html

Comments

2. An article about Apollo 8 40th Anniversary in the Houston AIAA Section newsletter:

http://www.aiaa-houston.org/

Comment by Al Jackson — December 21, 2008 @ 11:37 am

Moon Water – Again

December 23, 2008

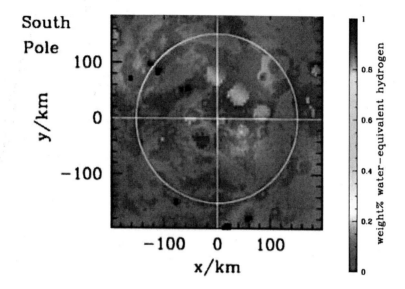

Model of the distribution of ice at the lunar south pole

The question, "Is there water on the Moon?" is still with us. Although water is not stable on the lunar surface in vacuum, the poles of the Moon contain deep craters whose floors are in permanent shadow. These dark areas are extremely cold – only about 50° above absolute zero. If a water molecule gets into one, no known physical process can remove it.

Where would such water come from? The Moon is constantly pelted by meteorites and comets, many containing water, either as ice or bound into mineral structures. This water is mostly lost to space during an impact, but some molecules may hop around on the surface for an extended time (minutes to hours). If by chance a water molecule fell into a polar dark area, it would be trapped there forever. Of course, this would be an extremely slow process, but the Moon is old (over 4,500 million years) and has plenty of time.

We will soon be obtaining new information on the poles of the Moon from the ongoing Chandrayaan-1 orbital mission. On that spacecraft, the Mini-RF SAR experiment[1] will use radar to map the poles, including all of the dark regions. Unlike neutron spectroscopy, radar probes a couple of meters below the surface and is sensitive to the presence of ice, not hydrogen.

The first hint that there may be ice in these polar cold traps[2] came from a radio experiment[3] on the Clementine mission in 1994. Four years later, a small satellite called Lunar Prospector (LP) carried an instrument[4] designed to measure the amount of neutrons given off the Moon's surface. Hydrogen absorbs neutrons, so when the LP investigators saw a decrease in neutron flux near the lunar poles, they concluded that excess amounts of hydrogen are present there.

A problem with the Lunar Prospector data is that its maps of hydrogen concentration are low in spatial resolution; we cannot identify any structure in the data smaller than about 40 km. Thus, in the LP neutron data, we see a large, smeared out area of enhanced hydrogen. We cannot tell if this excess is confined to the dark floors of permanently shadowed craters (consistent with the presence of water ice) or just an overall enrichment of hydrogen near the poles (consistent with implanted solar wind protons.)

Last week, new models of the distribution

of water[5] ice near the lunar poles were published. These maps indicate that if the polar hydrogen is present as water ice, ice concentrations may exceed 1 weight percent in some areas. This sounds like a small amount, but when added over a large area, it could constitute hundreds of millions of tons of water ice on the Moon. Moreover, because the neutron instrument only senses the outer 30 cm of the surface, total concentrations could be up to ten times greater than these results.

Of course, a model is not new data, but merely an attempt to envision how hydrogen might be distributed over the lunar poles. In concert, both neutron and radar data sets will provide an abundance of information that may allow us to finally resolve this vexing question: Is there water on the Moon?

Topic: Lunar Exploration

Links and References

1. Mini-RF, http://www.nasa.gov/mission_pages/Mini-RF/main/index.html
2. Ice on the Moon, http://www.thespacereview.com/article/740/1
3. Clementine bistatic experiment, http://www.sciencemag.org/cgi/content/abstract/274/5292/1495
4. Lunar Prospector neutron experiment, http://lunar.arc.nasa.gov/results/neutron.htm
5. Water distribution models, http://www.sciencedaily.com/releases/2008/12/081217192743.htm

Comments

3. Perhaps an antenna could be left at a higher point, on the rim of a crater, as a relay much like the broad band antenna I use for my internet connection. That might cost a bit less than a satellite in lunar orbit. I have read that power can be beamed by microwave which might replace a nuclear power source. Of course that might be a little more complicated to set up than a nuclear power source.

Comment by James West — December 29, 2008 @ 8:35 pm

4. Wouldn't it be great if it was possible to put a large array of solar panels in synchronous orbit around the moon. We would then be able to transmit power to numerous places including some dark ones. That would enable us to conduct numerous experiments over a long period of time using the panels many times instead of having to send new power sources each time. It could even be combined with a relay station for transmission of information.

Comment by James West — December 30, 2008 @ 12:53 am 5. James,

A permanent ground-based radio relay is feasible, but only after you already have significant infrastructure on the Moon. My comments above were in regard to a near-term, robotic rover mission to characterize the polar volatiles.

For power, you just can't beat simple solar arrays, placed on a spot which receives (near-) permanent sunlight. Such places exist near both poles. But again, this is only for power after you have a lunar installation. For near-term robotic missions, you need a portable power source and a small radioisotope generator is the simplest (if not the cheapest) solution.

Comment by Dr. Paul D. Spudis — December 30, 2008 @ 6:16 am

6. Dr. Paul, thank you for replying. The idea for a relay antenna on the rim of a crater came from a recent article I read referring to a flat area on the rim of a crater that was possibly suitable for landings. I was not aware that a relay antenna could not be attached to a spacecraft transporting a robotic rover. I didn't know it had to be a permanent ground based antenna to relay information. So much for trying to simplify things.

The link I have included sort of answered my question about a reusable solar array in orbit around the moon, particularly the paragraph that refers to lunar orbits. It does look to me like NASA would be on the right track if they were to pursue this, apparently, more practical approach to space exploration. The article is by Henry Brandhorst, Jr. at the Space Research Institute, Auburn University, AL. Thanks again for answering.

http://pdf.aiaa.org/preview/CDReady-MIECEC08_1836/PV2008_5601.pdf

Comment by James West — December 30, 2008 @ 8:11 pm

7. *I was not aware that a relay antenna could not be attached to a spacecraft transporting a robotic rover. I didn't know it had to be a permanent ground based antenna to relay information.*

I'm sorry — I misunderstood your proposal. Yes, you could do this, although the rim of Shackleton crater is visible from Earth only about half the time, so you would be out of communication 50% of the month. But it is a feasible strategy. You would also have to operate within the crater such that it is always visible to the spacecraft on the rim, which means the spacecraft has to be right on the very edge of the crater. The easiest way to do that would be to land and then move the lander to the edge of the crater, possibly with wheels as landing footpads.

Thanks for responding!

Comment by Dr. Paul D. Spudis — December 31, 2008 @ 6:07 am

8. Then, at the crater's edge, the antenna could telescope or unfold up for better range. What fun!

I sure hope the economy improves so the space program doesn't get pushed aside. Also, I hear a lot of grumbling from people who think the whole program is useless, even though they talk on cell phones every day.

Happy New Year!!!

Comment by James West — January 1, 2009 @ 1:06 am

9. Wow! Amazing no idea how new the discovery of Moonwater is currently. October 9th, 2009 was very recent. I am so flabbergasted at the new technology that was created by Mr.Spudis, it interesting that this equipment can identify ice meters below surface. I am so glad my teachers assigned me a project to look at recent events in the space race. These are concepts that reach out side of the box my only ? is what will this lead too. What does this mean for the future in ten years that water molecules exist on the moon? Is there going to be space shuttles and permanent bases? That I have not been able to find a clear answer. I'm also open for answers.

Comment by Kenny — March 23, 2010 @ 11:25 pm

10. Kenny,

Yes, it turns out that there is an enormous amount of water on the Moon, enough to support a small colony for many years. Whether we take advantage of that fact to establish a foothold there remains to be seen.

Comment by Paul D. Spudis — March 24, 2010 @ 4:30 am

Space Goals – One More Time

January 9, 2009

Yet another "study"…

It would appear that we are in the midst of yet another attempt to define the goals and objectives of our national space program. This time, the National Academy of Sciences is conducting a study on the Rationale and Goals of the U. S. Civil Space Program[1]. After completion, this study will no doubt be consigned to the large pile of previous studies gathering dust on the bookshelves of space students everywhere.

The study group is asking for public comment and input, in 600 words or less. This is what I have submitted:

The U.S. space program must serve national scientific, economic and security interests. Science has been well served by the space program and space exploration has revolutionized understanding of the universe and our place in it. Commercial opportunities in space have followed paths blazed by government, including launch services and operations in LEO to GEO Earth orbit. The next goal should be to expand the extent and capability of human "reach" beyond this zone first into cislunar and then into interplanetary space.

The ultimate object in space is to go anywhere, at any time, with whatever capabilities needed to do any task or objective. This ability is still far away; current spaceflight opportunities are mass and energy limited and will always be so if everything needed in space must be lifted from the deep gravity well of Earth's surface. To create greater capability, the resources of space must be harnessed to build, extend and operate a transportation system in space. The initial goal is to create a permanent infrastructure that can routinely access the entire volume of cislunar space (where all current space assets reside) with machines and people. As capabilities grow with time, such a system would be extended to interplanetary space.

To this end, the goal for next couple of decades should be to learn the skills and acquire the technologies needed to use the material and energy resources of space and to access, inhabit and work productively on the surfaces of extraterrestrial bodies. The Moon is the first target for research and use. It is both a school and a laboratory to learn how to get to, live on and explore other worlds. This task requires extended (ultimately, permanent) presence on the Moon with both machines and people.

Reconnaissance to explore, map and characterize work and habitat sites on the Moon can be done with robots and teleoperated machines. Demonstration experiments should be conducted to explore resource extraction techniques and processes, handling of materials, and create expanded capabilities and to emplace assets prior to human arrival. People will extend these capabilities and use the new infrastructure to understand the trade-offs, paybacks, difficulties and choke points of various resource extraction options. Humans will learn how to emplace, operate, maintain and expand planetary surface habitats.

A permanent human presence on the Moon creates new and exciting scientific opportunities. The

Moon is a complex, miniature planetary body and preserves both its own history and – uniquely – Earth's early history. The Moon records the output and history of our Sun and high-energy galactic particles for the last 4 billion years. Its surface environment enables the construction and emplacement of unique observational systems that can map in unprecedented detail the Earth and its environment, the local space neighborhood and the universe beyond.

To become a true space faring nation, the "umbilical cord" of space logistics must be cut to create a permanent, flexible and extensible transportation and habitation infrastructure beyond low Earth orbit. It is a difficult task, appropriate for government technical and financial support. It will open up the frontier of space for many and varied purposes, the fundamental objective of American space policy.

Please feel free to go and add your own two cents at the web site above. If repetition really is the mother of learning, perhaps we can repeat ourselves enough so that eventually, the right thing will be done.

Topics: Lunar Exploration, Lunar Resources, Space Politics

Links and References

1. Rationale and goals of the U.S. Civil space Program, http://www7.nationalacademies.org/ssb/rationale_goals_civil_space.html

Comment

1. I submitted the following, which I think in spirit is close to your ideas. ——— A civil space program should have this single minded long term goal as one critical ingredient – the enabling of routine, safe and affordable access to space. This goal should be communicated in the civil space programs documents, by its managers and leaders, to it's workforce of civil servants and principal contractors. If it is research and development, or the other extreme, the operation and ownership of a space transportation system, any effort should have to measure its relevance by the degree to which it furthers this goal.

When the Science enterprise at NASA sponsors a planetary probe it should be considered how this encourages launch vehicle providers, current or potential? Do extremely heavy/expensive probe projects create less encouragement for launch vehicle providers by needing fewer launches or the development and use of a very specific vehicle with little use to anyone else? Does technology from such probes have use on commercial satellites? Were manufacturing methods or volumes developed that encourage more satellites and space applications outside the civilian space programs? If all we are doing is internalizing the space business with direct revenue, and amortizing capabilities, then the longer term goal of furthering the development of space is lost.

Similarly, for our human presence in space, any vehicle development should measure its success by the degree to which that development or eventual operation can grow the market and capabilities for human access to space. Can a commercial developer benefit from the technology developed to also provide human access to space? Not in place of, but in addition to that of the civilian agency. Unless the thrust of the civil space program is toward creating advances in safety, affordability and responsiveness (many, many launches at a lower cost with great reliability) then again the investment only has direct benefit internal to the chosen industry players. The market does not expand.

Eventually, without a stated, emphasized goal to create routine, affordable access to space, and the selection of a mix of investments (technology, spaceports, probes, space transports, etc) and policy (breaking up vertical industry business structures, separating manufacturers from operators, as with airliners and airlines) that live by that goal, the civil space program every day becomes more and more an exclusive club of monopoly players. This club becomes more and more incapable every day of generating the incremental improvements (and occasional breakthroughs) that will make civilian space relevant on a national level. Over the near term any civilian national endeavor may gain supporters simply due to novelty. But over the long term a market that does not grow, and a culture that is closed to growth, will make subsequent achievements eventually appear repetitive until public support wanes to below critical levels. At that point a recovery would be impossible as the essential culture of progress, that civilian space should be a growing industry, will not be achievable at all due to years of stagnation of the entire US industry in a non-competitive closed market of select players.

Earth is a closed system, with the exception of the energy of the Sun arriving on Earth's surface. While living systems often adjust themselves to limited resources (also called reducing the population) it's also observed in nature that environmental change and limitations can cause once vibrant societies to perish. The first steps toward harvesting resources beyond Earth may be looked upon in the future as having been taken in our lifetimes. Or the period may be looked back upon as a lost opportunity due to an industry culture focused on maintaining a status quo of select players rather than on innovating and competing.

Comment by Edgar Zapata — January 10, 2009 @ 11:46 am

Radar Mapping the Moon

January 17, 2009

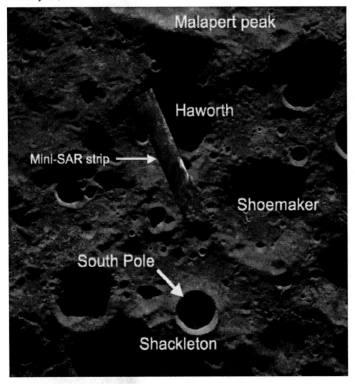

Mini-SAR image strip near the lunar south pole

The first images obtained by the Mini-SAR radar instrument[1] aboard the Indian Chandrayaan-1 spacecraft[2], currently orbiting the Moon, were released yesterday[3]. Although the spacecraft arrived last November, we are only now getting ready to map the poles of the Moon. The data released are test images, designed to make certain the instrument is working properly, that we can command it properly and that the end-to-end data stream from the Moon is configured correctly.

Imaging radar[4] works by sending a series of pulses which hit the Moon and are reflected back to the spacecraft. The range to the surface (measured by the time lag between send and receive pulses) and velocity of our instrument relative to the target (indicated by the slight Doppler shift[5] in frequency) are determined and stored as data. These data are sent down to Earth, where we reconstruct the information into an image that shows the radar reflectivity of the Moon's surface as a function of position. The result appears similar to

an optical image but is very different.

Unlike a photograph, radar provides its own illumination – the pulses sent to the surface are reflected, received back and converted into an image. It is this effect that allows us to "see" into the permanently dark areas near the lunar poles. Another difference is that radar images slightly penetrate the lunar surface to illuminate and image the subsurface structure. The depth of penetration depends upon the wavelength of the radar system, typically a few tens of wavelengths. In the case of Mini-SAR, our wavelength is 13 cm and we are looking into the Moon to depths of about 1 to 2 meters, the uppermost portion of the lunar regolith.

Our objectives are to map the poles of the Moon, examining the permanently dark areas and characterizing their surface properties. These dark areas never see sunlight and are extremely cold, only a few tens of degrees above absolute zero. Because the Moon is bombarded with water-bearing objects such as comets, we think it is possible that water might be trapped in these dark zones[6] near the poles. Radar is reflected from various surfaces in different ways. The reflections from rock and soil are very different than those from ice. By mapping all the dark regions and determining their surface back-scattering properties, we hope to identify any possible ice deposits, even though they cannot be seen with visible light.

The image released shows the interior of the lunar crater Haworth, a 35 km diameter feature just north of the crater Shackleton, at the south pole of the Moon. Because Haworth is both deep and partly shielded by the bulk of Malapert Mountain just north of it, its floor is in both Sun shadow (never being illuminated by the Sun) and in Earth shadow (meaning that radar from Earth-based radio telescopes cannot see into it). Mini-SAR has given us our first look into the dark interior of this crater. We see that it has a relatively flat floor, with numerous other craters formed on it. We have not yet analyzed its back-scatter properties[7], but the data are of sufficient quality that this analysis is possible. The first Mini-SAR images mostly show that the instrument is working properly. Future collects will image large, contiguous areas and gather high fidelity information that will allow us to identify and map the extent of anomalous reflective properties.

Despite some sensationalistic and inaccurate news reporting[8], we have not yet found any deposits of ice, or even any "extremely hard surfaces." (By the way, hard surfaces are not absent from the Moon – they can be found wherever bare rock is exposed, for example, in outcrops along crater and rille walls or the surfaces of very young features, like the solidified pools of shock impact melt[9] in King crater). I've been down this road before: a quick look at some data, someone shoots off their mouth, and the press takes it from there. Not this time. If we find something interesting, we will repeat our observations, probably several times, to make sure that the data is good before any interpretations are announced. So beware of what's reported in the news.

I am very pleased and excited to get new data from an instrument never before flown to the Moon. I'm looking forward to the next few months, when we'll get our full allocations of mapping time to build up our polar mosaics. The search for lunar ice has only just begun.

A Postscript

Emily Lakdawalla has noted the public release of Mini-SAR images on her blog at The Planetary Society[10]. She seems somewhat disappointed in the appearance of the images, noting that they seem "less crisp" than other SAR images she has seen. She attributes this lack of "crispness" to the smoothness of the Moon itself (features being eroded by the constant rain of micrometeorites) and to the compact miniaturization of our SAR instrument (Mini-SAR is less than 10 kg in mass, smaller than any previous space radar).

Neither explanation is correct. The principal reason for "crispness" of SAR images is the presence of cast shadows. Mini-SAR uses an incidence angle of about 35 degrees; as one of our main objectives is to search for ice on the Moon, we chose this imaging angle deliberately to enhance the back-scatter signal from the surface, making the identification of ice easier. In fact, for our purposes, cast shadows are not desirable because we want to map all of the dark areas near the pole — a cast shadow represents an unmapped area.

It is difficult to judge context from a narrow strip image. As we build up our contiguous mosaics of the poles, the radar images should be easier to examine and interpret.

Topics: Lunar Exploration, Lunar Resources

Links and References

1. Mini-SAR, http://www.nasa.gov/mission_pages/Mini-RF/main/
2. Chandrayaan-1 mission, http://www.chandrayaan-i.com/
3. First radar images, http://www.nasa.gov/mission_pages/Mini-RF/news/2009-01-16_radar_first_look.html
4. Imaging radar, http://en.wikipedia.org/wiki/Synthetic_aperture_radar
5. Doppler shift, http://en.wikipedia.org/wiki/Doppler_shift
6. polar darkness, http://www.thespacereview.com/article/740/1
7. back-scatter, http://www.thespacereview.com/article/740/1
8. Chandrayaan eyes ice on Moon, http://tinyurl.com/9akdum
9. shock melt, http://history.nasa.gov/SP-362/ch5.5.htm
10. Planetary Society blog, http://planetary.org/blog/article/00001812/

Comments

1. Thank you, Paul. This is a timely and useful response to the media reports, and much appreciated. I'm looking forward to the Jan. 29th meeting and initial results from Chandrayaan, and any comments you will have at that time.

Comment by Phil Stooke — January 25, 2009 @ 2:23 pm

2. Thanks Phil! By the way, great work and thanks for your locating the Chandrayaan MIP images on the Moon. I worked on that when I was in Bangalore last November — very tough!

More on Chandrayaan soon.

Comment by Dr. Paul D. Spudis — January 26, 2009 @ 6:54 am

What Apollo Was …. and Wasn't

January 25, 2009

Lost glory?

Miles O'Brien[1], late of CNN, recently wrote a column reflecting on the accomplishment of the Apollo program and the space program since then. He believes that Apollo was a great leap forward in space, a capability and step from which we then walked away. O'Brien asks why the country has turned its back on the promise of space and what it will take to re-establish the resolve we once showed in reaching for the Moon.

These thoughts are common enough in the space community. Why doesn't the nation see the conquest of space as critical? Why didn't wonderful accomplishments of the Apollo program propel us onward to the planets? Why do we continue to condone a space program of such apparent mediocrity?

None of these questions are the right ones because they misunderstand the true purpose of Apollo. The O'Brien column is only the latest in a long series of missives that long for a new renaissance of space travel, to pick up the mantle of space greatness promised by Apollo but abandoned afterwards.

Apollo was not about the Moon, or even about space. It took place in space and ultimately, on the Moon. But Apollo was a battle in the Cold War. John Kennedy did not say, "Go to the Moon and press onwards to the planets." He challenged America to show the superiority of its economic and political system by landing a man on the Moon and returning him to Earth "before this decade is out." The key objective was not going to the Moon – it was to beat the Soviets to the Moon. This objective was attained with profound consequences, critical to our Cold War victory[2] to a degree still not fully appreciated.

Most space program observers acknowledge this distinction, but they have only accepted it intellectually, not emotionally. To them, Apollo was a miscarriage of Wernher von Braun's dream of interplanetary flight. Indeed, von Braun himself thought this. As a firm believer in large engineering projects in service of national goals, he was willing to postpone his incremental, stepping stone architecture to the imperative of reaching the Moon before the Russians. But after the battle was won, he wanted to return to the classic, sequential framework[3] he had always advocated: shuttle, station, moon tug, Mars craft.

The political will for such a program did not exist then and doesn't exist now. We stopped going to the Moon for a very simple reason – after you win a battle, you don't keep fighting it. We beat the Soviets to the Moon – the reason for Apollo's existence. It required a significant fraction of the national wealth to pull off the Moon landing (at peak, almost 7% of the federal budget was spent on Apollo) along with the personal commitment of thousands of engineers and technicians across America, many of whom destroyed their marriages working double hours and weekends to meet tight deadlines. This kind of effort on behalf of a government program is not expended lightly; the Apollo program had the mentality of a war effort. And indeed, that's exactly what it was.

So what does this mean for our future? Has America lost its way? Is it simply that we no longer dare to do great things in space? I think not. What's needed is a space program that doesn't require an industrial war footing. The opportunity given to NASA called the Vision for Space Exploration[4] tried to foster a return to the Moon using small, incremental and cumulative steps. Such a program is congruent both with fiscal realities and with our aspirations to explore.

Miles O'Brien challenges us to drum up a heightened level of public enthusiasm, one that would support a new program that rivals Apollo in scale and ambition. But the experience of the last 30 years shows that in the absence of a credible external threat, the fiscal and human cost of such a space program is simply more than America is willing to pay.

I argue that the challenge for space advocates is to craft a program that matches the level of public support. Much can be done with existing resources if they are obtained and deployed in a rational and ingenious manner. Humanity instinctively gravitates toward exploration. People sense that engaging in these pursuits is about much more than politics, diplomacy or prestige. It's about human survival.

Are we willing to build a lasting space faring infrastructure – to change wartime resolve and bravery into an enduring legacy of human achievement? If so, the greatest days of our space program may yet lie ahead.

Topic: Lunar Exploration

Links and References

1. Miles O'Brien, http://milesobrien.wordpress.com/
2. Apollo Cold War victory, http://www.spudislunarresources.com/Opinion_Editorial/Apollo_30_op-ed.htm
3. von Braun architecture, http://en.wikipedia.org/wiki/Man_Will_Conquer_Space_Soon
4. Vision for Space Exploration, http://www.spaceref.com/news/viewpr.html?pid=13404

Comments

1. There are two issues that need to be addressed if you are talking about "drumming up more support for space" from citizen's enthusiasm for more government programs:

(1) Most ordinary tax payers think we are still paying the 10% of the federal budget (give or take a little bit) for NASA. I'm talking Baby Boomers here and older senior citizens, not the younger folks… but keep in mind these are the folks that actually vote, run for office, and chair all of the committees from local precinct-level parties to the congressional science committees that directly fund appropriations. When you point out that the amount of the Federal Budget for NASA is about 5% of what they think it is, they literally drop their jaw and wonder what happened. I guess the rest of us ask that same question too.

(2) There is more than one "space agency" in the U.S. Government. Indeed, NASA isn't even the largest in terms of budget or personnel. Other space agencies of the U.S. Federal Government include the NRO (National Reconnaissance Office), FAA-AST (Office of Commercial Space Transportation), and even NOAA's Space Weather Prediction Center. I'm sure a full review of the federal government would indicate a few more… and I wouldn't put it past all of these group to eventually have professional astronauts in one capacity or another. The predecessor of the NRO did in fact have professional astronauts with the "Manned Orbiting Laboratory" project that were eventually transferred to NASA.

Of the various "space agencies", the one I wouldn't mind seeing expand a little bit and certainly take on a mission of being an advocate for space development is the FAA-AST. This at least would be the "best bang for the buck" in terms of what would encourage further development and infrastructure for the development of space.

Comment by Robert Horning — January 26, 2009 @ 12:18 pm

3. Thank you for this article – you nailed it.

All the U.S. Presidents after Johnson have

wanted to give Kennedy's Man to the Moon speech. Many of them have tried, touting various programs. None of them have been willing to pay the true costs to fund their proposals.

NASA plays along with this game. Senior Agency officials promise Congress relatively cheap, re-usable Space Shuttles, a Space Station, and other programs for a fraction of their true (known) costs. In their eyes, the ends justify the means.

It's time for some honesty. Unless there is a return to the Cold War climate, there will be limited tax dollars available for NASA. Let's build a long term space infrastructure and exploration effort based on a realistic budget.

It's also time to make some hard choices instead of funding the status quo, which is primarily a high-tech jobs program supported by specific members of Congress with a vested interest.

Comment by Mike Nottle — January 27, 2009 @ 6:18 am

4. Of course Mr. Spudis is right–intellectually. Beating the Soviets to the Moon and back was what drove government policy and the money. But I think you can make the counter argument that below that level, when he talks about the double duty and breaking marriages, as many people or more did it for emotional or visionary reasons as for the "cold war" aspect of it. And that is what drives them today. I can't back that up other than providing anecdotal thoughts gained from working at NASA for six years.

I don't think too many voting citizens deal with space travel intellectually. If they deal with it at all, they deal with it emotionally. At some sub-level they may understand the science and engineering gains we have received from space exploration, but at the center of their emotional sweet spot they want somebody to go out there and break through those barriers for them. To do what they dreamed of as kids sitting up watching the moon out of their bedroom window. However, the cost is another reality and one that inescapably brings us back to an intellectual discussion. Because unlike in the 1960s cost matters very much.

So while I don't disagree with Mr. Spudis, I would rather read and dream along with Mr. O'Brien. It's just where my right brain leads me. My left brain, while understanding the intellectual reasoning, willingly follows.

Maybe we need to interweave both discussions into one. Could we combine the emotional and mystical with the logical and pragmatic? Maybe when we do that we can energize political leadership to lead us into a new generation of sustainable space exploration. Maybe we can catch an LED light in a bottle (lightning might be asking too much). Maybe.

Comment by Dan Carpenter — January 27, 2009 @ 2:51 pm

5. One thing that the public — and all too often supposedly knowledgeable historians — forget is that the Apollo program yielded incredible spin-off benefits that has harvested profits in the form of taxes collected by the federal, state and local governments across the country.

It was a technological stimulus that has yet to stop giving. It greatly jump-started the microprocessor revolution, it advanced materials science, it evolved communication, it evolved navigation systems in the form of inertial guidance, it evolved energy science in terms of fuel cell and battery technology and it even evolved medical science to boot.

Sitting in front of your computer, you are using a machine whose internal parts' provenance goes straight through the Apollo program. Kilby and Noyce may have built the first integrated circuits, but it was Apollo's need for small electronics packages, along with the Minuteman program, that fueled the first mass-production of them. At one point Apollo consumed over 60% of the world's IC supply. The computer you are using right now has dozens of IC's, including its CPU. Knowing that, it's fair to say that Apollo jump started the development that led to the Personal Computer Revolution and their ubiquity.

As far as energy goes, take a look at how NASA and the space program took the laboratory curiosity of fuel cells and advanced the technology such that it could stand the rigors of spaceflight. Those very same technologies sit at the heart of modern fuel cells, the ones we hope to power our future Hydrogen Economy. At the very least, Apollo advanced our knowledge of fuel cells and gave a very good basis for further research. This, of course, will help shorten our lead time towards energy independence.

Apollo also greatly advanced our knowledge of the moon and its creation — which of course advanced our knowledge of the history of our own planet. As one lunar geologist remarked recently, "seventy-five per cent of what we know about the moon came from the contingency sample Neil Armstrong put in his pocket when he stepped onto the lunar regolith."

Moreover, it was Apollo samples that proved the presence of Helium-3's abundance in the lunar regolith. This rare material on the Earth holds great promise to power nuclear fusion power plants economically, because it has lower radiation output than does Hydrogen. That would mean plants would last longer and of course be more cost-efficient. This discovery is relatively recent, and once again proves that Apollo return samples are gifts to science that keep on giving, decades past their collection.

That is the true legacy of Apollo, and it is one that seems to be almost completely ignored by historians and even NASA. While it is more than fair to say that Apollo was a Cold War propaganda battle, it is folly to ignore the long-term ramifications of it, and it is also suspect to think that even more ambitious projects wouldn't yield similar benefits for generations to come.

Comment by Charles Boyer — January 28, 2009 @ 9:49 am

6. Charles,

I don't disagree that Apollo produced a lot of technical innovation, much of which has produced enormous wealth. The problem with the spin-off argument is that you can never prove that a given development wouldn't have happened anyway in the absence of the Apollo program. Thus, the spin-off line of reasoning has no power as a convincing argument in favor of large, government space programs. I could easily argue that much of the spin-off benefits claimed for Apollo were actually produced by military R&D during the Cold War (e.g., integrated circuit chips). Of course, such a distinction is meaningless as both were part of the same effort, but that fact lessens its value as an argument for space.

Comment by Dr. Paul D. Spudis — January 28, 2009 @ 10:16 am

7. Dr. Spudis,

Your points are fair, but I will point out in riposte that as someone who held clearance that I have noticed military R&D's dissemination into the public sector is far slower than the same process from the civilian space program. Carbon-fiber tech comes to mind, but that may not be the best example.

Indeed, you could argue the integrated circuits point, however, the counter-argument would be the Charles Draper-MIT Apollo Guidance Computer development which used the lion's share of IC's, far above and beyond Minuteman.

Back to your central question — would it have happened sans Apollo? Certainly.

But would it have led to the creation of Intel in 1968, which quickly went to work on mid-scale integration and eventually large-scale integration? Probably not. IC's would have been kept black because the Soviets were still using vacuum and relay technology, and indeed, badly trailed the US in computational ability. That would have been an advantage our military would have been loathe to surrender.

That's why I posit that Apollo greatly stimulated the PC revolution. I seriously doubt we would have what we have on our desks, cell phones, televisions and even microwave ovens had that original IC tech been confined to our armed forces.

Comment by Charles Boyer — January 28, 2009 @ 10:45 am

8. Silicon ICs were invented by Robert Noyce at Fairchild Semiconductor in 1958. They were a private invention, using no public R&D money. They were in large scale use by the early 60's, well before the Apollo program existed. They have as much to do with the space program as velcro.

Comment by Chris — February 1, 2009 @ 5:12 am

The Strange Story of Lunar Magnetism

February 8, 2009

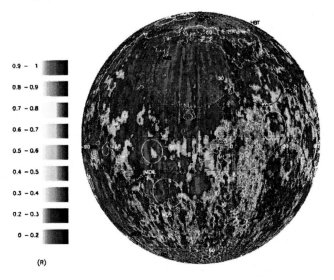

Magnetic field strength on the near side of the Moon

We've known since the beginning of the space age that the Moon has no global magnetic field. Before we returned samples from the Moon, this was thought to be well understood – compared to Earth, the Moon is a small body (1% the mass) and it rotates very slowly (almost 30 times slower). The large magnetic field of Earth is generated in its very large, rapidly rotating, molten nickel-iron core. This process, called a core dynamo, is thought to be the principal explanation for global magnetic fields in planets throughout the Solar System. Thus, it's not surprising that the Moon has no global field; for surface navigation, you can forget about packing a compass.

Rocks formed from cooling lava retain a memory of the strength and orientation[1] of any magnetic field existing during their solidification. The magnetic domains in minerals align themselves with the direction of the prevailing magnetic field. As the rock cools below a critical point (the Curie point[2]), the magnetic field is recorded as something called thermal remnant magnetism, which can be measured in the laboratory. Determination of remnant magnetism in Earth rocks has documented the existence of plate tectonics (continental drift), contributed to the unraveling of complex geological histories, and enabled the solution of many problems in terrestrial geology.

So it was quite the surprise to find that although there is no magnetic field around the Moon now, apparently there was one over 3 billion years ago[3].

The Apollo samples were studied by every known technique, so measuring their remnant magnetism was one of many tests to which each sample was subjected. To everyone's amazement, some of the lunar rocks were quite intensely magnetized – they cooled in the presence of a strong, steady oriented field. At first we thought this result simply must be wrong; perhaps the rock acquired it's magnetism by traveling back from the Moon, through the strong magnetic fields that surround the Earth and give rise to its radiation belts and polar aurora. We even took one lunar sample back to the Moon, to be sure that its magnetism was not induced by the transport to Earth.

In conjunction with these sample results, some very strong local magnetic fields were recorded in place on the Moon's surface by instruments carried by the Apollo astronauts and measured from orbit by robotic missions. John Young, Commander of the Apollo 16 mission, measured one of the strongest magnetic fields[4] near his Descartes landing site, about one-hundredth the strength of Earth's global field. It wasn't until the flight of Lunar Prospector more than 26 years later that we found that his Descartes landing site is one of the strongest magnetic anomalies on the Moon. These magnetic anomalies have no clear or obvious distribution on the Moon, except many of them are associated with unusual bright swirls that appear to drape themselves over the landscape of the Moon.

Recently, a team from MIT has found remnant magnetism[5] in a very old (almost 4.3 billion years) lunar rock. This rock is not a surface sample, created by impact and its associated shock, but rather, a deep-seated igneous rock, formed by the slow cooling of a magma at many kilometers depth within the Moon. The presence of a magnetic field in such a rock argues for 1) a stable, constant field, not likely to be achieved by impact; and 2) this field was in place early in lunar history. Both observations support the idea of a lunar core dynamo, one that somehow started early (sometime in the first few hundred million years of the Moon's existence), operated for awhile, and then quit (around 3.5 billion years ago.) Is it really plausible for a body as small as the Moon to have had a molten core, generating a global field?

To say that this is scientifically troubling is an understatement. Scientists typically don't like these "Just So" stories because they are simply contrivances that fit the presently known facts, and have little or no predictive power. A useful hypothesis not only explains the phenomenon in question, it also has specific implications or predictions that can be explored and tested. If subsequent work shows the idea to be a false one, it is discarded. In fact, we need some mechanism to discard useless hypotheses, lest we be drowned in worthless trivia.

We appear to have a non-magnetic Moon, made up (at least in part) of magnetized rocks. In theory, there is no reason why a magnetic field should start, then abruptly stop operating. On Earth, reconstructing the magnetic story requires hundreds of carefully selected, oriented bedrock samples, something which we totally lack for the Moon. Resolving the paradox of the origin of lunar magnetism will be a high priority of lunar return.

Topics: Lunar Exploration, Lunar Science

Links and References

1. paleomagnetism, http://en.wikipedia.org/wiki/Paleomagnetism
2. Curie Point, http://en.wikipedia.org/wiki/Curie_point
3. ancient lunar field, http://www.sciam.com/article.cfm?id=explaining-the-moons-anci
4. Apollo 16 magnetic field, http://www.lpi.usra.edu/lunar/missions/apollo/apollo_16/experiments/lpm/
5. New results on lunar magnetism, http://www.nytimes.com/2009/01/20/science/space/20moon.html?ref=space

Comments

1. Out of curiosity: How was "a deep-seated igneous rock" obtained?

I've seen it said that the nearly spherical shape of the moon implies it was once molten, so why is it implausible that a lunar core dynamo may have existed that later (around 3.5 billion years ago) solidified and ceased to create a magnetic field?

Comment by Jim Taylor — February 19, 2009 @ 5:08 pm

2. I guess the only way to navigate on the moon would be as we do at sea. (By the stars)?

Comment by Very Interesting. — February 19, 2009 @ 11:41 pm

3. Jim,

Like all lunar samples, the "deep seated" rock was found on the surface. Large impact craters and basins dig up very deep material; this is how we sample the various depths of the Moon's crust.

There's no particular reason to expect a body as small as the Moon to even have a core, let alone a core dynamo. It also rotates very slowly, in contrast to the Earth. Also, thermal models suggest that a molten core at 4 billion years would have too much heat to solidify at 3.5 billion years. Why the core dynamo switched off (if it indeed existed) is unknown.

Comment by Dr. Paul D. Spudis — February 20, 2009 @ 9:07 am

4. Stellar navigation is definitely an option for the Moon. On the Apollo missions, the lunar roving vehicle used "dead reckoning", which means that after initialization at the Lunar Module, the rover computer just kept track of how far and in what direction the LRV went. It could then calculate the heading and distance to the LM at any point in its traverse. This system worked well.

The last option involves putting a network of positioning satellites around the Moon, similar to Earth GPS systems.

Comment by Dr. Paul D. Spudis — February 20, 2009 @ 9:10 am

5. How many GPS satellites would it take to provide coverage on the moon to all latitudes 90% of the time?

For Mars, how many GPS satellites would it take to provide coverage to mid latitudes 90% of the time?

Comment by George Kossuth — February 23, 2009 @ 10:34 am

6. *How many GPS satellites would it take to provide coverage on the moon to all latitudes 90% of the time?*

I don't know (this isn't my specialty) but during study of a far side sample return mission I was involved in a few years ago, just 2-3 small satellites provide nearly constant communications relay coverage of the far side. I think you could extrapolate that to a GPS system (which needs basically two signals and a very precise time). The Moon's extremely slow rotation (708 hrs.) helps you a lot.

I don't know about the Mars case, but it's likely more complicated, since Mars has the same rotation period as Earth (~24 hrs.)

Comment by Dr. Paul D. Spudis — February 23, 2009 @ 7:46 pm

Another Strategic Plan Misfires

February 20, 2009

How interested in space exploration are we?

There seems to be no end of new "strategic plans" designed to "save" our nation's space program from the purgatory of mediocrity. The latest entry[1] into the strategic planning sweepstakes comes from the Baker Institute at Rice University. Originally, I had planned to say nothing about this report, out of deference to my old friend from the Stafford Synthesis Group[2], George Abbey, who is listed as an author. But recently, another author (Neal Lane) has made some public statements[3] that are so egregiously ignorant that I cannot remain silent.

Briefly, the Abbey/Lane report[1] urges the new administration to direct NASA to: 1) continue flying the Shuttle until 2015; 2) abandon the Moon as a goal because, "People don't care about going back to the Moon and there's no rationale for going back to the Moon"; and 3) focus NASA research on energy development and global climate change.

Aside from the idea of continuing to fly the Space Shuttle (not a very good idea for many reasons[4]), none of this is particularly new but rather a re-statement of the Apollo-era meme that, "If we can go to the Moon, we can solve the (fill-in-the-blank) crisis." Since energy and climate change are the current crises du jour, some seek to capitalize on the public's fondness for the NASA of old ("The Right Stuff")

with the frantic cry that it should be redirected to make these "fixes."

Although there are good reasons to question[5] the Apollo problem-solving template, I want to focus here on the argument Lane makes that people don't care about the Moon. It may surprise you to learn that I agree in part with his assessment but believe it is irrelevant to the determination of national space goals.

When I was on the Aldridge Commission[6], we received a presentation from NASA Public Affairs which showed 50 years of polling data[7] on the question, "Do you support the American space program?" The numbers on this question have bounced around through the years, ranging from as high as about 60 to as low as around 40. Surprisingly, no matter what the agency was doing, how it was faring, what disasters it endured or triumphs it achieved, the typical breakdown was roughly 50-50, plus or minus 10. This result is as rock-solid as almost any polling number in existence over a similar time span.

Needless to say, NASA wrings its hands endlessly over this result: "How can we excite the people? If we could just come up with the correct PR plan, the public and Congress will shower us with money and support!"

I think we should look at these numbers dif-

43

ferently. If your poll results are always around 50-50, then in a fundamental sense, people are "indifferent" about what you're doing. So, in one sense, Lane is right – the public really doesn't "care" about going to the Moon. What he leaves unspoken is the fact that at least half of the country doesn't really "care" about anything NASA does. True enough, many do have a fascination with spaceflight; attendance at the National Air and Space Museum is consistently the highest of all the museums on the Mall in Washington DC. But as with any museum visit, their curiosity is easily satiated and few dwell on national strategic goals and objectives in space.

Although NASA sees 50-50 polling as a problem, I see it as an opportunity. In broad and vague terms, people support our space program – they don't want to see NASA on the chopping block. They like the idea of going to new places and making new discoveries – they just don't focus and orient their lives around the "sausage making" of space policy, like we in the business do. What they want from their government is a space program that does interesting things (and not too many dumb things) with programs that will make and keep the country smarter, inspired, proud and hopeful.

Given such an attitude and with a funding level almost literally in the noise compared with other federal programs (at less than 1% of the federal budget, much smaller than most believe it to be), what should NASA's strategic direction be? I think that it should be the incremental build-up of our capability to go farther, stay longer, and to develop and increase human "reach" beyond low Earth orbit, first into cislunar (where so many national assets reside) and then into interplanetary space.

So what does this have to do with the Moon? The Vision for Space Exploration (VSE) has exactly these objectives[8]. The plan is to fund NASA at a politically sustainable level (in constant dollars, the agency's current budget has been at more or less the same level for the last 30 years) and give it the authority to create a growing space faring capability, assembled in small, incremental, cumulative steps. Our Moon plays a key role as it is the first place beyond low Earth orbit with the building block resources needed to develop and expand our space faring capability. Initially, this means oxygen and hydrogen, which provide consumables to support human presence and rocket propellant for re-fueling spacecraft.

Lane claims that the public doesn't "care" about the Moon and he may be right. However, I note with some amusement that in a recent poll on critical issues[9] the public was worried about, concern for man-made global warming came in dead last. Maybe turning NASA into EPA in orbit isn't any better at inspiring people than creating new capabilities to explore ever more distant reaches of space.

The Vision has the promise to give us the flexibility to pursue a set of long-term goals in space that ultimately will allow us to go anywhere, for any amount of time, to do almost any job we can imagine, as well as many more that we can not yet imagine. Is this not why we have a national space program?

Topics: Lunar Exploration, Space Politics, Space Transportation

Links and References

1. Baker Institute report, http://www.bakerinstitute.org/publications/SPACE-pub-ObamaTransitionAbbeyLaneMuratore-012009.pdf
2. Stafford Synthesis Group, http://history.nasa.gov/staffordrep/main_toc.PDF
3. Neal Lane statement, http://www.spacepolitics.com/2009/02/19/lane-on-his-report-and-the-nasa-administrator-search/
4. Shuttle retirement, http://caib.nasa.gov/news/report/default.html
5. Apollo template, http://pajamasmedia.com/blog/energy-independence-shooting-for-the-moon/
6. Aldridge report, http://www.nasa.gov/pdf/60736main_M2M_report_small.pdf
7. polling data on space, http://www.hq.nasa.gov/office/hqlibrary/pathfinders/opinion.htm
8. The Vision and the Mission, http://tinyurl.com/aqer2t
9. Policy priorities, http://people-press.org/report/485/economy-top-policy-priority

Comments

3. Absolutely dead on. The Rice University "study" is an example of the almost daily proof that anyone can come along and release a "Study," and the publicity it received is in direct proportion to its purported support of the Frankfort School's toxic Politically Correct agenda.

The Space Shuttle is a dangerously out of date concept that was a mistake when Congress and the Nixon administration shoved it down NASA's throat in 1971. We are playing Russian Roulette each and every time it flies. And not just with the lives of astronauts but with the political will. We are now, today, falling behind ESA. The Vision was a practical plan.

Those who wrote this study must not have read the National Academies' Scientific Context for the Exploration of the Moon (2007). Had they done so, they would realize the necessity of coming to grips with the hazards of lunar exploration before attempting a trip to Mars that is guaranteed to exceed the present lifetime limits on the probability of Radiation Exposure Induced Death (4 percent per individual). Indeed, the rest of the planet is interested in Lunar exploration for the most practical reasons, and not to test their engineering or to "catch up" with the United States. The moon is central to the future survival of billions. By 2050, we will realize that.

Are there not now enough resources and whole forests being wiped out to deal with Climate? Bugger redundant concepts like "climate change" and elaborate mythologies based on apparently magic words like "sustainability."

Comment by Joel Raupe — February 21, 2009 @ 12:38 pm

4. 100% agree with Paul – and Joel. The shuttle is an engineering marvel, but completely impractical, and every flight carries a substantial risk. There's lots of great science still to be done on the Moon, and a huge amount to learn about living and operating in space. We are nowhere near ready to cope with Mars yet. We can tackle other problems without abandoning space – as if NASA's budget could save the world anyway, a ludicrous notion. The Vision was well thought out and should remain US space policy.

One thing that hurt the Vision was a perception that it was under funded. That's not as true as it may seem. The original plan was to finish the Station and retire the Shuttle before getting started on the next step. The money freed up by ending the other programs (except for Station operations would go to Ares and Orion. That was logical, but would cause a long gap. It was sensible to try to close the gap with lots of prior planning and design work, as NASA has tried to do, but naturally people said there was not enough money for the new projects. If a 'stimulus' would help here, then beefing up the new program makes lots of sense. Prolonging the Shuttle's lifetime really doesn't.

Comment by Phil Stooke — February 22, 2009 @ 1:24 pm

Human Spaceflight: What Value to Science? (Part 1)

February 28, 2009

Luis and Walter Alvarez at the KT boundary. Dinosaurs on the bottom right; little critters on the top left.

There is a brief but vociferous debate about the value of human spaceflight[1] over at Space Politics, under a discussion of the new NASA proposed budget. An often expressed opinion is that in general, humans contribute little to the scientific exploration of space. Indeed, my scientific colleagues often make this argument, largely in the hope that some cancelled human space program cash will magically work its way into their programs.

I'm sometimes asked what we learned by going to the Moon with Apollo (1969-1972). People vaguely remember that we brought back some rocks and maybe some remember that experiments were laid out on the surface of the Moon. But they don't remember any great discovery on the Moon, like mountains of solid platinum or the uncovering of some new "unobtainium" element with anti-gravity properties.

Indeed, such was not found on the Moon. The Moon is rather ordinary in composition[2]. It's made of the same elements found in Earth rocks, although there are some interesting differences, such as the lunar samples are depleted in the so-called "volatile" elements, i.e., those with very low boiling temperatures. So the conventional wisdom is that we found nothing of value on the Moon.

In truth, we found true scientific gold. The rocks brought back from the Moon told us the story of the Solar System's early history, details both surprising and astonishing. It was a time when planets collided and giant asteroids blew holes in planetary crusts hundreds to thousands of kilometers across. The outer part of the Moon completely melted, forming a global ocean of liquid rock. Our ideas about planetary formation and evolution had to be re-written from scratch after Apollo.

What does this have to do with human exploration? Because people went to the Moon, we now have a completely different view of how life has evolved on Earth. That's a bold assertion, but I believe it to be true.

The lunar rocks are shaped by the process of hypervelocity impact and shock wave mechanics. These events leave tell-tale physical and chemical signs[3], learned through extensive laboratory experiments, field studies of terrestrial impact sites (like Meteor Crater, Arizona), and complex numerical modeling. Because we had done this work before Apollo, when the lunar rocks were returned, we knew how to read and interpret these signatures.

Now fast-forward to the early 1980's. Geolo-

Human Spaceflight: What Value to Science? (Part 1)

gist Walter Alvarez studies sediments in Italy, trying to figure out the rates at which limestone accumulates. Luis Alvarez, his father and Nobel-prize winning physicist, suggests to his son that he should use the concentration of the element iridium, which is rare in the Earth's crust but common in meteorites, as a sedimentation "clock." Because meteorites constantly rain down on Earth's surface (including oceans) at a known rate, this meteoritic "sedimentation" is used to measure how fast the limestone accumulated.

When they applied this technique to the rocks, the Alvarezes got a big surprise. A huge spike in iridium concentration was found at the boundary (formed 65 million years ago), between the older Cretaceous rocks and the overlying Tertiary rocks, a boundary coincident with the extinction[4] of more than half of all fossil species, including most famously, the dinosaurs.

To everyone's astonishment, the iridium spike at this boundary is found worldwide. But the story gets even better: other iridium spikes are found elsewhere around the world at other rock sequence boundaries and many, if not all, are associated with mass extinctions, as evidenced in the fossil record. In fact, impact mass extinctions are a driving mechanism of evolution, as the disappearance of species opens an ecological niche for their replacement by new creatures, such as the rise of the mammals after the Cretaceous-Tertiary extinction of the dinosaurs.

What does all this mean? The lunar rocks melted by impact show enhanced amounts of certain iron-loving elements, including iridium. These elements are added to the rock during high velocity impact. It took years of painstaking study by geologists, chemists, and physicists to understand these distinctive features and diagnostic properties of the impact process. This knowledge was first applied to the Apollo lunar samples, which led to deciphering the impact history of the Moon. Later, this experience allowed us to uncover a wholly new and unexpected page in Earth's geological history.

By going to the Moon, we gained a new perspective on the history of life on Earth and new insight into how the process of evolution actually works. Not a bad scientific return from a program whose real motivations were geopolitical, not scientific[5].

Why couldn't this have been done with robot spacecraft? There are lots of answers to this question[6], but fundamentally, it boils down to the fact that we would never have gone to the Moon strictly for science. However, because we went to the Moon for other socially compelling reasons, exploration enabled science. Only human explorers, trained in the methods of terrestrial field science, were capable of finding and selecting geologically controlled samples, rocks for which we could reconstruct a context that made their stories understandable.

I'm not done with this topic. Next time, I want to look at the value of human spaceflight from a philosophical perspective.

Topics: Lunar Exploration, Lunar Science

Links and References

1. value of human spaceflight, http://www.spacepolitics.com/2009/02/27/reacting-to-the-budget-proposal/
2. composition of the Moon, http://www.psrd.hawaii.edu/Dec04/LunarCrust.html
3. indicators of impact, http://www.lpi.usra.edu/publications/books/CB-954/CB-954.intro.html
4. KT extinction, http://rainbow.ldeo.columbia.edu/courses/v1001/impact23.html
5. What Apollo was, http://blogs.airspacemag.com/moon/2009/01/25/what-apollo-was-and-wasnt/
6. capabilities of robot spacecraft, http://encarta.msn.com/encnet/refpages/RefAuxArt.aspx?refid=461576326

Comments

1. Great Article!

Comment by Scott Bass — February 28, 2009 @ 4:12 pm

2. Tell it, Dr. Spudis. Explain what we mean when we talk about "Ground Truth." The use of the K-T Boundary was an excellent beginning and an illustration.

I salute you again for never growing weary of having to explain this Ground Truth to any who will listen, tirelessly, and for many years.

The K-T Boundary, in particular, is an excellent beginning because, ultimately, human spaceflight is about survival, and exploration and survival on the Moon an essential first step that cannot be ignored.

I've often wondered whether any robot capable of being built at the time would have noticed the orange glass that caught Dr. Schmitt's eye at Taurus-Littrow, almost forty years ago.

Comment by Joel Raupe — February 28, 2009 @ 5:17 pm

3. Scott and Joel,

Many thanks for your comments! Check back later for Part 2.

Comment by Dr. Paul D. Spudis — February 28, 2009 @ 5:38 pm

5. "Perspective." Nailed it. OH YES Paul thank you for that!

Whilst robot pioneers can provide the data, humans can give perspective. And without the human perspective the data is largely without context. Especially for those outside of the relevant specialty… like most taxpayers!

Which see many images of Earth taken by various weather satellites, Lunar Orbiter, etc. But it was the Earthrise of Apollo 8 that had MEANING.

Comment by Vacuum.Head — March 14, 2009 @ 9:49 pm

Human Spaceflight: What Value to Science? (Pt. 2)

March 1, 2009

People can do field work better than robots

The discussion at Space Politics[1] got me thinking about the scientific value of human spaceflight. Although there are many reasons for humans to go into space, I also believe that humans bring unique and non-duplicative skills to scientific exploration as well.

Last time[2], I discussed how the capabilities and experience of the Apollo missions led to a revolution in our understanding of the history of both Moon and Earth and gave new insight into the process of evolution and in turn, our origin. In this post, I want to look ahead to the value humans bring to future exploration, particularly to the scientific exploration of the Moon and their critical role in lunar return, resource development and ultimately, settlement.

A common article of faith in many academic and space circles is that robotic spaceflight is the preferred method of scientific exploration. Many famous space scientists (including James Van Allen[3] and Carl Sagan) preached the superiority of unmanned missions to human ones. Indeed, many phenomena in space (such as plasmas and magnetic fields) can-

not be sensed directly by humans or in some cases (e.g., detecting the tenuous lunar "atmosphere"), the presence of people interfere with the property being measured. I agree that some scientific activities cannot or should not be done by people. But in other areas of science, human presence is not merely beneficial, it's critical.

The Moon is a natural laboratory where we will answer critical scientific questions. The "field" is the world in its natural state, where the phenomena we study are on display and where we learn the key facts that permit us to reconstruct past processes and histories. Field work is not merely a matter of picking up rocks or taking pictures. It is the conceptual visualization of the four-dimensional (three spatial dimensions plus time) make-up of planetary crusts.

A good example of the differences in capabilities between humans and robots is illustrated by the experience with the Mars Exploration Rovers[4] (2003-present). These machines have traversed many kilometers of terrain, examined and analyzed rock and soil

Human Spaceflight: What Value to Science? (Part 2)

samples, and mapped the local surface over the course of five years. Many gigabytes of data have been returned by these rovers, giving us an unprecedented view of the martian surface and its geology. They are truly marvels of modern engineering.

Yet after all this exploration, we are unable to draw a simple geological cross section through either of the two MER landing sites. We do not know the origin of the bedded sediments strikingly shown in the surface panoramas, whether they are of water-lain sedimentary, impact, or igneous origins. We don't know the mineral composition of rocks for which we have chemical analyses; such information is crucial to determine processes and origins.

After five years of Mars surface exploration, we do not know things about the field site that a human geologist could determine after an afternoon's reconnaissance. In contrast, we have an incredibly detailed conceptual model, albeit incomplete, of the geology and structure of each of the Apollo landing sites. The longest stay on the Moon for these missions was three days, most of which was spent inside the Lunar Module.

A robotic rover can be designed to collect a sample, but it cannot be designed to collect the correct sample. Field work involves posing and answering conceptual questions in real time, when emerging models and ideas can be tested in the field. It is a complex and iterative process; we sometimes spend years at certain field sites on the Earth, asking and answering different and ever more detailed scientific questions. Our objective in the geological exploration of the Moon is knowledge and understanding. A rock is just a rock, a piece of data. It is not knowledge. Robots collect data, not knowledge.

It has been argued that planetary exploration robots are controlled by people so human intelligence guides the robot explorer. Having done both types of exploration in the field on Earth, I contend that remote teleoperated robotic exploration is no substitute for being there. All robotic systems have critical sensory limitations – important sensory aspects as resolution, depth of field, and peripheral vision. They have even greater limitations in physical manipulation, an extremely important aspect of field work, where picking a sample, removing some secondary over coating and examining a fresh surface is an important aspect of work in the field. The makers of the

MER rovers recognized this need by including the RAT (Rock Abrasion Tool) to create fresh surfaces; it became worn down and unusable after a short period of operation.

Ultimately, we need both people and machines to explore the Moon and other planets. Each has their appropriate skill base and limits. Machines can gather early reconnaissance data, make preliminary measurements, and do repetitive or exhaustive manual work. But only people can think. And thinking – and acting and working based upon the results of that thinking – is what field work is all about.

Topic: Lunar Exploration

Links and References

1.Value of human spaceflight, http://www.spacepolitics.com/2009/02/27/reacting-to-the-budget-proposal/
2. Value to science of human spaceflight, http://blogs.airspacemag.com/moon/2009/02/28/human-spaceflight-%e2%80%93-what-value-to-science-part-1/
3. James van Allen, http://en.wikipedia.org/wiki/James_A._Van_Allen
4. MER, http://marsrovers.jpl.nasa.gov/overview/
5. Mars exploration program, http://marsrovers.jpl.nasa.gov/mission/wir/index.html
6. people and machines, http://encarta.msn.com/encnet/refpages/RefAuxArt.aspx?refid=461576326

Comments

2. What you really mean is an astronaut geologist and a spectrometer.

No, that's not what I really mean and if you had read what I wrote, you would know that sample (and spectra) collection is only the manual labor portion of field science. Most of field science consists of thinking, in the field, where the rocks are. Machines cannot do this now and will not be able to in the near future. Apparently, it's difficult for some people as well.

Comment by Dr. Paul D. Spudis — March 2, 2009 @ 6:20 am

3. It seems that this is an interesting debatable topic, but I think it misses a more fundamental point; that is, the rationale for such a large expenditure for purely exploration and increasing human scientific understanding is not a sufficient justification for the proposed endeavor. That being said, the idea of human return to the Moon and to Mars and to the Asteroids can and should be justified by the more important desire/need to expand human presence for the purpose of increasing (existing) wealth, and new wealth (category) creation. It was, I believe, the primary reason for human expansion and exploration in our past, and it will and should continue to drive us in the present and the future.

Comment by SpaceFusionGuild — March 11, 2009 @ 6:43 pm

49

Of Science and Cathedral-Building

March 15, 2009

The Smithsonian National Air and Space Museum. Photo by Eric Long/NASM, National Air and Space Museum, Smithsonian Institution

The Daily Planet[1], my new companion blog here at Air & Space Magazine, highlights a speech[2] recently given by my good friend Dr. Neil Tyson at the Space Foundation breakfast. Noted is Neil's oft-mentioned concept that historically, three drivers[3] are responsible for societies or nations undertaking great collective endeavors, such as pyramid- or cathedral-building, New World exploration, or Apollo Moon programs. These drivers are waging war, seeking wealth, and the praise or worship of a deity. Neil believes that the first two are still operative in today's world, but the last one is a quaint relic of some bygone era.

I feel the need to reconsider this elimination of the third driver. On many very fundamental levels, humanity never really changes. However, our terminology changes over time and I think that all three rationales are as valid today as ever. For the last category, we still construct monuments, have an elite priesthood to preside over it, and constantly revise and extend the current dogma. But we call it "science."

I can already hear howls of protest from my scientific colleagues, but let's step back a minute and examine the role and nature of science in today's society. Science is commonly portrayed as the objective search for truth, as opposed to the superstitious or faith-based (depending upon which side of the fence you are) approach of religion. But science is not really as objective and pure as it is often portrayed and probably never was.

Modern science is a collective, social experience and as such, is subject to human failures and imperfections as is any societal institution. Moreover, science is most often funded by government, adding a layer of political accountability to what should be the objective search for truth. Most problems that we study in modern science are driven by social pressures and direction. Important problems are defined as those that the scientific community thinks are important. Often, these problems are deemed politically important and Congress will vote significant sums of money for their study.

Science deals with verifiable facts and observations. Scientists invent hypotheses to explain phenomena; the media often confuse the concept of hypothesis with that of theory, but in science, we have lots of hypotheses (which are a dime a dozen) but very few theories

Of Science and Cathedral-Building

(which are well established edifices of knowledge, verified and confirmed by observation and experiment over a long period of time.) Scientific concepts and ideas should be subjected to vigorous question and debate.

But in fact, no modern scientist can investigate all phenomena to verify the results of others; we rely on peer review and the opinions of experts to evaluate the validity or lack thereof of new claims and discoveries. Thus, faith is an element of modern science too. The scientific establishment deals with heretics and mavericks, sometimes with amusing condescension and at other times with ostracization. Sometimes, particularly for scientific issues that have serious political or economic effects, the current paradigm becomes a catechism and woe to the scientist who dares to dissent[4] from the "consensus[5]" (which in science, sometimes merely reflects the absence of definitive evidence.) As in religion, we scientists also have our fundamentalists, ecumenicalists, and schismatics.

Science is indeed self-correcting in the long run, but the time constant of this corrective mechanism is sometimes longer than a given scientist's career span. Thus, some scientists who were "right" never lived to see their vindication; Alfred Wegener[6], a German geologist who first articulated the evidence for continental drift early in the last century, was widely considered a crank during his lifetime. Only after the sea-floor had been mapped and sampled after World War II did we have the evidence that the continents had moved relative to each other.

Neil Tyson's third motivator is still with us. We call it "Science" and it is a key driver for many social activities, including and especially for space exploration. The current focus of the Mars program is the Search for Life[7], a goal many scientists believe should be the motivator for the entire space program. Some pursue this idea with an intensity I can only describe as "religious" – it is there and we will find it!

Cathedral building is alive and well in the 21st century.

Topics: Lunar Exploration, Lunar Science, Space Politics, Space and Society

Links and References

1. The Daily Planet blog, http://blogs.airspacemag.com/daily-planet/
2. speech by Neil Tyson, http://blogs.airspacemag.com/daily-planet/2009/03/13/neil-degrasse-tysons-broad-vision/
3. three drivers, http://www.haydenplanetarium.org/tyson/read/essays/nathist/reachingforstars
4. Michael Crichton, http://www.michaelcrichton.com/speech-alienscauseglobalwarming.html
5. scientific consensus, http://en.wikipedia.org/wiki/Scientific_consensus
6. Wegener, http://en.wikipedia.org/wiki/Alfred_Wegener
7. Life on Mars, http://www.space.com/scienceastronomy/080627-mars-life-question.html

Comments

1. Problem is, in the old days, fear of the power of deity (or royalty) led to major funded projects in their honor. In modern times, fear of science leads to budget cuts.

Comment by Neil deGrasse Tyson — March 15, 2009 @ 9:21 pm

2. *In modern times, fear of science leads to budget cuts.*

You mean like climate change research? Only some science gets cut; others get increases. Look at NIH's budget over the last 20 years.

Comment by Dr. Paul D. Spudis — March 16, 2009 @ 5:33 am

3. I was expecting at least a nod to the Church of Anthropogenic Climate Change.

Comment by Sissy Willis — March 16, 2009 @ 7:07 pm

4. Thank you for a very thought provoking article on something I always seem to get into conversations with my fundamentalist (Christian) friends and fundamentalist (Atheists/Anti-Creationists) nearly every day.

Science as a deity is a pretty interesting concept. It provides the Hope the traditional religions provide for something better than what currently exists.

It provides fear in the form of exclusionism and the threat that you will be left behind. A prime example was the push it lent to the Manhattan project. Fear that the Axis would find out how before the Allies did was as instrumental as the fact that Allies had the personnel and resources they poured into it.

It provides something that is awe-inspiring (anyone who's actually personally seen a shuttle launch live can attest to this).

Science = Religion. Personally I think its a better one than the older versions, but its a religion just the same. We seem to be made to believe in something....

Comment by Don D. — March 17, 2009 @ 12:55 pm

Mini-SAR Nears Completion of Its First Mapping Cycle

March 29, 2009

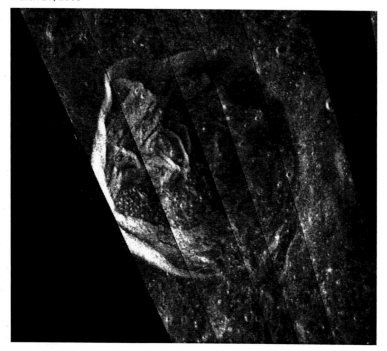

Rozhdestvensky K, near the north pole of the Moon

The Mini-SAR imaging radar[1] aboard the Indian Chandryaan-1 spacecraft[2] currently orbiting the Moon has been sending back some amazing images for the last couple of months. We are nearing the end of our first radar mapping season (which occurs when the sun illumination conditions on the Moon are unfavorable for normal surface or mineral mapping) and I think it's an appropriate time to look at and evaluate the data in hand.

To refresh your memory, Mini-SAR works by sending radio pulses to the Moon from the orbiting spacecraft and then very precisely recording the radio echoes bounced off the surface along with their timing and frequency. From this information, we construct images of the Moon that not only show the terrain in areas we could not otherwise see, such as permanently shadowed areas near the pole, but also contains information on the physical nature of the surface covered, specifically, the presence of terrain with unusual scatter-

ing properties. Such properties can be caused by many different things, such as composition, particle size, and physical configuration. Most famously, radar reflections can indicate the presence of water ice[3], based upon the distinct signature of planetary polar caps and the icy moons of Jupiter.

Typically, radar instruments are very massive and use copious amounts of power, but Mini-SAR is less than 10 kg (22 lbs.) and uses less power than the reading light in your living room. It does, however, generate large amounts of data and only a certain amount can fit in the data recorder aboard the Chandrayaan spacecraft. As imaging cameras and spectral instruments also produce large data volumes, we operate Mini-SAR during times when it does not compete against these instruments. Fortunately, because radar imaging provides its own illumination, we do not need the Sun and all of Mini-SAR's operations occur during lunar night, when the cameras cannot operate anyway.

Our mapping season began in mid-February, 2009. Over the past six weeks, we've mapped about 85% of the polar areas. A curious result of SAR imaging is that we never image the areas directly beneath the spacecraft (the nadir[4] groundtrack), so the poles themselves end up as an excluded zone[5]. There are ways to mitigate this effect and get images of the poles, but orbital mechanics dictate that it will take many months to fill in this gap in coverage. In the mean time, we have lots of data of the near polar regions to analyze, including areas in permanent darkness that may contain water ice.

The initial images look very clean, with a few collection artifacts and some missed orbits. Some of the mosaics have mismatched, offset features, not because of any fault in the instrument but because we still do not have a precise global cartographic control net for the Moon, a missing data set that will be filled by the mapping currently taking place by Chandrayaan, the Japanese Kaguya, the Chinese Chang'E and soon, the American Lunar Reconnaissance Orbiter missions. Much of the shadowed terrain covered by Mini-SAR shows a surface much like the surface of the Moon not in shadow, with small craters of a variety of shapes and sizes present. Some images show spectacular surface features, including wall slumping, central peaks and flat, smooth floors.

A particularly interesting and unusual feature was imaged by Mini-SAR almost by accident. Because of a timing error, we started a few mapping passes of the south pole early, before the scheduled start at 80° south latitude. Good thing we did! We covered the fresh, spectacular Schrödinger impact basin[6], on the lunar far side. Schrödinger shows an unusual, keyhole-shaped crater along a long fissure on the basin floor. This crater is surrounded by optically dark material, which has been interpreted as volcanic ash deposits. The new Mini-SAR image shows that this material is also dark in radar reflectivity[7], exactly what would be expected from a fine-grained, block-free deposit. Thus, our radar images confirm the geological interpretation first derived in 1994 from Clementine images.

The new radar images are not only visually arresting, but they will be extremely useful in unraveling the complex geological history of the Moon as a whole. We are hard at work finishing the calibration of our instrument, which

is required in order to make definite statements about the nature of the radar back-scatter signature, the tell-tale sign of the presence or absence of water ice. This determination is extremely important and one of our major experimental goals, so you'll appreciate that we want to get it right and be as certain as we possibly can be before we pronounce on it. I'll keep you posted on this blog of our progress.

Topics: Lunar Exploration, Lunar Science

Links and References

1. Mini-SAR, http://www.nasa.gov/mission_pages/Mini-RF/main/index.html
2. Chandrayaan-1 mission, http://www.chandrayaan-i.com/
3. Ice on the Moon, http://www.thespacereview.com/article/740/1
4. nadir, http://en.wikipedia.org/wiki/Nadir
5. polar excluded zone, http://www.nasa.gov/mission_pages/Mini-RF/multimedia/1st_map_cycle_1.html
6. Schrödinger, http://en.wikipedia.org/wiki/Schr%C3%B6dinger_(crater)
7. radar dark material, http://www.nasa.gov/mission_pages/Mini-RF/multimedia/1st_map_cycle_6.html

Comments

2. A historical question: is there any reason that bi-static radar of the moon's polar regions could not have been done way back in the 60s by those Lunar Orbiters which flew to polar orbits? Might they have produced the same evidence for lunar water eventually provided by later spacecraft? I know that b-static radar was done be at least on Lunar Orbiter in an equatorial orbit.

Comment by Nels Anderson — August 16, 2009 @ 8:27 pm

3. Nels,

In principle, no. But in the 1960's, we didn't understand planetary radar very well and its use to probe the geological deposits of planetary bodies, or even of the Earth, had not yet achieved any significant level of insight. It's not just getting the radar data — it's having the intellectual background to understand what it's telling you that was required, a state not fully reached even today.

Comment by Paul D. Spudis — August 17, 2009 @ 5:40 am

moon vs. Moon:
A Study in Arrant Pedantry

April 2, 2009

Pedant's paradise – The Moon.

When you write, do you capitalize the word "Moon?" And by this, I mean Earth's Moon, Luna, the natural satellite of our home planet. Well, believe it or not, some of the longest, most vociferous, and yes – the dumbest – arguments I've ever had were over this issue.

In the preface of my book, The Once and Future Moon[1], I argued over a decade ago that the Moon was one of the largest satellites in the Solar System, our first destination off the Earth and mankind's future home and thus, deserved the dignity of capitalization. I proceeded to capitalize the word "Moon" ever afterwards, except when I write for the press, which obstinately insists that it should be "moon" and ruthlessly proceeds to change all my brilliant text.

Why do so many editors insist upon this obnoxious practice? Apparently because the Associated Press (AP) Stylebook[2] says so. To quote the black book of AP style directly:

AP capitalizes the proper names of planets, including Earth, stars, constellations, etc., but lowercases sun and moon.

Uh, OK. I guess that settles that. Ordinarily, I like ex cathedra pronouncements about language (Fowler was famous for them), but usually, they tend to have some reasonable basis in grammatical or linguistic fact. If there is such a basis for the "rule" given above, I don't know what it is. I can speculate on one.

All of the major bodies of our Solar System have Roman (Latin) names – Mercury, Venus, Jupiter, etc. The only exceptions are the objects Earth, Moon, and Sun, whose names are derived from Germanic languages (the Latin equivalents are Terra, Luna, and Sol, respectively). Interestingly, the AP stylebook says to capitalize the Earth but not the Sun and Moon. My guess is that some classically educated nit-picker who was forced to sit through endless hours on the joys of the ablative absolute in Latin class decided that the Roman-named objects of the universe were worthy of linguistic worship, but the vulgar, barbarian Germanic names given to those other three bodies did not deserve to be capitalized.

I beg to differ. All three words are proper nouns; they refer to definite objects, one of which is home to humanity itself and another that soon will be. If these objects do not deserve capitalization, what does?

Simply put, the AP Stylebook is wrong. When referring to "the Moon" – that is, our Moon, Luna, site of Neil Armstrong's landing in 1969 – the word should be capitalized. When referring to any moon, such as in "the moons of Jupiter", it becomes a generic descriptor and hence, should not be capitalized. Our Moon is a world with its own history, one intimately entwined with our own. It has the material and energy resources needed to help us bootstrap a true space faring capability. It will one day become a second home for humanity.

After being criticized for ending too many sentences with a preposition, Sir Winston Churchill supposedly responded: "This is the sort of arrant pedantry up with which I will not put." Whether Churchill said this or not (it is disputed[3], but it certainly sounds like him), it nicely captures my thoughts on this "controversy."

Now, if we could just get the BBC to stop writing "Nasa[4]" for "NASA"……

Topics: Lunar Exploration, Space and Society

Links and References

1. The Once and Future Moon, http://tinyurl.com/23qjfat
2. Associated Press Stylebook, http://www.apstylebook.com/
3. Churchill quote, http://itre.cis.upenn.edu/~myl/languagelog/archives/001715.html
4. "nasa" http://www.bbc.co.uk/topics/nasa

Comments

1. I absolutely agree! The Moon is a specific place, whose beauty and uniqueness deserves capitalization far more than all the little runt satellites of the outer planets!

Comment by Chuck Wood — April 2, 2009 @ 6:30 pm

2. Most of the questionable decisions I've found in AP style (e.g. lower-casing "the President") appear to be based not on grammar or other logic but simply upon type size. A capital _M_ takes up more room in the line of type; newspapers want to pack as many words into the column-inch as possible; ergo they'll use the smaller alphanumeric character, in this case the little _m_.

And although it could be argued that there is only one true "moon" and all others are merely satellites (since their period of revolution is not a month), I would still like to see this exclusivity emphasized with a capital _M_.

Comment by Michael Spence — April 2, 2009 @ 8:56 pm

3. I capitalized Moon in my atlas, and (if my memory is correct, it was a while ago) I checked the preface to your book before deciding to do so. As it coincided with my own preference, it had to be right!

Comment by Phil Stooke — April 4, 2009 @ 2:01 pm

4. Well said. I always capitalize "Moon" and find it pet-peevy if I see it with a lower-case m. All other satellites have proper names — just as ours used to. First it was the Greek Goddess Selene, then the Roman (Latin) Luna. The idea that we've added an article to it, making it "the Moon" doesn't make it any less a prominent satellite deserving of a proper name.

Comment by Pillownaut — April 4, 2009 @ 2:26 pm

5. Thanks for this well-stated post. I got into an argument with a colleague once on the subject of whether "martian" should be capitalized. I argued that it should not because it is an adjective, just like "lunar." What is your position?

Comment by Brian Shiro — April 10, 2009 @ 7:52 pm

6. Brian,

I agree with you. A lowercase "martian" is an adjective. When "Martian" is capitalized, it is a noun and refers to an inhabitant of Mars, as in "Marvin is a Martian."

Comment by Dr. Paul D. Spudis — April 11, 2009 @ 5:16 am

Those Were The Days....

April 10, 2009

The TRW Capistrano Test Site

An item caught my eye[1] this morning as I scanned the space news of the day. A famous aerospace facility, the TRW Capistrano Test Site[2] in southern California is closing. The closure of a space facility is hardly news. In fact, such a headline could have been written any time over the last 20 years or so. But it got me thinking about the nature of the spaceflight business and why things seem to be more difficult to accomplish now than in the "good ol' days" of Apollo.

I have argued elsewhere[3] that for many reasons, the Apollo program is not a good template for new space endeavors. Yet the idea that we have somehow lost something – something vital – since the days of Apollo persists in the minds of many space advocates. They're not wrong. Among other things, we have lost a significant and critical component of space exploration: a robust aerospace industrial infrastructure.

It was inevitable that with the end of the Cold War in the 1990's, many high-technology companies with lucrative defense contracts would either re-focus their attention elsewhere (such as consumer products), be bought out by other larger companies, or simply go out of business. This was and is a critical issue

for national security. Many of these companies made parts and subsystems vital to the proper functioning of national defense systems. I always thought that the real motivation of the Space Exploration Initiative[4] of President George H. W. Bush in 1989 was to give this industrial base something difficult yet achievable to do, keeping them occupied with challenging work until they were needed for future defense production.

President Bush never articulated this reason for SEI and with his defeat in the election of 1992, it was easy to write off the SEI as a failed political "stunt" (a criticism I've never understood as no President has ever reaped any political gains from making a major space declaration and that includes John Kennedy, who was in serious political trouble in the south in 1963, even after shoveling tons of cash into Dixie from the Apollo program.) In the post-Cold War era, the fight against terrorism soon emerged as a national defense priority, but this war was different and largely fought with existing space assets and technologies.

A few isolated projects in the 1990's were able to draw on the legacy technical base of the Cold War. The 1994 Clementine mission[5] was a test of very low mass, low power sensors originally developed for defense, but applied to space exploration. Some of those instruments were manufactured by small companies that no longer exist. Over the course of the last five years, as we built the Mini-SAR instrument now orbiting the Moon on Chandrayaan-1, our progress was impeded because many of the parts and subsystems we needed were unavailable or only available from a small pool of vendors.

There were only two major bids to NASA in 2006 for the development of the new Orion spacecraft, the replacement for the Space Shuttle now under development. Contrast this with the 12 bids NASA considered for the Apollo spacecraft in 1961. The difference is not simply a consequence of corporations teaming with each other; joint proposals were common during the Apollo program. The difference is that there are few aerospace companies left. Many of the small high-technology companies and the highly skilled workforce that made up our industrial base during the Cold War are gone or have been absorbed into the body of Boeing-LockMart-Grumman.

The net effect of all this is less innovation and resourcefulness in the execution of space projects. As they age, the tendency is for organizations to continue to do business in a certain way because "that's the way we've always done it." Institutions become more risk averse and less enterprising with time. It's no surprise to read headlines about new delays in the Orion CEV program and how our return to the Moon will take longer and cost more because we have come to expect that from our space program.

It's not surprising. Just depressing.

Topics: Space Politics, Space and Society

Links and References

1. Closing of Capistrano site, http://sciencedude.
freedomblogging.com/2009/04/09/oc-factory-that-put-men-
on-moon-to-close/24657/#comment-15125
2. TRW Capistrano site, http://www.aoainc.com/capabilities/
space/propulsion/capistrano/capistrano.html
3. What Apollo was, http://blogs.airspacemag.com/
moon/2009/01/25/what-apollo-was-and-wasnt/
4. SEI, http://history.nasa.gov/sei.htm
5. Clementine mission, http://www.cmf.nrl.navy.mil/
clementine/

Comments

1. I think you may have this somewhat backwards. Less aerospace companies does not (necessarily) result in less innovation. The loss of innovation looks to me to be more a product of less small programs. All too many

aerospace programs are huge with all-encompassing scope and excruciatingly detailed requirements. It is very hard to be both innovative and successful with such enormous programs. True innovation is risky.

Small programs can be much more risky. Failure of them is not as expensive or embarrassing. Lessons learned from them can be applied more quickly. Some of them will be very successful. Others will fail. They can all be learned from. Small programs can (usually) be handled by small companies.

Unfortunately, small programs also look more expensive. Single all-encompassing programs usually appear to have efficiencies of scale. That seems to usually be a mirage.

Large programs generally take large companies to manage and coordinate, even though they nearly always subcontract out much (or most) of the work to small companies. The corporate culture and knowledge base needed for a successful large company is difficult to form and takes time to gel.

NASA's current Constellation Program (composed of Orion, Ares I, Ares V, Altair, CSSS Spacesuit, etc.) unfortunately looks to be a textbook example of a large program. NASA's much smaller COTS (Commercial Orbital Transportation Services) program is an interesting contrast. COTS competes almost directly with the Orion part of Constellation (at least it could if the COTS-D contract is pursued). It would be very interesting to see which succeeded.

I believe that we need more small programs.

Comment by Tom D — April 10, 2009 @ 5:20 pm

2. Tom,

Thanks for your comment. I am a big fan of small programs (like Clementine), but my piece was focused more towards the deterioration of our national technical support base, which for the last 60 years has been supported mostly by defense spending. Although that's a "big program" in the macro- sense, it actually consisted of a myriad of programs and projects of a wide variety of sizes, all of which required many capabilities and technologies to be developed. What I am concerned about is the erosion and extinction of these capabilities.

Comment by Dr. Paul D. Spudis — April 11, 2009 @ 5:22 am

3. I think you have it exactly right. When I was in high school, I visited the Bethpage Long Island Grumman plant several times where I saw the manufacturing underway of Apollo Lunar Modules (also F-14 fighters and other aircraft). I had been active at YMCA and Boys Club in woodworking and metal working, and here were the same kinds of planers, lathes, and other machines but on a huge industrial scale – building sized. They would start with solid blocks of metal and machine it down to the needed components – the shell of the ascent module.

Today, much of the area where the plant lay is a shopping mall. Along with the torn down or shuttered buildings, where are all the engineers and technicians ?

In the early 1980s I worked for Rockwell and regularly visited the Downey plant where the Shuttle orbiters were manufactured and assembled, and before that Shuttle, Apollo, and before that various aircraft going back at least to the P-51 Mustang of WWII. Today that plant lies in mothballs – shut down. No capability.

Fifteen years ago I was assigned to work on NASA Mir, leading the redesign and integration of the Ikas module of Mir, Priroda. NASA-JSC was just about to go to full cost accounting but it had not quite happened yet. So when we wanted to design and build and test and certify hardware, we had the machine shops, the test facilities, and the technicians.

We redesigned the Russian module, built a US secondary mechanical, electrical and data infrastructure, and we did, most of it in-house at NASA's JSC. Because we needed the hardware fast, to meet Russian schedules, we built an enormous amount of flight, training and back-up hardware-racks, lockers, electrical systems, computers and data systems, and all of the utility routing. We did it on an expedited schedule working around the clock. In several cases hardware went from conception to flight in a matter of months. We were able to move quickly because we had an experienced cadre and the required facilities were open and available.

A good example was the COSS computer training system. The requirement for this orbital system was realized during Norm Thagard's long duration Mir stay in July, 1995. The hardware and software was needed in time for Shannon Lucid's mission in March, 1996, less than a year. We were able to design the system, procure, manufacture, modify, test, certify and integrate the hardware and software for flight. The hardware flew on schedule. Today as a result of full cost accounting many of the facilities required to support these kinds of DDT&E have been shut down and the experienced workforce is no longer in place.

An important part of the Nation's Vision should be to ensure the nation has the required facilities and a trained and experienced workforce that is kept in place and working towards future missions on a continuing basis.

At one time, aerospace manufacturing was one of the greatest and most profitable manufacturing industries in the US. We cannot afford to lose the facilities, capabilities, expertise and personnel or the nation's industry.

Comment by GaryK — April 30, 2009 @ 5:11 pm

4. Hi Gary,

Many thanks for your thoughtful comment. I neglected to mention in my piece the long-term deterioration of NASA's in-house engineering capabilities. As we lose these assets in both the private sector and the government, who will pick up the slack? I suspect foreign entities, a result not necessarily good for us as a nation. Thanks for dropping by and come back often!

Comment by Dr. Paul D. Spudis — May 1, 2009 @ 8:58 am

The Deadly Dust of the Moon

April 24, 2009

An inhabitant of the lunar outpost deals with electrostatic charging of dust

Lunar dust sticks to everything! It's electrically charged! It causes silicosis[1] – astronauts on the Moon will get "black lung" disease, just like coal miners on Earth! It's so abrasive that under its obnoxious influence, moving parts slowly grind to a halt! We can't possibly cope with it! So much for our plans to live and work on the Moon. Guess we better stick to low Earth orbit.

To see the headlines[2] and read some[3] of the articles[4] written on this topic[5] over the last few years, one might get these impressions. Is any of it true? We actually have some hard data on this subject from measurements made on the Moon, experience dealing with dust during the Apollo missions, and from studies of returned lunar samples. Being a unique product of its environment and the processes that made it, the dust of the Moon is fascinating material. But is it a "show-stopper" for lunar return?

The surface of the Moon is covered by a fine powder, the ground-up reside of surface rocks produced over eons by micrometeorite bombardment. The lunar soil or "regolith[6]" covers everything and as a surface layer, its thickness increases with time. Because lunar soil is formed by impact disaggregation, the edges of the dust grains are angular and sharp; no water or wind erosion is present to round off the grain edges as on the Earth. The highly angular nature of lunar dust makes it self-compacting. The upper few centimeters of soil are loose and fluffy, but the soil becomes considerably denser and well packed below this depth. The dust is very fine grained; the mean grain size of lunar regolith is about 40 microns, roughly half the width of a human hair and a little coarser than talcum powder (about 10 microns), but hard and abrasive instead of soft.

The experience of the Apollo astronauts with lunar dust was that of the unprepared meeting the unknown. We didn't really understand the nature of the dust when the Apollo EVA suits and equipment were designed. The crews went to the Moon with brushes to clean off dusty surfaces, but the adhesive nature of the dust (caused by the high degree of angularity of each grain) was not appreciated. We also didn't (and still don't) understand the electrostatic charging properties that makes the dust "cling" to things, just as your clothes do when you first take them out of the dryer.

Dust got into everything during the Apollo missions. Plastic bags refused to seal properly. Fenders fell off the lunar rover, spraying dust all over the astronauts and their equipment. Metal seal rings on space suits because clogged with dust and refused to seat properly. As if all this weren't bad enough, the astronauts themselves seemed to revel in getting down and wallowing in the stuff, covering their lower bodies with black, charcoal-like smudges. After a hard day exploring the Moon, the crews noted the acrid smell of dust[7] in the LM cabin (Buzz Aldrin said that it smelled like gunpowder.) They breathed it into their lungs during the rest periods.

So, will lunar dust prohibit long-term habitation of the Moon? Hardly. Lunar dust has some unique properties that require careful consideration during the design of surface systems, but many of its alleged hazards can be avoided or mitigated. One of the simplest things to do is to avoid contaminating things with dust. The Apollo astronauts got covered with dust largely because they could not bend at the waist in their suits. They would fall forward on their faces[8], stop the fall with one hand, grab a rock or tool with the other and then do a one-handed push-up to stand up. This technique worked great – except for covering their suits with dust. New suits will allow crews much greater flexibility, including the ability to bend at the waist.

Another idea is to keep the suits outside[9], leaving the interior of the habitat completely dust free. But no matter how carefully we avoid it, some dust will get into places we want to keep clean. Brushing only seems to grind it into porous surfaces. But amazingly enough, we have found that much of the dust is magnetic[10]. Vapor-deposited metallic iron coats the surfaces of many mineral and glass dust grains. This so-called "nanophase iron" (from its extremely small size) makes the dust easily attracted to a simple magnet. A brush made with magnetic bristles will clean surfaces of most of the dust. Incidentally, this same property permits the lunar soil to be fused into glass using low-energy microwaves, allowing us to "pave" roads and landing pads near and around the lunar outpost and to make bricks for construction and radiation shielding.

Although we do not fully understand the electrical charging properties of lunar dust, several experiments on current and future robotic missions[11] will characterize these properties thoroughly. The idea that charged dust can levitate and coat exposed surfaces is widely believed, but there is no solid evidence that this process occurs to any significant degree and there is considerable evidence that it does not. The Surveyor 3 spacecraft, exposed on the lunar surface for over 30 months before examination by the Apollo 12 crew, was not covered by any significant amount of dust, other than that thrown up by the nearby landing of the Lunar Module.

Dust is both an asset and an issue. We will process the dust of the Moon to extract useable products, like hydrogen and oxygen. At the same time, we will learn how to live with it. We'd better

– dust occurs throughout the Solar System. On Mars, it may have toxic chemical properties. It's not a deadly hazard –just another property of the new worlds to which we journey.

Topics: Lunar Exploration, Lunar Resources, Lunar Science

Links and References

1. silicosis, http://en.wikipedia.org/wiki/Silicosis
2. dust, http://www.space.com/scienceastronomy/090421-st-moon-dust-sunangle.html
3. dust, http://www.newscientist.com/article/dn11326-lint-rollers-may-collect-dangerous-moon-dust.html
4. dust, http://www.msnbc.msn.com/id/30334967/
5. dust, http://www.redorbit.com/news/space/1570115/the_trouble_with_moon_dust/
6. regolith, http://en.wikipedia.org/wiki/Regolith
7. Apollo dust smell, http://science.nasa.gov/headlines/y2006/30jan_smellofmoondust.htm
8. dust fall, http://www.moondaily.com/reports/Dust_Busting_Lunar_Style_999.html
9. suits outside, http://www.newscientist.com/article/dn15047-no-spacesuits-needed-in-new-lunar-rover.html
10. magnetic dust, http://www.newscientist.com/article/dn11406-magnetic-elephant-trunk-sucks-up-lunar-soil.html
11. future robotic missions, http://www.space.com/missionlaunches/080410-ladee-moon-dust-mission.html

Comments

1. Dr. Spudis: Thanks for giving the talk at Lunar & Planetary Institute last night. We are doing experiments to ensure that lunar dust is not an obstacle but a resource for lunar missions.

Comment by L Riofrio — April 24, 2009 @ 1:50 pm

2. Great! I have no doubt that the dust can be handled. I just don't think it's the serious problem that it's often portrayed.

And many thanks for coming to my public lecture at the LPI last night!

Comment by Dr. Paul D. Spudis — April 24, 2009 @ 2:24 pm

3. [...] Dr. Paul D. Spudis put an intriguing blog post on The Deadly Dust of the Moon | The Once and Future Moon. Here's a quick excerpt. This so-called "nanophase iron" (from its extremely small size) makes the dust easily attracted to a simple magnet. A brush made with magnetic bristles will clean surfaces of most of the dust. Incidentally, this same property permits ... The Surveyor 3 spacecraft, exposed on the lunar surface for over 30 months before examination by the Apollo 12 crew, was not covered by any significant amount of dust, other than that thrown up by the nearby landing of the Lunar Module. ... [...]

Pingback by Topics about Electromagnetic-pulse | The Deadly Dust of the Moon | The Once and Future Moon — April 30, 2009 @ 9:42 pm

Return to the Moon: Outpost or Sorties?

May 5, 2009

Outpost on the Moon: Too expensive?

Recently, the acting Administrator of NASA testified[1] before Congress on his agency's implementation[2] of our National Space Policy, previously known as the Vision for Space Exploration[3] (VSE). In the question and answer period, he made a rather startling statement to the effect NASA was still trying to understand what "lunar return" means – that an outpost would be "expensive" and that lunar return might instead entail a series of smaller scale sortie missions, similar to the later Apollo expeditions of the early 1970's. He added that people should remember that the "original purpose" of the VSE was to prepare to go to Mars and other destinations.

I found this exchange fascinating because it suggests that NASA, as an executing entity, still doesn't fully understand the nature of their mission to the Moon[4] and to the extent that it is understood, they have transformed it into something very different from what the VSE actually said and what was intended.

To begin with, what did the Vision actually say about lunar return? The Vision consisted of both documents and speeches (all linked on this page[5]) that included the following points:

1. The purpose of the VSE is to serve national scientific, security and economic interests. 2. The Moon is a source of material and energy resources that we can access and use to create new space faring capability. 3. We return to the Moon to explore it scientifically, to learn how to live and work on another world and how to extract and use lunar resources. 4. This experience on the Moon will allow us to journey beyond the Earth-Moon system, first to Mars and then to other destinations. 5. We undertake this journey with small, incremental, cumulative steps, all designed to fit under NASA's current budgetary envelope.

Apparently, the view of NASA's acting Administrator is that the Moon is a box to be checked-off on the way to Mars. Hence, we don't really need to establish an outpost because we're just satisfying a political requirement in implementing policy, not conducting a technical experiment to use the Moon to prepare for journeys beyond.

Why is an outpost on the Moon necessary? Or more pressingly, why are sortie missions undesirable? Each time you go to the Moon, launch and spacecraft assets (in the form of equipment and vehicles) are expended. This equipment has significant cost; it is currently estimated that a single Orion sortie mission to the Moon might cost upwards of $ 4-5 billion dollars. A sortie mission goes to a single, specially designated landing site, the crew

61

explores the local area, sets up some experiments and then departs. All equipment brought to the Moon is abandoned. Thus, a sortie mission's high costs are pure expenditure, not investment. The return from sortie missions is scientific knowledge, something which as a lunar scientist, I certainly desire. But a sortie mission does little to advance the goals of learning how to live on another world or for extracting and using local resources, except for relatively short duration stays and technology demonstrations.

In contrast, an outpost at a single, optimally placed location can be built up from delivered hardware as each flight incrementally adds some equipment or facility. Thus, the cumulative capability of the outpost increases with time. This seems fairly obvious and in fact, has been recognized for many years; Arthur C. Clarke's classic book The Exploration of Space (1951)[6] notes:

> For a considerable time all flights to the Moon would be directed to the same spot, so that material and stores could be accumulated where they would be most effective. There would be no scattering of resources over the Moon's twelve million square miles of surface – an area almost exactly the same as that of Africa.

An outpost allows infrastructure to be built up rapidly and over the long run, permits and enables more exploration than a series of sortie missions. In other words, you build up capability on the Moon first, then stage the sortie missions from the Moon rather than from the Earth. Ultimately, this permits longer and more capable exploration than would otherwise be possible. The idea that an outpost is "more expensive" than sortie missions is ludicrous; an outpost can be built at any rate that budgetary limits permit. And it represents more value for expenditure because equipment gets re-used or at a minimum, used once to the maximum extent possible.

Much of this is so obvious as to be beyond debate. Could the agency's focus on sortie missions be because those implementing the program don't believe in the primary mission, which is to use the Moon to learn how to live off planet[7]? Do they think it unlikely that this will be possible or do they believe that activities on the Moon will "bog them down?" Is the strategy to advocate short duration sortie missions to "get data" and conduct their experiments before ending the program? The Vision's purpose is clear, so one must conclude that they

are eager to "check off the box" for the Moon and get to their mission of going to Mars. To do that, the first three parts of the VSE must be curtailed. As others have pointed out, the Constellation program is not optimized for the establishment of an outpost on the Moon.

A return to the Moon using the Apollo exploratory template is truly a "been there, done that[8]" exercise in space futility. The Vision was never intended to be a repeat of Apollo – the idea was to use the Moon to create new space faring capabilities. This is a task that's never even been attempted in space, let alone accomplished. It is the antithesis of "been there, done that."

It appears that some at NASA have been successful in obfuscating the purpose of the Vision.

Topic: Lunar Exploration

Links and References

1. NASA testimony, http://www.flightglobal.com/articles/2009/04/30/325885/nasa-reconsiders-outpost-as-2020-moon-goal.html
2. VSE implementation, http://www.newscientist.com/article/dn17052-nasa-may-abandon-plans-for-moon-base.html
3. Vision for Space Exploration, http://www.spaceref.com/news/viewpr.html?pid=13404
4. Vision and Mission, http://www.spudislunarresources.com/Papers/The%20Vision%20and%20the%20Mission.pdf
5. VSE links, http://www.spudislunarresources.com/Links.htm
6. Arthur C. Clarke, The Exploration of Space, http://www.amazon.com/Exploration-Space-Fawcett-Premier-Book/dp/B000VONR3W/ref=sr_1_7?ie=UTF8&s=books&qid=1241455107&sr=1-7
7. Use the Moon, http://www.spudislunarresources.com/Papers/Spudis_NLSI_July_2008.pdf
8. been there, done that, http://spudislunarresources.blogspot.com/2008/07/been-there-done-that-space-policy.html

Comments

1. Outpost or nothing! an Apollo-remake has little or no sense! however, also talk now about that is a nonsense, since the Orion/Ares-1 will fly only in 2018 to the ISS and the first/new manned lunar landing should happen in 2022 or later... now, I believe it's much more urgent to talk about the RISKS of the Hubble SM4 that should start next week

After posting several comments on US' space forums and blogs (with links to my article about the Hubble SM4 risks) my question is: "Does the American Press take care of the Atlantis' astronauts lives?"

Well, while waiting for an answer about this question, I've UPDATED my article with other concerns regarding the STS-400 rescue mission, the space-junk problem and the Atlantis'

Return to the Moon: Outpost or Sorties?

radiator issue, also, in the same article, I give some suggestions to increase the Atlantis' astronauts chances to survive to this risky mission (if "something goes wrong" of course)

http://www.ghostnasa.com/posts/044sm4risks.html

Comment by gaetano marano - ghost-NASA.com — May 5, 2009 @ 1:53 pm

2. This does a very good job of demonstrating that NASA's purpose these days is very little more than keeping themselves employed. Government space programs exist to put government people into government-operated space; and they have no interest in anything that threatens that. Colonization of space means people living places and doing things outside the government's purview, and that seriously threatens the paradigm.

One does not need to believe in any conspiracies to explain this; it's just the result of the economic incentives at work. It does explain why there's been no meaningful colonization efforts (or even travel beyond LEO since the Apollo days).

Comment by Carl Soderstrom — May 5, 2009 @ 2:30 pm

3. Well said, Dr. Spudis!

Comment by James Antifaev — May 5, 2009 @ 2:38 pm

5. Good points all Paul.

Comment by Dennis Wingo — May 5, 2009 @ 5:41 pm

6. I completely agree with Paul, and I know a little about what NASA is doing, since I work for NASA and have been intimately involved in Constellation activities. What this points out is the very real need for some entity OUTSIDE of NASA to set strategic focus and priority for NASA, along with objectives.

When we were working on the first Lunar Architecture Team in late 2006, NASA attempted to gather an international group to help define the rationale for going back to the Moon (publicly called Themes for returning to the Moon). These Themes formed the original basis for the work that we did on the lunar architecture, but later these Themes were increasingly left by the wayside and did not drive the thinking nor the strategic decisions formulating the general shape of the lunar return. As a consequence, lunar architectural decisions were increasingly made based upon hollow strategic and parochial logic as well as cost, and not driven by more specific goals like Paul points out such as "learning to live off-planet".

Case in point: Original need for ISRU investment was strong, and would be strong for a goal or Objective to "Learn to Live Off-Planet", but subsequent emphasis on it has diminished. Further, instead of forcing NASA to look at local lunar resources for oxidizer for the Lunar Ascent Module, the emphasis now is on hypergolics, which cannot be generated on the Moon.

While NASA does get strategic input and priority for the Science missions, it did not receive this on the human missions, and we have been suffering because of this lack for 25 years.

The best thing that someone can do is advocate to this Administration that Strategic Objectives need to be provided to NASA, not just a general framework, since significant decisions will be relegated to short term cost or parochial interests.

Comment by Tony Lavoie — May 5, 2009 @ 9:20 pm

7. Yes, these are good points. I also believe the point by Mr. Soderstrom has great value as well.

Mr. Krukin latest post raises an interesting question about President Obama's understanding of the "commercial settlement and development of space". The 5 points above should be the foundation of dialog between Americans and their president. How do we create this dialog?

Comment by Evon Speckhard — May 5, 2009 @ 10:13 pm

8. WE (talking about that) and NASA (adding more life support time to the SM4) have just FIVE days to give more chances to survive to the Atlantis' crew

Comment by gaetano marano - ghost-NASA.com — May 6, 2009 @ 9:44 am

9. Gaetano,

I appreciate your passion on issues in regard to the Hubble mission, but please keep this forum for on-topic comments and discussion. Thank you.

Comment by Dr. Paul D. Spudis — May 6, 2009 @ 11:35 am

10. *Each time you go to the Moon, launch and spacecraft assets (in the form of equipment and vehicles) are expended.*

Of course this needn't be true, even for sortie missions…

http://www.tallgeorge.com/projectcon-stellation.php#The%20Altair%20Moon%20Lander:2007%20Lockheed%20Martin%20Proposal

http://www.nasaspaceflight.com/2007/04/lockheed-studies-centaur-applications-for-lunar-lander-concepts/

http://www.slideshare.net/niceguyted/reus-able-lunar-lander-1308693

...but in an architecture that doesn't use Earth orbital assembly and/or refueling (and/or RLVs), it may be unavoidable.

If reusability isn't designed in at the start, it's no surprise that it's not there at the end. Sorties *and* outpost building can only benefit from it.

Comment by Frank Glover — May 6, 2009 @ 11:39 am

12. This whole thing is nothing but an attempt by the Obama Administration to make it easy to kill any space program that entails flight beyond Low Earth Orbit. First they'll kill the idea of a lunar outpost, then they'll kill the Orion heavy lift vehicle and the whole idea of even the sorties. Then we'll be left with a vehicle with about the same capabilities as the Russian Soyuz and Chinese Shenzhou vehicles and still be stuck in LEO for the next 25 to 35 years. This plain sucks.

Comment by Ross Warren — May 7, 2009 @ 2:19 pm

14. Guys –

I mostly agree with Paul; and totally agree with Tony Lavoie. Even when given an original specific direction, our Agency, once faced with the inevitable challenges (budget, etc.) keeps defaulting to a mode where we forget what the true strategic goals are – or at least, need to be. We did the same with shuttle as budget times got harder and harder; made it unaffordable to own without telling anyone that's how we were handling the challenges.

The more correct answer to Congress' question might be something like this:

Establishing a long-term base at one good location, like we've had in Antarctica for decades, is justified on a scientific basis alone – Antarctica. But in addition, we know we have

to learn to live off the land in space to afford to stay there, and a place only three days away is the best place to do that. Plus showing we can live, work, learn, and reduce the cost of staying there over time is exactly the type of experience we need to show you in Congress and the American people we can go beyond the moon.

This sort of rationale means you don't put ISRU in the back seat for years- you put it closer to the front. You use scenarios that already exist, like Antarctica; and you also make it clear that we know we won't get permission or funds to go beyond the moon unless we prove to Congress et al that we can sustainability stay on the moon – and get something out of it in the process. And once we establish a base, others will, too. Commercially (for tourism, first), as well as bases by other nations – again, like Antarctica.

However, if we can't have, establish, and then operate in a way we can afford to own a simple Antarctic-style station on the moon, then we won't be able to sustain anything more grandiose beyond it. The moon – and I would maintain even more so, NEOs, with their potential danger and great resources – presents not just technical and operational learning experiences we need; but the needed demos for economic and political sustainability as well.

Comment by Dave Huntsman — May 8, 2009 @ 9:34 pm

16. So if it is to become a field trip why not then re-engineer a Soyuz for lunar reentry, re-make the Grumman LM with modern electronics and send them over using a DIRECT launcher? maybe there wont be as much storage space for rocks but remote sensing equipment is adequate enough these days, right?

You could also build a Skylab-sized can and send some 'big brother' contestants to Mars on a free-return trajectory.

All this could still be cheaper and quicker than the Orion/Ares project, have a similar feel-good factor as well as rate well on TV...

Comment by Phil Thomas — May 15, 2009 @ 6:43 am

What the Augustine Committee Didn't Know in 1990

May 15, 2009

Earth set over the south pole of the Moon, seen from the Kaguya orbiter

A newly formed commission[1] led by Norman Augustine will review NASA's human spaceflight program with the aim of determining if we are on the "right track." This is familiar territory for Augustine, who led the 1990 Advisory Committee on the Future of the US Space Program[2]. Now, 19 years later, it may seem that he's treading across similar ground, but the landscape has changed.

Compiled in the wake of the Challenger accident, the first loss of an American crew in flight, the Commission was charged to consider 1) Should America have a human space program; and 2) if yes, what should be its goals? The 1990 report concluded that human spaceflight was an essential part of the program and that the long-term goal of human exploration is Mars, preceded by a LEO Space Station which emphasizes life-sciences and microgravity research, an exploration base on the Moon, and robotic precursors to the Moon and Mars (Recommendations 4-7[3]). The view of the Moon in Augustine 1990 was to explore the Moon as a test-bed for future missions to Mars. The report emphasized that the lunar base was to be built such that it could be human-tended, but not require permanent staffing. Surface activities were worded carefully because as far as it was known in 1990, the Moon offered little or nothing to sustain a long-term human presence.

Sustainability on the Moon revolves around two principal issues: power and consumables. Destinations in space typically lack one or both of these vital commodities. That's why we bring all our consumables (air, water, rocket propellant and electrical power) with us. For some time now, we've been able to generate electrical power in space using solar arrays, but these have the drawback of being unusable during night and periods of eclipse.

The Moon was viewed as a barren desert. True, it has no atmosphere or flowing water. Lunar samples are exceedingly dry and contain no water-bearing minerals. Moreover, the Moon has a slow rotation rate (28 days), meaning that after a scorching 14 Earth days in sunlight, you freeze for 14 days of cold, no-solar-power nights. These properties led to a concept of operations that emphasized transient stays on the Moon, science exploration and use of the Moon as a test-bed for Mars.

However, in the intervening twenty years since that Augustine report, several robotic missions have changed the way we perceive the Moon. We found that the poles are very different from the rest of the Moon. The 1994 Clementine mission[4] found large areas in permanent shadow near both poles; the sun never reaches the bottoms of craters here because Moon's spin axis is almost perpendicular to the plane of Earth's orbit around the Sun (the ecliptic). Such areas are extremely cold, possibly only a few tens of degrees above absolute zero. Water added to the Moon through bombardment by water-bearing meteorites and comets for billions of years could be retained in these dark areas. Additionally, we found areas in close proximity to these dark regions on mountain peaks rising above the local horizon that are nearly continuously illuminated by the Sun. In 1998, the Lunar Prospector mission[5] found elevated amounts of hydrogen in the polar regions, consistent with the accumulation of excess volatiles (including water).

So what do these discoveries mean for lunar return? We now know that sustained human presence on the Moon is possible, largely because we've found a source of near-constant power (permanent sunlight) and a source of sustenance and rocket propellant (volatiles, including water). The robotic Clementine and Lunar Prospector missions showed us that the poles, almost completely unknown in 1990, are inviting oases on the lunar desert. There, we can extract hydrogen and oxygen to make air and water for life support and propellant to fuel rockets. The sunlit areas can generate near continuous electrical power, with regenerative fuel cells providing power for the short duration eclipse periods. Locally obtained power and consumables means that continuous human presence is possible, without the enormous expense or unproven technology of large nuclear reactors and the delivery of massive quantities of material from Earth.

The new Augustine committee should be made cognizant of these facts. The more we learn about the true nature of the Moon, the more the goal of learning to live there on a quasi-self sufficient basis appears feasible. This opens up wholly new areas of operations and commerce in space, undreamed of as little as twenty years ago. It has the potential to change the entire paradigm of spaceflight, from a narrow, government-run, science-oriented program, completely dependent upon the caprice Congressional largess to a self-sustaining, free-market program, in which NASA develops and demonstrates new technologies that open up space faring by many different passengers and payloads for a wide variety of purposes.

Much of the original 1990 Augustine report is directly applicable to today[6], including a gradual movement of humans beyond LEO and a "go-as-you-pay[7]" paradigm for agency

funding. By using the Moon, where expanded dimensions in exploration can be developed and tested, we will transform and enhance the business of spaceflight.

Topic: Lunar Exploration

Links and References

1. 2009 Augustine committee, http://blogs.airspacemag.com/daily-planet/2009/05/08/nasa-needs-direction-call-norm-augustine/
2. 1990 Augustine committee, http://history.nasa.gov/augustine/racfup1.htm
3. recommendations, http://history.nasa.gov/augustine/racfup1.htm
4. Clementine mission, http://www.lpi.usra.edu/expmoon/clementine/clementine.html
5. Lunar Prospector mission, http://www.lpi.usra.edu/expmoon/prospector/prospector.html
6. norm v. norm, http://rocketsandsuch.blogspot.com/2009/05/norm-90-v-norm-09.html
7. go as you pay, http://history.nasa.gov/augustine/racfup1.htm

Comments

1. With the greatest of respect I feel I must counter your suggestion that lunar Hydrogen be utilized as rocket propellant! I am of the opinion that barring the discovery of vast quantities of lunar Hydrogen and/ or water ice; that any LUNH be conserved as a "Common Heritage" in a life support loop rather than thrown away. Even on an international mission to Mars! For local 'hops' a Al/LUNOX monopropellant rocket is a viable alternative and the Moon is an ideal launch pad and shipyard for "brute force" Nuclear propulsion systems. Perhaps, if the corrosion problem could be solved, these could just use LUNOX as reaction mass.

Once we are mining NEOs and Main belt asteroids (Ceres!) for water we could return to LH as a propellant. (But we probably won't need to!)

With regards to other lunar resources LUNOX; Al; Si; Ti;... extraction costs will mean that a profit must be had and thus exemption from Common Heritage rules will be required. As a fig leaf for international assent some form of levy: land deeds, taxation, etc should be in place to raise a common International Space Fund for those nations without any space infrastructure and used to bootstrap their off world activities.

In this way the Moon Treaty can be upheld in principle and become the future foundation of the development of Lunar ISRU rather than a roadblock.

Comment by brobof — May 16, 2009 @ 8:17 am

2. First the U.S is not required to follow the Moon Treaty, thanks to the foresight and efforts to stop the U.S from joining it 30 years.

Second, there are no Common Heritage "Rules". The concept of Common Heritage just means that ALL nations have the right to use lunar resources as they wish. No single nation may stop another from having reasonable access to it, but there is no requirement for any nation to share what they use with other nations. The legal principle is the same as fishing on the open ocean, you are free to catch what you want. But you are not allowed to stop others from fishing as well. And of course there is no legal requirement to share what you catch with the world. Its yours to keep.

And this legal principle has already been established by the Russia selling some of their lunar samples into the private market and the U.S. government suing to recover stolen Apollo samples under the laws that govern stolen government property.

Comment by Thomas Matula — May 16, 2009 @ 2:04 pm

3. If lunar water is present but not vastly abundant, more important than drinking it is studying it and other frozen comet gases for a compositional chronology of the projectiles that penetrated the inner solar system over time. We may not be returning to the Moon for science, but to willfully destroy a unique scientific resource would be uncivilized.

Comment by Chuck Wood — May 16, 2009 @ 3:31 pm

4. Dr. Wood,

Granted. And knowing its composition before you use it for any purpose will be important as a safety issue. Trace elements could create a lot of problems. But with modern techniques the amount needed for sampling for scientific research would not be great.

Comment by Thomas Matula — May 16, 2009 @ 6:01 pm

6. *If lunar water is present but not vastly abundant.*

Define "vastly abundant." The best current estimates for the amount of water at each pole is between 1 and 10 billion metric tonnes. If it's even the lower limit, that's enough hydrogen and oxygen to launch the equivalent of one Space Shuttle per day for over 200 years.

Comment by Paul D. Spudis — May 17, 2009 @ 7:51 am

8. As a follower of the Lunar Ice Deposits ever since Clemmie, I too share your optimistic view of an abundance of Lunar Hydrogen. Somewhere! It certainly makes things that much easier and for certain posters to continue to ignore the PRINCIPLE of the Moon Treaty!

However I feel that we must prepare for another disappointment when LCROSS reaches ground zero. The fact that Lunar Prospector failed to kick up a signature in Shoemaker (nee Mawson!) ...was a grave disappointment and the "Arecibo Radar Mapping of the Lunar Poles: A Search for Ice Deposits" by Stacy, Campbell & Ford puts forward a pretty convincing argument that lunar ice may be largely artifactual. Alas, academic arguments aside, nothing will be resolved until after an exhaustive search for the Ground Truth. On the Ground.

Finally, you are no doubt aware of the most recent model by Dr Vincent Eke (Institute for Computational Cosmology, Durham University); "Moon's polar craters could be the place to find lunar ice, scientists report" (Icarus, December 18th, 2008, http://www.physorg.com/news148805928.html) His estimate is 200 MegaTonnes ("metric tons") and sufficient (by extending his simile to that of the Northumbrians it supports!) ... for a Lunar Population of roughly 300,000!

That will do nicely until Ceres sets up its export business!

Comment by brobof — May 18, 2009 @ 6:04 pm

9. brobof,

I am well aware of the scientific argument about polar ice, having been personally involved in it for the last 15 years. The Arecibo results are not contra-indicative of ice, they are ambiguous. We are mapping the poles now using radar in lunar orbit and will soon have data from a better viewing geometry that will allow us to put constraints on the presence and quantities of polar ice. But you are correct — ultimately, we must land on the surface to unambiguously measure the amounts and physical states of polar ice.

Comment by Paul D. Spudis — May 19, 2009 @ 5:59 am

Can We Be "Resourceful" on the Moon? (Part 1)

May 30, 2009

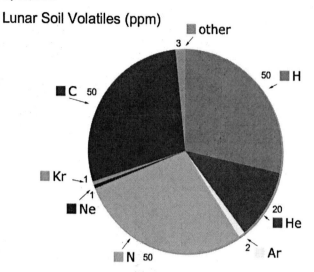

The solar wind implants gases into the lunar soil

While the resources of space have the potential to revolutionize spaceflight—giving us a much wider range of activities than are now possible, including habitation of other planetary bodies—discussions on various internet forums show that there is a lot of confusion and lack of knowledge about space resources in general and lunar resources in particular. To some, the idea of harvesting and using resources from a body other than Earth has a science-fiction aura—for them, air, propellant and food manufactured for people belongs on the silver screen, not in space or on other worlds.

So, what is a "space resource[1]?" In broad terms it is needed materials or energy derived from space itself. The value of space resources is immediately obvious; at transport and delivery costs exceeding $20,000 per pound to low Earth orbit, everything we find and use in space is one more thing we do not have to pay exorbitant costs to transport. For human spaceflight, consumables are heavy but absolutely necessary. This category includes air, water, electrical power, and rocket propellant. On a long space journey, we could save millions to billions of dollars by re-fitting consum-

ables in space rather than dragging them all up to orbit with us from the deep gravity well of the Earth.

Unfortunately, the materials and energy we need in space are not there in the form we need them. Thus, our task is to convert what we find into what we want[2]. There's nothing magical about this – modern industrial chemistry is largely concerned with this very topic. All you need to convert chemical substances in one form to another one is time and energy. Fortunately, both are available in quantity in space, especially if you use automation and robotics to perform a lot of this work. Creating systems to harvest and utilize space resources has the benefit of boot-strapping a self-sustaining, space faring capability instead of remaining tethered to never-ending, one-off expenditures and Earth's gravity well.

For many years we have used energy provided by the sun to generate electrical power in space but we have yet to use any material resources. The Moon is the nearest object offering usable resources[3]. Contrary to common belief, the lunar surface contains virtually all the elements one needs to create usable

Can We Be "Resourceful" on the Moon? (Part 1)

products for human space faring, including air, water and rocket propellant.

It's often said that the Moon is resource-poor. That is inaccurate; the Moon is resource *different*. It is depleted in volatile substances (those that have very low melting points). The most important rare resource on the Moon is hydrogen. The Moon itself has very little of this element, but the soils have a great deal of it; because the Moon has no atmosphere or global magnetic field, the stream of protons from the Sun (the solar wind) implants hydrogen onto the surface of the dust grains on the Moon. This solar wind hydrogen can be released through heating of the dust. When you have both hydrogen and oxygen, you have air, water, and rocket propellant.

The typical hydrogen concentration in most soils is 20 to 100 parts per million. This is enough quantity to extract and use, especially if much of the mining and processing work is done through robotic machines operated from Earth. Hydrogen appears to be present in higher quantities in soils that have high titanium content, which are abundant on the lunar near side (the Apollo 11 landing site has one of the highest titanium contents found on the Moon to date).

Now there are even more exciting resource prospects. The Moon has abundant hydrogen at the poles[4], enriched by more than a factor of three over the global average. Some of this hydrogen, present in the permanently dark and cold floors of polar craters, may be in the form of water ice[5]. Additionally, with the spin axis of the Moon perpendicular to the plane of its orbit around the Sun, some peaks near the poles appear to be in near-permanent sunlight, permitting continuous collection and use of solar electrical power, as well as the important benefit of a near constant surface temperature.

For these reasons, recent international exploration of the Moon has focused on the poles of the Moon, where extracting and using lunar resources is easiest and where humans have the greatest potential to learn how to live off-planet and exploit space resources to create routine access in cislunar space and into the Solar System.

Next: Changing our approach to spaceflight.

Topics: Lunar Exploration, Lunar Resources, Space Transportation

Links and References

1. space resource, http://www.psrd.hawaii.edu/Nov00/mining.html
2. convert what we find, http://www.psrd.hawaii.edu/Dec04/spaceResources.html
3. useable resources, http://www.spudislunarresources.com/Images_Maps/soil.htm
4. hydrogen at the poles, http://www.spectrum.ieee.org/jun09/9329
5. ice on the Moon, http://www.thespacereview.com/article/740/1

Comments

3. Here's a wild idea for the use of lunar resources. It may have no merit, but I found it interesting and would like to hear if any of it makes sense.

It seems that most of the elements present on the moon, not just volatiles implanted by the solar wind, may be used as part of various propellants, though not necessarily in the correct proportions and not without external hydrogen. Use of advanced and unconventional propellant combinations may lead to much more benefit from lunar ISRU and maybe even more efficient ISRU itself.

Two combinations are of particular interest: Silane + H_2O_2

Higher silanes + H_2O_2 are expected to be a dense, space storable and hypergolic propellant combination with better Isp and lower toxicity than MMH/NTO. This is useful for science missions, but since both silicon and oxygen are plentiful on the moon, this propellant combination would also benefit substantially more from lunar ISRU.

Prediction of Performance of (Higher) Silanes in Rocket / Scramjet Engines.

MMH/Al/NTO gels

Under the ISTP project NASA is developing an advanced MMH/NTO engine called AMBR for science missions. The goal is to increase the Isp to about 375s in the next ten years or so. One of the more advanced techniques under consideration is use of gelled propellants, with significant amounts of metal powders combined with the MMH. Depending on the mass fraction of the metal, Isp and density can be increased.

AMBR* Engine for Science Missions

Again, a second effect is that a larger mass fraction of the fuel can be sourced from ISRU. MMH consists entirely of elements that are rare on the moon, whereas aluminium is plentiful. Mass fractions of up to 70% metal are a serious possibility. For high metal fractions, the Isp drops but this is likely more than compensated for by the increased potential for ISRU.

Preliminary Assessment of Using Gelled and Hybrid Propellant Propulsion for VTOL/SSTO Launch Systems

Theoretical Effects of Aluminum Gel Propellant Secondary Atomization on Rocket Engine Performance

The following article gives the average composition of the lunar regolith: The average chemical composition of the lunar surface The main elements are oxygen, silicon, aluminium and various other metals. Oxygen is of course an oxidiser. Silicon can be turned into silane if external hydrogen is provided or it can be used in powder form as an additive to a gel propellant. Aluminium and the other metals can similarly be used as metal-loading for gel propellants.

It looks as if nearly all of it could be turned into propellant, provided external hydrogen is supplied. This may seem no improvement over simply using hydrogen from Earth together with lunar oxygen, but it does mean that all of the regolith/ore could be processed into propellant, producing no slag. If the Isp of the resulting propellant is good enough, it may allow processing of the regolith to be moved to L1 instead of having to do it on the lunar surface.

Establishing the initial infrastructure there is cheaper than doing it on the lunar surface, because of the lower delta-v. Also, with uninterrupted sunlight and cheaper solar panels it would be much easier to generate massive amounts of power. You would still need some propellant production on the surface, but the scale could be reduced. This would work especially well for an architecture that wants to go to L1 anyway.

Comment by Martijn Meijering — June 1, 2009 @ 9:21 pm

4. Dr. Spudis,

Thank you for sharing your thoughts on this blog. Your ability to state the importance of exploring the moon in the 21st century may be unmatched. I hope you have an opportunity to testify before the Augustine Commission.

Comment by Jason — June 2, 2009 @ 12:40 pm

5. "the Moon is resource different"

Absolutely! 'Skimming' the regolith (think way big robotic combine harvesters) will also be a useful way of getting Helium too, as this is a non renewable. On the Earth

http://www.energybulletin.net/node/34563

"Peak Helium". And this may be a viable industry after we are mining NEOs and other cometary debris for CHON. However equating "Unobtanium" to He 3 is a moondoggle too far, as we are far from the technological readiness of MkI Fusion let alone MkII. But is a useful pretext for a film!

A speculative left brain thought: could our regolith fields be 'seeded' with materials designed to retain the wind blown volatiles and make the process more efficient? Or would skimming fresh ground be more productive? A useful near term experiment would be to discover how quickly an area is replenished. And a wild right brain dream: if we are going to create big furrows in Tranquility, why not draw a picture! Mega graphics! Something big enough to see with a telescope but not the naked eye would seem to be appropriate! Note "E=MC^2" has only two curves:)

/speculation

Finally I echo Jason hopes. The commission needs the testimony of someone who has quietly but passionately advocated a return to the Moon over (too) many years!

Fingers Crossed! Dave

Comment by Vacuum.Head — June 4, 2009 @ 5:47 am

Lunar Resources (Part 2): Changing Our Approach to Spaceflight

June 5, 2009

Carnegie-Mellon's Scarab rover: Prospecting for resources

Last time, I outlined some of the basic principles of lunar resource utilization. The Moon is our nearest source of material resources in space and learning how to extract what we need from the Moon is a key skill[1] in our expansion into the Solar System.

All this is very well and good, but how do we go about using the resources of the Moon and of space in general? Many people tend to think of huge industrial factories, similar to oil refineries, built on the Moon, with large mining communities similar to those depicted in the movie *Outland*[2]. In fact, the beauty of space resource utilization is that it's possible to start very small and build up capability with time. The "factory" needed to produce a metric tonne of oxygen on the Moon is the size of a typical office desk.

Mining is a very old and venerable field and has some simple precepts[3]. First, you must find and characterize the prospect. Next, you need to understand the concentrations and physical states of the "ore body" that you wish to mine. Then you must collect the feedstock, convey it for processing, extract the desired element or compound, discard the waste and store the product. For lunar mining, we are now in the process of finding and characterizing the prospect through remote-sensing and mapping of compositions from space and analysis of returned samples. This characterization has been going on for many years, giving us a first-order understanding of the compositions and physical states of lunar materials.

The Moon is rather ordinary in composition[4], having a crust (like the Earth) rich in oxides of aluminum, iron, and silicon. The oxide portion is key: the Moon is over 40 % by weight oxygen. This oxygen is tightly bound to its host metals and breaking these chemical bonds is one way to produce oxygen, which serves both human life support (air to breathe) and transportation (oxidizer for rocket propellant). A variety of processes can accomplish these tasks, including electrolysis (melting the soil into a liquid and then passing an electrical current through it) and chemical reduction (using hydrogen or fluorine brought from Earth as a reducing agent). None of these techniques are in any way "risky" in a technical sense – reduction as a chemical process dates from medieval times. Abundant solar energy provides virtually unlimited power; some areas near the poles of the Moon are in near-constant sunlight.

The "long pole" in the tent is getting started. Right now, the architecture for lunar return[5] has no requirement or provision for resource utilization. NASA's efforts to date have focused on rocket-building and planning for scientific sortie missions. Yet learning how to gather, process and use the resources of the Moon is major goal of the Vision for Space Exploration. The idea is to use what we find in space to create new capabilities. This goal has the promise of freeing us from the "tyranny of the rocket equation" – we would no longer be mass and power-limited in space.

The key to bootstrapping this capability is the judicious use of robotic precursor missions[6]. Robotic spacecraft are now orbiting the Moon, mapping the distribution of elements such as hydrogen and ascertaining the nature of the environment near the poles. The next steps are to measure the composition and physical properties of the polar deposits from the surface; this requires soft-landers capable of landing payloads on the order of a few to tens of kilograms. Small surface rovers would be able to map out the elemental concentration of volatiles and determine the best places to mine.

After the prospects are mapped, we must experiment with different techniques for harvesting and processing. Again, this work can be done by modestly sized robotic missions, landing small excavators and trucks (Mars rover-sized) and using laboratory bench-scale processing equipment. Landing and experimentation with this equipment will allow us to find out which techniques are most effective, what processing methods use the least amounts of energy and have the highest yields, and determine where the choke-points are in the processing and production stream.

These small initial steps allow us to begin extracting and storing resources immediately. Over time, we can increase these capabilities such that when people finally return to the Moon, they have at their disposal a cached accumulation of consumables, including air, water and rocket propellant. In effect, we are creating the initial phases of self-sufficiency even before human arrival through the emplacement and use of automated, robotic infrastructure.

No one knows if lunar resources can be extracted and used in the manner described here. But that's why we're going to the Moon in the first place – to answer these questions. We are using the Moon as a laboratory to learn how to live and work productively on other worlds. The skills and technologies developed here will serve us well wherever else we go in the Solar System. And the sooner we get started on this path, the sooner we will develop a true space faring infrastructure.

Topics: Lunar Exploration, Lunar Resources, Space Transportation

Links and References

1. key skill, http://www.space.com/businesstechnology/technology/moon_mining_041110.html
2. Outland, http://www.imdb.com/title/tt0082869/
3. mining precepts, http://en.wikipedia.org/wiki/Mining
4. Moon composition, http://www.spudislunarresources.com/Images_Maps/soil.jpg
5. lunar return architecture, http://www.nasa.gov/exploration/news/ESAS_report.html
6. robotic precursor missions, http://tinyurl.com/2croecd

Comments

2. Bravo, Dr. Spudis. As we have come to expect, you have set the case clearly before us, and without the bothersome reminders of lost opportunities or being explicit as to the real hazards standing in our path to deeper destinations, like Mars.

I seem to have to always bring these things up when trying to communicate the necessity of the Moon, as the University of Space Exploration. The United States dropped out in 1972, and merely flirted with the admissions office in 1994 and 1998.

Thank God people like you kept the fires lit over decades.

It's particularly exciting now that we can see LRO and LCROSS stacked and packed, ready to return us to Freshman Orientation, with the other precursor missions waiting in the wings. Let's hope. rather than drop out, the United States doesn't simply flunk out this next time.

Comment by Joel Raupe — June 6, 2009 @ 9:16 pm

First, Nail Down the Mission

June 25, 2009

A continent spanned: Is cislunar space next?

The new Augustine Commission[1] met for the first time last week (June 17). The one-day agenda was filled with presentations[2] on rocket-building, including reviews of NASA's current efforts along those lines, followed by briefings on a number of possible alternatives. Suddenly, the space blogosphere was filled with speculation on the possible demise of the new Ares I launch vehicle and its replacement by either a commercial or some alternative Shuttle-derived rocket.

An early focus on rockets is perhaps inevitable, given the cost, schedule and technical issues that the Ares program has experienced. But in fact, all this rocket talk is quite beside the point. The real issue is, as it has always been, "What is the mission?" Why are we going to the Moon[3]? Why should we send people into space? Can't robotic missions explore the universe more cheaply and easily?

Such questions about the space program are answered repeatedly, but the discussion never advances. Recognizing that I am rushing in where space angels fear to tread, let me give it yet another go.

There are many motivations for a national space program. Scientific knowledge is an important objective, but it is not the only one and perhaps not even the most important one. The Vision for Space Exploration[4] is being undertaken "to advance U.S. scientific, economic and security interests." The Vision, proposed by the President and endorsed by two Congresses, was carefully crafted to give logical, long-term purpose and direction for expanded possibilities and opportunities in space. In a speech on the Vision[5] given a couple of years after its announcement, Presidential Science Advisor John Marburger said, "Questions about the vision boil down to whether we want to incorporate the Solar System in our economic sphere, or not."

Here is the problem. Leaving Earth means escaping from a very deep gravity well[6]. It is very costly to lift mass out of this well; current estimates vary widely, but $20,000 per pound to low-Earth orbit is a commonly cited cost for delivery by the Shuttle. As long as we must lift everything we need in space from the surface of the Earth, we are mass- and power-limited. Thus, we are also capability-limited. And under the existing rules of spaceflight, we always will be.

So, let's change the rules. Rather than lifting all the water, air and propellant we need up from Earth, let's find and make those commodities in space. Once we do that, our capabilities multiply many fold. We will be able to go anywhere we want, for as long as we want, to do any job or task we can imagine.

Why the Moon? Because the Moon is the closest, most easily accessible place beyond low Earth orbit that has the resources we need. Water is the currency of space faring – we need it for life support, energy storage and rocket propellant. The Moon has abundant supplies of both hydrogen and oxygen; no matter what form those two elements may take, we can extract and make these needed commodities from lunar materials.

Making propellant from lunar material allows us to access not only the Moon's surface, but any other point in cislunar space (the volume of space between Earth and Moon) on a routine basis. This zone is where all our commercial and national strategic assets reside. Rather than building custom spacecraft, launching them on an expendable rockets, using them for a few years and then abandoning them in place, we would be able to create maintainable and extensible space systems. Spacecraft can be refueled in orbit instead of launched whole cloth from Earth. The VSE asks NASA to find and use what's out there to create a wholly new, sustainable space faring capability.

This is our "mission" on the Moon: learn the skills and develop the technologies needed to live and work productively on another world. Creating a space transportation infrastructure is akin to building the first transcontinental railroad; it will open up the frontier of cislunar space. And a system that can access cislunar space will take us to the planets.

NASA's task is to probe beyond low Earth orbit—opening the space frontier for sustained exploration. The agency's job is not to industrialize the Moon, but to answer the question, "Can the Moon be industrialized?" This new direction is far removed from the geopolitically driven Apollo template[7] of "flags and footprints." The multinational fleet of probes scouting the Moon is testament to mankind's boundless curiosity and a timely reminder that those who explore, excel.

A mission statement must be clear and simple. When the mission is understood, debate about rockets and architectures take place in an information-rich environment. The launchers used and the way mission elements are put together is optimized based on the requirements of the mission. Developing those requirements cannot begin until you know the mission.

One hundred and forty years ago, the mission was understood — to span the continent with a transportation system, opening up the frontier to development. That mission created a modern industrial nation. We seek to do the same with cislunar space. And then, the planets.

Topics: Lunar Exploration, Lunar Resources, Space Politics, Space Transportation, Space and Society

Links and References

1. Augustine committee, http://tinyurl.com/23wkmwp
2. presentations, http://www.nasa.gov/offices/hsf/meetings/06_17_meeting.html
3. why go to Moon, http://www.spudislunarresources.com/Papers/The%20Vision%20and%20the%20Mission.pdf
4. Vision for Space Exploration, http://www.spaceref.com/news/viewpr.html?pid=13404
5. Marburger speech, http://www.spaceref.com/news/viewsr.html?pid=19999
6. gravity well, http://en.wikipedia.org/wiki/Gravity_well
7. Apollo template, http://blogs.airspacemag.com/moon/2009/01/25/what-apollo-was-and-wasnt/

Comments

1. I've advocated the idea of a Earth-Moon-Earth "Space Ferry" from 2005 posting on a space forum and I agree that we need as much as RE-USABLE vehicles to cut the costs of Space exploration.

About the Augustine Commission, I've just posted my suggestion #4 (of about three dozens I've in mind to write) for the Human Space Flight Plans Committee and NASA:

"put the spacecrafts and spaceflights SAFETY as FIRST concern" http://ow.ly/f3vQ

Comment by gaetano marano - ghostNASA.com — June 25, 2009 @ 12:29 pm

2. You want safety as the "first" concern? Then don't fly a space ship. It's worth remembering that if someone chooses to fly into space, then they must have some concern ahead of safety else they wouldn't do it.

Comment by Karl Hallowell — June 25, 2009 @ 9:59 pm

3. Great post, Paul. Kind of reminds of a rant I did over at the Selenian Boondocks a few years back:

http://selenianboondocks.blogspot.com/2005/12/were-on-road-to-nowhere.html

What's frustrating is that there are sensible ways to move forward and start on this now, but no one wants to see them because their vision is bound by institutional interests, or trapped in an Apollo mindset.

Comment by Ken Murphy — June 26, 2009 @ 12:31 am

4. @Karl Hallowell

I'm aware that, travel in Space, never will be as safe as swim in the home pool (although many are drowned in their pool ...) but I just want that, all future spacecrafts, will be not always designed to put the astronauts' lives under serious risks every time a small problem happens

Comment by gaetano marano - ghostNASA.com — June 26, 2009 @ 9:35 pm

5. How about the ISRU potential of Near Earth asteroids? I'm fond of L1/L2 as a cheaper initial destination than the Moon and one that also supports Moon, Mars and NEO's.

Comment by Martijn Meijering — June 28, 2009 @ 12:53 pm

6. Good essay. It's amazing how many people assume that space is about science, when really it is (or should be) about expanding our economic sphere, and ultimately spreading humanity throughout the solar system.

Comment by Joe Strout — June 28, 2009 @ 6:48 pm

7. This is my first visit to this blog Mr. Spudis and I agree with Mr. Strout; this is a good essay and you have made clear some aspects about going to the moon that I have not considered.

I am not sure I am in agreement with the last part of Mr. Strout's expression – considering the way humanity currently functions here on earth.

Nevertheless I will come back to read more of your thoughts.

Comment by Rob — July 7, 2009 @ 9:40 am

Would More Money Improve NASA?

July 8, 2009

How much money should we spend on space?

How much should we spend on America's space program? Does NASA's budget need an infusion of billions of dollars? The way these questions are answered gives some indication of why one believes we have a space program, what it should be doing and whether money is the key needed to unlock the barriers hindering our access to space.

Former NASA administrator Michael Griffin recently opined[1] that, "we're going to have to spend what it takes." If we can't pursue space goals "with sufficient robustness," he hopes that the newly formed Augustine Commission recommends that "we just not do it." Additionally, Norm Augustine himself recently said[2] that, despite our current technology and knowledge, ultimately, "It boils down to what we can afford."

As we all know, spending money is easy. Spending money wisely is something else entirely. The Apollo era, when money supposedly flowed freely, is often cited as the glory days of NASA. Early Apollo spending[3] was high primarily for two reasons.

First, in its early days, NASA had little infrastructure – few field centers, test equipment, space vehicles and the people to design, build, test and fly the spacecraft. A lot of NASA's early funding went toward building up the facilities needed to go to the Moon: the KSC Moonport (the VAB and Launch Complex 39), the Houston MSC (now JSC) campus, the

Deep Space Network and the several other installations around the country.

Second, there was a perceived political imperative that required rapid progress in space and this urgency expressed itself as high rates of expenditure. Apollo was not a journey to the Moon—it was a race and the Soviets were thought to be ahead of us. They orbited the first satellite and the first human. They did the first spacewalk and were the first to hit the Moon with a robotic probe. Being behind was a jolt and a wakeup call for Americans who believed our country was and should remain the world's leader in technology.

With the lunar landing accomplished, the urgency of space dissipated and NASA adjusted to being just another federal agency, seeking to retain what it already had while expanding its sphere of activities to the extent that it could. However, institutionally, NASA never abandoned its business model of "racing to somewhere." This way of thinking is manifest as NASA again pins its hopes for a viable space program on getting more money.

In 1989, President George H. W. Bush (Bush 41) outlined what became known as the Space Exploration Initiative[4] (SEI). It called for a permanent lunar base and a manned mission to Mars. NASA's response to the new mission directive was tepid; it produced a 90-Day Study[5] that concluded we could do SEI if the agency budget was increased substantially. In other words, the agency response to the Presidential directive was "Give us more money."

Flash forward 15 years. In an attempt to set a long term strategic direction for space and to assure that we maintain a productive, technological workforce, President George W. Bush (Bush 43) outlined the Vision for Space Exploration[6] (VSE). It directed NASA to return humans to the Moon and learn to use lunar resources, followed by manned Mars missions, all the while integrating private industry into the architecture.

NASA's response to the Vision was the Exploration Systems Architecture Study[7] (ESAS), which outlined an approach using Shuttle-derived hardware. However, as work unfolded, Shuttle heritage was diluted, costs rose rapidly, the date of lunar return receded, and the idea of incrementally developing a sustainable space infrastructure using lunar resources[8] was abandoned. The lunar surface mission was warped into a "touch and go" demonstration followed by an Apollo-style Mars mission, staged entirely from the Earth. Once again the agency's response to a new exploration challenge was to close ranks and follow the Apollo template, repeating the refrain, "Give us more money."

Now NASA's former administrator, Mike Griffin tells Norm Augustine to "do it right or don't do it at all[1]." In effect, Griffin is playing the "Washington Monument" game, a form of budgetary blackmail that threatens to terminate something believed to be strongly supported by Congress and the public (in this case, human spaceflight) unless some increased budgetary threshold is reached (in this case, more money to implement the ESAS.) But this "game" only works when you're holding the high cards, in this case that the public won't stand for the termination of human spaceflight.

In a previous blog[9] I discussed the issue of public support for space exploration. While people don't often think about space (and virtually no one casts their vote based on how the space program is funded), they still like the idea of having a space program. The fact that we've had the same space budget[10] for thirty years (in constant dollars, between 0.5 and 1.0% of federal spending, more or less) suggests that this level of funding is politically sustainable.

As celebrations for the 40th anniversary of Apollo 11 begin, the agency has been unable to create a sustainable architecture for lunar

return, thereby bleeding the life out of exploration efforts. There is still no plan for lunar surface activities. In their urgency to exit the Moon as rapidly as possible and get to Mars, NASA is side-stepping the principal reason they were to go to the Moon in the first place – to learn the skills needed to live and work productively on another world. Is it any wonder that Congress and the public are uneasy about their space agency and its plea for more money?

The response to the questions I asked at the beginning should not be, "Give them more money." The question we need to ask NASA is, "Given a constant level of funding over time, can you create a program that incrementally and cumulatively builds up a real space faring capability?"

If the answer to that question is "No," then we need to ask, "Why not?"

Topics: Lunar Exploration, Space and Society

Links and References

1. Griffin op-ed, http://tinyurl.com/p48pff
2. Augustine afford, http://tinyurl.com/32eoq3w
3. Apollo spending, http://en.wikipedia.org/wiki/NASA_Budget
4. SEI, http://history.nasa.gov/sei.htm
5. 90-day study, http://history.nasa.gov/90_day_study.pdf
6. VSE, http://history.nasa.gov/Bush%20SEP.htm
7. ESAS, http://www.nasa.gov/exploration/news/ESAS_report.html
8. sustainable space infrastructure, http://www.spaceref.com/news/viewnews.html?id=1334
9. public support, http://blogs.airspacemag.com/moon/2009/02/20/another-strategic-plan-misfires/
10. space budget, http://en.wikipedia.org/wiki/NASA_Budget

Comments

3. The current NASA budget request, just under 19 billion, comes out to about .54 percent of the over all budget request for 2010, about 3.5 trillion, on the low end of what the article says is sustainable. 1 percent would create a 35 billion dollar NASA, which is quite outside the realm of possibility.

Two or three billion more, which is what people like Griffin is claiming is needed, on the other hand, brings the share of NASA to a little more than .6 percent, which is well within the range of sustainability stated in the article.

So I'm not really sure what the complaint it about.

Comment by Mark R. Whittington — July 8, 2009 @ 2:18 pm

4. I agree on give to NASA an higher and

known budget over the next years and this is, exactly, one of my suggestions I want to give to the Human Space Flight Plans Committee and NASA here:

http://ow.ly/f3vQ

Comment by gaetano marano - ghost-NASA.com — July 8, 2009 @ 2:32 pm

5. *Two or three billion more, which is what people like Griffin is claiming is needed,....So I'm not really sure what the complaint it about.*

Hi Mark,

First, I'm not complaining — NASA is. They are the ones who claim to need more money. Second, I simply do not believe that $2-3 billion more per year will make all of the ESAS problems go away; what has been crafted is essentially an Apollo-style architecture (with a sortie mission profile no less, but that is another story) that has large recurring costs and leaves no legacy hardware in place on the Moon or in cislunar space. Third, it doesn't matter that doubling the agency budget keeps it under 1% — what's clear is that there is no political will to raise it to that number. We'll be lucky to keep it at current levels of ~0.5%. Fourth, my point is that such a number is still a lot of money (!) and it behooves the agency to attempt to devise something a bit more clever and innovative than Apollo on Steroids.

Comment by Paul D. Spudis — July 8, 2009 @ 3:45 pm

6. I agree with some of the other comments posted here. Mr. Spudis you have portrayed NASA as being some kind of budgetary sinkhole that whines for more money to keep it afloat. The reality is that NASA's budget has declined or flatlined for years when inflation is taken into account. NASA could not even meet the demands of its existing programs because of rising costs and inflation forcing the agency cut a number of programs. How can anyone expect NASA to establish a lunar based infrastructure when their budget is less than half of the Apollo era budget adjusted in today's dollars. Senator Bill Nelson remarked on this budget shortfall in today's confirmation hearings.

Comment by Gary Miles — July 8, 2009 @ 6:06 pm

7. Paul – I don't think that anyone is suggesting that an infusion of cash will make the return to the Moon's problems go away. But what such an infusion would allow the people working on those problems to better able ad-dress them and would relieve NASA of the temptation of cutting corners to save cash, a practise that in the past has come back to bite them.

If we want the return to the Moon program to expand to create a lunar settlement, more money is definitely required. Mind, that can be had through commercial (my preference) or international (I suspect the administration's preference) partnerships as well as increased appropriations.

Finally, I think that what will there is to put more money in the return to the Moon project depends on what Augustine 2.0 has to say. My take from how people on the hill are talking is much depends on what it recommends.

The thing I do not want to see is the practice NASA has done before when costs have become an issue, which is to start jettisoning capability (already started) and stretching out the schedule. Both have had pernicious results when resorted to in the past.

Comment by Mark R. Whittington — July 8, 2009 @ 6:08 pm

8. *If we want the return to the Moon program to expand to create a lunar settlement, more money is definitely required.*

Mark, I simply do not agree with this. It all depends on what your mission on the Moon is and how you approach it.

Comment by Paul D. Spudis — July 8, 2009 @ 6:37 pm

9. *you have portrayed NASA as being some kind of budgetary sinkhole that whines for more money to keep it afloat. The reality is that NASA's budget has declined or flatlined for years when inflation is taken into account*

It has not declined — it has been static for the last 30 years. In aggregate, NASA has received the same cumulative amount of money in the last 30 years as it received to do the Apollo program (a point made by no less than Dr. Michael Griffin himself.)

You used the term "budgetary sinkhole" not me. My point is that if NASA is waiting for a significant increase in its budget before it implements the VSE, we can wait until Doomsday and end up with nothing. The current path is not affordable now nor is it sustainable in the future. Stay the course and we'll get nothing, or at best, another 30 years in low Earth orbit.

Comment by Paul D. Spudis — July 8, 2009 @ 6:42 pm

10. *It has not declined — it has been static for the last 30 years. In aggregate, NASA has received the same cumulative amount of money in the last 30 years as it received to do the Apollo program (a point made by no less than Dr. Michael Griffin himself.)*

Please check out this Wikipedia link on NASA's annual budget since 1958 and check out the nominal and adjusted 2007 columns. Clearly over the last 30 years there has been several declines in NASA's budget. The differences become glaringly obvious when comparing the nominal budgets to the 2007 adjusted. The %G of Fed Budget column is also worth noting. As a percentage of federal budget, NASA's budget is 10 times less today than it was in 1968.

But you also miss the essential point here. While NASA's budget has been static or declining, NASA's operational costs have increased virtually every year due to inflation, higher facilities costs, higher material costs, and higher labor costs. Since the budget did not increase, NASA was forced to cut programs and unable to meet many of its goals. While I support the VSE and ESAS, without an adequate raise in budget, it amounted to little more than an unfunded mandate.

That comment about Dr. Griffin was taken out of context. His point at the time was that NASA spent the last 37 years stuck in LEO spending the same amount of money that NASA received for the first decade developing the Apollo program and going to the Moon.

Comment by Gary Miles — July 8, 2009 @ 8:45 pm

11. Gary,

But you also miss the essential point here. While NASA's budget has been static or declining, NASA's operational costs have increased virtually every year due to inflation, higher facilities costs, higher material costs, and higher labor costs. Since the budget did not increase, NASA was forced to cut programs and unable to meet many of its goals. While I support the VSE and ESAS, without an adequate raise in budget, it amounted to little more than an unfunded mandate.

The funding for NASA at the 0.5-1.0% level is in constant dollars. The level of inflation is already accounted for in such a reckoning. So my claim is valid: NASA has been funded at (more or less) a constant level since Apollo ended.

I know that this "unfunded mandate" charge is a common theme in the blogosphere in regard to the VSE but it is simply untrue. The VSE was specifically designed around the idea that the agency's budget would NOT increase more than inflation. NASA was specifically told to craft an architecture doable within existing funding limits (the "sand chart"); this directive was ignored. So how is that an "unfunded mandate"?

That comment about Dr. Griffin was taken out of context. His point at the time was that NASA spent the last 37 years stuck in LEO spending the same amount of money that NASA received for the first decade developing the Apollo program and going to the Moon.

No, I did not take it out of context — that was exactly my point: what you do in space and how you do it is more important than how much you spend.

Comment by Paul D. Spudis — July 9, 2009 @ 5:45 am

12. Nice essay, Paul. And I agree with you that what you do in space and how you do it is more important than how much you spend. NASA is by far the most well-funded and unified space agency in the world, and sometimes its apologists seem to forget that.

But the way you're poo-poo'ing the idea that NASA is the victim of unfunded mandates seems frankly disingenuous. As you have so clearly pointed out, the agency's funding has been essentially static for the past 30 years. Yet you have entirely failed to mention that NASA's mandated missions have *vastly* expanded since the days of Apollo, largely through acts of Congress and the President. The end result is that the agency is being asked to achieve a very ambitious and diverse dossier of tasks with a budget that realistically can support only a handful of them. It would be nice if you would acknowledge this obvious fact in your critiques of the agency and its requests for more funding.

Comment by Lee Billings — July 9, 2009 @ 12:18 pm

3. Lee,

Thank you for the kind words.

My comment on "unfunded mandates" was specifically in relation to implementation of the Vision for Space Exploration. My point is that the VSE was supposed to be accomplished under the existing budget envelope from the beginning. In other words, NASA was tasked to construct a lunar return using the existing levels of money in the human spaceflight program. The way this was to happen was to 1) retire the Shuttle; and 2) complete ISS, followed by American withdrawal from ISS operations in ten years. The money freed up by this was to be used to build a CEV and return to the Moon.

As far as NASA's "full plate" goes, every space mission and program that the agency is "assigned" by Congress and the White House gets specifically designated funds for their completion by the Congress. If there are shortfalls, it's usually because the cost estimates for these missions have been low-balled.

Comment by Paul D. Spudis — July 9, 2009 @ 1:03 pm

14. Thanks for the reply, Paul, and your keen mention of the significant problem of low-balled estimates of mission costs.

This is a contentious topic, but it seems NASA's stuck between a rock and a hard place when it comes to its funding and cost estimations.

For instance, in this post you've endorsed the popular opinion that NASA's response to Bush 41's SEI plan was tepid, because the agency's cost estimations were too high. The cynical thinking goes that NASA thus deliberately sabotaged the 1989 plan to go back to the Moon and on to Mars.

But now you're also saying that the real cause of NASA's budgetary woes is that the agency's cost estimates tend to be too low.

So which is it? I know this isn't an apples-to-apples comparison; SEI was just one instance, and the damaging trend of low-balled cost estimates is very clear. But it seems that when NASA responds to ambitious mandates with more "realistic" (that is, rather expensive) cost estimates, the outcome is no better, and arguably even worse: Programs never get off the ground in the first place because of sticker-shock.

Surely we can all agree that space exploration is very expensive, and that more efforts need to be made, publicly and privately, to make getting out of Earth's gravity well more affordable. The halcyon days of cheap, routine space exploration have not yet arrived. And so at present the case remains that, as the phrase goes, no bucks, no Buck Rogers.

I still think NASA is simply being asked, again and again, to do too much with too little.

In this light, public statements by current and former agency officials that NASA needs more money are entirely appropriate and not necessarily just cynical machinations.

Comment by Lee Billings — July 9, 2009 @ 2:06 pm

15. Paul

My comment on "unfunded mandates" was specifically in relation to implementation of the Vision for Space Exploration. My point is that the VSE was supposed to be accomplished under the existing budget envelope from the beginning. In other words, NASA was tasked to construct a lunar return using the existing levels of money in the human spaceflight program. The way this was to happen was to 1) retire the Shuttle; and 2) complete ISS, followed by American withdrawal from ISS operations in ten years. The money freed up by this was to be used to build a CEV and return to the Moon.

NASA did submit a development program that would fit within the originally proposed budget when VSE was released. Then the Office of Management and Budget slashed some $15 billion from that budget over a five year period. Those funds were critically needed up front during the early development phase to iron out engineering problems before testing and construction began. Also, it is not realistic to not expect some hidden costs not foreseen in the original budget.

Comment by Gary Miles — July 9, 2009 @ 2:13 pm

16. *NASA did submit a development program that would fit within the originally proposed budget when VSE was released.*

At that time, no architecture had been chosen so there was no basis for anything like a real program budget. What NASA did for the first two years of the VSE was to generate stacks of viewgraphs and institutional blither about "spiral development" and all that required was the existing organizational overhead. The only real mission hardware designed in that time period was LRO (which is a whole other story).

OMB has always acted as the executive brake on NASA while Congressional appropriations acts as the legislative brake. The agency was authorized to come up with an architecture to implement the VSE and they chose to come up with one that was already known to require more money than was expected.

Comment by Paul D. Spudis — July 9, 2009 @ 2:46 pm

17. In 2006, Congress endorsed and approved the changes that NASA made to VSE based on the ESAS in support of the Constellation program. Thus with Congressional approval, OMB had the necessary authorization to increase NASA's budget.

Comment by Gary Miles — July 9, 2009 @ 3:53 pm

18. I think we're also forgetting that return to flight costs for the shuttle fleet after Columbia were much higher than originally anticipated. Those costs came out of other NASA accounts, including exploration.

Comment by Mark R. Whittington — July 9, 2009 @ 5:17 pm

19. *For instance, in this post you've endorsed the popular opinion that NASA's response to Bush 41's SEI plan was tepid, because the agency's cost estimations were too high. The cynical thinking goes that NASA thus deliberately sabotaged the 1989 plan to go back to the Moon and on to Mars. But now you're also saying that the real cause of NASA's budgetary woes is that the agency's cost estimates tend to be too low. So which is it?*

The two are not mutually exclusive; you're mixing up two different things.

NASA sabotaged SEI because at the time, they had not launched even one piece of Space Station hardware and much of the 90-day study used Station-derived hardware. Alternative architectures proposed then (e.g., Lowell Wood's inflatables) were dismissed out of hand (even though later, JSC developed Transhab, an inflatable hab module for deep space exploration.)

The low-balling of costs is a strategy undertaken to sell missions or programs that the agency wants. The increase in the agency's portfolio is a chosen strategy to diversify and stay in business. This is fundamentally different from avoidance of a mission entirely.

One more point about "costs." In the case of SEI, the money issue was only partly NASA's fault — the commonly cited media cost number for SEI of $600 billion (for which see this) was the aggregate cost for a 30-year program, which in fact, would have been spent anyway. But NASA didn't help matters any by devising an architecture whose purpose was mostly to "feed the beast" rather than develop new capabilities.

Comment by Paul D. Spudis — July 10, 2009 @ 8:40 am

Space Program vs. Space Commerce

July 16, 2009

Museum piece: 40 years ago, on its way to the Moon

"Your job is not to envision the future, but to enable it." – Antoine de St. Exupery

Originally, I had not planned to write anything for the blog today; the web is already inundated with retrospective-, perspective-, nostalgia-laden, crying-in-my-beer pieces on today's 40th anniversary of the Apollo 11 launch. One shudders to think what's coming in just a few days, on July 20, the actual anniversary of the lunar landing. However, a news piece caught my eye because it casts the lunar landing in stark perspective (for me, at least) and makes me wonder whether some historical amnesia is at work within the space enthusiast community.

Ned Potter of ABC News has cast some cold water[1] on the warm and rosy glow of the good old Apollo days. In brief, he's looked at some of the old polling data and finds that, lo and behold, the public was split on the value of going to the Moon. In fact, a plurality of people polled back then opposed the lunar voyage, thinking that it cost too much money and delivered too little. This actually jibes pretty well with my recollections. As a young space enthusiast (I was 16 years old during the Apollo 11 mission), I can remember constantly defending spending money on the space program. Mostly, I did this with my high school classmates but sometimes, I even argued with my parents about it!

The simple fact is, the public is now and has always been split on the value of a federal government space program. It received its largest margin of support when viewed in the perspective of national security, but the lunar program was under fire as a waste of tax money from the beginning. Even John F. Kennedy, widely credited today as a "visionary" President, desperately wanted to find some other arena in which we could compete with the Soviets and win. At one point, he favored desalination of seawater[2] as an appropriate scientific and technical challenge for America.

I have previously written[3] on the nature of public support for space. In my opinion, it's a mile wide and an inch deep. People are at best lukewarm in their support and virtually no one casts a vote on the basis of whether or not a candidate supports the space program. Yet there is amongst the space community a feeling that somehow, getting people enthused about a space goal is essential to accomplishing that goal (putting the cart before the horse).

If the ABC News story is even partly correct, such a belief is unwarranted, both then and now. In retrospect, we remember the successes and the triumphs, but forget about the naysayers. How many remember that on the day of the Apollo 11 launch, there was a "poor people's march" on Kennedy Space Center, led by Rev. Ralph Abernathy[4]? He told reporters that he had came "not to protest the moon program, but to remind America of its unfulfilled promises." When the rocket actually launched, he too was swept up with the thrill and emotion of it all and cheered on the mission as loudly as anyone. Imagine! Success drawing in a doubting public! When you accomplish exciting things in space, you capture the imagination—people begin to believe that they too can own a piece of this great adventure.

Others look on the issue of public support for space as improving[5] or static[6]. Andy Chaikin asks whether anyone "still cares[7]" about the Moon. But the real issues are who supports a lunar return and for what reasons. As a people, Americans tend not to support endless public spectacles. Today, people who still follow NASA are asking, "Where's the beef?" They're looking for more than a postcard from space or a PR extravaganza. They want some legacy from their investment and they are right for doing so.

Just saying that you're doing something exciting isn't the same as doing something exciting. And hearing about how exciting something is sure as hell isn't the same as doing it. To launch an enduring legacy built on the accomplishment of Apollo 11 and begin the return on our investment, NASA needs to enable the private sector's ability to capitalize on space resources, starting by demonstrating this capability[8] on the Moon. Then the public can join in the construction of a road to space as opposed to being mere spectators in a museum to the past.

Start the countdown to true space exploration freedom—the one that brings all of humanity into the business and the adventure of advancing to the stars. Put the horse in front of the cart and enable the future. Humanity's imagination and resourcefulness will expand in the process and the benefits will be many.

Topics: Lunar Exploration, Lunar Resources, Space and Society

Links and References

1. Apollo public support, http://abcnews.go.com/Technology/Apollo11MoonLanding/story?id=8090280&page=1
2. seawater desalinization, http://www.dow.com/liquidseps/prod/sw.htm
3. public support of space, http://blogs.airspacemag.com/moon/2009/02/20/another-strategic-plan-misfires/
4. Rev. Ralph Abernathy, http://en.wikipedia.org/wiki/Ralph_Abernathy
5. improving, http://www.associatedcontent.com/article/1949028/since_apollo_11_changing_public_attitudes.html?cat=15
6. static, http://www.space.com/news/090716-apollo11-40th-chaikin.html
7. Chaikin, http://www.space.com/news/090716-apollo11-40th-chaikin.html
8. Demonstrate resource capability, http://www.spaceref.com/news/viewnews.html?id=1334

Comments

1. The problem today is that people don't just complain about the government "wasting" its money on space travel, they even complain when private citizens "waste" their money on space travel. The argument is, how dare you spend large sums of money on frivolous pursuits when we haven't solved [global warming, world hunger, HIV/AIDS, universal healthcare...pick your favorite crisis]. I think these people are being hypocritical, but I'm not likely to convince them of that. (Let's see, did you go to a movie in the past year? Why didn't you donate that $10 to Greenpeace?)

The bottom line is those who want to go to space are going to have to do it in spite of the naysayers. That crowd isn't going away, no matter how much the private sector accomplishes. They simply have different values (and a faulty set of economic principles, too).

Comment by Bill Hensley — July 17, 2009 @ 3:13 pm

2. Exactly right. Nobody cares about the airline. It's all about were it takes you. Apollo was great, but what we need now is a dull, boring, flight-now-boarding-at-gate-nine, spaceline.

Comment by Frediiiie — July 18, 2009 @ 12:02 am

3. Good points Paul. Probably Americans were more ambivalent about the Moon Shot than the rest of the world. One need look no further than that classic Gil Scott Heron tune for evidence of that!

Just found your blog today, sir. It's awesome! Keep up the good work!

PS: We need to get you on coasttocoastam!

Comment by Warren Platts — July 18, 2009 @ 1:43 pm

4. Paul, I agree. What is needed now is not a new NASA program, subject to all the political whims that buffet NASA constantly, but a public/private partnership outside of NASA. One that would stay focused on infrastructure and sustainability.

Comment by Thomas Matula — July 18, 2009 @ 7:21 pm

Can You Legally Own a Piece of the Moon?

July 24, 2009

A Moon rock on Mt. Everest: Not for keeps

Mr. Ian Sheffield of Edinburgh Scotland is miffed. He claims to have not one, but two[1] dust samples of the Moon—one from the Apollo 11 mission and another from the Apollo 15 mission. He explains that he bought these lunar samples "from a dealer" about 3 years ago. The article does not indicate how much he paid for them, but he does allow that each is valued at "around £2000" (about $3300) each.

A problem arose when he planned to display his samples to the public. He apparently wrote to NASA asking if he could exhibit them. To his astonishment, NASA refused to give him permission and demanded the return of the samples, claiming that the lunar dust in his possession was property of the United States government.

Mr. Sheffield's story of how the samples came into his possession[2] is interesting. He states the dust came off a camera film pack to which a technician in the Lunar Receiving Laboratory was accidentally exposed. Because no one was sure the lunar samples would not contain some possible primitive (and pathogenic) organisms when the Apollo 11 crew first returned to Earth, they had to spend three weeks in quarantine. Anybody in the LRL exposed to lunar material was compelled to join the astronauts in their quaran-

tine. The technician who was exposed went into isolation and (the story claims) upon his release, "was given the dust as a memento[1]."

My antennae went up at this point. No lunar samples are "given" to private individuals. Each piece of the Moon returned by the Apollo astronauts is carefully accounted for and resides in the Lunar Curatorial Facility[3] in Houston, where they are kept in two separate hurricane-proof vaults. Many lunar samples are loaned to scientific institutions for study. The only lunar samples given away (of which I am aware) were to about a hundred national leaders during President Nixon's 1969 world tour. The beautiful "Space Window[4]" in the Washington National Cathedral, honoring man's landing on the Moon, holds a 7.18-gram basalt from Mare Tranquillitatis, on loan to the Cathedral. Other moon rocks were presented to the Apollo astronauts (and Walter Cronkite)[5] in 2004. However, each plaque came with a catch: the lunar samples can not be personally held by the recipients, and must be displayed at a local school or museum. Recently, Astronaut Scott Parazynski was loaned a sample of the Moon's regolith that he carried to the summit of Mount Everest[6].

Some diplomatic gifts of lunar samples have found their way onto the black market. A notorious case is a sample presented to the people of Honduras[7] back in 1969. This sample turned up during a NASA Inspector General "sting" which was designed to catch dealers of fake lunar samples. To the agents' surprise, they were offered a genuine lunar rock: asking price, $5 million. A meeting was arranged and the rock (and presumably, the seller) was seized. Another lunar sample was stolen from a museum in Malta[8] between 1990 and 1994; it was recovered in another sting operation in 1998.

The federal government forbids private ownership of any Apollo sample. Yet, such samples show up every now and then. The most common form they take is dust stuck to adhesive tape (an easy way to "clean" the surface of some exposed sample container, tool, or space suit used on the lunar surface). Mr. Sheffield's sample is likely to be one of these pieces. Its status, I was surprised to find out, is legally uncertain[9]. Although NASA has sued in court to recover any such bootleg sample, no prosecution has succeeded, except for those caught (literally) in the act of theft. In an embarrassing incident for NASA, a summer intern and two companions carried a safe full of lunar samples[10] out of a building at Johnson Space Center (as Dave Barry would say, I am not making this up). They were apprehended while trying to sell them at bargain basement prices and subsequently prosecuted.

It was rumored for years that several of the Apollo astronauts held samples from their respective missions. If they did, it was probably inadvertent—the lunar dust is extremely adhesive and it is possible that smudges of lunar dust clung to personal items returned from the Moon in their Personal Preference Kits. Alan Bean, who documents the Apollo experience through his oil paintings, is said to add ground-up patches retrieved from his lunar space suit to his works. His reasoning is that because his suit was dirty with lunar dust, some of that dust must find its way into his paintings[11], giving them a true "lunar" ambiance.

So Mr. Ian Sheffield of Edinburgh may be home free. I might suggest to him that given their quasi-legal status, he is probably better off not calling attention to his possession of these unique artifacts. In fact, although NASA frowns on owning stolen Apollo lunar samples, there are dozens of lunar samples available for sale on eBay. A number of meteorites recovered on Earth, came from the Moon. Although most of them belong to national governments that sponsor the recovery of meteorites from Antarctica, several are in

private hands and can be bought and sold, just as any commodity. Right now, there is a very nice anorthositic breccia from the lunar highlands for sale. Better hurry though – the sale only lasts another day. Oh yes, the asking price: a mere $144,000.

By the way, over the years, I have been asked to look at a few "lunar" samples that were in fact, lunar fakes. *Caveat Emptor!*

Topics: Lunar Exploration, Lunar Science, Space and Society

Links and References

1. Sheffield samples, http://edinburghnews.scotsman.com/topstories/Houston-has-a-problem-.5491120.jp
2. how he got samples, http://www.eastlothiancourier.com/news/roundup/articles/2009/07/23/389965-one-giant-leap-for-local-astonomer/
3. Lunar Curatorial Facility, http://www-curator.jsc.nasa.gov/lunar/index.cfm
4. space window, http://www.museummasterworks.com/Catalog.php?page=7
5. presented lunar samples, http://www.msnbc.msn.com/id/5431472/
6. sample to Mt. Everest, http://www.onorbit.com/node/1027
7. Honduras sample, http://www.geotimes.org/nov04/trends.html
8. sample theft in Malta, http://www.msnbc.msn.com/id/5031216/
9. legal status, http://www.msnbc.msn.com/id/5431472/
10. JSC sample safe, http://gizmodo.com/5242736/how-an-intern-stole-nasas-moon-rocks
11. Alan Bean dust in paintings, http://www.dailycamera.com/news/2009/jul/17/astronaut-alan-bean-moon-artist/

Comments

1. This is interesting. The market price for Moon rocks seems to be about $1,000 USD per gram. So if a private consortium could land on the Moon for say $200 million, all they would have to do is bring back about 200 kg of cool Moon rocks to break even. . . .

Comment by Warren Platts — July 30, 2009 @ 4:06 pm

2. Warren,

A difficult proposition. Each Apollo mission brought back between 60 and 110 kg; this was possible with the large ascent stage of a capable system. Current studies of robotic sample return from the Moon typically return a couple of kilograms of soil and rock, tops. Thus, a robotic mission would find it difficult to bring back the amount of sample you imagine.

Keeping such a mission below $200 million cost is a whole other story.....

But I like your idea. There may well be a future market in lunar samples as gem stones.

Comment by Paul D. Spudis — July 31, 2009 @ 5:21 am

3. Who says lunar samples have to be carried back on the same vehicle? It wouldn't be hard to carry a catapult to the Moon and repeatedly launch lunar material into Earth orbit, from which they could be collected and returned to Earth for sale.

Comment by Bob Carver — July 31, 2009 @ 2:24 pm

4. Bob,

Good idea! But then you have the problem of finding the lunar rocks somewhere in space after they have been hurled off the Moon. Space is very big.

However, your idea is a good one. People have been looking at the idea of a "mass driver" that would electromagnetically catapult things off the Moon, into orbit around the Earth. If we ever make rocket propellant on the Moon, this would be a good way to get it into space, where it could then be used to re-fuel spacecraft.

Comment by Paul D. Spudis — August 1, 2009 @ 8:14 am

5. So, Dr. Spudis, what do you know about kimberlite deposits on the Moon? Have any precious or semi-precious stones been found on the Moon so far? What about things like GOLD! Yes, I know that gold deposits on Earth are hydrothermally produced. But isn't it true that there used to be a shallow water ocean on the Moon? In which case, we could expect water to diffuse down into the lunar mantle, and then percolate back up and leave behind deposits of all sorts of useful things.

Just a thought.

Comment by Warren Platts — August 1, 2009 @ 2:01 pm

6. Warren,

Most deposits of gems and precious metals on Earth typically involve volatiles in their emplacement — usually water (steam) but also carbon dioxide, halogens and other exotic species. Most all of these are largely absent from the Moon and always have been. The "ocean" that used to be on the Moon you refer to is a "magma ocean", which refers to the ancient time when the outer few hundred kilometers of the Moon was completely molten. In other words, an ocean of molten rock, not water. Indigenous lunar water is present only in vanishingly tiny amounts.

Comment by Paul D. Spudis — August 1, 2009 @ 3:27 pm

7. The "ocean" that used to be on the Moon you refer to is a "magma ocean", which refers to the ancient time when the outer few hundred kilometers of the Moon was completely molten. In other words, an ocean of molten rock, not water. Indigenous lunar water is present only in vanishingly tiny amounts. I can't remember where I thought I had read somewhere that there might have very briefly, geologically speaking, been some water on the surface of the Moon right after it first formed. . . . Maybe it was just my imagination.

Be that as it may, we know that the standard theory is that the water deposits posited at the poles are the result of allochthonous comets; yet there are alternative explanations:

http://news.brown.edu/pressreleases/2008/07/moon08

"In a paper published in the July 10 issue of the journal Nature, the team, led by Alberto Saal, assistant professor of geological sciences at Brown, believes that the water was contained in magmas erupted from fire fountains onto the surface of the Moon more than 3 billion years ago. About 95 percent of the water vapor from the magma was lost to space during this eruptive "degassing," the team estimates. But traces of water vapor may have drifted toward the cold poles of the Moon, where they may remain as ice in permanently shadowed craters."

Then there are those mysterious "gas leaks":

http://lunarscience.arc.nasa.gov/articles/erupting-gas-may-cause-lunar-flashes

"If you tie all this together in one package, you can convince yourself there's a story here," says Paul Spudis of the Lunar and Planetary Institute in Houston, Texas. "At one time the Moon had volatiles; it might still have some remnant of those in the deep interior," he says."

So can we say for sure that there haven't been hydrothermal processes analogous to those on Earth capable of concentrating valuable minerals like GOLD?

Therefore, I respectfully request a blog post on primordial volatiles, please, sir. The blog post should discuss the ISRU significance of such primordials, especially at the gas leak zones.

You know, basalt flows are capable of forming both seals and storage zones for commercial natural gas deposits right here on Earth. Might it be possible that similar formations on the Moon might also contain valuable gas pockets that might be drillable using conventional oil and gas techniques?

Comment by Warren Platts — August 1, 2009 @ 11:09 pm

8. Warren,

It all goes to my statement above — yes, there are volatiles in the Moon, but they are present in extremely small amounts.

Maybe I'll do a blog post on the details of this topic sometime in the future.

Comment by Paul D. Spudis — August 2, 2009 @ 6:18 am

9. In 1970, NASA sold medallions commemorating the 1st Lunar Landing. The medallions were made from metal from the spacecrafts Eagle & Columbia; so from the theory of Moon dust sticking to the Eagle, and anything that came in contact with it, these medallions likely contain some Moon dust.

I purchased one of these medallions in 1970, but it isn't exactly clear how one could determine if it contains any traces of the dust (also, since the Eagle was left in orbit, it isn't clear how the metal from the Eagle was collected, but the inscription on the medallion confirms the sources).

Comment by J. Todd — August 21, 2009 @ 3:24 am

10. Interesting article. In 1949, Robert A. Heinlein published a science fiction story called "The Man Who Sold The Moon". His concept was that the purchase of property on earth included rights to land, minerals, et cetera, and therefore should include rights to air and lunar property, given that earth's moon belongs to the earth as it is held by earth's gravity in an earth orbit. The main character in the story then proceeded to purchase the rights to the moon from every person whose land was passed over along the moon's orbital path. It's an interesting concept, and an interesting story. To read the story, look for it in 'The Past Through Tomorrow' by Robert A. Heinlein, pp.121-212.

Comment by Carl N Graves — August 31, 2009 @ 1:59 pm

Next Step or No Step

August 3, 2009

Target and Distraction: Which is which?

The Moon versus Mars controversy has reared its ugly head yet again. For the newcomers, this is the perennial "debate" among space buffs about what the next destination in space should be. I do not mean to suggest that all possibilities are encompassed by these two options; it just seems that most advocates fall into one or the other of these two camps.

In part, this argument has arisen because the Augustine Commission, currently deliberating the future of NASA's human spaceflight program, has resurrected the debate with an architectural option they call "Mars First[1]" (a.k.a. Mars Direct[2], Direct to Mars, Apollo to Mars and Mars-in-MY-lifetime), beloved of the Mars Society and ex-astronauts[3] everywhere. Briefly, this plan calls for sending people to Mars as soon as possible – no Moon, no asteroids, no L-points: do not pass "Go," do not collect $200. In such a scenario, all pieces of the Mars mission are launched directly from the Earth; this roughly one-million-pound on-orbit mass includes all the propellant needed for the trip, which makes up about 85% of the mass of the spacecraft.

The Mars First option follows the "Apollo template[4]." In 1961, faced by the political ne-

cessity to get men to the Moon and back within a decade, Wernher von Braun designed the biggest rocket he could imagine – basically a scaled-up, clustered V-2 – to lift all of the parts he needed into space. This super heavy lift vehicle was actually a family of rockets (Saturn class), whose ultimate behemoth was the Nova, a vehicle with a lift-off weight exceeding 13 million pounds. Fortunately, the choice of lunar orbit rendezvous for the Apollo mission mode made Nova unnecessary and a self-contained mission was launched by a single, smaller (7 million pound) Saturn V.

The Apollo template makes use of maximum disposability. As the mission proceeds and each flight element is thrown away, unused and unusable, the vehicle gets smaller and lighter. For some items, such as fuel tanks and structural elements, this doesn't introduce unwarranted penalties, but some parts of the vehicle are high in cost and value. Within the Apollo template, however, their loss is inevitable.

A significant part of the Apollo template is the lack of infrastructure legacy, i.e., the elements brought to a destination that are available for use by the next crew. We need to develop an architecture that leaves equipment in place for future use and expansion by subsequent visitors. This is one reason why sortie missions are inferior[5] to establishing an outpost or a base; sortie missions spread surface assets over a large area where they cannot mutually support each other.

Much of the support for Mars First comes from the belief of its advocates that we will get "stuck" on the Moon or somewhere else, sort of like we have been "stuck" in low Earth orbit for the last 40 years. In their minds, Mars is THE destination. To hear the pitch, one might believe Mars has it all – atmosphere, water, a 24 hour day, and possible ancient fossil life. Adventure! Thrills! What else could a space cadet want?

Although the "Mars First" advocates vigorously present their position each and every time the direction of our space policy is debated, they have never won the argument. Why? Is it some evil conspiracy to keep them from their Mars dream? Is it just the stupidity of policy makers? Some simple facts suggest otherwise.

We do not now have the technology we need to support multi-month, self-sufficient human space travel. The International Space Station needs nearly constant servicing and re-supply from Earth. In fact, one of the missions of ISS is to learn how to live in space without such service and re-supply, closing the various life-support loops and thereby developing sustained human presence. This is experimental technology and not nearly mature enough upon which to rest the lives of a Mars mission crew. Regardless of claims, a Mars mission is at least one (and possibly two) order(s) of magnitude more costly than any alternative mission.

There isn't the will in either the Congress or the Executive to significantly increase the amount of money allocated to our national space program. Spectacular claims about "exciting the public" with a human Mars mission, regardless of their veracity (which is doubtful[6]), do not translate into higher budgets for NASA. To go to Mars using existing technology, with an Apollo-style business model, is both unachievable and unaffordable.

The Vision for Space Exploration[7] makes Mars a goal – along with every other space destination – after we go to the Moon to learn how to live and work on another world. Moreover, the VSE implicitly states that such is to be accomplished under existing budgetary envelopes. In contrast to the Apollo template, time rather than money is to be the free variable. The Moon can be reached with existing launch assets; although NASA is currently bogged down in a debate about rocket development, the real issues are how you go back to the Moon and what you do there. The Moon offers the material and energy resources[8] to develop the technology and skills necessary for sustained, long duration capability in space.

Mars First advocates worry about getting "stuck on the Moon." In fact, it is their obsession for Mars that has kept us in low Earth orbit for the last 40 years. By relentlessly pushing for a space goal that is well out of our technical and fiscal reach, they have gotten an undesired (but not unexpected) result: stasis. There is no choice. You use the Moon or you get nothing. Right now, Mars is a bridge too far – we need the stepping-stone of our Moon to reach it.

Topics: Lunar Exploration, Lunar Resources, Space Politics, Space Transportation, Space and Society

Links and References

1. Mars First, http://www.nasa.gov/pdf/368722main_Beyond_LEO_07_12_09.pdf
2. Mars Direct, http://en.wikipedia.org/wiki/Mars_Direct
3. ex-astronauts, http://tinyurl.com/lgfjq8
4. Apollo template, http://history.nasa.gov/SP-350/profile.html
5. sortie missions inferior, http://blogs.airspacemag.com/moon/2009/05/05/return-to-the-moon-outpost-or-sorties/
6. doubtful, http://blogs.airspacemag.com/moon/2009/07/16/space-program-vs-space-commerce/
7. Vision for Space Exploration, http://en.wikipedia.org/wiki/Vision_for_Space_Exploration
8. Moon resources, http://www.spaceref.com/news/viewnews.html?id=1334

Comments

1. I couldn't agree more. What we need is not an out of reach mission but one that simplifies and industrialises access to space. And actually, I really cannot understand why NASA are spending billions (3Bn$ spent already from what I read) on developing new rockets, especially the Ares I, when many are available commercially. These 3 Bn$ could have been allocated to development of say lunar infrastructure instead. And time would have been saved. And the programme would be more advanced. And further support would have been granted to private launch companies. I may be wrong, or ill-informed, but what is the added-value of reinventing the wheel at a much higher cost than existing alternatives?

Comment by matchad — August 4, 2009 @ 4:06 am

2. Dr. Spudis: another excellent post, sir. You should have been on the Augustine committee. (You would be NASA administrator if I was the President.) Do you have a link to the white paper that you gave to Augustine?

I must say that I was disappointed that the Lunar Base option was given such short shrift by the Committee. I guess it sort of lives on in the "Program of Record" option. But if President Obama decides to keep flying Shuttle and ISS beyond previous commitments, that's not going to leave a lot for other expeditions. Which means he will tilt toward the Flexible Path option.

What do you think Paul? Are near-Earth asteroids a viable target for exploration? I think Congress a while back did in fact mandate that NASA take steps to mitigate against the danger of asteroid collisions. But like you said, these would be disposable, easily cancelable missions.

Don't get me wrong, I am for Lunar Base, for several scientific, economic, and national security reasons. Also the cancelability factor for a Lunar Base would be low. Look at ISS and Shuttle. It's my hypothesis that they haven't been cancelled just because of their reusability. People think that they should not waste their "investment", despite the fact that it is irrational to make future decisions based on sunk costs. Once the first few modules on a Lunar Base were laid down, it would be difficult politically to walk away from our "investment".

I vote for sticking with the program of record, even if it takes time. But I could get behind asteroids as well. You've got to admit the planetary science and ISRU opportunities would be exciting. The main thing, I would say, is to just come up with SOMETHING. A single strategic plan around which tactical architectures can be designed.

Comment by Warren Platts — August 4, 2009 @ 11:47 am

3. I'm a member of The Mars Society, and you're simplifying their position quite a bit. Hardware that's built with Mars as the goal can fly to many destinations (asteroids, L-points, lunar orbit, and some even say lunar surface) without much alteration, and increasingly-stringent testing of Mars hardware before the first landing would be a sensible move. These types of missions would mirror the Apollo 7-10 testing in the 60s. Given a flight rate to Mars of every 26 months, there's no reason such missions couldn't continue in the time between launch windows.

That said, I personally favor the more organic approach of building infrastructure allowing us to go anywhere we want, but can't see that being a politically workable solution. These days the words 'less government control' are not winners.

Comment by Tom Hill — August 4, 2009 @ 12:54 pm

4. Right on Paul. I might quibble with the numbers a little, but I think you have it right. First, let's learn how to live off-planet, and in so doing, learn to live off the local land. After we have done that, a key by-product is the creation of new wealth, the magnitude of which we cannot predict now, as well as the creation of new commercial opportunities, including space tourism and commerce. The net sum of these 2 by-products may be zero, or small, but just as possible could be a demand and supply curve that has a large value and growth rate. Another by-product that will definitely occur while we are learning to live off-planet is the accumulation of new knowledge (i.e., Exploration!). Only after we have started on this path can we really afford as a nation to shift philosophy toward one of Exploration as the primary objective.

Comment by Tony Lavoie — August 4, 2009 @ 2:19 pm

5. My understanding is that NEOs and Phobos are the target of the "E" option not the surface of Mars. At least in the near to medium term.

Landers are expensive and the delta v to Phobos is less than the delta v to the lunar surface. No need to "land" on Phobos, you can kinda sorta just settle down, like docking.

An un-crewed gravity tractor mission to an NEO is also appealing — set up a forward operating base for future robotic and crewed exploration of that NEO, both for science and resource exploitation and use that base to deflect the asteroid's trajectory helping to develop systems to protect us in the event we find an asteroid on a collision course with Earth.

Lunar resources? Including those Dennis Wingo has been hinting about? NewSpace should team up with international partners and go exploit the Moon. Probably do it better than NASA would, anyways.

Comment by Bill White — August 4, 2009 @ 11:23 pm

6. The Mars/Moon debate is silly. Spudis is correct; Mars is not technically nor financially doable for a price the nation is willing to pay... but he then makes the assumption that the Moon is. The technical part is there, heck we have done it before...but there is no reason to go. None.

We will not go to the Moon or leave low earth orbit until there is a sustained space industry that can exist at least in part off the government dole. As it is right now, that does not exist.

But the future is coming. Here is a prediction...on the 50th anniversary of LtC Glenn's first flight...Elon Musk and Dragon fly go to orbit with the first non government crew ...

Until then Spudis, Zubrin and all the other folks are just wailing at the Moon with their various arguments.

Comment by Robert Oler — August 5, 2009 @ 12:23 am

7. Robert,

You're really arguing against NASA, not for or against a destination or activity in space. Take that up with Washington, not with me.

Comment by Paul D. Spudis — August 5, 2009 @ 5:05 am

8. As I have said before "Phobos First!" (By way of L1, ESL2, NEOs of opportunity, etc.) But this Space Cadet has a new mantra courtesy of HSF Augustine II: Flexibility!

Whilst an immediate lunar base (= Emergency Radiation Shelter) ...will be useful for ground truthing and proofing of 2nd Gen systems and hardware; Orbital Aggregation [of] Space Infrastructure Systems is the way to go. My take:: Roscosmos: LOS and Hypergol Depot; NASA: L1 "Gateway" L(UN)OX Depot; ESA: OTV infrastructure & Argon Depot; China: SEL2 Station Core (with a little help from everyone) plus an ISS Module of their very own! EVERYONE builds their flavour of Lander evolving to a Deep Space Vessel! And whilst we are dreaming: the UK fully funds SKYLON!

Fingers LCROSSED but failing the fabled "Ice Caves of Shackleton", I still think Tranquillitatis is the best place for tourism and the Mass Driver. If nothing else it will need to be crewed permanently to stop vandals and souvenir hunters!

The Moon is an essential next step *if nothing else* to resolve property rights; the inter-

nationalisation of space and eliminate once and for all the old style of Geopolitics:

"Planet Earth looks beautiful from space. There are no borders on the Earth," Sunita Williams. That's why we need the Moon. Perspective.

From Mars it's just a dot!

Comment by brobof — August 5, 2009 @ 5:45 am

9. This seems to be all, or mainly about NASA. What do you think of private initiatives of going to the moon to actually do something there. Do you think there is any chance we see that in the medium term?

Comment by matchad — August 5, 2009 @ 8:51 am

10. *What do you think of private initiatives of going to the moon to actually do something there.*

I'm all for them. But I assume that NASA will continue to get its $18 billion/year and as a taxpayer, I'd like to see something useful done with it. Tackling challenging, technically difficult things, like space resource utilization, is something a federal engineering R&D agency should be eager to do.

Comment by Paul D. Spudis — August 5, 2009 @ 9:03 am

11. Paul:

As a former and retired NASA MSFC employee and an old Army Ballistic Missile Agency employer I think we need to use the existing Shuttle based system minus the Shuttle replacing it with a heavy lift rocket system. To me this is using the KISS approach system(Keep It Simple Stupid. As you have stated the Apollo System was really an upgraded V-2.

Comment by Otha H Vaughan — August 5, 2009 @ 10:02 am

12. Paul.

I am not arguing for or against NASA any more then I would argue for or against the FAA. The issue in both cases (as well as any agency of the Federal government) is a basic one, " what is the agency suppose to accomplish?"

In the abstract, to make an agency of the federal government more then just a wealth transfer machine...the agency should do things that directly benefit the people of The Republic, the ones who make all federal agencies possible. To the extent that the agency consumes resources, its benefits should be more tangible and obvious. NASA consumes a fraction of the federal budget but all the money set up for civilian space flight. The people of The Republic should get some tangible benefits from those dollars, if the people do not, then the money spent is simply wealth transfer.

There is nothing in any plan to return to the Moon or go to Mars that has such a tangible benefit. Other then the ability to "learn how to live and work on another world", (a goal which could far more easily be accomplished at the bottom of our oceans) your op ed does not mention a single one. Stripped of the various "great country" or "how the west was won" arguments the reasons for going to the Moon or Mars or the Near Earth Asteroids or whatever boil down to "it is something to do and we are going to spend the money on NASA anyway so why not do (something)". That argument is what got us a multi decade effort to build a space station, which as it is completed, all the folks who argued for it are tired of.

The argument for (something else) replaces the ISS as a program but does nothing to ensure that the nation receives any tangible benefits from the goal. Figure out what has doomed ISS from returning any of the tangible goals that Ronald Reagan sold the nation on in his state of the Union speech and you will understand the fatal flaw of your piece. .

The good news is that after 40 years of stops and starts the government infrastructure which has felt no obligation to show value for its cost, has finally gotten so monstrous that it is collapsing of its own weight. The vacuum has allowed some private efforts, which seem to be "real" to fill the void. One day we will go back to the Moon (probably as a government effort) just as Lewis and Clark went out west. Not because it is hard, but because it is easy. The technology will not have to be developed; most of it will be in use. It will not be soon, but it will be sooner then we can get there today.

You are one of the bright folks in the field. Your heart is in the correct place and I applaud your integrity.

Comment by Robert G. Oler — August 5, 2009 @ 3:24 pm

13. Bob, you are wrong. There are many tangible benefits: one of them is US national security.

Consider that the Moon constitutes the 8th

continent of Earth; since the United States is a global superpower, the Moon is by definition of concern to the US. Moreover, the Moon's ultimate political status is still very much in the air, the Outer Space Treaty (OST) notwithstanding. It is not clear that the USA will remain a signatory to the OST in perpetuity. The desirability of remaining a signatory was seriously reviewed by the Bush administration in 2004. Logically, therefore, it would not be surprising if other space faring nations will question whether they want to remain signatories as well.

By building a permanently manned base on the Moon, the USA will ensure that the final political disposition of the Moon will be on terms favorable to the United States. In particular, the US would like to see a free and democratic lunar legal regime that is favorable to business. If another country established permanent lunar bases first, and then chose to withdraw from the OST, they could claim the right, through use and occupation, to dictate the terms of a new lunar legal regime, that we might not like.

Furthermore, historically, the Moon has been considered as a possible platform for missiles or directed energy weapons, or as a strategic command-and-control base. This is not to suggest that the United States ought to weaponize the Moon; however, a permanent American lunar presence would reduce any temptation for other nations to do so.

In addition, there is the following chilling scenario to consider that is not talked about much: once a nation develops ISRU techniques for extracting pure elements such as oxygen, titanium, aluminum, gallium and silicon from the lunar regolith, there is nothing in principle that would prevent that nation from constructing a satellite manufacturing facility. Such a facility could effectively "corner" the market for LEO satellites. This is not a matter of economics; it is a matter of physics. The delta-v requirement from the Earth's surface to LEO is ~10 km/s, whereas the delta-v budget from the surface of the Moon to LEO is only ~6 km/s; the use of a heat shield and aerobraking would reduce this figure to ~3 km s-1; the addition of a lunar maglev launch system (LMLS) could reduce this figure to < 1 km/s. In essence, an LMLS could essentially eliminate the marginal cost of placing satellites into orbit.

The implication is that the nation possessing a lunar satellite factory and an LMLS could build and cheaply launch a "Brilliant Pebbles" constellation consisting of thousands of satellites capable of a full-spectrum global dominance of the entire cis-lunar space and the upper atmosphere of Earth, and therefore of the Earth itself.

Given that a Delta IV Medium rocket costs ~$100 million USD, the cost savings would likely be measured in hundreds of billions–if not trillions–of dollars. Moreover, the marginal ratio cost (MRC)–the incremental cost of each countermeasure versus the incremental cost of each new Brilliant Pebble satellite–would not be favorable to the Earth-bound nation. If current trends continue, China will have an economy larger than the United States in a mere 20 years. If the US were on the losing side of such a MRC, that could prove decisive. The US would find itself in the same position as the former Soviet Union–financially run into the ground. On the other hand, if the US were in possession of the satellite factory and LMLS, that would be an important asymmetric force multiplier.

In sum, if America does not choose to master the lunar environment, some other nation will. That such a mastery of the lunar environment could translate into a political and/or military advantage cannot be ruled out from the comfort of your armchair. Seen in this light, a new NASA Moon program is pretty darn cheap insurance if you ask me. To consistently argue otherwise, you'd have to claim that the US Navy doesn't deliver tangible benefits to American citizens.

Comment by Warren Platts — August 6, 2009 @ 12:47 pm

14. Hello Warren.

Try as I might, I cannot see a single national security concern "now" or in the foreseeable future that would prompt a return to the Moon. I have no doubt that sometime in the future the Moon and its resources will play a significant role in national security; but that day is at best half a century away.

President Jefferson, when he bought the LA purchase sent Lewis and Clarke out west in part for national security reasons. A nation cannot legitimately claim what it does not even map or understand…and Jefferson knew that.

But in terms of the Moon we as a society and a technology are no where near where Jefferson was as L&C left. There were no significant technology hurdles that the effort had to clear. All the expedition had to do

was buy the hardware and go The folks who went already had the basic skills for "living off the land". There were no "resupply" shipment needed. The list goes on (and is quite lengthy) but concludes with one basic statement. There were no obstacles to Americans and American industry following L&C except their own initiative and grit.

NONE of those things exist right now in human spaceflight. No nation on this planet could "go" or "return" to the Moon and even start to scratch any of the assumptions that you have stated as reasons. Even if Dennis Wingo is correct and there are lots of "raw materials" on the Moon; from mining them to fashioning them into things like semiconductors and then forming those subcomponents into satellites is not going to happen with any of the infrastructure that the US (or any country) could put on the Moon in the next 50 years. It is not Delta V that is at issue, it is manufacturing economics.

Why? Because in the case of the Chinese even if they had the will, they don't have the technology that makes the effort affordable… and in our case not only is the effort not affordable; but there is no hint that other then government any other part of the American infrastructure is ready to follow such an effort.

I was having lunch with Jim Oberg the other day…and he told the story of the Russians in the 1980's trying to open Siberia. They found gold there and tried to mine it. The region was so tough that even with convict labor that was essentially working itself to death that the infrastructure was so expensive, it was not worth the effort.

That is where we are right now with humans going to other bodies in space. Until that changes (and the space station is the key to changing it) there is no other body in space that is not a recipe for a failed program.

If the Moon is the 8th continent, there is no history of civilization from one continent getting a toehold on another, without transportation cost that are manageable.

As I told the writer of the op ed, go read Reagan's SOTU where he plugs the space station. Figure out why NONE of those things have happened yet…and you will understand why the Augustine commission is wasting its time.

Comment by Robert G. Oler — August 6, 2009 @ 4:26 pm

16. Hi Bob,

You say that "no doubt that sometime in the future the Moon and its resources will play a significant role in national security; but that day is at best half a century away." Even if that's true, that doesn't entail that we shouldn't start planning a satellite factory right now. Like I said, if China were to continue to grow at 7% per year and the US only grows at 3%, then in 20 years, the Chinese economy will exceed the USA's in a mere 20 years. Therefore, if we get in a space race with China that starts 50 years from now, we cannot hope to come out on top. On the other hand, if we start out now, while we have a temporary advantage, the efforts we make now will pay dividends 50 years from now.

Furthermore, there is a danger that you overestimate what it takes to build a spacecraft on the Moon. The hard part is getting the refined raw materials from lunar rocks–this is the only new technology that needs developed. Once this is accomplished, assembling satellites is comparatively trivial. If we stay on schedule, we could get back to the Moon by 2020. By 2030, if we stay focused, we could have a permanent lunar base that is at least as functional as the current ISS. Thus, by 2040, you would have a decade of solid research on the Moon for developing ISRU techniques. By 2050, if not sooner, the first satellites could be launched.

Alternatively, if we get involved in a decades-long odyssey to Mars, and decide to "leave the Moon" for others, and China steps into the resultant vacuum and do what they do best–setting up factories–then the stage could be set for a true technological "strategic surprise" in the Clausewitzian sense, from which Western civilization and the values we hold dear may never recover.

Also you are wrong about Lewis and Clarke not needing resupply. They brought tons of supplies with them, which they cached in various locations to which they later returned, and they bought horses, food, and boats from Indians they encountered. Your Siberia example is also misleading: probably, the Soviet implosion in the 1980's had more to do with the failure of the gold mine that James Oberg mentioned. Certainly, Canadians and Alaskans have no problem extracting minerals from equally challenging environments.

I don't know why you think that the ISS is going to be the key to revolutionizing space travel. There are no game changing technologies that are going to come from Earth or the ISS. To be sure, the ISS has proved that manned outposts in space can be occupied for years at a time, and it also demonstrated the power of modular construction (and also the weaknesses: Skylab delivered 80% of the volume for a 10th of the price; a single launch from an Ares V could

launch a station with 2.5 times the volume of the ISS). We've basically maxed out the possibilities. Reusable SSTO's from Earth capable with any reasonable payload mass fraction are just not physically possible. (Incidentally, you might find it interesting that most SSTO research was funded by the SDIO.) The only way to escape the Earth's gravity well is to avoid it altogether by building our spacecraft off planet. The Moon is by far the easiest place to do that. The sooner we get started on that task, the sooner you will get your "manageable" transportation costs.

Whether the Augustine committee turns out to be a waste of time depends upon how they "spin" their final recommended options.

Comment by Warren Platts — August 7, 2009 @ 8:36 am

17. Hello Warren.

First let me say that I opposes Zubrin's plan as much as I oppose "going back to the Moon". As I made clear in my reply to the op ed writer… I think both plans have their minds stuck in an Apollo "big government/big plan" mentality which I would like to see buried in the dust bin of history.

I don't trust economic forecast more then 6 months ahead of time, and even those I sort of shake my head at. If I were to tell you in 1960 that in 2010 time frame we would be running multi trillion dollar deficits and have run massive deficits for sometime (several decades), then even I would think that the statement was nutty. So who knows what our economy will be like in 2015 much less 2050.

In any event along with not seeing the national security edge the Moon gives us, I don't see in any reasonable time frame the infrastructure on the Moon to make "satellites" of any sophistication, particularly ones whose components are completely "moon grown"…Much less a sky full of brilliant pebbles…and actually I think that in another 10 years brilliant pebbles will be an idea whose time has passed.

What I do see is ISS. We have spent hundreds of billions on it…and we need to make it the fulcrum of a new space age. One, where we eschew large government "explorations" and instead concentrate on building a true space industry. That is occurring. Musk has a very good shot at not only flying his vehicles, but flying them at prices and internal infrastructure that cost far less then anything Lockmart and Boeing can come up with. Think the airmail contract of the 1930's and you will see how ISS is the key to developing a real space infrastructure that can do just magnificent things in space both crewed and unscrewed.

We are on the brink of having the private industry (truly private industry not the equivalent to Soviet design bureaus) which will launch for an industry affordable price, be able to do on orbit assembly of massive payloads and even on orbit repair of those massive payloads as far as geo synch orbit. That is the capability the military wants now..and will find useful when there are lots of directed energy weapons in geo synch orbit...which is what is going to replace brilliant pebbles.

All a government return to the Moon or going to Mars or doing anything big ticket in human spaceflight does is suck up cash on useless adventures. The cash really ought not to be spent at all (we have this deficit) but if it is going to be spent, it should be on something that develops infrastructure that the rest of the Republic can use...not just something that keeps requiring government money.

As an aside...I don't fear the Chinese all that much. Comment by Robert G. Oler — August 7, 2009 @ 9:54 pm

18. Bob, you've got it completely backwards. The ISS is the drain on resources, and it always will be, until it is deorbited. It's worth was mainly in the building of it, and as an experiment to demonstrate both the strengths and weaknesses of international cooperation. At this point it is a drag that is slowing down innovation.

I understand your mail-contract analogy of the 1930's. But it pales in comparison to the effects that WWI and WWII had on the aerospace industry. All you want to do is shift government transfer payments from ATK to SpaceX. The effect would be a marginal improvement at best.

Your idiosyncratic theories about economics, technology, and the dynamics of the space industry are systematically blinding you to the potentially game-changing opportunity that a lunar base would represent. Infrastructure is not what counts. The EFFECT of infrastructure is what we are after.

For one thing, the ISS cannot grow itself. A lunar base, once a threshold of equipment is emplaced, will be capable of increasing its size and functionality largely by itself, using local materials. That is the game-changing reality of a lunar base versus a space station. Moreover, long before actual spacecraft manufacturing capability is achieved, lunar rocket fuel will be produced.

Meanwhile, the ISS cannot and will not ever manufacture anything. It is a heavy, unsafe behemoth that would require at least 7 space shuttle loads of fuel to move to a useful orbit. I can't see how the ISS will be the fulcrum for much of anything; but how about this: let's offer to sell the American components to Elon Musk. How much do you think he would pay? I wonder if he would

even bid $1 USD. Probably not, but if your theory is correct he would pay at least a dollar for it, and if he did, that would solve everyone's problems, wouldn't it? It would free up NASA resources, and give a big jumpstart to the new aerospace upstarts.

Paul has got it right: it's the Moon or nowhere–take your pick.

As for China and SDI, the SDIO looked very carefully at all sorts of SDI architectures, including the giant, Battlestar Galactica directed energy platforms you envision. The main problem is energy. To capture the energy required for repeatedly firing an effective weapon would require HUGE solar arrays. And their very size makes them vulnerable to stealthy kinetic kill vehicles. And they are expensive to launch because they would require multiple launches to assemble one.

But even if you are correct that a Battlestar Galactica system would be superior to a Brilliant Pebbles system, that doesn't affect my argument by one iota. The nation in possession of a lunar spacecraft manufacturing facility on the Moon is STILL going to have a huge strategic advantage. To paraphrase what I said earlier: it is not a matter of architecture, it is a matter of physics. The delta-v advantage that the Moon has over the Earth is inexorable. Here's a prediction for you: the strategic advantage is such that Earth will never allow a lunar civilization to exist as an independent political entity.

As for you lack of fear of China, I can't do better than quote from a recent paper on strategic surprise that I ran across recently:

"China, with or without a Russian consort, is by far the leading candidate to play the starring role in opposition to the U.S. hegemon. Predictable capabilities support this view, as does an unsentimental appreciation of China's political and strategic culture. Some among us believe that China will mature in its modernization into a contented and generally cooperative, profit-maximizing trading partner in a U.S. policed world order. People of that opinion would do well to ponder these words written by the eminent cultural historian, Adda B. Bozeman:

'[I]t is noteworthy that the Chinese themselves have traditionally conceptualized the Middle Kingdom not as one bounded state in the company of others, but as a civilization so uniquely superior that it cannot be presumed to have frontiers. This self-view spawned China's insistently Sinocentric worldview; sanctioned imperial schemes of military and political expansion; and sustained several politically and culturally potent ideas of imperial administration, chief among them the notion of the emperor's "heavenly mandate" and the concept of a family of unequal and inferior nations held together by the "Imperial Father"—images persuasively concretized throughout the centuries by the tribute system and the well-organized dependence on hedge-guarding satellites and surrogates.'"

http://www.strategicstudiesinstitute.army.mil/Pubs/display.cfm?pubid=602

Comment by Warren Platts — August 9, 2009 @ 1:23 pm

19. Hello Warren

Nice little discussion we have going on here, I certainly am enjoying it.

Our differences of the future aside, I think that you grossly overestimate the ease (or underestimate the difficulty) of setting up 1) any kind of mining operation on the Moon and 2) any kind of manufacturing system on the Moon for refined parts. I don't see the latter happening on any scale (other then maybe sustainables for a lunar base) in the next 50 years...heck I don't even see lunar base construction starting until 2030 at best... I have heard Dennis Wingo beat the same drum and he does a lot of blue sky waving...of course a few years ago he was all excited about assembling satellites at the space station.

Plus I see zero participation in such an effort by anything other then government and that is a foundation for failure.

ISS on the other hand is here. I think that you are completely underestimating what can be done in microgravity with the timely participation of private enterprise. The effort to utilize the resources of near earth space have floundered because of NASA requirements (which would be on any lunar facility) and because of long time lines. Joe Allen, who I am reasonable friends with made in the 80's a convincing case that a private facility run in a private manner...could do some cutting edge research and eventual product work...if there was private access to space.

To me ISS, which the nation has invested billions in, is such a "anchor partner". We are for national pride etc reasons going to continue to fly to ISS for another 10-15 years and probably longer. There is no reason that this effort is not the "anchor tenant" for private access to space... and eventually that will, as the airmail contract did start up the cycle of private infrastructure development.

That is my gripe with Obama's stimulus package...it did nothing to stimulate private capital all it did was sustain government infrastructure... and a misbegotten effort to head to the Moon (or Mars or anywhere) will only do the same thing.

If I were "King" we would immediately move to a Comsat like corporation to run ISS, move all lift to private concerns (think airlines) and get NASA out of any business but unscrewed exploration and crewed technology development. Eventually in my view that will let a real space infrastructure/private enterprise system develop which will eventually put together the "parts" that some government (or private concern) can use to go back to the Moon...and do it on an affordable basis.

That is what made air travel an American adventure...we developed through the airmail contracts and with some technology spin off from WWII a private American industry that could on its own build the Dash 80. it is essential for any space future that this cycle (private enterprise providing a product for services) start...otherwise we will go nowhere in space.

As for the military...I agree with Gates (who I admire a lot)...any "program" that is more then 10 years in the making doesn't work well in the real world. We are going to evolve our C3 systems in space which will include larger and larger platforms and eventual servicing of those. We are not in my view going to move much past theater ballistic missile defense systems. MAD will work well into this century.

As for the Chinese. While I read and understand what you say, I also remember the last administration beating the drums about Saddam and how he was going to kill us all. I spent a good chunk of my recent life searching for his WMD in Anbar...and found nothing. We don't need to invent enemies. They will come all by themselves.

Again this is quite a good conversation with quite a nice tone.

Comment by Robert G. Oler — August 10, 2009 @ 10:12 pm

Two Views of the Vision

August 11, 2009

The White House had a different view of the Vision for Space Exploration than NASA

Last week, the Augustine Commission[1] held another public meeting in Washington DC and Dr. John Marburger testified[2]. For those just joining our story in progress, Marburger was President Bush's Science Advisor and the Director of the Office of Science and Technology Policy in the White House between 2001 and 2009. He was a key player in the development of the Vision for Space Exploration[3] (VSE) and his comments on the intent and reality of the VSE were interesting and insightful.

Marburger described a split between NASA and the White House during formulation of the Vision. NASA (led by former Administrator Sean O'Keefe, Chief Scientist John Grunsfeld and an internal study group within the agency) wanted a manned Mars mission (as it has for the last 50 years) while the White House (led by Marburger, his OSTP colleagues and some members of the National Security Council) called for a new direction and orientation of the space program. They favored a return to the Moon with the "mission" of radically changing the rules of spaceflight.

This latter course involved learning how to use the material and energy resources of the Moon to produce life support consumables, electrical power and rocket fuel, thereby creating new spaceflight capabilities. The White House group was informed by an abundance of detailed studies[4] done over the past decade that demonstrated how the resources of the Moon could be tapped and utilized. Given the unlikelihood of significant new money for NASA, they believed that some kind of "game-changer" was needed – a way to step beyond low Earth orbit by incorporating innovative ways of conducting space business. A sustainable path, if you will.

Marburger's biggest concern was that by inserting Mars as a goal (not by any means an "ultimate goal") or even a date for lunar return, the path forward would become "burdened by deadlines and difficult budget issues." He believed that a program composed of small, incremental steps would gradually but continuously expand human "reach" into space beyond low Earth orbit—with economy provided by a template of bootstrapping. The key was to use robotic missions as pathfinders to understand, access and acquire products derived from lunar and space resources.

As these differing threads were woven into a policy statement, NASA viewed the VSE as the next "large space program" for the agency. NASA's traditional template dominated public discussion of the Vision, where gaps, arbitrary time scales and the long-desired human Mars mission as the "ultimate goal" became familiar talking points – not surprising, considering that the agency had sole custody of the VSE after it was crafted. Lunar return by 2020 was not meant as a deadline, but it is widely interpreted as such. Although the VSE is careful to mention trips to "Mars and other destinations," the latter part of that phrase seldom appears in NASA charts.

The subsequent Exploration Systems Architecture Study[5] (ESAS) is pure NASA. In classic agency fashion, "Apollo-on-steroids" (big giant booster, mega-capsule and gargantuan lander) was rolled out. The programmatic significance of Ares V in the architecture should not be overlooked – delivering 150 metric tones to LEO, it is a rocket designed for human Mars mission done in the Apollo-style, with everything needed for Mars dragged up from the deep gravity well of the Earth. It is overkill for almost any other space job, including missions to the Moon. Overkill can work, if you have the money (although it isn't good practice even if you do have the money). But even with the most optimistic assumptions, the ESAS doesn't fit into NASA's current or projected budget.

Marburger's concern is exactly what has happened. NASA thinks that its principal mission on the Moon is to conduct Apollo-style local site exploration and serve as a test-bed for the Mars flags-and-footprints extravaganza. The idea of building a spaceflight infrastructure using lunar resources was swept aside. An Apollo-like architecture was developed but with no political backing to pay for it. Now the agency finds itself subject to a protracted and embarrassing "public audit" of its mission and methods of doing business. The country is not disposed to a significant increase in spending on space, not just because of the poor state of the economy (although that doesn't help) but because they think we are already spending the right amount. The comfortable, old shoe cannot be resoled; you cannot conduct space business today using the Apollo model, whereby technical difficulties are bludgeoned into submission by cash and long hard (and expensive) man-hours of work.

The way forward involves approaching the problem differently. Marburger's take on the VSE[6] is adaptable to any budgetary level. It makes continuous progress, using small steps when times are tough and larger ones when things are flush. It sets no deadlines but it does set strategic directions – incrementally beyond low Earth orbit, using what we find along the way to create new capabilities and possibilities. It has intermediate milestones that map progress and provide societal payback. It brings commercial enterprise along,

with the aim of expanding our space economy and high-technology industrial base. In other words, it is sustainable. It is the antithesis of the conventional form of space exploration.

Given the dwindling amount of money for discretionary spending in the federal budget, perhaps the idea of using lunar resources to build a sustainable infrastructure in space should be embraced.

Topics: Lunar Exploration, Lunar Resources, Space Politics, Space Transportation, Space and Society

Links and References

1. Augustine committee, http://www.nasa.gov/offices/hsf/home/index.html
2. Marburger testimony, http://tinyurl.com/337zrb3
3. Vision for Space Exploration, http://history.nasa.gov/Bush%20SEP.htm
4. detailed studies, http://tinyurl.com/337zrb3
5. ESAS, http://www.nasa.gov/exploration/news/ESAS_report.html
6. Marburger and VSE, http://www.spaceref.com/news/viewsr.html?pid=19999

Comments

5. The BIG question (IMHO) is that given the above and the current exigencies that NASA faces; can it *in itself* CHANGE? Or will external agencies be required. This Space Cadet feels that NASA has become too ossified and has some ideas. (Follow link for Handwaving and Hyperbole.)

Looking forward to today's Meeting tho'. Should be a cracker!

http://www.nasa.gov/pdf/376015main_August%2012%20Meeting%20Agenda%20-%20Final.pdf

Ronald (Wilson) Reagan Building :)

Comment by brobof — August 12, 2009 @ 7:54 am

6. Let's face it, the O'bama administration wants to use money slated for space exploration and use it on his own pet programs. There's no support, vision or any leadership regarding manned space exploration at the White House anymore. Sure, O'bama will speak wistfully and eloquently about space exploration but he won't do anything substantial. Face it folks, we're going nowhere. Watch funding for all of NASA to be slashed as we continue to hitch a ride on Soyuz space capsules. All you space supporters and dreamers, you'll all go to your grave never seeing Americans walk on the Moon again, let alone Mars.

Comment by Dan Roberts — August 12, 2009 @ 9:14 am

7. In football it is hard to get one player, sized and trained to play one position, to play a different position. Can you imagine a 300 lb. defensive lineman trying to think and play like a cornerback, or quarterback? NASA is a 300 lb. rocket scientist. It only knows how to build rockets and spacecraft. Trying to get today's NASA to seriously think about working and living on another planetary body, developing surface based assets, living off the land, and promoting economic potential is about as hopeless as a "hail mary".

Comment by John G. — August 12, 2009 @ 9:41 am

9. Remember what the economic incentives of government programs are, and VSE/ESAS will make much more sense.

Private enterprise serves the consumer, because the consumer pays for it voluntarily and may easily switch to another service provider if service is relatively poor.

Government takes money from people by threat of violence (pay your taxes or we throw you in jail or shoot you). They have *no* incentive to serve the taxpayer. They have *every* incentive to serve those who give them money — the politicians who are trying to appease their constituencies. This is why the segmented SRB design was chosen in the first place, even tho it was by far the worst design out of 4 contenders. Government programs are chosen based on how well they favor certain groups with political 'pull'.

Also, government programs are rewarded for failure. If your program fails, that means you didn't have enough money and should be given more. If you succeed, that means you should be given more money so you can succeed further. There is *no* mechanism for readily inducing a shrinkage or efficiency increase, other than some vague and quickly forgotten promises to the taxpayers every 2, 4, or 6 years.

In short, government can't go out of business, so they have no reason to be efficient. They are several steps removed from the taxpayers whose labor supports them; so they have *no* incentive to get you or I into space. Because of this, we should give them as little money as possible, so they will waste as little as possible.

Comment by Carl — August 12, 2009 @ 11:10 am

10. First, Paul, you did an excellent and eloquent job summarizing Marburger's perspective for those of us who didn't know it. Second, I think that a little stronger oversight of NASA might help, and in particular NASA's Strategic Plan developed by OSTP as an example of what could be done. Let them give NASA the focus. Case in point: the emphasis and priority on ISS has been heavily skewed toward the facility and astronaut needs at the expense of payloads. A stronger mandate in a Strategic Plan, defined outside of NASA would have balanced the activity on ISS to get more of a response for payloads. The same for ISRU in early lunar outpost discussions. ISRU is practically gone now, even though this seems like more of the intent from Marburger's perspective... I don't think NASA is that good at setting its own priorities.

Comment by Tony Lavoie — August 17, 2009 @ 6:12 pm

11. Paul

Excellent exposition on Marburger's thoughts and directions. If NASA had taken that route, Augustine II would not have happened and we would be well on the way to the Moon by now.

Comment by Dennis Wingo — August 22, 2009 @ 2:33 pm |Edit This

Scientists vs. The Icy Commander

August 21, 2009

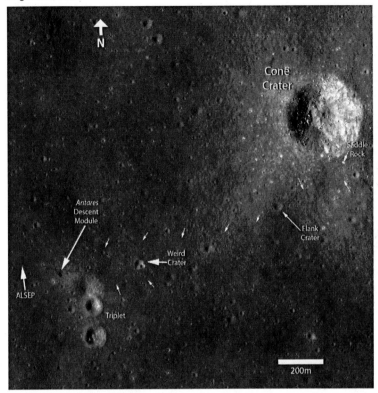

The path to Cone crater (LROC image, Ariz. State Univ.)

In 1961, Alan B. Shepard's[1] successful 15-minute sub-orbital hop gave President Kennedy the high cover needed to announce a reach for the Moon, "by the end of this decade." America's spirit was lifted and Alan Shepard became a national hero, getting ticker tape parades and White House receptions. Then, as in a Greek tragedy, he was struck from the flight list after developing Meniere's syndrome[2] (an imbalance of the inner ear). His flying days were over. Or were they?

Shepard, a smart, tough, no-nonsense aviator, took a job helping Deke Slayton (previously grounded by a heart murmur) run the Astronaut Office. Shepard and Slayton picked all flight crews for the Gemini and Apollo missions. Very early on, it became clear that you did not cross Al Shepard, lest your career come to a screeching halt. Shepard never stopped his Apollo training or flying in the T-38, even though he had to "backseat it" with another astronaut. His personality was memorably captured in Tom Wolfe's book, *The Right Stuff*[3], as "The Icy Commander."

After taking a chance on experimental surgery to correct his inner ear problem in 1969, he successfully returned to active flight status and looked ahead to an Apollo flight assignment. Rejected for the Commander's seat on the next available flight by NASA Headquarters (on the grounds that he needed more training time), he was named to command a subsequent flight, while Jim Lovell was named Commander of Apollo 13.

Geologists who worked on the Apollo training were ecstatic – Lovell was one of their favorite pilot astronauts, a smart, capable guy with a keen eye and an analytic mind. He was being sent to Fra Mauro, the first highland site to be visited on the Moon. This region

99

was considered a key locale to decipher lunar geological history, being located on the ejecta blanket of the Imbrium basin[4], the largest impact crater on the near side.

Jim Lovell was considered the right man to study this site and collect the key samples scientists needed to help unlock the secrets of the Moon. Unfortunately, with the failure of Apollo 13, Jim Lovell didn't land on the Moon. Still, the Fra Mauro site was considered so important, it became the designated landing site for Apollo 14[5], eighteen months later.

Uh-oh. Lunar scientists didn't have Jim Lovell to explore with—they had drawn the "Icy Commander," the guy who cheerfully admitted that, compared to aeronautics, he thought geology was a low-grade science. Nevertheless, Shepard assured the Apollo scientists he would try to do the best job he could for them.

While successful in almost every way, the Apollo 14 mission[6] was not without controversy. Cone crater, a large young impact feature, had apparently dug up rocks from deep within the Fra Mauro Formation, including it was hoped, ejecta from the Imbrium basin. During their second moonwalk, Al Shepard and Ed Mitchell trudged up steep slopes leading to Cone, dragging along their Modularized Equipment Transporter (MET), a small pull-cart designed to carry tools and samples with them, getting more winded and disoriented with each step. Getting to the rim of Cone crater was considered critical to the scientific success of the mission.

At 47, Shepard was the oldest man to fly to the Moon and many felt that he was out of shape and not up to the rigors of lunar trekking (which didn't explain why Ed Mitchell was also having problems.) Moreover, it seemed that Shepard was all too eager to abandon the trek and declare victory after he radioed to the ground that he thought they were already at the rim of Cone crater[7]. (Enough with the hiking trip! We're running out of time and consumables. Let's sample this area and call it the rim of Cone crater.)

Scientists in the back room were aghast. Getting Cone crater samples was critical to mission success. And now this old, panting geezer was destroying their chance to unlock a deep secret about the Moon. Although they put on a good face, scientists were resentful; after all their work on geological training, the "Icy Commander" simply declares victory and turns for home. Adding insult to their perceived

injury, back at the Lunar Module, Shepard pulled out a 6-iron and conducted a little sand trap practice. (He abandoned the quest for Cone crater – to play golf, no less!)

Now, thirty-eight years later, we've just received a magnificent picture of the Apollo 14 landing site[8] from the Lunar Reconnaissance Orbiter Camera (LROC). Its quality is so good we can see the path of the astronauts footprints and MET tracks on the Moon. It is even possible to follow their tracks all the way up to Cone crater—to the point where Al Shepard declared victory.

Oops. Al Shepard was right. He was at the rim of Cone crater. Terrain around the rim is so hilly that he and Ed Mitchell didn't know they had reached the rim; the deep crater interior is just over a slight rise, a few tens of meters north of where they were. The samples that Shepard and Mitchell collected do represent the deepest ejecta from Cone crater, thereby fulfilling that goal geologists set many moons ago. For almost 40 years, the "Icy Commander" was right. Yet his name lived in infamy in lunar geologic circles.

If there is a moral to this story, it could be that scientists should never state something is absolutely known and settled. It's likely they'll be proven wrong.

Topics: Lunar Exploration, Lunar Science, Space and Society

Links and References

1. Alan Shepard, http://en.wikipedia.org/wiki/Alan_Shepard
2. Meniere's Syndrome, http://tinyurl.com/4lwzmn
3. The Right Stuff, http://www.tomwolfe.com/RightStuff.html
4. Imbrium basin, http://en.wikipedia.org/wiki/Imbrium_basin
5. Apollo 14 landing site, http://www.lpi.usra.edu/lunar/missions/apollo/apollo_14/landing_site/
6. Apollo 14 mission, http://www.lpi.usra.edu/lunar/missions/apollo/apollo_14/
7. Cone crater, http://www.hq.nasa.gov/office/pao/History/alsj/a14/a14.tocone.html
8. LROC image, http://lroc.sese.asu.edu/news/?archives/91-Trail-of-Discovery-at-Fra-Mauro.html

Comments

2. We never doubted we were at the top. Photos proved it

Comment by Edgar Mitchell — August 22, 2009 @ 9:36 am

3. Hi Ed,

Thanks for dropping by and commenting. Indeed the photos did, but sometimes the objective truth gets buried by people's impressions and that was my theme.

Anyway, the post-mission photos have now been corroborated by the new LRO data.

Come back often!

Comment by Paul D. Spudis — August 22, 2009 @ 10:04 am

6. There's an even larger lesson to be learned from this. It's not just that scientists — even when they agree with each other completely after long discussion — can be wrong. It's that anyone can be wrong. There are sets of conditions when the Alan Shepard style of leadership is quite correct. To use one famous example, long, thoughtful discussion of what to do next would have been hilariously inappropriate on Normandy beach on June 6, 1944. Perhaps even under the conditions present on the Moon during the Apollo explorations. After all, if the astronauts had died on the Moon, then those samples would not have come back.

On the other hand, many people argue that the kind of leadership exhibited by Shepard in the astronaut office and on the Moon is today holding back progress in space exploration and development. Ever hear the phrase "Not Invented Here"? That kind of close minded behavior stifles further development.

Comment by Chuck Divine — August 24, 2009 @ 9:06 am

7. That, sir, is an open question IMHO— whether we suffer from too much or too little Alan Shepard-style management. There is a difference between true closed-mindedness, and a single-minded determination to succeed.

Comment by Warren Platts — August 28, 2009 @ 12:23 am

8. Glad I found this, since other outlets make it seems as if they didn't quite reach the rim. So was it close enough to the geologic features, just not close enough to say "I'm on the crater rim!"

They were 30 meters short of seeing into the crater itself, apparently.

http://www.universetoday.com/2009/08/20/latest-lro-image-solves-apollo-14-mystery/

Comment by Ralph H — September 30, 2009 @ 4:57 pm

9. *So was it close enough to the geologic features, just not close enough to say "I'm on the crater rim!" ?*

Yes. Their objective was to sample the rim of Cone crater, where the deepest ejecta presumably would occur. They did that. Mission accomplished.

Comment by Paul D. Spudis — October 1, 2009 @ 4:56 am

I Aim At the Stars...But Sometimes I Only Make Viewgraphs

September 9, 2009

The more things change....

Over the long holiday weekend, Turner Classic Movies[1] regaled us with a really obscure one – the 1960 biopic, I Aim at the Stars[2], starring Curd Jürgens. This movie is a biography of Wernher von Braun, the German rocket scientist who built the V-2 for Hitler and the Saturn V for America. Although no landmark in cinematic history, it was an interesting and reasonably well told story, even if it glossed over a few inconvenient facts about von Braun, like his nominal membership in Himmler's SS[3].

What fascinated me in this movie (which I had not seen) was not von Braun, but the character played by James Daly, Major William Taggert (an intelligence officer in the U.S. Army who, having lost his family to a V-2 hitting London, hated von Braun and all of the Peenemünde rocket group). After the war, Taggert follows the Germans as they relocate, first to White Sands and finally to Huntsville to continue their research into rocket flight. Taggert becomes a reporter (his civilian occupation) who beats his media pulpit about the ir-

relevancy of space flight. "All the money spent on space could build schools and hospitals instead!" he angrily harangues via television, a philosophical counterpoint to von Braun's plea for an American satellite program.

Watching the movie, I was struck that this debate has been ongoing for the last 50 years. Something about space exploration or human forays into new realms sticks in the craw of some people. Although the context of the von Braun-Taggart argument was Sputnik and a possible American response, much has remained the same over these last 50 years. The public still falls into two camps – those who believe that our survival depends on continued reach beyond Earth versus those who think it's a waste of money or that the money could be better spent. NASA spends most of its outreach efforts trying to win the hearts and minds of this latter group.

Case in point: a NASA "white paper[4]," clearly a rough draft, leaked to the press[5], describing the post-Augustine space program. Omitting the use of our Moon as the logical next step, "Generation Mars" is billed as the necessary pathway to keep NASA relevant, the public engaged and the required pipeline for sustainable product and group input cycles. No more idiotic fooling around with, or distractions from, lunar bases. The "exciting" destination is Mars – in about thirty years or so. In the mean time, keep flying Shuttle so as not to upset the applecart. Oh, and imagine, "use" the ISS for something (Just pull one or two studies—from the hundreds gathering dust—off the shelf of unfunded programs).

A key assumption here is that NASA's survival revolves around an excited and engaged public. The authors of this piece apparently think this will happen with Mars because the public doesn't care about the Moon; that the Mars Generation can become "emotionally engaged because they will become contributors to the Mars goal and part of the maturation process in achieving it." Great stuff that – "emotional engagement," not reason or logic. The system of taking incremental steps using lunar resources to make space faring routine is abandoned for a multi-decadal agency program to take an "excited" public to Mars, a program "owned" by its contributors. That's a lot of time and work needed to engage, excite and own something. It sounds like the description of an entitlement program, not a mission statement.

After 50 years of obvious benefits from space flight, many still are, at best, indifferent to it. But even more significantly, few feel the need to be emotionally engaged with it. People understand that along with our vast interstate road network, we have other vital economic infrastructure, such as railroad transportation, air traffic and more recently, a network of telecommunication satellites orbiting Earth. We depend upon this infrastructure on a daily basis, but except for buffs, we do not get emotionally engaged in their day to day operations.

As no significant additional money is likely to materialize, we must strive for achievable goals and a paced rate of advancement. A program that promises accomplishment thirty years in the future is not a program at all, but rather, an excuse to "study" the problem indefinitely. In other words, it means another thirty years like the previous thirty years – lots of swell viewgraphs, color artwork of astronauts climbing the walls of Valles Marineris[6], and bureaucratic blither about exciting students. But no actual spaceflight infrastructure.

I've touched on this issue[7] before; no one votes for a candidate based on their position on the space program. The net effect of this environment of public indifference is that NASA's budget (which comes from an ever shrinking slice of the tax-funded, discretionary spending pie) will remain at existing levels for the foreseeable future. What does this mean for NASA's Mad Men advertising campaign[8] for "Generation Mars?" Basically it means that a space agency dependent upon public excitement to enrich its budget is one that is not likely to prosper. With budgets devoured by countless cycles of viewgraphs, white papers and consensus management missives in the coming decades, what remains is an agency with no sustainable space exploration system.

To add space to our other national transportation networks, the kind that we take for granted but that contribute in so many ways to our prosperity and security, NASA needs to lay the groundwork for private industry to follow. NASA needs to be the driver of private sector technology as it explores. Without logical steps, NASA becomes the devourer of resources and not a technology driver.

As the next frontier is scouted, business will follow, as it always does. Business is eager to follow. NASA needs to finish laying the groundwork before moving on. The Moon is the next destination in space. Will America lead and have a stake in this new land or will we stay behind and watch the movie?

Topics: Lunar Exploration, Space Politics, Space Transportation, Space and Society

Links and References

1. Turner Classic Movies, http://www.tcm.com/tcmdb/title.jsp?stid=27789
2. I Aim at the Stars, http://www.imdb.com/title/tt0053440/
3. von Braun, SS, http://tinyurl.com/29l3mhu
4. Generation Mars, http://www.spaceref.com/news/viewsr.html?pid=32268
5. leaked to press, http://blogs.orlandosentinel.com/news_space_thewritestuff/2009/09/nasa-aims-for-a-mars-landing-in-30-years-.html
6. artwork, http://www.patrawlings.com/detail.cfm?id=975
7. public support, http://blogs.airspacemag.com/moon/2009/07/16/space-program-vs-space-commerce/
8. Mad Men, http://www.amctv.com/originals/madmen/

Comments

1. I too used to take for granted that public excitement was essential to reaching the next major milestone in space exploration development. But reading you post, I realise that, beyond the resources consumed in trying to engage the public, that very goal is sending us back to the old days. Trying to revive public

excitement is another way of trying to revive history. It actually makes perfect sense that the level of public excitement about space will probably never ever reach again the level it has back in the 60s, as space is no longer new. Likewise, people are no longer excited about say, the internet. BUT, the internet and NTIC at large now offer something more important : the prospect of an exciting career to young students. Excited students who will become excited researchers and engineers is what the space industry really needs, not an excited public. Once again, a most enlightening article, thank you !

Comment by matchad — September 9, 2009 @ 7:38 am

2. As far as a candidate, particularly at the national level turning support for NASA into a campaign issue, much of it is the candidates themselves. There rarely is a huge difference, and in the case of Obama he needed to be gently reminded that NASA is more than an engineering jobs program that could be cut to fund the Dept. of Education. I'd hardly call either George W. Bush's or John McCain's support of NASA to be that of a hardcore fan, although Bush being from Texas certainly was familiar with NASA programs.

Basically, I don't think the general public has been given a clear choice between the two camps in a visible race to really be able to express their feelings on this issue. Congressmen and Senators from districts and states with huge aerospace infrastructure are going to be whole hog supporters of NASA and spaceflight (Texas, Florida, Alabama, and northern Utah), or at worst apathetic about space exploration. Congressional delegations in other states might be mildly against NASA, but otherwise simply don't care either. It hasn't been an issue in part because there is no real difference in philosophical attitudes by the candidates.

It would be interesting to see what would happen if there was a significant political race with somebody firmly in each "camp" of support/opposition to spaceflight and see how that might play out as an issue.

Missing from this discussion is somebody like Walt Disney, who played a major role in selling space exploration to the generation of Americans in the 1950's and 1960's. While there are some like James Cameron (he wants to book a private spaceflight to make a film in orbit) and Tom Hanks (a major supporter of the X-Prize foundation) who are strong sup-

porters of spaceflight, I don't think anybody could match the handiwork of the master of the mouse. An old-fashioned sci-fi thriller filmed in orbit with actual astronaut-actors might just pull off something that would put actual views of Earth orbit onto the big screen could be one of those kind of major moments in history to allow ordinary folks to "get it" on what can be done in space. Yes, there have been Imax films done like that which are awesome, but those don't go everywhere or are seen by ordinary folks that aren't already space fans. I just don't know who would be able to pull off the kind of P.R. campaign that Disney did back in the early days of spaceflight.

Comment by Robert Horning — September 9, 2009 @ 11:53 am

3. It is my humble assertion that the previous generation are sated. Largely by the cinematic guile of Hollywood! Not only have we gone to the Moon; we've gone to 40 Eridani A and met the slightly green tinted, pointy eared hominids with hemocyanin for blood! Or Canopus and the silicon based Sand Worms. How can the orbital plodding of the ISS compete with Warp Drives and Guild Steersmen? Add the fact that [strike]Spice[/strike] Space has always been a political football: Proxmire anyone?

Unless NASA develops a program to re-engage the next generation. Say with the massively hands on lunar tele-presence exploreathon that I keep banging on about. (See Link) Then the drift away from the High Frontier will continue, leaving the field for less jaded nations: China, Russia, Japan and indeed even tired old Europe. "Festina Lente!" As one tired old European (Augustus Caesar) used to say!

Comment by brobof — September 10, 2009 @ 4:38 pm

4. You know if I were President I'd be at every shuttle launch. That would be one way to draw attention to the program. I would have made sure the press saw me watching the test today of the Ares rocket. Things like that would make a big difference. A few comments about just how much the economy has been repaid for what we spent in space in technology spin offs and lives saved by weather satellites. I'd also spend some time just evoking the gosh gee whiz wow factor. It's not hard. Not really. A statement like this "Investing in man space is an investment to our children's future and will materially make their lives better as it has ours."

The meek can inherit the Earth. The rest of us are going to the stars. It would be shame if those people were Indian or Chinese and not American.

Comment by Stacy Brian Bartley — September 10, 2009 @ 5:06 pm

5. Nations are just too overstretched to justify spending such large sums of money on something that gives no immediate return(in the wallet). We will return to the Moon, I think, if we get a working fusion reactor to generate electricity... but that's some while away. Perhaps this would lead to the merging of the military-industrial complex with oil and mining. Run out of oil? lets go to Titan!

Comment by Phil Thomas — September 11, 2009 @ 2:48 am

6. Generation Mars, GEEEZ! Once again NASA doesn't get it. A few quotes from the Augustine Committee summary, "Planning for a human spaceflight program should begin with a choice about its goals-rather than a choice of possible destinations." So what does NASA do, it picks a destination. Quoting again, "The Committee concluded that the ultimate goal of human exploration is to chart a path for human expansion into the solar system." Shooting for Mars is truly 'Apollo on steroids', just another big rocket program not even leaving bread crumbs for a path. Quoting again, "The Committee finds that Mars is the ultimate destination for human exploration; but it is not the best first destination." Hello, is NASA reading this stuff? (I do find it curious that the committee finds Mars the "ultimate destination". I'm glad our forefathers didn't think of the Appalachian Mountains as the ultimate destination. Such short-sightedness.) The plan is out there folks, its been there for 50 years! Read Arthur C. Clarke's 'The Exploration of Space' or Werner Von Braun's articles in Colliers magazine from the 1950s. Incremental steps, with the Moon being the first step, learning how to work and live on another planetary body, and using the knowledge gained on the Moon to venture to Mars, and other places in the solar system. Mars Generation, GEEEZ.

Comment by John G. — September 11, 2009 @ 11:36 am

7. The idea of excitement as the basis of NASA support was always contingent on there being little real opposition, and the ability of NASA-backers to ridicule their opposition as ignorant made them ignore it, till way too late. In fact, while in January of 1962 42 percent of the population thought we needed to spend more on spaceflight, and only 6 percent thought we should spend less, by January of 1972 43 percent thought we should spend more on spaceflight, but 38 percent now thought we should spend less. The political profit for any politician had dropped steeply. All NASA's attempts to win back that 30+ percent in opposition have failed, because they thought "excitement" and "spinoffs" claims could do it.

They cannot do it. That leaves a competent building of space infrastructure that really can allow our society to lower the costs of human spaceflight, so we can do many more missions for the same money we have today. That means refusing the lure of "excitement" which was always based on the ability to manipulate the great mass of people. They *know* that is being tried, and they resent it bitterly!

We must admit that human spaceflight is pointed at the settlement of the rest of the Solar system, and that making it cheaper to do this is NASA's primary mission, including lots of science to let settlers know what they will be getting for resources, and for problems in their new environments. This is a rational goal for the growth that rational people will realize is needed for our society.

*Forget*manipulation* and concentrate on competent lowering of costs, and we will then have a rational argument that we can take to people we admit are rational themselves, not the emotion-driven morons of academic and bureaucratic fantasy.

Comment by Tom Billings — September 11, 2009 @ 12:37 pm

9. I once saw an Australian TV story about helium-3 and how its the best fuel available for fusion reactors. This material is most abundant on the moon, as its regolith is soaked in solar particles.

I think that our lack of natural resources is eventually going to force us out of the ground and into space. In this TV program Harrison Schmitt was all for mining the moon while Ed Mitchell was flatly against it.

I just don't think we'll be able to raise the megabucks necessary to leave the earth in the near future unless there's serious commercial gain in it, or the war on terrorism ends.

Comment by Phil Thomas — September 12, 2009 @ 6:26 am

11. I agree that frenzied excitement isn't necessary for NASA survival; in any case, a significant proportion of the public were not

excited by the Apollo landings, (cf. Gil Scott Heron's proto-rap tune about it.) As it is, every shuttle launch rates a blurb in the mainstream media. I think going back to the Moon would increase interest levels.

That said, I don't appreciate the way the Augustine committee has not-so-subtly inverted the VSE priorities. The Moon was the "must have" destination, while the shuttle and ISS were "nice to haves" that would require extra, new monies to keep going; now it's the opposite.

PS: Paul, have you seen ULA's proposal to use EELV's as an alternative to the CxN scenario? It came out a couple of days after your post. It says we can get back to the Moon by 2018 under the current budget without heavy lift, and proposes a cool, "Space 1999"-like horizontally landing lunar lander architecture. Very refreshing after the depressing Augustine hearings. This is an important third alternative to Cxn and Augustine, something concrete that Congress can get behind. With any luck, the lunar base will still get constructed within our lifetimes.

http://www.nasaspaceflight.com/2009/09/ula-claim-gap-reducing-solution-via-eelv-exploration-master-plan/

Comment by Warren Platts — September 17, 2009 @ 8:14 am

12. To understand why NASA is where it is; one needs to look at the economic incentives it has.

Who is NASA's customer (i.e. who gives it money)? The U.S. Government.

Whom does NASA serve, in order to get that money? The U.S. Government.

I posit to you that *at best* government space programs exist to get government people into government-operated space. At worst, they're self-perpetuating jobs programs with marginal productivity when compared to an organization that has competition driving it.

NASA will not get you or I into space. They have *no* incentive to do so. Even if they tried they probably couldn't do it effectively because there are no economizing forces on them. They get their money from taxes, which are not given in voluntary exchange.

On the other hand, Burt Rutan and the like *are* getting a lot of interest from individuals and the media, because they *do* promise a way that non-government-affiliated individuals can get into space. It costs a lot of money right now; but it doesn't cost a lifetime, and it will get much cheaper and better in short order due to competition.

If you want to go into space, you do indeed need to engage them... and you need to engage their wallets. Offer them real value that they can see, and they will vote you loads of money straight from their pockets.

We like to justify the government space program with its spinoffs and money returned; but in a lot of cases it's just self-justification. It ignores the fact that a lot of those spinoffs would have happened anyway, because they were technologically ready to happen. The benefits of the space program to the average consumer have been largely due to market-driven forces; such as communication satellites, weather prediction (as consumed by commercial aviation and many other enterprises), mapping (again as consumed commercially), and others. Those benefits largely come to the consumers via commercial channels, which means money is available to fund their operation. The consumers won't give money voluntarily to a space program that pretty obviously returns very little value directly to them. They give it to the companies that provide them value and *then* those companies use space-based technology.

Comment by Carl — September 17, 2009 @ 12:25 pm

14. *"it was an interesting and reasonably well told story, even if it glossed over a few inconvenient facts about von Braun, like his nominal membership in Himmler's SS."*

This is slightly off topic, but speaking of Faustian bargains, I just now ran across the curious case of Gerry Bull, a rocket scientist who's life kind of mirrored Von Braun's. Von Braun was born in the German periphery in what is now Poland; Bull was from Canada. Von Braun worked on new-fangled liquid fueled rockets; Bull worked on the old-fashioned, Jules Verne technology of using big guns to launch things into orbit. Both would turn to any military that would fund their ideas, but Von Braun worked for the bad guys first, and then turned to the good guys, whereas Bull worked for the good guys first, and when that didn't pan out, he turned to Saddam Hussein and helped develop the infamous "Babylon Gun", that would have been by far the world's biggest gun ever built. (In theory, the gun could have delivered 100 pound payloads into orbit for ~$300 per pound–not bad when compared to the $5,000 per pound that current EELV's cost!) To complete the bizarre mirror symmetry, Von Braun died a natural death; Bull was shot in the back of his head 6 times, probably by the Mossad. Crazy . . . .

Comment by Warren Platts — September 22, 2009 @ 11:05 pm

Water, Water Everywhere....

September 25, 2009

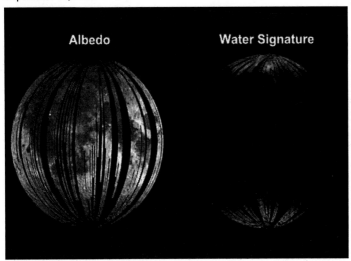

Water-bearing minerals on the Moon. (Chandrayaan M3 Team and NASA)

The extreme dryness of the Moon is established scientific dogma. The study of Apollo rock and soil samples pretty much had convinced scientists that the Moon has no water. Because its surface is in a vacuum and experiences extreme temperature swings at the equator (from -150° to 100° C), the Moon was believed to have a bone dry surface. Moreover, minerals that make up the lunar rocks not only have no water, but crystallized in a very reducing, waterless environment, indicating no significant water at depth.

Yet, some irritating facts suggested that the whole story was more complicated. Water is being added to the lunar surface. We know the Moon is bombarded with comets (mostly water ice) and meteorites rich in water-bearing minerals. Additionally, the solar wind (mostly hydrogen atoms or protons) constantly hits the surface, implanting itself into the dust grains and a possible source for the creation of water[1]. An experiment laid out on the surface by the Apollo astronauts observed water vapor[2] after the crew left the Moon. It was thought this vapor might be latent out-gassing from the Lunar Module descent stage, but scientists couldn't be sure.

So what happens to all this water? Most of it is thought lost to space by a variety of processes, including dissociation by sunlight, thermal loss from the extremely high daytime temperatures, and sputtering induced by the impact of high-energy particles from space. Some areas near the poles

of the Moon are permanently dark and cold, so if any of this stray water happened into them, they would be "trapped" forever in the dark areas. And although an extremely slow process, over millions of years a considerable amount of water ice might accumulate[3]. But we don't know how much water is made and how much might be present on the Moon.

Just published results from spectral mapping instruments[4] on three different spacecraft indicate the presence of large amounts of either water or the OH^5 molecule in the soils of the Moon. This water is present at high latitudes at both poles and occurs in sunlit areas (these instruments rely on reflected sunlight). Although the authors of these new results don't understand the source of this water, they favor the creation of water by the interaction of solar wind with surface minerals. Solar wind protons reduce metal oxides in the soil, creating free metal (usually pure iron, FeO) and water. The M3 Team suggested that this water might act as a source for the water believed to be trapped in the dark polar cold traps.

What's surprising about this new data is not the presence of water, but its pervasiveness. The published image (above) shows this water to be present from the poles down to about 60° latitude. This area subtends over 10 million square kilometers, or about one-third the surface area of the entire Moon! Although the water appears to be present only in the upper few millimeters of the

107

surface, its total mass could be enormous, greatly exceeding the several hundred million tones estimated to be present as ice in the dark areas of the poles.

As always with good science, the new results raise many more questions than they answer. In part, this is a "chicken or egg" issue – do the newly discovered deposits result from surface alteration by water derived from the polar ice, or do they serve as a source for such deposits? How does water form, move, get destroyed or get cold-trapped on the Moon? What are rates of water deposition and removal? What and where are the ice deposits and how pure might they be? Right now we can only dimly perceive the beginnings of a whole new sub-discipline of lunar studies: polar geoscience.

This exciting story isn't over. More developments in this field are on the horizon. Results from other experiments carried aboard the Chandrayaan-1 spacecraft, including my own Mini-SAR imaging radar[6], have yet to be fully reported. The American Lunar Reconnaissance Orbiter[7] (LRO) mission is settled into its mapping orbit and will be examining the Moon in detail over the next couple of years. Every time we get new data from the Moon or examine and map it with some new technique, we learn new and surprising facts.

In a future post, I'll examine the implications of large amounts of lunar water for human return to the Moon and the possibilities for a permanent sustainable presence[8] on our nearest planetary neighbor. Stay tuned – things are getting very interesting.

Topic: Lunar Exploration

Links and References

1. source for water, http://www.ifa.hawaii.edu/~meech/a740/2006/spring/papers/Vondrak-Crider-2003.pdf
2. observed water vapor, http://tinyurl.com/2vacbas
3. accumulate water, http://www.thespacereview.com/article/740/1
4. spectral mapping instruments, http://m3science.geo.brown.edu/?p=167
5. large amounts of water, http://www.sciencemag.org/cgi/content/abstract/1178658
6. Mini-SAR, http://www.nasa.gov/mission_pages/Mini-RF/main/index.html
7. LRO mission, http://lro.gsfc.nasa.gov/
8. permanent sustainable presence, http://www.spaceref.com/news/viewnews.html?id=1349

Comments

1. Quoting Arthur C. Clarke from 1951,

"The first lunar explorers will probably be mainly interested in the mineral resources of their new world, and upon these its future will very largely depend."

The Augustine Committee clearly showed that we cannot get humans to the Moon anytime soon with NASA's current budget and thinking. A huge increase in NASA's budget (though small when compared to stimulus bills and bailouts of the banking system and auto industry) is not likely. So, what do we do? While we spend the next decade extending the human use of the International Space Station, and coming up with a new US human space transportation system, we can also be sending robots to the Moon, to continue exploration of the lunar surface and begin lunar resource utilization and development in preparation for a human presence sometime in the late 2020s or early 2030s. Robotic exploration and development of the Moon during 2010-2020 is an affordable, incremental step along the path of human expansion into the solar system, that subsequent steps could benefit from. Does this last sentence sound familiar? It should, it was espoused by Clarke and von Braun in the 1950s, more recently by the Augustine Committee, and by plenty of other people in between. Going back to the Moon does not have to start out like another big NASA rocket program. It just requires some old thinking, and some new implementing.

Comment by John G. — September 25, 2009 @ 11:59 am

4. Very exciting developments. LCROSS can only yield new clues and greater understanding, I'm so anxious for that on October 9th. As for where the hydroxyl exists in which layers, or from whence it can be coaxed with new extraction technologies, it's also exciting to think that we don't know the whole story yet. Think how many lunar samples we've had since 1969, but only now have developed techniques to unfold this amazing story. I'm thinking this is the tip of the proverbial lunar iceberg…   :)

Comment by Pillownaut — September 25, 2009 @ 8:23 pm

5. @ John G.

The Augustine Committee clearly showed that we cannot get humans to the Moon anytime soon with NASA's current budget and thinking. Nevertheless, that does not entail that it cannot be done. Cf. the recent proposal by ULA that just came out. They can get boots on the ground permanently by 2018.

@ Dr. Spudis:

What about the fumaroles? Do you think they play a role in the lunar hydrology?

Comment by Warren Platts — September 27, 2009 @ 9:39 am

6. *What about the fumaroles? Do you think they play a role in the lunar hydrology?*

Although we cannot totally rule out anything at the moment, I doubt it. Fumaroles are associated with volcanic activity and the vast bulk of the Moon has been volcanically dead for billions of years. This water must be geologically young, otherwise it would have been lost to space and destroyed. In fact, its pervasiveness argues for its current, continual formation on the Moon. A more likely origin is solar wind reduction of silicates in the existing rocks. This newly formed water may eventually migrate to the polar cold traps.

Comment by Paul D. Spudis — September 27, 2009 @ 10:47 am

7. Dr Spudis,

I was trying to describe how much potential water was on the Moon to a friend of mine in terms he would understand. According to your article and others, I have read;

Assuming that H20 (I understand that research is talking about OH) is present at a concentration of 1 part per 1,000 by volume over 1/3 of the Moon's surface and we extracted the water with 100% efficiency by strip mining one meter deep, we would expect to collect approximately 12.6 km^3 of water. Approximately 1/3 the volume of Lake Meade (volume of Lake Meade is approximately 35.2 km^3).

Have I calculated that correctly or have I understated the potential volume somehow?

Comment by Russell — September 28, 2009 @ 2:44 pm

8. If the theory that THERA collided with the Earth, creating both a larger Planet and of course the Moon, are correct then I think this data provides more ammunition for the seeding of Earth from Space. Not only in terms of Organics from the Kuiper Belt and Oort cloud meteors, comets etc but also water for the Moon as well as it would seem the Earth during the period of maximum(sorry about this) "Velikovsky" activity.

That being so, perhaps the moon is a case of development in progress. The Earth contains an abundance of resources. After this discovery perhaps the Moon does too! I've always been askance at the attitude that: "Well we've been to the moon and there's not much there! On to Mars!" and felt it was somewhat premature; considering that the explorations made were very cursory, and then only to affirm manned landing viability.

Oddly, the same mindset goes for Mars. We now find that had Viking scooped just a bit deeper, Ice would most likely have been found! Less than 4 years after the last moon mission as well. What a jump start for REAL space exploration – at a time where we had both the money and manpower resources to do something about it!

One really gets a grasp of how negative suspect motivation can be for certain endeavors; not merely in terms of the right hardware for the wrong mission and canceling everything once the "crisis" is over, but the entire direction of the space program: most especially the manned sector.

Clearly the motivation has little to do with Exploration since it's now clear in Space that machines can do 95% of what is required with a minimum of manned participation: Hubble excepted.

Even so that was actually service and Maintenance. I'm sure most of the Astronauts do not like to be thought of as Space mechanics, but in truth, that's what they are!

The Gravity Well of Earth being what it is, if we wish to do more in exo-LEO Space then we need a permanent presence there. What better place than a resource-rich abode like the moon with its protective Regolith if we're prepared to go underground? The moon is superior in most respect than the Lagrange points or even LEO as a Space Base both in occupation and Launch terms. If normal and light water are there in useable(but probably @ largely recyclable level) quantities. what else there? We now have prototyping machines which would be ideal for creating an industrial infrastructure without transportation of heavy plant; and what better place than the moon for Fission based and later Fusion based power generation. Then there are Far-Side Astronomical observatories, laboratories…the list is almost endless and I'm sure there are ways to work with local resources so as not to haul hardware up the Earth's Gravity well. Once we're fully established, Interplanetary ships/craft can be placed in Lunar Orbit to shuttle between Mars Orbit and Back using Lunar resources rather than the much more expensive one of Earth. Not to mention the saving of some 8 m/s of Delta V, trading mass for Velocity. I say back on the moon by the 50th Anniversary: we can do if we apply ourselves – in any case we have greater resources to work with. Time to cease procrastinating!

Comment by Kit Hildreth — September 28, 2009 @ 5:32 pm

9. Russell,

Your estimate is as good as any other that I've seen. Well done!

Because this volume is spread over a very large area, it may be difficult to mine and extract. We calculated the amount of water possibly present as concentrated ice in the polar dark areas based on a variety of radar about as much water as in the Great Salt Lake.

Comment by Paul D. Spudis — September 28, 2009 @ 6:53 pm

10. Wow that's a lot less than my calculations, but I noticed that you are looking at concentrated ice. Thanks for clarifying that for me.

Volume of water in the Great Salt Lake: 0.0189200000 km^3

Comment by Russell — September 28, 2009 @ 8:30 pm

Space Exploration Sets
Sail on Lunar Water

October 4, 2009

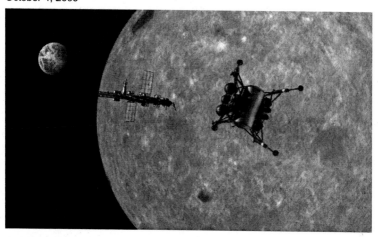

A cislunar transport system will revolutionize space travel (NASA artwork by Pat Rawlings)

Water is an extremely useful substance in space. The recent finding[1] of water on the Moon has generated considerable comment in the space community; a quick search on Google using the phrase "lunar water" returns over 7.66 million hits. Lunar water's significance lies not in its role as a medium for the presence of extraterrestrial life[2] but rather in its potential to support terrestrial life—ours—as humanity moves beyond Earth. The Moon is the port from where we will navigate—the safe harbor where we will learn how to live and work productively in space and from where we will set sail into our Solar System, thereby ensuring the survival of our species.

The three principal uses for this water are life support, energy storage, and rocket propellant.

We can easily imagine drinking water. We need about 2 liters of water per day under ordinary circumstances. Water is also a constituent of food, both unprepared and preserved, adding at least another liter to that total. In addition to consumed water, we can also use water to make oxygen, replenishing the air we bring with us to create a breathable atmosphere. Water is over 85% oxygen by weight and the liquid is easily broken into its constituent gases by passing an electrical current through it.

Another way that water supports life is by offering shielding and protection against solar and galactic cosmic radiation. Water harvested from the Moon can fill the outer jackets of surface habitats, protecting not only human life and technology within it, but also the plants that we will want to grow there, both for food supply and carbon dioxide scrubbing of the habitat air. Thus, water supports life on the Moon as both a consumable and as a building material.

A second main use of water is less often considered. We can break down water into its component gases[3] using electricity, but the process can also be reversed – hydrogen and oxygen gas can be combined to generate electricity in a device called a fuel cell. When these gases combine, they generate electrical energy and make water as a by-product. This technique was used in the Apollo spacecraft for power and water production[4]. When combined with another technique to generate electrical power (e.g., arrays of solar cells or a nuclear reactor), we make a completely reversible, self-sustaining power and water system. Thus, the water becomes a medium of energy storage – during lunar night, we combine hydrogen and oxygen to make water and electrical power while during the daytime, we reverse the process by using electrical power generated by sunlight to disassociate the water back into its constituent gases. Such a rechargeable fuel cell system enables permanent, sustainable human presence on the Moon.

The third important use for lunar water is for the production of rocket fuel. Liquid hydrogen and oxygen are the most powerful chemical rocket propellants known. By manufacturing rocket propellant from lunar water, we make the Moon a refueling station and logistics depot in space. The critical value of this ability is that such rocket fuel not only permits our routine access to and from the Moon, but also enables access to any other point in cislunar space (the volume of space between Earth and Moon.)

All satellites reside in cislunar space. Numerous remote-sensing satellites are found in low Earth orbit. GPS elements reside in moderately high (few hundred kilometer) orbits. Communication satellites are found at geosynchronous orbit, 35,000 km above the Earth. Other specialized satellites occur at different altitudes. At present, we cannot access these satellites with either human or robotic spacecraft. So we design, build and fly these space assets, use them for a time then abandon them, replacing them as needed with new satellites—at great cost. The ability to reach valuable space assets routinely with people and machines allows us to change the way we conduct business in space. Instead of the current "fly and throw away" template, we can build extensible, maintainable and upgradeable systems.

Very large, distributed space systems will enable new capabilities, such as global communications using hand held cell phone-sized equipment, anywhere in the world at any time. New remote-sensing platforms can be built to look at any corner of the globe at any wavelength in unprecedented detail. Telescopes built on the Moon's far side, where they will be shielded from Earth's radio noise, can scan the universe in new areas of the spectrum. These and many more capabilities are enabled by a cislunar transportation system[5] and will vastly improve life on Earth.

By understanding and using the resources of our Moon, we can push out to the stars. An abundance of water on the Moon fundamentally allows us to change the rules of exploration and spaceflight[6] to our advantage. We stand at the threshold of a new understanding of how the Moon evolved and works—and works to humanity's advantage.

Topics: Lunar Exploration, Lunar Resources, Space Transportation, Space and Society

Links and References

1. water on Moon, http://blogs.airspacemag.com/moon/2009/09/25/water-water-everywhere%E2%80%A6/
2. role of water, http://www.scientificamerican.com/article.cfm?id=water-lust-why-all-the-ex
3. breakdown of water, http://en.wikipedia.org/wiki/Electrolysis_of_water
4. Apollo fuel cells, http://airandspace.si.edu/exhibitions/attm/a11.jo.fc.1.html
5. cislunar transportation, http://www.spudislunarresources.com/Papers/Cremins%20and%20Spudis%202007%20Astropolitics.pdf
6. rules of spaceflight, http://www.spaceref.com/news/viewnews.html?id=1349

Comments

1. The amount of water discovered recently is tiny. You write enthusiastically in your book about using implanted solar wind hydrogen but over the whole Moon this is equivalent to the water in a single small lake. The recently discovered water is probably not much more abundant, especially if it sublimates away during the daytime and is replaced on the same timescale. This is not enough water to establish a self sustaining community on the Moon, much less to use as radiation shielding instead of the much more abundant regolith. As for using it as rocket fuel, this is really useful only if you can establish a self sustaining community to build, maintain and operate the spacecraft. In any case the amount of equipment you would need to transport to the Moon to establish that community would be well beyond the capability of a transportation system based on chemical rockets. If you use nuclear rockets the propellant can be whichever fluid is most convenient, which for the Moon would be liquid oxygen since you could extract oceans of the stuff from the lunar rocks.

Comment by Eric Paul Brunner — October 5, 2009 @ 7:05 pm

2. Hi Eric,

Wrong, on just about every one of your points. First, even before this new discovery, we estimated about 10 billion metric tonnes of water in the lunar polar areas on the basis of Lunar Prospector neutron and Clementine radar data. That's an amount of water roughly equivalent to the Great Salt Lake, hardly a minor amount. If converted to hydrogen and oxygen, it could launch from the Moon a rocket the mass equivalent of a Space Shuttle every day for over 2000 years. Second, you don't need a "community" and industrial plant to make propellant. Much of the preliminary work can be done with robots teleoperated from the Earth. Resource extraction can be demo'ed by small machines about the size of an office desk. The size of a water extraction machine that could make several tonnes of water per month is not much bigger than the payload of a single lander.

So, you're wrong about the quantity, wrong about the resource processing, and wrong on the difficulty of establishing a new capability.

Comment by Paul D. Spudis — October 6, 2009 @ 5:04 am

4. Whilst waiting for the LCROSS Post Impact Presser, I take time off (with metaphorical fingers crossed) to respectfully disagree with your third utilisation of any watery bonanza. Any 'low hanging' polar hydrogen will be too precious a commodity to throw away as rocket fuel. Indeed any particularly scenic "Ice Caves of Shackleton" should be protected as a (tourist) resource for future generations! (Or, shh, perhaps even constructed!) And as Dennis Wingo has pointed out; lunar food production is just as important. With the bonus that the hydrogen and nitrogen stays in the life support loop... in one form or another! Last, but not least, we will need DEEP swimming pools for the lunar high diving competitions! And which would you rather have; a personal zero gee swimming pool (and handy radiation shelter) or a single shuttle launch?

Whilst conceding that a renewable hydrogen supply by the 'daily' harvesting of solar protons could be a game changer. Although certainly a logistical nightmare. And, come to think of it, an ethical one too. Will future generations thank us if we disfigure Luna with strip mines? Even if they are on Farside. However there are alternatives: Al/LUNOX will get you off the Lunar surface and for heavy lift there is always the (original) nuclear option: the original Orion-Cargo can use the "Big Rock" throwing Lunar Authority catapult.

Once in space, most of our mutually agreed cis-lunar space faring architecture can take the form of tele-operated robotic vehicles, that use SEP and a gentler pace of living. Water and eventually rock shielded habitats in LLO, GSO and Lagrangian points will morph into proper O'Neillian settlements where humans can perform the tele-tasks that require zero time delay and repair the repair sats. Again water will be too precious...

In the short term hydrogen WILL be used as fuel, as quicker transits WILL be required. Leveraged by LUNOX as oxidiser. But surely the hydrogen (and nitrogen) will be initially sourced terrestrially via the new generation space companies and space gas stations. In the medium term, utilising the PROFAC concept, we can totally outsource these gases from out of the Well. In the long term we must get over the nuclear hangup and a develop a serious ceramic that will allow us to use LUNOX as the working fluid in a Nuclear Thermal Rocket.

By this time we should have a "Flexible Path" to CHON rich NEOs and have discovered that Phobos/Deimos is loaded with the stuff. But again, even with these sources, the water (hydrogen) will be too useful as water to a growing offworld population. Indeed I suspect that it won't be until we reach the Main Belt that water as fuel will no longer be an issue. Super heated steam powered rockets anyone!

Of course all this is moot if Robots are to be our only explorers. But look how popular the LCROSS "moonbombing" mission was: trending #1 on twitter no less.

One post presser comment (paraphrased from Marvin the Martian): "Where was the kaboom? There was supposed to be an moon-shattering kaboom!" The December conference will be a long time to wait but papers (and careers) come first; hopefully with some good news. But from an initial impression this Space Cadet still thinks we need 'Bots – lots of 'Bots – followed by humans for a real ground truth.

PS by my calculation 18,920,000 m³ and that doesn't support any human life at all!

Dave Comment by brobof — October 9, 2009 @ 11:19 am

5. Dave,

Our estimate is that about 10 billion metric tonnes (that's 10^9 tonnes, about 3 orders of magnitude greater than your estimate) occur within the upper few meters of each pole; this estimate is based on both the extent of shadow area, the strength of the enhanced Clementine radar signal, and the corroboration of excess hydrogen from the Lunar Prospector neutron data. That's only within the shadowed areas; more probably exists outside this zone, which is what the M3 instrument was seeing.

I'm not worried about using lunar water; there's plenty for everybody. Special areas of great beauty on the Moon can be set aside as a preserve or national monument, but I am not willing to declare a priori vast tracts of lunar land off limits to mining and resource harvesting, especially as we don't know yet exactly where and in what condition are most of the resources.

But, to each his own.

Comment by Paul D. Spudis — October 9, 2009 @ 2:07 pm

6. Paul,

Awesome post as always! Keep them coming please! The sooner the better :)

WRT using LUNH$_2$ as rocket fuel, I see both sides of the story. On the one hand, maybe we should just let the market decide on the best allocation of resources. But on the other, there is the recognition that markets may not always make the best decisions overall. E.g., economically, WRT to whale hunting, arguably it is more economical to hunt a whale species to extinction, literally, and then investing the profits in other ventures. But that's not right— we wouldn't want to go there.

No matter how you slice it, the Moon is Dune.

The Great Salt Lake seems like a bunch of water, but it's one of those places that are a mile wide and an inch deep. Volume-wise, the Great Salt Lake is roughly on the same order as Lake Mead, the USA's largest manmade reservoir.

Water to the Moon is like rice is to Japan. Economically, it would make more sense to farm out Japanese rice production to other places where land and labor are far cheaper. But I don't blame them for protecting their rice farmers from the effects of foreign competition.

It would be one thing if true market forces could work their magic. But the Moon will be a "command and control" economy during its initial stages. So the decision will come down to trade studies. I'm just guessing, but I would think that sacrificing food production capacity for rocket fuel in order to import food from Earth will not make sense economically.

Then again, there is the QWERTY phenomenon. The QWERTY keyboard layout was designed in order to slow down typing speeds, because the old typewriters would get jammed if you typed too fast. Now we are stuck with what was once a good market decision.

What we don't want to do is to get locked into a decision just because of legacy rocket hardware issues. That is, we shouldn't be burning up lunar hydrogen just because that's what our rocket engines are designed to use. There are practicable alternatives. A 450+ Isp isn't necessary to lift stuff off the Moon's surface. The old hypergolic propellant used on the LEM had an Isp of ~ 300s. Aluminum/LO$_2$ should have an Isp of ~ 280 s. The low Isp does not render the concept impractical for lunar gravities–though if used on Earth, it rapidly leads to absurd mass fractions.

I think we can all agree that a few million USD for John Wickman to develop his Al rocket designs a little more would be money well spent.

Comment by Warren Platts — October 10, 2009 @ 1:53 pm

7. Hi Warren,

Two points in response to your post.

First, I agree that other propellant possibilities should be examined and experimented with. In fact, I look upon all of lunar ISRU (at the moment) as an experiment — is it possible? If so, how difficult is it? And finally, is it worth doing? Once we have some real data from the surface of the lunar poles, we'll be able to answer such questions intelligently. I note in passing that lunar water is only a convenient form of hydrogen, not the only form of it on the Moon. Hydrogen is present as implanted solar wind and thus, trillions of tonnes are present on the Moon. A principal rule of mining is to get the easy stuff first, then move on to the harder (i.e., lower grade) ores.

Second, we want propellant for more than just lifting payloads off the Moon. Propellant export creates a market for propellant in cislunar space. And that is the real reason to go to the Moon — to create a long-lasting, reusable, refuelable space transportation system. We need propellant not only for orbit change but attitude control and minor maneuvers. The goal is to make a system that can routinely access not only the lunar surface, but all other points in cislunar space as well, for a variety of reasons — maintenance, construction, tourism, etc. We might need LOX-LH$_2$ systems for quite some time. My only point is that there is a lot of water on the Moon, certainly enough to keep us in business until we are able to access and mine the asteroids.

Comment by Paul D. Spudis — October 11, 2009 @ 5:58 am

8. [W]e want propellant for more than just lifting payloads off the Moon. Propellant ex-

port creates a market for propellant in cislunar space. And that is the real reason to go to the Moon — to create a long-lasting, reusable, refuelable space transportation system. We need propellant not only for orbit change but attitude control and minor maneuvers. The goal is to make a system that can routinely access not only the lunar surface, but all other points in cislunar space as well, for a variety of reasons — maintenance, construction, tourism, etc.

I think I see what you're saying. I foresee a day when the only commodity that will be lifted from the gravity well of Earth will be humans. Why pay to launch a satellite from Earth when the same satellite can be manufactured on the Moon and be launched for much less?

In this regard, I've been dying to ask you this question: :) You used to work on the Clementine project. As you know, it was partially underwritten by the old SDIO. That is, Clementine was meant to be a test of "Brilliant Pebbles" technology. But the Brilliant Pebbles concept would have required a constellation of a thousand or more satellites. So the SDIO was very interested in researching cheaper ways to launch payloads into LEO, as evidenced by their support of the Delta Clipper SSTO project, among others.

So my question is why was SDIO and DARPA so interested in the Moon? The Brilliant Pebbles technology itself could have been just as easily tested within LEO or GEO: places where real satellites are in orbit. Instead they combined the mission with a survey of the Moon's poles. But why?

I'm just guessing that they were thinking that the Moon might make a good place for a satellite factory that could cheaply place thousands of satellites into LEO. I cannot find any open source literature to support my contention, despite extensive searching. But it makes sense. They were researching cheap launch methods. They have a history of thinking outside the box. So they were curious about the amount of water on the Moon in case there was enough of it to make it economical to launch satellites from the Moon. Weren't they?

I'm not asking you to violate any Secret clearances or anything, or to give up any classified information. But any light that you can legally shed on this question would be very much appreciated.

Comment by Warren Platts — October 12, 2009 @ 9:45 am

9. Hi Warren,

So my question is why was SDIO and DARPA so interested in the Moon?

First, Clementine was a project solely of SDIO (later BMDO) and DARPA had nothing to do with it. And actually, they weren't particularly interested in the Moon. The Clementine mission came about largely at the instigation of my colleague, Stu Nozette, who came up with the idea of a lunar and asteroid flyby with a Brilliant Pebble (BP) spacecraft in 1989. If any single person deserves significant credit for getting Clementine as a lunar mission, it's probably Stu. A key event was getting NASA to agree to co-sponsor the mission, providing a science team (paid for by NASA). Clementine was initiated in 1992 and launched in 1994.

The SDIO interest in Clementine was to test a variety of small spacecraft technologies and sensors that had been developed as part of the SDIO BP program over several years. The Moon was selected as a target because we knew a bit about its surface properties, so it would offer a good test of the BP sensor suite. It was only after we got into the lunar orbit that we found the poles to be interesting and the bistatic radar experiment was improvised after we were already at the Moon (if we had known about the large shadow beforehand, we might have included a neutron spectrometer as part of the payload.)

Nothing classified or more conspiratorial than that. Sorry.

Comment by Paul D. Spudis — October 12, 2009 @ 2:07 pm

10. Please, give veracity to the calculations, comments, given references, that is formal scientific literature (and el al. documents)

Comment by Oscar M Brandt — November 15, 2009 @ 6:50 pm

11. Oscar,

Feel free to check the numbers yourself. I suggest that you start with the total water estimate we made in the Clementine bistatic radar paper:

Nozette S. et al. (1996) The Clementine Bistatic Experiment. Science v. 274, n. 5292, p. 1495-1498.

Comment by Paul D. Spudis — November 16, 2009 @ 5:43 am

LCROSS: Mission to HYPErspace

October 12, 2009

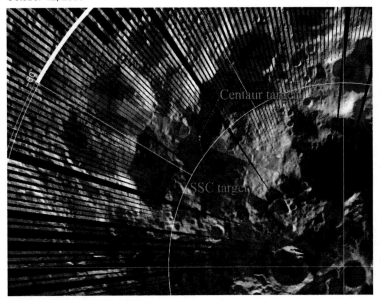

The LCROSS impact site seen from LRO

Early last Friday, the public and families of employees at Ames Research Center in California, where the LCROSS mission[1] was conceived, built and operated, camped on the lawn in an all-night vigil. NASA's educational outreach and public relations push[2] about the pending lunar impact event was very effective, having reached a wide audience in the weeks leading up to the much hyped event. Alas, the promised giant plume of impact debris was invisible from Earth, leaving a receptive public feeling cheated and disappointed[3].

The understanding that a high-velocity impactor can yield important information about planetary composition and state is very old. The first probes to the Moon (both Soviet[4] and American[5]) were impactors. We know that when something strikes a planetary surface at high speed, target material is thrown up into space, some of it vaporized by heat generated in the energy of the impact. By studying this impact ejecta, we learn about the composition of the target object.

I didn't post on it earlier, but as the LCROSS mission has successfully concluded[6], I think it is a good time to examine this mission, how it came about, and the lessons that hopefully it has taught NASA about public appeal and its involvement with space.

LCROSS was not originally a part of the robotic precursor program for lunar return. Initially, the Lunar Reconnaissance Orbiter (LRO)[7] spacecraft was to be launched on a Delta II. By the end of 2005 it had outgrown its booster and was forced onto the much larger Atlas V booster where it had surplus payload margin. The Associate Administrator for the Exploration Systems Mission Directorate (ESMD) Scott Horowitz, decided to use this margin to fly an additional small spacecraft (called a secondary payload) that would address the raging debate about whether water ice exists at the poles of the Moon. Horowitz looked to NASA's field centers for a small payload that would provide data about this contentious and nagging issue.

Although a variety of small missions were proposed, including survivable hard landers and small "hoppers," the idea of slamming the Centaur upper stage into the Moon and examining the resulting ejecta plume was selected as LCROSS in April 2006. It was considered a low-risk, low-cost concept, as the used Centaur upper stage had no value and would have been steered into a solar orbit anyway. A small satellite was built to track the Centaur impact, measure the properties of the ejected plume and with luck, would "settle" the issue of water on the Moon.

A serious defect in this mission concept was that it presupposed that we understood the Moon well enough to identify in advance the most likely site for ice on the Moon. Lunar investigators knew from previous data that water ice, if present, was not present everywhere – it had a patchy, heterogeneous distribution because the permanent shadow around the poles (where the ice would be stable) is itself patchy. Moreover, the remote sensing data of the time was ambiguous[8] as to which shadowed locales contained ice, if any.

In March of 2006, because of these uncertainties, those who had worked on the robotic precursor program laid out a sequential, incremental strategy to first map the deposits from orbit and identify the best candidate sites for ice. Following orbital mapping, we would softland with capable rovers and map and test the surface composition at a minimum of about 20 different sites. Although this strategy is more costly than a simple impactor mission, it would have provided us an unequivocal answer to the ice issue; we would know without doubt whether there is or is not water ice at the poles of the Moon. Moreover, rovers would collect information on the possible presence, physical nature and setting of other volatile substances (such as ammonia and methane) that have resource value. In other words, we would have collected the critical strategic information needed to locate, prospect, harvest and use lunar water.

Instead, the mission chosen and flown and heavily advertised by NASA as a citizen participation viewing event to find water on the Moon, could not answer key questions about polar water. If LCROSS detects water, we still won't know where all the ice deposits are located, what other species might be present, what its physical state might be, and how it is distributed laterally and vertically in the surface regolith. If LCROSS detects nothing, it won't prove that water doesn't exist on the Moon, only that the wrong site was selected. In other words, after this mission, we will still know next to nothing about the material that will enable and advance permanent, sustainable economic presence on the Moon.

An impact plume wasn't the only thing missing. Hopefully, NASA will recognize the real discovery of LCROSS – mission hype is a poor substitute for shortcomings in programmatic logic.

Topics: Lunar Exploration, Lunar Resources, Lunar Science

Links and References

1. LCROSS, http://www.nasa.gov/mission_pages/LCROSS/overview/index.html
2. public relations, http://www.nasa.gov/mission_pages/LCROSS/impact/event_index.html
3. cheated, http://www.mercurynews.com/news/ci_13530257
4. Soviet moon probes, http://en.wikipedia.org/wiki/Luna_2
5. U.S. moon probe, http://en.wikipedia.org/wiki/Moon_landing#Ranger_missions
6. LCROSS concluded, http://www.usatoday.com/tech/science/columnist/vergano/2009-10-09-nasa-lcross-moon_N.htm
7. LRO, http://lunar.gsfc.nasa.gov/
8. ambiguous, http://www.thespacereview.com/article/740/1

Comments

1. The only justification I can imagine for you being upset is that the survivable hard lander or small hopper path was not taken as an add-on to the LRO mission, instead of the impactor that was selected. The rest of your arguments about knowing WHERE first to send a future lander/rover depend on the results of the LRO mission, which has just begun. Thus they would likely had fared no better than the impactor if they too were a just shot in the dark. So do not fret, as nothing precludes us from still sending a dedicated lander or rover after LRO data come in. As you did not specify the cost or schedule issues of the alternatives to the impactor, this reader is left with insufficient data to know if any of these alternatives could have met the LRO schedule constraints (which may have been "firm" at that time) or cost constraints (which no doubt had a desired cap). While I agree that LCROSS may have been overhyped, if NASA had NOT hyped-it-up, it would have been harassed for NOT doing that — as NASA-Watch just did with regards to a recent sounding rocket mission from Wallops.

Comment by Ronnie Lajoie — October 12, 2009 @ 6:52 pm

3. I can tell you that LCROSS generated a lot excitement. Our blog gets anywhere from 30 to 70 hits per day. Last Friday, we had over 300. The public is interested in space exploration.

Comment by Dave — October 13, 2009 @ 12:41 am

4. Ronnie,

You completely missed my point.

It's not that NASA should have selected a hopper or other kind of hard lander (those had their own issues) — it's that they chose to fly a mission that basically adds nothing to our strategic knowledge of the polar deposits while discarding entirely the robotic flight program that would have given us such information. They didn't have to fly a secondary payload at all, so the LRO launch date (which was over a year late anyway) was not a constraint.

The mission has been advertised as big exploration event. I contend that it is not.

Comment by Paul D. Spudis — October 13, 2009 @ 5:13 am

6. The data from the LCROSS mission is still coming in, so hopefully the next weeks or months of analysis will find something useful. While the mission may not have been as scientifically useful as the science community may have hoped, it did raise interest in the Moon by the general public; interest that hopefully won't go away too soon in light of the lack of an obvious impact splash. As for the deferral of the Robotic Lunar Exploration Program (RLEP) program, now the much weakened Lunar Precursor Robotic Program (LPRP), that decision was indeed made to provide funds for near-term Constellation goals, and I too was not happy about that action. But we have a new NASA administration now and thus have an opportunity to seek to restore a viable robotic lunar exploration program (while waiting for human return). Let's get some good LRO data, then finally get a U.S. rover to the Moon and down to a pole where it can drill for ice. So do not despair; the NEW Golden Age of Space Exploration is just beginning. Ad Astra!

Comment by Ronnie Lajoie — October 13, 2009 @ 11:20 am

7. Paul,

You make excellent points — ones that, I suspect, all but the most informed are not even aware of. That is a weakness of the current culture. Scientists and engineers need to get better at communications — with each other and with the larger public. There also need to be ways for scientists such as yourself to get greater attention from the larger world. Too many scientists and engineers dismiss this kind of work out of hand. You, thankfully, seem to be an exception. There are others. We need more.

Comment by Chuck Divine — October 13, 2009 @ 1:07 pm

10. What's your take on Arlin Crotts' suggestion (particularly http://thespacereview.

com/article/1485/1) that lunar water may be primordial?

Comment by Nels Anderson — October 14, 2009 @ 2:18 am

11. Yeah, it was rather disappointing. I was watching the live coverage and hoping to see *something*.

Still, apart from the broad criticism and uninformed complaints of 'bombing' of the Moon, I am hoping that soon someone will actually step up and try to explain why there was no impact ejecta. Because it does seem to me that if we expected something and didn't get it, then that is of interest in itself.

If there was no impact plume, then what did we hit?

Comment by Paul — October 14, 2009 @ 3:33 am

12. Nels,

While I appreciate Arlin's enthusiasm and tenacity in following up on the TLP problem, my suspicion is that the vast bulk of lunar water is from external sources, not the Moon itself. But the whole issue should be re-opened and examined afresh.

Comment by Paul D. Spudis — October 14, 2009 @ 5:19 am

13. *If there was no impact plume, then what did we hit?*

There probably was an impact plume — it was just much smaller and much darker than had been supposed beforehand.

Comment by Paul D. Spudis — October 14, 2009 @ 5:20 am

14. Paul,

I'm surprised at your attitude. As you will soon see LCROSS is a spectacular scientific success. The fact that the plume was not as pronounced as predicted IS significant. The data we are obtaining from these missions WILL be the justification for the landers etc. we all agree on. While I admire your forceful commentary, I believe you just may become an advocate.

Comment by Pete Worden — October 16, 2009 @ 7:36 pm

15. Pete,

Thanks for reading the piece and for your thoughtful comment. We simply disagree on the value of LCROSS, which I believe to be a "bargain basement" mission that ESMD used and is using as an excuse to ignore their duty to craft a robotic program capable of obtaining the data we need to use lunar resources. My position is that NASA as an agency has never embraced the "mission" of the Vision for Space Exploration — to go to the Moon and learn to use its resources to create new space faring capability. Because they have never accepted that charge, they do not see strategic knowledge about the Moon (particularly regarding resources and ISRU) as in the critical path. Hence, a real robotic program to get the strategic knowledge we need was dumped.

Comment by Paul D. Spudis — October 17, 2009 @ 5:49 am

16. I was listening to Science Friday and it got me thinking, why hasn't anyone planed a mars rover like mission to take soil samples. If those rovers lasted 5 years with mars level dust storms and weaker solar energy a similar rover could give scientist 10 years of functionality.

Comment by Matt — October 19, 2009 @ 10:55 am

17. Matt,

Your question is excellent and should be directed to the Associate Administrator of NASA's Exploration Systems Mission Directorate.

Comment by Paul D. Spudis — October 19, 2009 @ 11:39 am

18. Unfortunately, this all may not matter anyway. The Augustine report is about to come out, and the current Presidential administration and NASA leaders spend a lot of their time talking about the 'flexible option', one that mostly ignores the Moon and instead flies around space on "a three hour tour" to near-earth asteroids or the moons of Mars. It doesn't help when an anonymous administration official asserts that options that include returning astronauts to the moon "are not sellable to the public or to the president." Politics will trump any logical, methodical approach to space exploration as it always has. I suppose one could say the word 'flexible' could include the Moon, but I'm concerned that 'flexible' will be used by NASA to look a little at a lot, and come up with such an amorphous, open-ended do all for everybody plan that we wind up with nothing but view graphs for another 20 years. The path is so obvious and it goes through the Moon. It is painful to watch politics make a mess of it all. Maybe a more appropriate name for the new crew exploration vehicle should be the S.S. Minnow.

Comment by JohnG — October 21, 2009 @ 10:23 am

Paradigms Lost

October 23, 2009

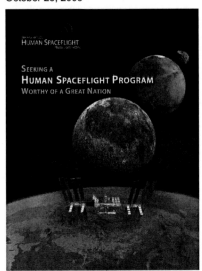

New report - same old assumptions?

There is nothing more difficult to take in hand, more perilous to conduct, or more uncertain in its success, than to take the lead in the introduction of a new order of things. – Niccolo Machiavelli, The Prince[1].

In his famous book *The Structure of Scientific Revolutions*[2], Thomas Kuhn described two types of science: normal science, the everyday background work, where constant, steady but unspectacular advances occur in our knowledge, and revolutionary science, where fundamental assumptions and ways of conducting business are unalterably changed forever. Kuhn called such a change a paradigm shift; a new paradigm (i.e., a framework of knowledge, including the assumptions, worldview, approaches and techniques to conduct business under a given set of circumstances) replaces the existing one and the new approaches and attitudes become the norm.

The paradigm model might also be applied to conducting business in other fields, in particular, the business of spaceflight. Since it arose more than 50 years ago, the paradigm of spaceflight has largely remained unchanged. In short, we conceive a mission (robotic or human), then design, build and launch a spacecraft to conduct that mission.

This satellite or spacecraft operates for a time in space—gathering information or providing a service—until it breaks down or becomes obsolete and is abandoned. We then imagine the next mission—going back to the drawing board to design the next spacecraft—a process repeated continuously and a major cost of space exploration.

Is a paradigm shift – a "revolution" in space travel possible? One would think that with 50 years of experience under our belts, we would have already exhausted all the possibilities. Indeed, the imminent development of warp drive or "Cavorite[3]" does not seem likely, but then, that's the nature of truly revolutionary breakthroughs, isn't it? On the other hand, is there something missing – something that could be done right now using existing knowledge to change the rules of spaceflight and possibly spur additional breakthroughs?

As long as we're chained to the existing spaceflight paradigm, we must continue hauling from Earth everything we need in space. For human missions this includes all the air, water and other consumables needed for life support. The cost to lift all this mass (which includes the weight of a massive amount of fuel needed to escape from Earth's very deep gravity well) is budget busting. So for "normal" space exploration, costs will never be lower except at the margins and we will always be mass-limited in space. And when you are mass-limited, you are capability-limited as well.

I've argued here[4] and elsewhere[5] that there is a method that is already well understood in principle, but its practical application and viability is completely unknown. If we could use what we find in space to create new capabilities, we would change the rules of spaceflight, thereby ushering in a true paradigm shift in space travel.

Such was the original intent of the Vision for Space Exploration (VSE)[6]. The desire for fundamental change in perspective was behind the program's specific direction to study and experiment with using the material and energy resources of the Moon. From the moment it was announced, the true purpose of a lunar return was misunderstood, both inadvertently and deliberately. Constellation is a rocket program; the VSE is not.

No one knows if using space resources[7] is possible but we can find out by pursuing innovative technology. In theory it works. We've never attempted high-risk mining on the Moon and it may have significant practical difficulties but potentially, it could become a highly leveraging activity.

If we can extract and make rocket propellant on the Moon, we can create a completely reusable, refuelable transportation infrastructure in cislunar space. If we can extract the oxygen and the hydrogen, we can live in space. Of course, such an outcome would change and transform the business model of space—something that fascinates and attracts many but repels others and hence, its mixed reception in aerospace circles.

This would truly be a revolution, a paradigm shift in the same sense as we understand it from Kuhn's description of scientific progress; as a vast new expanse is opened to us and we are free to move about the universe, the world changes and things are never the same again.

In order to mitigate risk and to ensuring our economic and national security, government often steps in to develop technology that the private sector cannot or will not take on. A government push to learn how to use the resources of space will break the cycle of launch and discard. Instead of having a short "shelf-life," our indispensable and unprotected systems in space become maintainable, reusable, exten-

sible and affordable.

While reading the newly released Augustine report[8], keep in mind its background and its assumptions. It is based solidly on the traditional models of conducting business in space – design, launch and abandon, along with the accompanying plea for more money to ensure a "robust" program of space exploration.

As long as such assumptions prevail, advances never will.

Topics: Lunar Exploration, Lunar Resources, Space Politics, Space Transportation, Space and Society

Links and References

1. Niccolo Machiavelli (1532) The Prince, http://tinyurl.com/28hj7xj
2. Kuhn, http://en.wikipedia.org/wiki/The_Structure_of_Scientific_Revolutions
3. Cavorite, http://www.technovelgy.com/ct/content.asp?Bnum=717
4. resources, http://blogs.airspacemag.com/moon/2009/05/30/can-we-be-%e2%80%9cresourceful%e2%80%9d-on-the-moon-part-1/
5. resources, http://www.spaceref.com/news/viewnews.html?id=1349
6. intent of VSE, http://www.spaceref.com/news/viewsr.html?pid=19999
7. using space resources, http://www.spaceref.com/news/viewnews.html?id=1334
8. Augustine report, http://www.nasa.gov/offices/hsf/home/index.html

Comments

1. Mr. Spudis, I fully concur with this post, especially in light of the clear goal of 'expansion into the solar system' the Augustine commission has given to human spaceflight.

I was just wondering how realistic it is to expect a major paradigm change like this to originate in a politically restrained NASA or a presidential research panel. Especially a paradigm change that would certainly require major workforce and money shifting...

Maybe initiatives like the Google Lunar X Prize could provide some concrete results. A firm like SpaceX went from scepsis to resupply contracts, and now may be called upon to transport astronauts. At least I'm hopeful that these private efforts could provide a wake-up call that a paradigm change like the one described is feasible and a good thing.

Comment by Simon Vanden Bussche — October 24, 2009 @ 9:34 am

2. *I was just wondering how realistic it is to expect a major paradigm change like this to originate in a politically restrained NASA or a presidential research panel*

Simon,

A fair point. However, several good solid steps in the direction of a different approach were taken early in the implementation of the Vision, by people inside and outside of NASA. Unfortunately, old business methods and attitudes die hard.

Look at it another way. If the agency cannot reform itself, it will simply drift off into irrelevance.

Comment by Paul D. Spudis — October 24, 2009 @ 1:00 pm

3. I think there is one paradigm shift we need to make before the one you're talking about, Paul. It is to standardize and reuse the vehicles we build. NASA will want to continue to pour most of their money into the old way of doing things, but if they can be persuaded to put a fraction of their investment into the commercial/reusable paradigm, it may help to birth an industry that could better support ISRU research. In short, we still need to get the cost of access to LEO down before the next steps are possible.

Comment by Bill Hensley — October 24, 2009 @ 3:18 pm

4. Well said. I couldn't help but be reminded of the ancient days of the First Bush White House, when some clever White House wag was quoted as saying, "Brother, can you Paradigm?"

After reading the much ballyhooed and perhaps unnecessary Final Report I was struck by a couple of things. It didn't strike me as looking like the result of 90 hour work weeks, the release has received almost no press attention, and what little I've seen is headlined by things not in the report, perhaps based on interviews with committee members.

And finally, an appalling understatement of one of the central conclusions from the NAS "Scientific Context for the Exploration of the Moon" from 2007, namely equating the sticky problem of mitigating the dangers posed to deep space travel by galactic cosmic rays, equating this with the physiological challenges of micro-gravity and the psychological challenge of long-term travel.

News accounts of the committee's conclusions are headlined as a recommendation to skip the Moon and proceed with their, so-called, "Flexible Path." That's not what I read in the report at all.

While I disagree with the committee's conclusion that "Mars is clearly the most interesting long-term destination," I don't think even the Flexible Path option is a viable way around the Moon.

Cosmic rays and an accelerated risk of Radiation Exposure Induced Death make the Moon as necessary as undergraduate work before entering the Graduate courses. Sure, I can understand why folks would like to skip high school and proceed directly to college, but you can't say you've been to high school if, as an elementary school

student, you visited the auditorium there a few times.

Comment by Joel Raupe — October 24, 2009 @ 3:32 pm

5. In short, we still need to get the cost of access to LEO down before the next steps are possible.

Bill,

We can start understanding how to do ISRU, what the feedstocks are like, how difficult it is, how to extract and store product all before the cost of launch to LEO is lowered. True, inexpensive access to LEO is a good thing, but once we've got refueling depots in space, you have a fully functional space transportation system, one that can access all of cislunar and whose existence will naturally increase flight rates to LEO.

I've heard about "lowering" launch costs for years, but it's a "chicken or the egg" issue. We already know why LEO access is expensive — the "marching army" required to assemble, prepare and launch rockets. Automate most of that and you'll lower launch costs.

Comment by Paul D. Spudis — October 24, 2009 @ 4:46 pm

6. Paul,

I'm all for going full speed ahead on ISRU research, but it can't really become operational until we can afford to actually get to the moon. You indicate that the chief weakness of the current mission-oriented paradigm is that we have to lift everything up from the earth's surface. I'm saying its chief weakness is having a government agency conduct a multi-year program to design and build special purpose hardware that is used once and thrown away. That's an exceedingly expensive way to proceed and the sheer cost of it limits our forward progress.

You mention fuel depots. I expect that would be an element of the commercial/reusable paradigm I'm suggesting needs to come first. I guess what I'm saying is that what you envision as a single paradigm shift is really two paradigm shifts. The first step is to convince NASA to focus on defining standard interfaces and letting commercial contracts for spaceflight services. That in itself is a major paradigm change for space exploration. In fact, it's too large for NASA as a politically driven organization to embrace quickly. After reading all the back and forth on the Augustine Committee over the past few months, I think about the best we can realistically hope for is a partial step in that direction. Perhaps NASA will be able to embrace spending $5 billion or so on commercial crew services as long as they get to keep building Orion and some big heavy lift vehicle. The latter two will be obscenely expensive and will provide cover for the former.

Comment by Bill Hensley — October 24, 2009 @ 5:43 pm

7. *I'm all for going full speed ahead on ISRU research, but it can't really become operational until we can afford to actually get to the moon.*

Bill,

We can afford to get to the Moon right now — robotically. Moreover, much of what we need to know, at least initially, can be done via robotic spacecraft and rovers, operated from Earth via teleoperations. This is one area in which NASA should be able to perform. This is also a program that doesn't require significant new money, only the will to execute it.

Comment by Paul D. Spudis — October 25, 2009 @ 5:02 am

8. Well, I agree with that. We may not be so far apart in terms of our ideas about practical steps forward. For instance, if anything actually comes of the Google Lunar X-Prize, I hope NASA will find a way to leverage it and to encourage further advancement. Another prize, perhaps. Or a COTS-like program to fly NASA experiments to the lunar surface aboard GLXP-heritage spacecraft.

Comment by Bill Hensley — October 25, 2009 @ 10:34 am

9. Dr. Spudis -

Given the realities of today's NASA (as well evidenced by the Augustine Report) perhaps you and Buzz Aldrin could make common cause, at least to the extent of his October 12th call to develop the Moon – much along the lines you propose – but as a public sector & private sector partnership in cooperation with the entire world.

http://www.huffingtonpost.com/buzz-aldrin/a-different-kind-of-moon_b_317786.html

Archimedes famously said that with a place to stand, he could move the Earth. If we are to move (change) NASA perhaps we need a fulcrum outside the United States if we are to obtain sufficient leverage to achieve that goal.

Both Senator Shelby (R-AL) and Representative Giffords (D-AZ) are representative of numerous staunch and influential defenders of the "Program of Record" and it appears so is much of NASA's middle management. Should we attempt to batter down the Maginot Line, or go around?

Going forward, should we bemoan opportunities (and paradigms) lost or seek the most effective route for achieving the vision you (and Dennis Wingo) have articulated?

I believe Buzz Aldrin is 100% spot on, here:

"New space powers such as China and India have dedicated and complex space programs now under development, with the Moon as their target. Trying to "win" a Moon race with them would be foolish. They would eventually reach the Moon, with or without our help. What would be our policy then? Try to deny them access to the Moon's bountiful resources in minerals — and maybe water as well? Such an attitude is more appropriate for the Cold War era that has been over for more than two decades."

Comment by Bill White — October 25, 2009 @ 10:47 am

10. *you and Buzz Aldrin could make common cause...to develop the Moon – much along the lines you propose – but as a public sector & private sector partnership in cooperation with the entire world.*

Bill,

Two problems with what Buzz proposes.

First, not everyone shares our economic and political values, including the respect for contract law that for over 200 years has been largely responsible for creating the wealth that America now enjoys (and seems hell bent on destroying.) One of the principal arguments for America "leading" in space endeavors is to ensure that our values become the values of the new frontier.

Second, in the piece you reference, Buzz defends the existing paradigm — he simply wants to direct us elsewhere, specifically to Mars. If the idea of using space resources to enable new capability makes any sense at all (which you seem to acknowledge), why does it make any sense to continue the old ways of doing business with human missions using the Apollo template to destinations other than the Moon?

Comment by Paul D. Spudis — October 25, 2009 @ 1:15 pm

11. Paul, the link I gave does offer a new paradigm for lunar exploration.

He calls for a public/private partnership "to build the communication and navigation satellites needed by future lunar travelers, develop fuel depots using lunar LOX — perhaps derived from the recently discovered lunar water — and construct habitats that will shelter space travelers while on the surface. It will enable a sustainable human presence on the Moon that will be accessible to all the nations on Earth."

Lunar ISRU is central to this approach.

Also, as for a capitalist (free market) business structure he writes:

"Unlike the International Space Station (ISS), which is governed by complex treaties, the LIDC will have the same flexibility as an NGO in working with different nations and private entities to finance build and operate the facilities and equipment needed for lunar exploration. Using a corporate structure, the LIDC will allow nations to join through the purchase of shares and enable them to contribute at a level that is sustainable for their economies. Intelsat, the international corporation that bought the benefits of communication satellites to the nations of the world is an example of the potential benefits of a focused NGO in developing global space infrastructure."

That sounds more "free market" than a NASA led program could ever be and if the Indians and Japanese and Europeans and Chinese are all participating then it will be far easier to demand a significant role for U.S. NewSpace ventures as well.

That said, perhaps the NGO/Intelsat model is not the best model and could be improved upon. But surely it is a better model than anything we can expect from NASA in the next few years.

Comment by Bill White — October 25, 2009 @ 2:18 pm

12. *the link I gave does offer a new paradigm for lunar exploration.*

My point is that Buzz's "new approach" does NOT carry to his preferred destination — Mars. If it makes sense to develop commercial lunar ISRU, why then does it make sense to do a big government Mars mission?

Buzz's idea is simply a bone thrown to the lunar community to divert their criticism of his tenacious clinging to the old Apollo-template in his preferred humans-to-Mars flags-and-footprints program. Sorry — I will not be misled or diverted.

Comment by Paul D. Spudis — October 25, 2009 @ 3:12 pm

13. Dr. Spudis,

How do we go about changing the Paradigm? Find some other agency besides NASA to do space development? Pure commercial?

Comment by Karen Cramer Shea — October 25, 2009 @ 6:23 pm

14. Karen,

I think that NASA could lead the way, followed by a transition to the private sector. That's what the Vision for Space Exploration was all about, although the agency misinterpreted it as a human Mars mission. Learning to use space resources to develop new space capabilities is the mission on the Moon.

Needless to say, many in the space business have never gotten and will never get this concept. But I know many who do, both inside and outside the agency, and are doing their best to educate their associates.

Comment by Paul D. Spudis — October 26, 2009 @ 4:59 am

15. Dr. Spudis -

I also believe NASA "could" lead the way, followed by a transition to the private sector. But if they won't, doesn't a "Plan B" become necessary?

Even if Buzz Aldrin's idea is not your first choice, how about as a "Plan B" alternative?

Comment by Bill White — October 26, 2009 @ 7:46 am

16. I personally think our best bet to cause a paradigm shift at NASA is to find some other agency, possibly DOE to fund space solar power research. Space solar power will at some point become necessary for our technological civilization and profitable, as alternatives fail to meet demand. So far only $80 million has been spent on its development, so anyone who writes it off is doing so very prematurely.

A space solar power industry would naturally develop lunar resource utilization because of the physics. So long term without NASA this path would develop the Moon and Earth Orbit. The effect could work on a much shorter term by NASA feeling threatened by the competition and getting involved in actual space development.

It is possible that NASA might change its ways but currently it doesn't look promising that it can. If NASA is forced to extend the life of the shuttle, a paradigm shift maybe impossible until after the next shuttle accident.

Comment by Karen Cramer Shea — October 26, 2009 @ 10:20 am

17. [...] Part of the reason to set up a human presence on the Moon is to learn how to tap the resources found there to do better stuff in space. Oxygen is the most-cited export for Moon industry, and with good reason. The Moon is composed, in bulk, of about 40% oxygen, primarily bound up into the minerals. The reason oxygen is such a valuable commodity in space is that it makes up the bulk, by mass, of propellant loads. When you're launching from Low-Earth orbit to near-Moon space, some 75-80% of your 'wet' mass (payload, vehicle and propellant masses) is composed of propellant. Of that propellant, 7/8ths of the 'weight' (the mass you're moving out of the gravity well) is oxygen. If that oxygen can be delivered to LEO from somewhere other than the surface of the Earth then the potential exists for significant cost savings. That kind of paradigm is visited by Paul in his latest Once and Future Moon post, "Paradigms Lost". [...]

Pingback by Carnival of Space #126 (& Carnival of the Moon) - Out of the Cradle — October 26, 2009 @ 11:38 pm

Caves On the Moon?

October 27, 2009

Collapse breccia near a lava tube entrance (Photo[1] by Dr. Harmon Maher, Univ. Nebraska)

The science team of the Japanese Kaguya mission[2] have just published a paper claiming to have found an opening to a cave[3] on the Moon. Such a discovery is a potentially important development for future lunar habitation. Lava tubes[4] are large caves created during the volcanic eruption of a very fluid, highly effusive lava. They are common on Earth, especially in iron-rich basaltic lavas, such as those that make up most of the Hawaiian islands.

The idea that caves occur on the Moon has been around for a long time. We have long known that the lunar maria (the dark, smooth, relatively uncratered plains of the Moon) are made up of old basaltic lava flows. Looking at orbital photographs[5], we find many narrow, winding channels (or rilles) in the maria. These channels cannot be the product of water erosion, as flowing liquid water cannot exist in the vacuum of the lunar surface. So workers looked for another explanation. They found it in lava channels and tubes.

On Earth, volcanic terrains often show small channels within young lava flows. Lava tubes form when hot lava erupts and pours out onto the surface. The lava immediately begins to cool, with the outermost edges cooling first. As the lava cools and hardens from the outside edges inward, the flow of still-molten lava becomes constricted to a central, narrow, inte-

rior conduit. When the eruption stops, the still-liquid lava drains out, leaving behind an empty cave-like tube-shaped segment. In some instances, the roof of a drained tube collapses, exposing the tube interior as a channel or, if less extensive, creating a "skylight[6]" or a hole that allows access to the cave interior. Lava caves are quite common on volcanoes made up of runny (low viscosity) lava, such as the shield volcanoes of Hawaii.

Caves found on the Moon[7] would be very useful[8]. Because they form in dense basaltic lava, the space inside a tube is protected from both the hard radiation of the lunar surface and the constant micrometeorite bombardment the Moon experiences. Moreover, the temperature of the subsurface of the Moon is very stable; below the zone which experiences the extreme temperatures of night and day, lunar temperatures are fairly constant at about -20° C. On Earth, lava caves can be quite roomy, with diameters tens of meters across and hundreds of meters long. On the Moon, these dimensions may be much larger – the low gravity of the Moon results in much bigger lunar lava tubes and channels than their terrestrial counterparts, being hundreds of meters across and many kilometers long. Thus, they offer many potential advantages to future lunar inhabitants.

Before we pack our bags for the Marius Hills, we should take note of some other properties of lava tubes. Many lava tubes partly[1] or completely collapse immediately after their formation. If the roofed segments are weakened by flowing lava, earthquakes, or are very thin, they cannot support their own weight and after the lava drains out, the roof falls into the void. This is seen on both the Earth and the Moon. Hadley Rille, visited by the Apollo 15 astronauts[9] in 1971, is a lava channel, parts of which were roofed over as a tube. The crew landed near a channel portion, but a roofed segment is only about 12 km from the site. High resolution images of that segment show no entrance to an underground cave there or elsewhere along the rille (channel). That doesn't mean that there is no cave portion of Hadley Rille, but it does suggest there is no entrance to a cave there.

Other candidates on the Moon look more promising. Numerous lava tube "skylights" have been noted[5] in association with many lava channels on the Moon. These skylights are typically unconnected to each other or any nearby feature and are found as individual tube segments that appear to start and stop along the trend of a rille. It is impossible to identify lava cave entrances because most of the images we have for these features are low resolution and have near-vertical viewing geometry.

The new Kaguya pictures[10] show a circular, rimless pit[11] on the floor of the projected segment of a rille. Collapse pits are not uncommon on the Moon and many of them are not associated with lava channels or tubes. So while the new Kaguya images are intriguing, they are not definitive evidence for a cave.

There are other issues in regard to the use of lunar lava tubes. Many (if not most) terrestrial lava tubes are not void; they are either filled with late-stage lava, which plugs up the cave, or by collapse debris, which buries it. Finding a new void lava tube is celebrated by the caving community simply because void tubes are rare. But even if a void tube formed on the Moon, it may not remain that way for all time. Lunar volcanism was active over 3 billion years ago. Since then the Moon has been constantly bombarded by debris, initiating landslides, infilling craters, and generating seismic waves. Such a bombardment could well act as a leveler to collapse and fill in void lava caves that might have existed on the Moon.

But the biggest problem with lunar caves is even more fundamental – they aren't where we want them. Sustained human presence on the Moon is enabled by the presence of the material and energy resources needed to support human life and operations around the Moon. After over a decade of study and exploration, we now know that these locations

are near the poles of the Moon. Unfortunately, both poles are in the highlands and finding a lava tube in such non-volcanic terrain is highly unlikely, regardless of the imaginative ramblings of certain science-fiction authors[12]. If a lunar cave were present there, we would certainly consider using it. But it makes no more sense to locate a lunar base near the caves, than it does to build a water-park in the Sahara desert.

The formation of lunar lava tubes and caves is an interesting scientific topic, but their utilitarian value is uncertain, at least until we have established a permanent presence on the Moon. Ultimately, we may be able to use them to live on the Moon, but first, we need to follow the Willie Sutton principle and go where the money is.

Topics: Lunar Exploration, Lunar Resources, Lunar Science

Links and References

1. terrestrial lava cave, http://maps.unomaha.edu/Maher/brecciacourse/brecciapictures/lavatubecollapsebreccia.jpg
2. Kaguya, http://www.kaguya.jaxa.jp/index_e.htm
3. cave found on Moon, http://news.nationalgeographic.com/news/2009/10/091026-moon-skylight-lunar-base.html
4. lava tubes, http://en.wikipedia.org/wiki/Lava_tube
5. orbital images, http://lpod.wikispaces.com/October+23,+2009
6. skylight, http://volcanoes.usgs.gov/images/pglossary/skylight.php
7. caves on Moon, http://www.nss.org/settlement/moon/library/LB2-208-LavaTubes.pdf
8. useful, http://www.oregonl5.org/lbrt/l5isru1.html
9. Apollo 15, http://www.lpi.usra.edu/lunar/missions/apollo/apollo_15/landing_site/
10. new cave, http://www.newscientist.com/article/dn18030-found-first-skylight-on-the-moon.html
11. rimless pit, http://planetary.org/blog/article/00002173/
12. science fiction, http://www.spudislunarresources.com/Moonwake/mw.htm
13. Willie Sutton, http://en.wikipedia.org/wiki/Willie_Sutton

Comment

1. [...] Emily over at the Planetary Society Blog notes that researchers at JAXA appear to have discovered a cave entrance on the Moon (or a least a really deep pit too deep to be a crater) in the Marius Hills, which is great news for those who consider going underground to be the solution to the radiation pelting the surface of the Moon. Coupled with light-pipes, large underground spaces are a boon on the Moon for those who want to set up a human presence. [Update: Paul Spudis reminds us of the real estate maxim: location, location, location] [...]

Pingback by Carnival of Space #126 (& Carnival of the Moon) - Out of the Cradle — October 27, 2009 @ 6:50 pm

A Rainbow On the Moon

November 14, 2009

An ice rainbow seen in cirrus clouds on Earth. (UCSB Dept. Geography)

Five weeks ago a crater from the LCROSS impact[1] formed on the Moon. The pre-impact build-up[2] had been sensational, but the actual event was largely invisible to observers on Earth. It was a different story on the Moon. The slowly growing impact ejecta curtain threw water ice particles and vapor far out into space. When the crater formed, flying ice particles could have refracted the glare of unfiltered sunlight into an "ice rainbow," similar to those seen through very high altitude clouds on Earth. For a very brief time, a rainbow might have been visible to an observer standing on the lunar surface. And like its namesake[3], this rainbow is a promise – a promise that the Moon is habitable. It is an invitation to humanity to extend man's domain to our nearest planetary neighbor.

The LCROSS science team's initial analysis of ejected impact plume data found evidence for water[4]. It appears that several other species, particularly some carbon substances also found in the cores of comets, may be present. The new results suggest that some lunar polar volatiles may have their origins from outside the Moon, deposited there over millions of years by the impact of comets and asteroids.

Over the last 50 years, the idea of water ice at the lunar poles[5] has generated as much angst as excitement within the scientific community. Ice on the Moon was suggested by Watson, Murray and Brown in 1960. They recognized that, regardless of the fate of such substances elsewhere on the Moon, the dark, cold floors of polar craters might retain volatile substances. Rock and soil samples returned by the Apollo missions were not only bone-dry, but crystallized in a very reducing environment, suggesting that any indigenous lunar water, if present, must have been a very minor component. Apollo scientist Jim Arnold resurrected the Watson et al. hypothesis forty years ago, concluding that their original proposal of water ice at the poles was still feasible and that a polar lunar orbiter was needed to search for such deposits.

We know that over geologic time, the Moon was bombarded by water-bearing objects. Meteorites contain water, and just as they've landed on Earth, they've also hit the Moon. Moreover, we've detected water vapor in the tails of comets with Earth-based telescopes. But it was widely speculated that all this water must be lost from the Moon, which left the issue of polar ice unresolved.

Fifteen years ago, the 1994 Clementine orbiter mission revived our interest in the Moon's polar regions. When Clementine's images of the Moon's poles revealed large areas of shadowed terrain, it reminded Gene Shoemaker and the science team of the Watson and Arnold papers. Large shadowed areas suggested that polar cold traps might really exist, so an experiment was improvised using the spacecraft transmitter to beam RF energy into the shadowed areas. Analysis of the radio echoes[6] suggested the presence of ice in shadowed areas near the south pole. This result was questioned, largely because our team couldn't repeat the passes using the improvised experiment.

In 1998, Lunar Prospector found evidence for excess hydrogen[7] in the surface soils of both lunar poles. These data could not show what form the hydrogen was in and had very low spatial resolution. The issue, as to whether the observed polar hydrogen represented water ice in the dark cold traps or elemental hydrogen implanted by solar wind protons, was vigorously debated. The preponderance of evidence[5] in the years since Lunar Prospector, suggests that water ice is present in the polar areas, but its form, distribution and physical state are completely unknown.

The current flotilla of lunar orbiting spacecraft carry several advanced sensors, all designed to better characterize the environment and deposits of the polar regions of the Moon. We have seen extremely low temperatures in the polar dark regions using the DIVINER instrument[8] on the American Lunar Reconnaissance Orbiter (LRO)[9] spacecraft. The Japanese Kaguya[10] mission mapped the topography and terrain of the polar areas and showed us the extent of the shadowed areas. The Indian Chandryaan[11] mission sent a probe into the south pole, mapped the extent of sunlight and carried two NASA instruments – the Moon Mineralogy Mapper (M³)[12] and Mini-SAR radar[13]. In September, the M³ instrument found significant amounts of water bound into mineral structures at high latitudes. The Mini-SAR instrument has made maps showing the interior of dark polar craters. These maps are being analyzed for scattering characteristics to determine whether water ice might be present there; our initial results will be announced soon.

Now, the LCROSS impactor – sent to kick up the dust of the polar dark regions – has shown us that water ice does exist there. We still don't know how much water ice in total may be present; from Clementine, we estimated there are billions of metric tones of water ice present in the south polar area. Complete analysis of all of the remote sensing information in the next couple of years will ultimately give us a good estimate of the total amount of water available. Clementine also revealed peaks of near-permanent sunlight in proximity to regions of permanent darkness at the poles (where the sun's circular rotation keeps temperatures benign).

If you don't know where you're going, any path will get you there.

The Moon has the resources needed to bootstrap a sustained, permanent human presence. It is the place where we can learn how to live and work productively in space. The Moon has put out a welcome mat. What are we waiting for?

Topics: Lunar Exploration, Lunar Resources, Lunar Science, Space and Society

Links and References

1. LCROSS, http://lcross.arc.nasa.gov/
2. build-up, http://blogs.airspacemag.com/moon/2009/10/12/lcross-a-mission-to-hyperspace/
3. namesake, http://www.biblegateway.com/passage/?search=Genesis+9%3A8-17&version=KJV
4. evidence for water, http://www.spaceref.com/news/viewpr.html?pid=29613
5. ice at poles, http://www.thespacereview.com/article/740/1
6. radio echoes, http://www.psrd.hawaii.edu/Dec96/IceonMoon.html
7. excess hydrogen, http://lunar.lanl.gov/
8. DIVINER, http://diviner.ucla.edu/
9. LRO, http://lunar.gsfc.nasa.gov/
10. Kaguya, http://www.selene.jaxa.jp/index_e.htm
11. Chandrayaan, http://www.isro.org/Chandrayaan/htmls/home.htm
12. Moon Mineralogy Mapper, http://m3.jpl.nasa.gov/
13. Mini-SAR, http://www.nasa.gov/mission_pages/Mini-RF/main/index.html
14. water, http://www.sciencemag.org/cgi/content/abstract/1178658

Comments

3. Paul Spudis,

You appreciation of the lunar 'stuff' had been very consistent and you have pushed my enthusiasm to a great desire to better understand the Moon. I am currently (re)reading your 'Once and Future Moon', Smithsonian Institution 1996, upon the exciting news on 'water on the moon' based on the LCROSS's data.

Yes, indeed, the Moon is the "place where we can learn how to live and work productively in space'. In one article published on 9/14/09, 'Objectives before Architecture' you clearly articulated the main difficulty created by the 'tyranny of the Rocket Equation' and I do agree with the idea that our primary mission is to locate, access and process lunar resources in order to 'learn how to use the lunar material and energy resources'; in order 'to create the ability through the use of space resources to go anywhere and do everything'. We truly need to start designing missions that will attempt to understand the feasibility or the unfeasibility of such endeavor. That answers the question posed in this lunar water related article: "the Moon has put out a welcome mat.

127

What are we waiting for?"

Comment by Ernst Wilson, Elizabeth City State University — November 14, 2009 @ 2:16 pm

5. Ernst,

Nice to hear from you again. Thanks for your comments; I appreciate them. Glad you're enjoying the book (again) — tell your students!

Comment by Paul D. Spudis — November 14, 2009 @ 3:43 pm

7. With the confirmation of bucketfuls of water and, it would seem, tasty hydrocarbons too! Ethanol? Headline: "Scientists discover Moonshine!"

My main concern is over the nature of the Vision... post LCROSS. Will the unilateralist Vision of Bush/ ESAS/ Cx re-emerge to trounce the multi-lateralist Vision of Augustine? After all what's at stake here: "Most Valuable Real Estate in the Solar System" (Peter Diamandis Huffington Post http://www.huffingtonpost.com/peter-diamandis/most-valuable-real-estate_b_357177.html)

Now where have I heard that before! Ahh Yes:

http://www.space.com/businesstechnology/technology/moon_next_020923-2.html

And they spelt Shackleton wrong!

I earnestly hope that you are correct and that not only are there 730,500 shuttle launches worth but a bountiful re-hydration cycle too. This space cadet reckons that revisiting the Moon Treaty may be the only way to preserve some of the science for future generations before it gets used to: "develop capabilities, plans, and options to ensure freedom of action in space, and, if directed, deny such freedom of action to adversaries."

Time will tell!

Comment by brobof — November 14, 2009 @ 5:45 pm

8. brobof,

Well, we are talking about the most valuable piece of real estate in the Solar System, in the near field anyway. A good sales phrase has its own life!

In regard to preserving the scientific integrity of the ice deposits, there is more than enough lunar ice for that. We could take hundreds of core samples and archive them

permanently to preserve that volatile record without any impact whatsoever to mining and extraction activities.

There's plenty of water on the Moon for everybody. Its role is to allow us to begin routine space operations and habitation. It doesn't have to last us forever; by the time we get our foothold established there, we will have other opportunities and sources for volatiles, metals and energy.

Comment by Paul D. Spudis — November 15, 2009 @ 4:45 am

9. Amen to that!

"Lunar polar orbiters, using remote sensors to search for water-ice in the permanently shadowed frozen craters near the lunar poles. If ice is found, the Moon can become a refueling base not only for oxygen, which constitutes 86% by weight of rocket propellant and is already known to be the most abundant element on the lunar surface, but also for hydrogen, which is the remaining 14% by weight of rocket propellant. The first such mission, if started now, could be designed, built, flown and return information within five years, around 1995."

Alternative Plan for U.S. National Space Program Gerard K. O'Neill 1989 (For goodness sake!)

I remember well when Clementine ('94) Kudos! and later Lunar Prospector hinted at lunar water reserves. Why do these things have to take sooo long? Perhaps things will speed up a bit now!

Comment by brobof — November 15, 2009 @ 8:46 am

10. Hi Paul,

Of all the talk I have heard over the years on where we should go and what we should do in space, I want to tell you that your position – the exploitation of lunar resources in support of human expansion into the solar system – seems the most rational, the most objective.

Given the near certainty in knowledge that water is there, do you think the moon will become the focus of NASA's future human space flight program?

Comment by Philip Backman — November 15, 2009 @ 8:45 pm

11. Hi Phillip,

In a rational world, yes it would. But we don't live in that world, so the Moon's future is uncertain. NASA is wedded to the Apollo template of spaceflight (design, build, launch and abandon), so reusable vehicles, incremental steps and using space resources are alien concepts to them. It remains to be seen whether they can change the way they approach their business.

Thanks for your encouraging comments!

Comment by Paul D. Spudis — November 16, 2009 @ 4:40 am

12. Paul Spudis,

I will not be surprised to see a NASA shift of paradigm, based upon the new lunar dis(un) coveries, as I will not be surprised to see astrolawyers revisit the 1979 Moon Agreement in order to come up with a new legal regime governing the principles of lunar activities.

regards, ernst

Comment by Ernst Wilson, Elizabeth City State University — November 16, 2009 @ 2:05 pm

13. Paul,

I want to share a quote and an observation.

"We have to learn again that science without contact with experiments is an enterprise which is likely to go completely astray into imaginary conjecture." — Hannes Alfven

With LCROSS, NASA kept contact with an experiment and took a calculated risk with a relatively small investment. The question is "WHEN will that investment pay off by affecting space exploration decisions."

Thanks for stimulating these discussions!

Comment by Steve at NASA — November 16, 2009 @ 9:49 pm

14. Wondering is there a mineral that we can gain from probing the moon.

Comment by Spence — November 18, 2009 @ 6:57 pm

15. Spence,

There are a variety of minerals that make up the Moon's crust. Almost all of them are common species that have no economic value on the Earth. The real value of lunar materials is their use in space. Extracting what's needed and can be used from the Moon enables us to live and work in space easier and more cheaply. This is what makes lunar resources extremely valuable.

Comment by Paul D. Spudis — November 19, 2009 @ 4:44 am

Thanksgiving On the Moon: A Lunar Feast

November 22, 2009

Lunch from the lunar dirt (Photo courtesy of Dr. Jeff Taylor, Univ. Hawaii)

We often hear the Moon described as a lifeless desert, a barren rock in space where nothing can survive. Although the Moon is certainly different from the Earth, it is hardly barren. From the 1970's through the 1990's (largely before we knew about the presence of water and other volatiles in the lunar polar regions) the late, lunar scientist Dr. Larry Haskin[1] set forth some basic facts about the chemical composition of the Moon. Larry was a chemist by training and his view was that the Moon has all that we need – just not in the form in which we need it.

Larry wrote a very interesting paper[2] for the 1988 Second Symposium on Lunar Bases. Over the years, I heard him give several different versions of this talk. Initially, he called it "Wine and Cheese from the Lunar Desert" but after deciding that he didn't want to drive away or offend any teetotalers in his audience, he changed it, first to "Cola and Cheese" and then "Water and Cheese from the Lunar Desert." Although the liquid varied, the cheese stayed.

Haskin's argument is very simple. Take a cubic volume of soil (about 1 meter in dimension) from anywhere on the Moon. In that volume of soil (weight about 1600 kg), there is enough hydrogen, carbon, and nitrogen – the principal volatile elements implanted by the solar wind – to make lunch for two[2]. Larry's menu was modest, but satisfying: two cheese sandwiches, two glasses of wine (or cola, with real sugar), and two plums. Chemical atoms needed to make up this meal are all present in that relatively small volume of soil; they are just not arranged in the form that we need them. But the task is possible, given time and energy.

Because the Moon has no atmosphere and no global magnetic field, the highly energetic stream of particles from the Sun (the solar wind[3]) implants its atoms directly onto the dust grains of the soil. This material is mostly hydrogen and helium, but other light atoms such as carbon, nitrogen and other noble gases are also present. These volatile elements seem to correlate with a property called "maturity" which means the amount of time a soil has been exposed to the space environment. The amount of solar wind gas also correlates inversely with grain size – the finest fraction of the soil contains the most solar wind. Another unusual correlation is with titanium; the highest quantities of solar wind hydrogen are found in very high titanium soils. It's not clear why this should be true, although it is postulated that the crystal structure of ilmenite (an iron- and titanium-oxide mineral) acts as a "sponge" for solar wind atoms.

Thanksgiving on the Moon: A Lunar Feast

Given these properties, the best soil on the Moon to process and extract these important volatile substances would be very fine-grained, high-titanium soils. In fact, this soil occurs as the dark pyroclastic ash[4] that sometimes covers mare and highlands areas on the Moon. They are very fine-grained (typical mean grain sizes of a few tens of microns) and some are rich in titanium. The tiny black glass beads returned by the Apollo 17 mission have up to 13 wt.% titanium dioxide (among the highest found on the Moon). However, these Apollo 17 samples were buried by a landslide for millions of years so we do not know how much volatile material a mature, exposed surface ash deposit might contain. A robotic mission to such an area to measure the amount of solar wind gas could answer these questions.

Extracting the volatiles from soil[5] is very simple: just heat the soil to about 700° C. Although simple in concept, in practice this may be a very difficult job. We need to find a way to process the lunar soil in a continuous stream. Batch processing is much less efficient and expensive. Soil roasters that continuously roam the surface, heating the soil using solar thermal power and collecting and storing the emitted gas, is likely to be at least part of the ultimate solution.

During recent discussions about using lunar polar ice[6], some expressed concern that we would too rapidly devour what they perceive to be a limited resource. Although the Moon has hundreds of millions of tones of water around the poles, ultimately, we will need to learn how to use the lower grade ore present elsewhere around the globe. In this case, it will be the bountiful lunar regolith – the meters-thick outer layer of the Moon. This resource can truly last a lifetime – the lifetime of humanity in space. Wine and cheese[2] (or beer or cola and cheese, if you prefer) is there for the taking and the making. We are limited not by the intrinsic resources of the Moon but only by our own imaginations.

Something else to be thankful for this season—a Moon that has what we need to survive and thrive in space.

Topics: Lunar Resources, Space and Society

Links and References

1.Larry Haskin, http://epsc.wustl.edu/admin/people/haskin.html
2. water and cheese, http://www.nss.org/settlement/moon/library/LB2-504-WaterAndCheeseFromLunarDesert.pdf
3. solar wind, http://en.wikipedia.org/wiki/Solar_wind
4. pyroclastic ash, http://astrogeology.usgs.gov/Projects/LunarPyroclasticVolcanism/lunpyroWebimages.html
5. volatiles from soil, http://blogs.airspacemag.com/moon/2009/05/30/can-we-be-%e2%80%9cresourceful%e2%80%9d-on-the-moon-part-1/
6. polar ice, http://blogs.airspacemag.com/moon/2009/11/14/a-rainbow-on-the-moon/

Comments

4. Good post, Paul. I've read a bit about various researchers, Larry Taylor in particular, using microwave ovens to heat Apollo 17 regolith samples to well over 700° C. May it be possible that the lunar feast is cooked in a microwave?

Comment by James — November 23, 2009 @ 5:26 am

5. James,

You'll have to heat up the soil to get the volatiles out of it anyway. As you'll have the microwave handy, why not?

Comment by Paul D. Spudis — November 23, 2009 @ 5:46 am

6. Hello, The conference was held in 1988, not 1989 as noted in the article. :) I remember, I was there. :)

Comment by Bryan — November 23, 2009 @ 10:55 am

7. Bryan,

You're right — I "misremembered it"! I've fixed the text. The Proceedings of the conference did not appear until 1992 — I remember that because I was an Associate Editor! Thanks for the correction.

Comment by Paul D. Spudis — November 23, 2009 @ 11:04 am

Another Moon-Forming Collision?

December 7, 2009

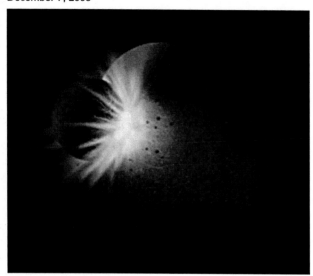

Another time, another place

A recent discovery from the Spitzer Space Telescope[1] may yield new insight into the origin of our own Moon. Although this discovery was in the news some time ago, the advent of the Augustine report and the LCROSS mission results have eclipsed it.

The Spitzer Telescope found evidence for a planetary collision[2] around the star HD 172555, about 100 light-years away from our Solar System. This evidence was a heat signature associated with spectral evidence for silicon monoxide gas[3] (a fairly rare substance) and glassy silica dioxide, a common form of silica glass found associated with volcanoes on Earth. These substances were found associated with a large cloud of silicate debris: the ground-up and pulverized parts of the outer portions of two rocky planets. The evidence suggests that two planets collided with each other at relative speeds exceeding 10 kilometers per second.

This set of circumstances is (more or less) the same that we expect in the aftermath of the currently favored model for the origin of our own Moon. Traditionally, Earth's Moon was considered to have formed in one of three different ways[4]. One model called for Earth and Moon to accrete (assemble) from a collection of small bodies simultaneously; with great

imagination, this model was called "co-accretion." The second idea called for the Moon to form somewhere else in the Solar System and then be "captured" into orbit by a near-miss encounter with the Earth. The last model of "fission" suggested that the Moon was ripped from the body of the Earth at a very early stage of our history when it was molten (or nearly so) and spinning very quickly. This spun-off piece of molten slag then became our Moon.

Although each of these models had its proponents in the days before and immediately after the Apollo missions, none of them seem to simultaneously satisfy all the constraints those program results provided. The general composition of the Moon is very similar to the mantle of the Earth, suggesting some variant of the fission model might be the answer. The problem with fission was its physical implausibility, as the early Earth was not like binary stars cited as analogs by the model's proponents. The near-identical oxygen isotopes of Earth and Moon indicated that co-accretion might be correct, but why would such planetary formation create two different objects (Earth and Moon) instead of a single body? Capture was an attractive way of explaining the subtle differences between Earth and Moon, but not their similarities and it was not an easy model to reconcile dynamically with the Moon's orbital properties.

The advent of the "Giant Impact" model[5] of lunar origin supposedly severed this Gordian Knot of lunar origin. Although first proposed in the mid-1970's, it seemed to emerge fully grown from the forehead of Zeus at the 1985 Origin of the Moon conference in Kona, Hawaii. In one fell swoop, the "big whack" explained all the salient features[6] of the Earth-Moon system, including its oxygen isotopes (both Earth and proto-Moon formed at the same position near the Sun), its depletion in volatiles (such a large impact would vaporize the planets' mantles and this material would be depleted in volatiles), and the high degree of angular momentum in the Earth-Moon system (an off-center impact would speed up the Earth's rotation while the Moon was spun off into orbit around it).

Although many scientists embraced the Giant Impact model, a few dissidents remain. Much of the model's attraction stems from its apparent ability to explain any particular fact about or constraint on the Moon. Lack of volatiles troubling? No problem – the big whack would drive them away. High spin rates on the early Earth? The impact hit off-center. Moon's bulk composition like the Earth's mantle? The Moon came from the Earth's mantle. Is it too unlike the Earth's mantle? No problem – it comes mostly from the mantle of the (now destroyed) impactor planet. In other words, the model is largely unconstrained and elastic enough to fit any fact or observation.

Scientists tend to be uncomfortable with such models; not only do they lack predictive power, they seem too much like "Just-So" stories[7]. However, finding something unexpected that matches the predictions of a model tend to give that model veracity and "completes" the jigsaw puzzle. The new Spitzer findings might be that missing piece of the puzzle. They suggest that planetary collisions do happen (we didn't really doubt that, but it's nice to have a concrete example). The presence of silicon monoxide gas indicates very high temperature, non-equilibrium processes – exactly what would be expected from a giant impact.

We may well find systems in various stages of evolution as we continue to observe the nearer stars and examine their planetary systems. Having other examples of multiple planetary systems allows us to "field check" our suppositions about the history of our own Solar System. Some of the apparently contrived "Just So" stories may just be correct.

Topics: Lunar Exploration, Lunar Science

Links and References

1. Spitzer telescope, http://www.spitzer.caltech.edu/
2. collision, http://www.washingtonpost.com/wp-dyn/content/article/2009/08/19/AR2009081902033.html
3. silicon monoxide, http://www.spitzer.caltech.edu/Media/releases/ssc2009-16/release.shtml
4. three different ways, http://en.wikipedia.org/wiki/Moon#Origin_and_geologic_evolution
5. giant impact, http://en.wikipedia.org/wiki/Giant_impact_hypothesis
6. salient features, http://www.psrd.hawaii.edu/Dec98/OriginEarthMoon.html
7. Just-so stories, http://en.wikipedia.org/wiki/Just_So_Stories

Comments

1. Dr. Spudis,

I have enjoyed your blog posts. They are very informative and well written for lay people like myself to understand.

If you could elaborate more on why the fission model of lunar formation is implausible I would appreciate it. I don't know enough about binary star systems to see how using them as an analog to lunar formation is not plausible.

Respectfully,

Comment by Jason — December 8, 2009 @ 5:26 pm

2. Hi Jason,

Thanks for reading the blog.

In brief, binary stars are rotating systems of gravity controlled nuclear explosions. Protoplanets — even if molten — are silicate objects, not undergoing nuclear fusion and not analogous in terms of viscosity, rotation rates, and angular momentum. One end member of the big whack model is effectively "fission" by deriving most of the Moon from the mantle of the Earth. It provides a mechanism to "spin off" a piece of the Earth as the Moon.

Comment by Paul D. Spudis — December 8, 2009 @ 6:34 pm

Arguing About Human Space Exploration

December 16, 2009

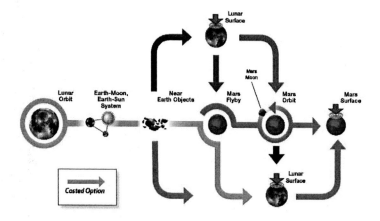

Flexible Path -- Metro Map to Nowhere (Augustine Report)

Hot rumor has it that, like Christmas, the Obama Administration's response to the Augustine Committee Report, Seeking a Human Space Program Worthy of a Great Nation[1], is imminent. Much excitement is discernible in the space blogosphere that a major change is at hand.

The Augustine Committee report concluded that NASA cannot execute the existing Program of Record (POR) of moving humans beyond low Earth orbit (LEO) to the Moon at existing or projected levels of funding. The report offered up "Flexible Path" (FP)[2], an alternate beyond LEO mission architecture of sending humans to asteroids and other destinations.

Flexible Path is billed as a low-cost alternative to the POR. It avoids going to the Moon, a destination viewed by the Augustine Committee chairman as a repeat of Apollo (a defensible position, at least in terms of the current plans by NASA's Exploration Systems Architecture Study[3]). FP describes a trans-LEO architecture that uses fuel depots and visits Lagrangian-points, asteroids and the moons of Mars. By not building a "costly" lander spacecraft for descent into the gravity well of the Moon, NASA can "save" money. The Augustine Report envisions human trips (in terms of total mission duration and remote-from-Earth operations) to L-points and asteroids as intermediate steps to human missions to Mars, their chosen "ultimate destination" of human spaceflight.

Many space advocates see great advantages[4] to FP. It relies on the idea of propellant depots, where fuel is cached at staging locations (e.g., Earth-Moon L-1; see below) and human vehicles are re-fueled in space for voyages beyond – trips destined for places at which no additional significant propulsive maneuvers are required. Thus, its targets are theoretical points in space or objects with very low surface gravity, such as near-Earth asteroids (rocky objects with orbits between Earth and Mars, not those in the asteroid belt, between Mars and Jupiter) or the small, asteroid-like moons of Mars. The latter are particularly interesting in that they could allow humans to control (teleoperate) robots on the surface of Mars with a near-instantaneous response, eliminating the tens-of-minutes time delay of radio communication with Earth. Such operations might permit true field geological exploration of the surface of Mars without the necessity of descending into the planet's relatively deep gravity well.

The Lagrangian points[5] (also called libration or L-points) are quasi-stable spots in space that are stationary with respect to two or more objects. For example, if you draw a line between Earth and Moon, it will revolve like the hand on a clock around the center of the Earth. If a satellite is put at a point on that line such that its period of revolution is identical with the Moon's, it will appear to be stationary in space relative to both Earth and Moon, even though it is flying through space just as fast as the Moon. L-points are found in relation to any two bodies, including Earth-Moon, but also Earth-Sun. They have many advantages as observation points, where satellites or telescopes can point and stare at targets in space for long dwell times and as staging areas for trips to other destinations.

So what's the problem with FP? At first glance, it appears to be an innovative way to move people beyond low Earth orbit at a relatively low cost. Advocates claim that human trips to destinations never visited by people are more exciting than repeating what we did 40 years ago and that by investing in things like depots, we get a flexible, extensible space infrastructure that will ultimately permit routine access to all destinations. Why complain about such a thing? Especially as many have advocated exactly such an approach[6] for the return to the Moon, an approach sorely lacking in NASA's existing plans.

To be blunt, the difficulty with Flexible Path lies in its motivations, assumptions and likely implementation. Development of FP by the Augustine Committee was driven largely by their determination that NASA's chosen architecture for lunar return (ESAS) is unaffordable. Assuming that movement of people beyond LEO is desirable, FP offers an allegedly low cost path to accomplish such. But to what end? The Augustine report is a bit vague as to the objectives and goals of the various FP missions. Mentioned is the servicing of telescopes at the L-points; the problem is there are none, at least at the moment. The James Webb Space Telescope[7] is not yet launched, nor is it designed for human servicing. The L-points are empty spots in space; there's nothing there except what we put there. In that sense, as a destination for people, it is no different from low Earth orbit, except that being outside the Van Allen belts (which protect astronauts on the ISS and Shuttle,) the radiation hazard is much greater.

The Augustine Report indicates that human missions to asteroids—Near Earth Objects (NEO)[8], could yield valuable information, including gathering strategic information for the

possible mitigation of asteroid collision with the Earth. Yet such targets are potentially dangerous. Some NEO asteroids have very high rates of rotation (on the order of an hour or less) making close approach very hazardous, except near the poles. Many asteroids are loose piles of rubble and co-rotating pieces of debris in the near field of such bodies could pose a hazard to a human vehicle. The Orion spacecraft[9] must be completely depressurized to allow astronauts to egress and explore an asteroid with EVA, so having all the crew in suits would be required. Exploratory capability would be very limited, on a scale similar to that of Apollo 11, the first lunar landing.

The several weeks-to-months duration of such mission will likely require the development of a "mission module," with significant life-support, power, and environmental control. Thus, selecting this path over the lunar surface does not relieve the agency from the task of developing and qualifying a completely new spacecraft, so the alleged "savings" of not building a lander are gobbled up by expenditure on the mission module. Lockheed-Martin, contractor for the Orion spacecraft, has come up with a clever "kissing Crew Modules"[10] concept whereby two Orion spacecraft are docked together, doubling the available interior space and consumables. However, this arrangement complicates the mission design. With two separate vehicles (requiring two dangerous re-entries), this configuration (with no airlock or docking module) has little exploratory flexibility. With such architectural difficulties and the "safety first," risk-aversion-at-any-cost sensibility expressed in the report, a NEO mission is probably a non-starter, or more likely, it will morph into a multi-year "study."

The programmatic assumption behind FP is that lunar return is "boring" and trips to new destinations will somehow "excite" the public and sustain the NASA budget. The problem with this reasoning is that it's not public excitement that is needed – it's public support. People will support things that aren't exciting, if there is some perceived value to them. A return to the Moon to develop and utilize its material and energy resources and create new space faring capability may not be "exciting" but it certainly is productive and useful. It allows us to build an extensible, maintainable, expandable, affordable transportation system, serving many purposes, thereby moving beyond the existing spaceflight paradigm: Design, Launch, Use, Discard (and Repeat.) The Committee's assertion – that building a lander to get into and out of the Moon's gravity well makes lunar return unaffordable – is ludicrous. It's not the Altair lunar lander[11] that's eating NASA's lunch; it's the development of two entirely new (and arguably unnecessary) Ares launch vehicles[12].

Repeating what we did 40 years ago is not the reason for lunar return, although, I understand the confusion. Here was a great missed opportunity for the Committee – they could have pointed out that NASA flubbed the implementation of its lunar mission from the beginning, largely because the agency never really grasped the rationale behind going to the Moon[13], thereby leaving others unable to embrace or articulate the mission. Perhaps the Committee didn't point this out because they didn't understand it either. Or perhaps because so much money has already disappeared down the black hole of Ares development, it was deemed easier to frame the report in the familiar terms of hardware procurement rather than focus on the objectives of the mission.

The above dissection of U.S. space policy leads us to the question of how NASA would implement Flexible Path. The lesson to be drawn from NASA's Exploration Systems Architecture Study – whose fiscally unsustainable architecture led to the creation of the Augustine Committee in the first place – is that the agency only knows one way of conducting business: the Apollo template[14]. This business model calls for big rockets, big infrastructure, a large marching army, and big budgets. Who believes NASA would implement FP differently than how they've planned a return to the Moon? The Augustine Committee itself (largely made up of former NASA employees and agency contractors) claims that "heavy lift launch vehicles are essential for human trips beyond LEO," a clear expression of the Apollo mindset, yet a statement that, objectively speaking, simply isn't true.

In other words, changing the focus of our destination from the lunar surface to Flexible Path doesn't solve anything. Building big rockets and throwing away 95% of the vehicle leaves no lasting infrastructure in space and prevents access to the material and energy resources of the Moon, negating the original intent and beauty of the Vision for Space Exploration (VSE) – to learn how (or whether, if you prefer) to use those resources to build sustainable, extensible space faring infrastructure. The chorus of approval[15] that you hear in the space press for FP is based largely on

wishful thinking. Flexible Path, if implemented by NASA, will reflexively follow the existing panem et circenses paradigm, abandoning any hope of ultimately changing an outdated spaceflight business model.

A $3 billion vaccine won't rid NASA of this disease. Only a renewed sense of purpose can save the patient.

Topics: Lunar Exploration, Lunar Resources, Space Politics, Space Transportation, Space and Society

Links and References

1. Augustine report, http://www.nasa.gov/offices/hsf/meetings/10_22_pressconference.html
2. Flexible Path, http://www.msnbc.msn.com/id/32767421/ns/technology_and_science-space/
3. ESAS, http://www.nasa.gov/exploration/news/ESAS_report.html
4. advantages, http://selenianboondocks.com/2009/12/why-not-just-fund-the-program-of-record/
5. Lagranian points, http://en.wikipedia.org/wiki/Lagrangian_point
6. approach, http://www.spaceref.com/news/viewnews.html?id=1334
7. James Webb Space Telescope, http://www.jwst.nasa.gov/
8. NEO, http://en.wikipedia.org/wiki/Near-Earth_object
9. Orion spacecraft, http://www.nasa.gov/mission_pages/constellation/orion/index.html
10. kissing crew modules, http://www.space.com/missionlaunches/090902-orion-asteroid-mission.html
11. Altair lander, http://www.nasa.gov/mission_pages/constellation/altair/index.html
12. Ares launch vehicles, http://www.nasa.gov/mission_pages/constellation/ares/index.html
13. rationale, http://www.spudislunarresources.com/Opinion_Editorial/Why_we_are_going_to_the_Moon.htm
14. Apollo template, http://blogs.airspacemag.com/moon/2009/10/23/paradigms-lost/
15. chorus of approval, http://www.spacenews.com/civil/091106-bolden-flexible-path-attractive.html

Comments

1. Paul,

FWIW, I'm not super-wedded to the flexible path approach, I was just using it to compare what you get with the money that way versus the PoR. That article was tossed together a bit quickly, and I probably could've taken one of the Moon First options and compared it just as easily, because after all, all of the non-PoR options the A-com reported on in their last report included robust advanced R&D funding and commercial crew funding. I'm a Moon nut myself, and have been so for all of my adult life. That said, part of why I'm somewhat more interested in the Flex-Path route is that I'm pretty sure NASA will botch a direct lunar return, and I actually want the Moon to have a chance.

I'm sure you've seen it too, but so many people are in such a darned rush, that whenever you propose something like depots that need to be done if we ever want to get the cost of lunar access affordable enough for real sustained activity, the answer always is "We can't put a risk like that on the critical path!" And so we're stuck with technologies like orbital propellant transfer that we could've de-moed back in the late 60s or early 70s that we're still afraid to do because every time "It's unproven, we can't risk the critical path on unproven technologies!" By taking the lunar return off the critical path, it might just allow NASA the flexibility to do stuff with the Moon that it couldn't risk if it were the direct focus. Stuff like allowing a much larger commercial or international component be involved (who says the lander has to be NASA designed and operated?), developing stuff like propellant depots and ISRU that actually make this affordable enough that someday you can close a business case on lunar travel. Stuff like that.

So, while I'm a Moon nut, I think that the Flex-Path proponents do have a valid point that quickest route back to the Moon might not be the direct approach.

Comment by Jonathan Goff — December 16, 2009 @ 10:38 am

2. BTW, having read the rest of your article, I do agree that the standard NASA approach to implementing FP is likely going to be a disappointment compared to its potential. NASA will just hear what it wanted to hear from your report–WooHoo! We get to build ourselves an HLV!–and ignore the inconvenient parts.

That said, how is a more Moon direct approach going to avoid the same fate? I mean, in the end, is a bloated, overly expensive, underperforming Moon program really that much better than what you think we'd get out of FP?

Comment by Jonathan Goff — December 16, 2009 @ 10:46 am

3. Wonderful post! But it's far too logical and reasonable to be taken seriously…

Just kidding. Thank you for continuing to address "The Why".

Comment by Itokawa — December 16, 2009 @ 11:07 am

4. Jon,

I am completely sympathetic to your programmatic concerns. And anyone who's read my stuff knows the problems I have with the POR. We have an agency that will not think "outside of the box." The few within it that try to

change things for the better get slapped down very rapidly.

There is no guarantee that a "Moon first" program won't turn into a dead end. It all depends on whether someone with decision authority (eventually) understands the real objective of lunar return: to learn how to use its resources to create new space faring capability. At least on the Moon, we have the inherent ability to experiment with doing that. At Earth-Sun L-1, we don't.

Comment by Paul D. Spudis — December 16, 2009 @ 11:14 am

5. I would agree with Jon that Lagrange points and NEOs should come before the moon, simply because they are easier. The radiation hazards at L1/L2 make them excellent places for testing radiation shielding. Using Orion for such experiments has the benefit of not needing to develop separate vehicles and of testing the actual hardware that will be used for exploration.

I'm less certain whether the moon or Phobos/Deimos should come first, I think both are reasonable. On balance I think the moon should come first.

About unproven technologies on the critical path: as Jon has heard me say hundreds of times :-) I would advocate using non-cryogenic lander propellants (either toxic or non-toxic) to avoid putting cryogenic propellant transfer on the critical path while at the same time devoting substantial resources to maturing them as soon as possible.

EELV Phase 1 + noncryogenic propellants can give us the moon and perhaps Phobos, Deimos and Ceres. Add SEP tugs and we'll certainly have Phobos, Deimos and Ceres and perhaps the surface of Mars. All this is proven technology, although for the Martian surface and perhaps Ceres you would want surface nuclear power too.

And all this would give most of the benefits of propellant depots for commercial development of space.

Comment by Martijn Meijering — December 16, 2009 @ 11:47 am

6. It's going to be difficult to plant a flag at the Lagrange points. It's going to be just as heartbreaking to watch a future Spirit get stuck in the sand from Phobos as from Caltech.

And I'm pleased to hear someone echo the unmitigated concerns about GCRs (and secondary showers) as well as the clearly stated rationale in the Space Studies Board's "Scientific Context for the Exploration of the Moon."

I can understand the never-ending problem of carrying out long-term projects within the two-year horizon of the congressional election cycle, but the "flexible path" creates no real infrastructure anywhere but on Earth.

It appears that one day our astronauts will need to be functional in Hindi or Mandarin in order to get landing clearance to visit the relics of our own pioneering exploration.

Comment by Joel Raupe — December 16, 2009 @ 12:25 pm

7. The sad thing about all this is that we have existing launchers that could easily launch sophisticated probes to the surface of the Moon, to the various Lagrange points, to Phobos and elsewhere. It seems to me entirely feasible to consider developing lunar LOX ISRU using telerobotics, all controlled from Earth without any humans on the Moon. We could even prove out solar cell production on the Moon using telerobotics.

Indeed, if I could get brave, I could envision it being feasible to develop lunar LOX ISRU telerobotically and to establish an EML1 LOX depot telerobotically, all using existing launchers. Robotic vehicles would deliver LOX from the Moon to EML1 until the depot was stocked. That way, future landers would have access to descent and ascent LOX supplies.

But if we did that, then we wouldn't need a huge launcher like Ares V when it eventually came to launching humans. Hmmm... that might be a problem.

Comment by Itokawa — December 16, 2009 @ 12:43 pm

8. Gee the A-panel said we can't do a the Constellation moon/Mars exploration due to being short on Funds. So they give us flex path...a journey going no-where as in Look but don't touch. All those "experts" want us to believe Flex is the only answer. But then along comes ULA and shows how to do Constellation within the current budget and have man back on the moon by 2018! ULA does it all no HLV needed, Fuel depots established, commercial flights, saves billions using existing Delta and Atlas boosters, moon base, rovers then on to Mars. DON"T LET THOSE SHORT SIGHTED EXPERTS PULL THE WOOL OVER YOUR EYES!!!! Check it out...this is the program that puts NASA back in game and encourages commercial involvement. LET THE PUBLIC KNOW THERE ARE OTHER OPTIONS!!!!!! MUCH BETTER OPTIONS!!!

http://www.ulalaunch.com/docs/publications/AffordableExplorationArchitecture2009.pdf

http://www.popularmechanics.com/science/air_space/4330793.html

Comment by Doug Gard — December 16, 2009 @ 7:40 pm

9. Paul, I agree with you in a large part about Flexible Path. I also support development of lunar resources in order to facilitate long term space development. The ISS has created a paradigm shift in commercial human space development in LEO. NASA took the long expensive road to build the ISS because of the space shuttle program. The space station concept evolved over a decade before construction began. But nevertheless, that development has spurred a new round of commercial space development both on the suborbital level and LEO.

Ironically, NASA has abetted this movement by building an architecture which cannot possibly compete with commercial spaceflight in LEO. The Constellation program may not be perfect, but it includes two necessary components for space flight. A crew vehicle and a heavy lifter. By the time both are develop, there will be ample opportunity to refocus the Constellation onto lunar development and ISRU. Once the lunar base is establish, then a paradigm shift in space travel beyond LEO can occur.

Comment by Gary Miles — December 16, 2009 @ 10:47 pm

10. One of the problems with the Constellation Program right from the start was NASA's only half hearted commitment to a lunar base which made many people wonder if this was only an Apollo redux program.

But why build a permanent base on the Moon?

1. We'll finally determine if living long term under a 1/6 gravity environment is deleterious to humans and other animals as far as health, reproduction, and growth. The results of this alone could have profound implications for space exploration and potential colonization.

2. We'll finally discover if food can be efficiently grown under artificial conditions under a 1/6 gravity environment.

3. We'll finally determine just how much and how effective regolith shielding is as protection against galactic and solar radiation and micrometeorites on the lunar surface.

4. Unmanned Earth operated solar powered rovers could travel all over the Moon collecting rocks and soil, returning their samples to the lunar base for return to Earth.

5. Unmanned Earth operated solar powered rovers could travel all over the Moon prospecting and collecting lunar meteorites for their platinum resources returning to the Moon base with their precious goods.

6. With emerging commercial manned space programs on the horizon, and potentially thousands of wealthy individuals on this planet willing to pay for a trip into space, a Moonbase could be a future destination for hundreds or perhaps thousands of wealthy space tourist and lunar lotto winners annually.

7. In the long run (perhaps just 20 or 30 years after we return to the Moon), a large lunar colony might be a perfect place for satellite manufacturing and launching since its much cheaper to launch satellites from the Moon into Earth orbit than from the Earth's surface. The Moon would then dominate the satellite manufacturing and launch business and be at the core of a currently $100 billion a year satellite telecommunications industry.

8. The burial of cremated remains on the lunar surface might also be a lucrative venture perhaps generating billions of dollars annually. Perhaps a round trip of a loved one's ashes sprinkled with lunar dust and returning to Earth in a beautiful lunar manufactured urn might also be a lucrative venture.

Comment by Marcel F. Williams — December 17, 2009 @ 2:04 am

11. Oh! I probably should have added that lunar base will finally determine if we can efficiently extract oxygen from the lunar regolith and mine the ice and hydrocarbons at the lunar poles.

Comment by Marcel F. Williams — December 17, 2009 @ 2:18 am

12. Even if manned missions to Lagrange points and NEOs precede manned missions to the moon, as I advocate, there is no reason why robotic precursor missions to the moon should wait. Manned NEO missions and perhaps Lagrange point missions should probably be preceded by robotic precursors, but robotic precursor missions to the moon could

come first. It would probably be wise to do that just to keep some attention focused on the moon.

Sending a few remotely controlled roomba bots to sinter or melt landing pads would be great. Navigation beacons would also be a good idea. ISRU experiments and prospector missions are another possibility. Both would keep the link with later manned exploration clear.

Comment by Martijn Meijering — December 17, 2009 @ 4:25 pm

13. *Even if manned missions to Lagrange points and NEOs precede manned missions to the moon, as I advocate, there is no reason why robotic precursor missions to the moon should wait.*

Yes there is — there won't be any money for anything else. The current ESAS architecture has gutted the lunar robotic program; how would a NASA "big rockets to nowhere" implementation of FP change that arrangement?

Comment by Paul D. Spudis — December 17, 2009 @ 5:43 pm

14. It wouldn't. If all we get is big rockets to nowhere, we're screwed, even if we do get a moon base. You want one, I get that and I want one too. But what I want even more is commercial development of space. If I had to choose between that and exploration I'd choose commercial development of space.

Of course, the only reason you'd have to choose between these two is if NASA insists it needs to build big rockets and if Congress and the Obama administration agree. If that happens, any hope of NASA helping with opening up space is gone for a generation. If we can't have commercial synergy in the next couple of years then I'd rather choose killing off the Shuttle stack and having no moon base and even no exploration than doing exploration with an SDLV. That would at least leave open the possibility of commercial synergy a few years down the road.

You would perhaps rather settle for the moon base and I respect that. Of course, if they blow all the money on big rockets there will be nothing left for a moon base either.

Comment by Martijn Meijering — December 17, 2009 @ 8:39 pm

15. *the only reason you'd have to choose between these two is if NASA insists it needs to build big rockets and if Congress and the*

Obama administration agree

And this is exactly the situation we're in and that's why I wrote this column. By the way, in the list above, you left out the "Augustine committee." So they are all in agreement.

Comment by Paul D. Spudis — December 18, 2009 @ 4:04 am

16. Good point about the Augustine committee. But isn't the moon base lost anyway with an SDLV? A moon base with EELV might not even be affordable, let alone with SDLV. If SDLV is the choice as seems likely my only hope is that MSFC will screw up again between now and 2018 and the SDLV will be canceled for good. A five to ten year delay is better than a twenty five year delay. Of course, even if that happens it could still end up taking out the ISS.

If we want to see commercial development of space in our lifetimes all our hopes are now riding on SpaceX, whatever comes out of CCDev and suborbital efforts. SDLV is a travesty.

Comment by Martijn Meijering — December 18, 2009 @ 9:28 am

17. There are SDLVs and SDLVs — we could make a Shuttle side-mount using the existing SRB, ET and left-over SSME (there are enough to make between 15 and 19 SDLVs) for minimal investment. Each could put ~60 mT into LEO and that capability is enough to emplace a lunar outpost and associated equipment. Before human arrival, we could start resource prospecting and experiment with processes using EELV launched robotics and teleoperated machines; these assets would be left in place on the Moon for future use by humans.

I simply don't buy the NASA/Augustine line that there's not enough money to do this. They need to craft an architecture that permits them to make incremental progress under existing budgets. If they're the smartest people and the finest engineering organization in the world, that shouldn't be an insurmountable obstacle.

Comment by Paul D. Spudis — December 18, 2009 @ 10:17 am

18. *I simply don't buy the NASA/Augustine line that there's not enough money to do this.*

Are you assuming the ISS is scuttled in 2020 or perhaps earlier? The Augustine committee assumes you can have two out of the following three: SDLV, exploration, ISS. And

exploration would exclude a moon base. I'm having a hard time believing you could have Shuttle-C + ISS + moon base.

I would consider abandoning LEO an even bigger disaster, as what some NASA insiders call yielding LEO to commercial players translates to making off with the loot to safely beyond LEO and making sure there is nothing left in LEO.

If they're the smartest people and the finest engineering organization in the world, that shouldn't be an insurmountable obstacle.

Surely you don't believe that is true?

Comment by Martijn Meijering — December 18, 2009 @ 2:45 pm

19. *The Augustine committee assumes you can have two out of the following three: SDLV, exploration, ISS*

I know they do. And they are wrong.

The one aspect everyone forgets is that schedule is the real free variable. We do all these things and we take as long as we need to do them. Build the program with small, incremental, but cumulative steps. We get there eventually. But we make constant, steady progress, even when budgets are tight.

Comment by Paul D. Spudis — December 18, 2009 @ 3:28 pm

20. *Build the program with small, incremental, but cumulative steps.*

Amen to that. I'd say that that is in fact one of the potential strengths of the Flexible Path, even though you are right in pointing out NASA would likely find a way to mess that up.

But doesn't the high fixed cost of the Shuttle stack throw a spanner in the works here? Sure, you can develop the SDLV slowly, but if there isn't the money to both use it for exploration and maintain the ISS you'd have to sacrifice one capability for another. That's not incremental progress, although it is incremental change.

You have argued that the Flexible Path as implemented by NASA is unlikely to yield the space faring capabilities that people who want to see commercial development of space would hope for. You are probably right about that, or at least it is unlikely to yield much.

A moon base (if that is possible budget wise) is a perfectly honourable alternative goal to commercial development of space. It is a goal space faring enthusiasts might find less exciting, but it is certainly better than nothing. That part of your argument I find convincing.

But I don't think it's true that there would be no further pain for advocates of a space faring civilisation. Losing ISS (or another permanent presence in LEO) would be a high price to pay. Not having the opportunity to delay exploration in the hope that a later administration might be willing to go down the commercial route with depots and such is another non-zero opportunity cost. ISS and Lagrange point infrastructure are arguably of greater benefit to commercial development of space than a lunar base. In my opinion switching from NASA-style Flexible Path to a lunar base would be a step backward from the point of view of commercial development of space. In other words, although your argument has some merit, it likely won't and in my opinion shouldn't sway advocates of a space faring civilisation to switch their allegiance to a lunar base. Perfectly honourable position to take, but a different set of priorities.

BTW this little discussion we're having shows that it is actually possible for reasonable people who have different priorities (exploration vs commercial development of space) to have a civil, honourable and honest discussion without resorting to lies and distortions as happens so often on online forums.

Comment by Martijn Meijering — December 18, 2009 @ 4:16 pm

21. *But doesn't the high fixed cost of the Shuttle stack throw a spanner in the works here?*

The high costs of the Shuttle "stack" are mostly associated with the people costs of refurbishing the Orbiter after each flight, not by stacking launch components in the VAB. Shuttle-side mount won't have those costs.

But I don't think it's true that there would be no further pain for advocates of a space faring civilisation.

I didn't say there wouldn't be any "pain." I've been in the business over 30 years and it's been one continuous pain. But I happen to think it's important enough to endure it.

Comment by Paul D. Spudis — December 18, 2009 @ 5:53 pm

22. Paul,

A program built on developing human spaceflight through incremental steps within existing NASA budget would seem logical and ideal. For one, NASA's budget has not even been increased to match the inflation rate over the last four decades. Any kind of program, no matter how incremental, would have to be cut back over time due to the inflation rate unless NASA budget is given parity.

Two, as you are keenly aware, NASA is subject to the changes in political winds whenever a new White House administration takes over. From Nixon, to Reagan, Bush I, Clinton, and Bush II, NASA's policy has changed drastically. The sustained construction of the ISS is probably the most long term project NASA has endeavored to build and even then it evolved from the Space Station Freedom to International Space Station. So how do you propose to insulate NASA from Presidential politics in order to guide a long-term strategic vision?

Three, historically, human development rarely occurs incrementally. In fact, technological development tend to spur rapid changes in the market and human society. Who for instance could have predicted the impact of the Internet and the changes that it has created in information technology? Cellular communications? Even microwave technology in food preparation. The guiding factor seems to be creating or establishing new markets either through technological innovation or geographical distribution.

So getting to the Moon and staying there to develop its resources to create new markets could expand human society beyond LEO. Once these markets are established then development of better means of space travel can proceed more rapidly. Whether any of us like it or not, the decision on the architecture was made almost 5 years ago by Congress at the behest of Michael Griffin. What is relevant is the Augustine panel report stated that the Constellation program is executable and can succeed in getting to the Moon. If Obama changes this course now, what is to stop the next President changing Obama's decision 4 or 8 years from now? It is like a perpetual game of ping pong where nobody is winning and we are all losing. Staying the course and tweaking the Constellation program to succeed in getting to the Moon and establishing a permanent lunar base (fingers crossed) is a worthwhile goal. Once the lunar base is there, that is when your paradigm change will occur.

Comment by Gary Miles — December 19, 2009 @ 12:21 am

23. *What is relevant is the Augustine panel report stated that the Constellation program is executable and can succeed in getting to the Moon. If Obama changes this course now, what is to stop the next President changing Obama's decision 4 or 8 years from now?*

Gary,

The Augustine report is wrong. The current architecture is not executable for the simple reason that there is not enough money to do it. The Vision did not direct NASA to create an unaffordable architecture and then whine about not having the money for it.

As far as changes in strategic direction go, I am arguing against that — I want NASA to carry out the direction it was given on January 14, 2004. They are the ones who changed the mission, from one using lunar resources to establish a sustainable and useful human presence in space to a rocket-building entitlement program. They were given clear programmatic direction and chose not to do what they were told. So exactly where is the problem with program execution?

Comment by Paul D. Spudis — December 19, 2009 @ 4:59 am

24. Paul,

Sustainable government programs are a rarity given the changes in political discourse in the federal government over time. Social Security and Medicare have survived this long due to their immense popularity with a solid political base to protect them. Human spaceflight has no such political base. Commercial markets are what provides sustainability to transportation systems. So establishing a market on the Moon is what will sustain space travel beyond LEO.

You have pointed out that the Flexible Path option has no identifiable markets on which to develop and generate revenue to sustain the development of human space travel. Moreover, under NASA, the same government contractors that have been providing services for NASA for the last 40 years, namely Boeing, Lockheed Martin, ATK, Northrop Grumman will remain the same players for development Flexible Path option. The dates these companies have provided for operational launchers and spacecraft are no more reliable than the dates given by NASA.

Given the unlikelihood of NASA implementing an incremental, sustainable program, which strategy would you choose to support. The Moon First with either Constellation program or possibly Ares V Lite variant where there remains the possibility of establishing a market for real commercial expansion? Or Flexible Path which leads to nowhere and depends on developing technology that does not exist yet with no to little hope of establishing new markets beyond LEO?

Comment by Gary Miles — December 19, 2009 @ 3:38 pm

25. *Sustainable government programs are a rarity given the changes in political discourse in the federal government over time*

On the contrary, my impression is that no government spending program or agency ever goes

away, unless by a specific act of Congress. We had a tea-tasting board funded by the federal government for over 200 years. The only question is whether money will be spent wisely or not.

which strategy would you choose to support. The Moon First with either Constellation program or possibly Ares V Lite .... Or Flexible Path...?

Your choice is a false one. Neither will lead to the kind of long term space faring capability that we need. Either NASA will change the way it conducts business in space or it will go nowhere. It's as simple as that.

Comment by Paul D. Spudis — December 20, 2009 @ 4:33 am

26. I am a little surprised and disappointed that the Augustine commission didn't promote the idea of a Lunar space elevator, which (unlike a terrestrial space elevator) could be built with existing fiber. There is highly unlikely to be anything already at the Lagrange points, but a lunar space elevator would start at L1, and would seem to be a good use of human presence and heavy lift capacity.

The Pearson report on a Lunar Space Elevator http://www.niac.usra.edu/files/studies/final_report/1032Pearson.pdf

gives some background here, but I think is ambitious for initial goals, with elevator masses in the kiloton range. I would argue for an initial elevator that was just capable of bringing 1 kg of mass up from the Lunar surface. This would both provide a new lunar sample, presumably from the pre-Imbrian Sinus Medii mare at the sub-Earth point, and would enable us to gain experience with the workings of an actual elevator. I think that the cost to do this would be comparable, and the benefit far greater, than developing another LEM.

Comment by Marshall Eubanks — December 20, 2009 @ 6:01 pm

28. As Paul mentions, one of the reasons we are in the mess we are in when it comes to human space flight beyond LEO is because of the way NASA chose to implement the strategic plan it was given in January 2004. Here are a few excerpts from President's Bush's speech: "Beginning no later than 2008, we will send a series of robotic missions to the lunar surface to research and prepare for future human exploration. . . Also, the Moon is home to abundant resources. Its soil contains raw materials that might be harvested and processed into rocket fuel or breathable air. . . We can use our time on the Moon to develop and test new approaches and technologies and systems that will allow us to function in other, more challenging environments." POP QUIZ: How many robotic missions does NASA have in work for the lunar surface? (Answer:

ZERO). How has NASA included the use of lunar resources in their current architecture? (Answer: a few small demonstrations only). NASA obviously chose to ignore this whole thought pattern in the strategic plan that was given to them.

NASA also dropped the ball in adequately describing the strategic plan to the public. A few more excerpts from President Bush's speech: "The Moon is a logical step toward further progress and achievement. . . Returning to the Moon is an important step for our space program. Establishing an extended human presence on the Moon could vastly reduce the cost of further space exploration, making possible ever more ambitious missions. . . With the experience and knowledge gained on the Moon, we will then be ready to take the next steps of space exploration; human missions to Mars and to worlds beyond." So, what does Mike Griffin do? He calls the program "Apollo on steroids", and from then on whenever the press reported on human exploration beyond LEO it was always "NASA's Moon Program". And when the Ares-1X launched, headlines read "NASA launches new Moon rocket". The Ares-1 can only make it to LEO, of course. The moon was only supposed to be the first step, not the entire plan.

Finally, we are also where we are because of politics. Every time the United States get a new President, NASA gets a new direction. Congress too has its pet projects that it directs NASA to execute, many times at the expense of existing projects. There's no doubt that President Obama's Administration is very 'political', and by definition anything associated with former President Bush is BAD. So here is a couple of excerpts from Arthur C. Clarke in 1951: "The first lunar explorers will probably be mainly interested in the mineral resources of their new world, and upon these its future will very largely depend. . . The human race is remarkably fortunate in having so near at hand a full-sized world with which to experiment: before we aim at the planets, we will have had a chance of perfecting our astronautical techniques on our own satellite . . . the conquest of the Moon will be necessary and inevitable prelude to remoter and still more ambitious projects." Sound familiar? President Bush was only re-telling the story of a methodical, incremental, and practical approach to expanding human presence into the solar system, that has been told now for over 50 years. It's not 'Bush's plan'.

Trying to be an optimist, I'm hoping that reason will ultimately win the day, and we do embark on the expansion of humans into the solar system, using the Moon as the gateway that it truly is.

Comment by John G. — December 21, 2009 @ 12:24 pm

29. Paul,

A more careful reading of the Committee's "Flexible Path" option will show it *includes* Lunar return capability; but by *first* developing the deep-space capability (i.e., a restartable space transport stage), it lowers the technical difficulty of a lander to the point where a commercial procurement could make one; thus following that course would likely yield a lunar delivery capability at far lower cost/flight and cost/lb to lunar surface than a "get there now!" Lunar architecture.

Speaking for myself I'm well aware of the promise of Lunar ISRU and the advantage of incremental development. I could find no path for incremental development by NASA without incremental destinations, because the organization is so "destination driven". I think in terms of the capabilities needed, personally. Lunar development needs depots at LEO and a cislunar staging point, a reusable transfer stage between them, and a reusable lander with ISRU propellant, as long as you're using chemical propellants. The Flexible Path alternative has NASA developing the bulk of those things.

I question whether you *want* NASA as the organization developing Lunar surface infrastructure for cost-effective resource development. That is not playing to their strengths. If they develop enough transportation infrastructure to get you most of the way there, it's a help.

Comment by Jeff Greason — December 22, 2009 @ 12:02 am

30. Hi Jeff,

Thanks for reading my post and for your comments. The problem with FP's inclusion of the "capability" for lunar surface return is that it is optional and in NASA-speak, that means that you don't get it.

I could find no path for incremental development by NASA without incremental destinations, because the organization is so "destination driven". I think in terms of the capabilities needed, personally.

So do I. And we need a path that creates capabilities incrementally — the key is to emplace and use robotic assets first to demonstrate and begin to make what we need on the Moon. NASA claims that they want to build, launch and fly hardware. Let them do so by pre-emplacing ISRU equipment on the Moon and practice extracting resources through remote teleoperation. We can survey, prospect, and conduct demonstrations first. Then small payloads can extract and store product before human arrival. There is no destination short of the Moon where this can be done.

I do not buy the committee argument that this

is unaffordable; in the spring of 2004, Code M (now SOMD) had a tiger team examine an affordable architecture built around robotic outpost emplacement and a side-mount SDLV made from leftover STS pieces, which could put 60 mT into LEO. It had a lunar outpost in place and operating by 2016.

I question whether you *want* NASA as the organization developing Lunar surface infrastructure for cost-effective resource development.

You're right — their job is NOT to develop the surface infrastructure but to determine how difficult lunar ISRU is, where the choke points are, and how to mitigate any unforeseen difficulties. In other words, they do engineering R&D, not infrastructure creation.

That is not playing to their strengths. If they develop enough transportation infrastructure to get you most of the way there, it's a help.

Given what's transpired in the last 5 years, how is developing Earth-Moon transport playing to their "strengths"? The whole reason your committee was called into existence was because of the perception (accurate or not) that the agency had blown the task of transportation development.

Having no particular destination in mind (FP) is much worse; it intensifies and buttresses NASA's tendency to excel in pointless "busy work" and Powerpoint engineering rather than cutting metal and conducting flight test. The real problem is the agency's "Apollo" mindset, in which big rockets and big money are the sine qua non of space exploration. They need a major change in their business model, not a re-direct away from the Moon.

The Vision was about changing the paradigm, of finding out whether we could use space resources to change the rules of space faring. It was not a "ticket out of LEO for humans" or a "human Mars mission" or a "rocket building entitlement." It was NASA's last chance to do something significant and game-changing. Instead, they dusted off and trotted out "Apollo on steroids." I do not want to appear cynical, but it makes one question whether they are capable of significant change.

Comment by Paul D. Spudis — December 22, 2009 @ 5:13 am

31. Paul,

I can tell that the passion is flowing… keep it going! Those of us within NASA who believe it is the "Why" question (Why should NASA do the things that it does in its Human Space program?) that is the most critical aspect of the debate will do our part to honor the bigger vision context

and advocate to those within who don't fully understand it. We do what we do so that we learn how to live off-planet as we progress to become unbounded from Earth. By the way, I was also disappointed that the Augustine folks did not adequately address the "why" question.

Comment by TonyL — December 23, 2009 @ 11:29 am

32. Hi Tony,

Thanks for reading and commenting. The Augustine committee actually did say that human expansion into space was the "long-term goal" for human spaceflight.

They then proceeded to lay out a plan using the Apollo-template of big rockets and PR stunts that is the very antithesis of sustainable human presence.

So, like most committee reports, it's a mixture of the good and the bad — the sublime and the ridiculous.

Comment by Paul D. Spudis — December 23, 2009 @ 12:37 pm

33. If NASA is not protecting us from asteroids and comets, NASA is waste of money. The survival of human civilization being a compelling thought? Protect us instead of doing ivory tower make work projects.

And if Earth does get impaled by some rock or nuke itself, it might be nice to have some colonies on nearby planets. The survival of human civilization being a compelling thought?

You don't have to worry about the gravity well if you are not coming back. Send people on one way trips. If I were terminally ill I would go and work myself to death setting up house. I might even go healthy and try to eke out long term survival on a hideous algae diet. Many people would.

Mars has aerodynamic braking so it takes less fuel to land there. No point in a sample return mission from Mars, better to send people up there and see if there is some bug that kills them rather than bringing a bug here.

We will have to dig underground or pile up dirt to survive radiation. Might as well send the robots to get started. Is that taboo terraforming? might ruin the sacred science museum up there? Terraforming! now that's a goal? The survival of human civilization being a compelling thought?

Comment by Francis X. Gentile — January 9, 2010 @ 3:44 am

Cataclysmic Events On the Moon

January 9, 2010

The Imbrium basin immediately after formation, ~3.85 billion years ago. Artwork by Don Davis.

NASA recently announced that it has down-selected three New Frontiers mission concepts[1] for additional study. One of these missions, *Moonrise*[2], proposes to return rock and soil samples from the floor of the largest impact crater on the Moon, the South Pole-Aitken (SPA) basin[3], centered on the southern far side. Not only is this the largest basin on the Moon, it is also the oldest, as evidenced by a high density of impact craters superposed on top of its deposits. But knowing that it is the oldest basin does not tell us exactly when it formed. Samples collected from its floor could potentially determine exactly when, during the early history of the Moon, it was created.

Why is the absolute age of this feature important? When lunar samples are returned to Earth as they were by the Apollo missions forty years ago, they are subjected to virtually every conceivable chemical and mineral analysis we can imagine. From those studies we have reconstructed a rough outline of lunar history and the processes that have shaped it. From this work (which produced reams of detailed data on the elemental composition

and make-up of the Moon) came the startling discovery that, not only is the heavily cratered crust of the Moon very old (older than 3.8 billion years), but that the largest impact features of that crust seemed to have formed at the very end of that early period.

The Moon's molten crust solidified 4.3 billion years ago. Virtually all highland rocks assembled in large impact events date from around 3.8 billion years ago. Because it was believed that many of these impact events had been sampled by the Apollo missions, it was inferred that the Moon underwent a massive bombardment of large body projectiles, all closely sequenced at that time. This time period is referred to as the lunar "cataclysm[4]" or the "late heavy bombardment[5]." As both Earth and Moon orbit the Sun at the same distance, an impact cataclysm affecting the Moon also would have affected the early Earth.

A late, cataclysmic bombardment of the Earth-Moon system was not predicted by the then-existing models of lunar formation and growth. In an unexpected turn, this time period (3.8 billion years ago) was significant in another respect, as it is the oldest epoch from which we have preserved fossil (bacterial) life in Earth's rock record. Does this mean there is a connection between the end of the early, heavy impact bombardment and the emergence of life on Earth? It is very tempting to make this connection but when missions to the Moon ended, our ability to continue down this path of scientific inquiry also ended. If we are ever able to draw such a paradigm-changing conclusion, we must first determine if the lunar cataclysm really happened.

Two things about the Apollo samples must be considered. First, all were collected from six landing sites in the vicinity of the central, equatorial near side. The large Imbrium basin is one of the youngest on the Moon—it subdivides lunar history[6] and determining its absolute age was a top priority; at least two of the Apollo mission sites were chosen to address the composition and age of the Imbrium basin[7] and two (and possibly the remaining four) additional sites are well within the possible influence of this large feature. Thus, although we cannot be certain, many of the Apollo highland samples may contain the imprint of this single, watershed event, obscuring the record of earlier impacts.

Second, the nature of the Moon itself works against our understanding of the geological context of the returned samples. The Moon has had a complex history, whereby rocks were thrown hundreds of kilometers across its surface, mixed up with other deposits thrown out from other craters and basins, along with periodic lava flooding. The continuous impact bombardment of the Moon for billions of years has "sandblasted" the crust into a crushed mixture of rock and fine powder (called regolith[8]) that covers the surface. True rock outcrop is hard to find and virtually none of the Apollo missions sampled it[9].

Although a lunar sample can be subjected to excruciatingly detailed measurements and age determinations, such data are valueless unless you are able to relate the sample to some larger, regional geological unit. In the case of the impact cataclysm, how many and which basins did the Apollo missions sample?

Unless we can answer that question with some certainty, we cannot be sure that a "cataclysm" occurred – we may be looking at only the last (or last few) largest basin-forming impacts.

To resolve the issue of the cataclysm, we must collect samples from older, distinctly different impact basins[10], preferably far removed from the zones where Apollo explored. Hence, we desire to collect samples from an area exactly opposite to the near side Apollo sites—the far side's South Pole-Aitken, the largest and oldest basin on the Moon. If we sample impact melt from this event, we could determine the age of SPA with a degree of confidence and understand whether all basins formed at nearly the same time (which would be the case if SPA is the same 3.85 billion year age as the near side Imbrium basin, sampled by Apollo) or if the formation of the basins was spread out over 400 million years, as would be the inference if SPA is 4.3 billion years old.

Obtaining this key sample would be monumentally important for lunar science. Does this sound too good to be true? Perhaps, but it is exciting to anticipate the possibilities of such a discovery.

In my next post, I will discuss some of the difficulties in deciphering the cratering history of the Moon from its rocks.

Topic: Lunar Exploration

Links and References

1. New Frontiers, http://discoverynewfrontiers.nasa.gov/news/New%20Frontiers/2009/news_123009.html
2. Moonrise, http://news-info.wustl.edu/news/page/normal/15260.html
3. SPA basin, http://www.psrd.hawaii.edu/July98/spa.html
4. cataclysm, http://www.psrd.hawaii.edu/Jan01/lunarCataclysm.html
5. late heavy bombardment, http://en.wikipedia.org/wiki/Late_Heavy_Bombardment
6. subdivides lunar history, http://ser.sese.asu.edu/GHM/ghm_10txt.pdf
7. Imbrium basin, http://ser.sese.asu.edu/GHM/ghm_10txt.pdf
8. regolith, http://en.wikipedia.org/wiki/Regolith
9. outcrop sampling, http://www.hq.nasa.gov/office/pao/History/SP-4214/cover.html
10. older basins, http://www.moondaily.com/reports/Moon_Mission_In_Running_For_Next_Big_Space_Venture_999.html

Robotic Sample Return and Interpreting Lunar History: The Importance of Getting it Right

January 11, 2010

A robotic mission sends samples back to Earth: What can we learn from them?

Deciphering the cratering history of the Moon is an important scientific problem. My previous post discussed early lunar cratering history, the apparent impact "cataclysm" 3.8 billion years ago, its significance to Earth's early history and how remaining questions might be resolved by collecting and returning new samples from the Moon. Here, I will describe the scientific difficulty and critical importance of planetary sample collection and analysis. With so many demands on NASA's budget, we need to approach this problem carefully, making every effort to maximize the prospect that we obtain not just samples but the right samples to answer the question of the Moon-Earth cataclysm.

NASA has announced that the proposed New Frontiers *Moonrise*[1] robotic sample return mission is one of three selected for detailed concept study[2]. The objective of this mission is to sample, date and analyze the composition of the impact-generated rocks produced by the largest and oldest crater on the Moon, the South Pole-Aitken (SPA) basin.

The return of surface samples has the po-

tential to answer many important scientific questions. How do we reconstruct the history of a planet from rocks returned from its surface? What are some of the difficulties in such a reconstruction? How well do we really understand the history of the Moon from returned lunar samples? Because context is vital to the correct interpretation of sample return data, these questions must be understood and considered, and underlie the mission strategy.

A great deal can be learned from remote surface measurements, but some properties can only be measured to very high degrees of precision by using returned samples[3]. One key piece of information that is difficult to measure remotely is a rock's age (measured by its radiometric isotopes). This determination requires a significant sample preparation, handling, and precision measurements; in some dating methods, we must literally take the rock apart, grain-by-grain. The machinery needed to measure isotopic composition tends to be big, massive, and power hungry, all undesirable properties for lunar and planetary payloads.

Geologists collect samples because they cannot bring into the field all the complex and sophisticated equipment used to analyze and describe the physical, chemical and mineral properties of planetary crusts. Samples allow them to conduct many different kinds of measurements in a controlled environment, eliminating external factors that can contaminate results. In addition, samples have long-term value in that they can be stored, archived and examined in detail (sometimes by newly invented techniques) as concepts and understanding change. It is for this reason that lunar sample studies continue to unravel new aspects of the complex history of the Moon 40 years after Apollo 11.

We are able to design a spacecraft to collect rocks and soil[4] and return them to Earth. After analysis, we have lots of data and numbers, but not necessarily any new understanding. Context is important in translating sample data into knowledge.

The geologist in the field must collect samples carefully; field work is not just picking up rocks – it is the attempt to unravel and comprehend the spatial and temporal make-up of planetary crusts. Samples must be representative of the larger, regional geological units they come from. A sample must be of the appropriate size (coarse-grained rocks need larger samples than fine-grained rocks to be representative of their parent units). If possible, we must collect rock samples from outcrop[5] (in place bedrock); rocks obtained from loose pieces on the ground (called "float" by geologists) have uncertain or unknown context and hence, the conclusions we draw from such samples may not apply to regional units. And when done on the Moon or another planetary body, all of this activity must conform to the constraints imposed by the flight system, such as total mass and volume limits for returned samples.

Recently, many countries have flown sensors that have yielded compositional information and globally mapped the Moon. From these data, we can determine chemical and mineral compositions of the geological units of the Moon (which are delineated by extent, morphology and physical properties). When this information is combined with data from returned samples, we can characterize the unit and its history even more fully than traditional field work, where intense, protracted ground study is possible. This is the promise of the new approach – allowing us to combine the low fidelity but broadly distributed data of remote sensing with the highly detailed but narrowly restricted information provided from samples.

However, due to the very nature of the Moon, there are significant geological complications that must be taken into account. Exposed bedrock is rare. A thick cover of regolith[6] is everywhere on the lunar surface. In the highlands (the oldest geological units on the Moon), there may be no bedrock at all, the surface having been thoroughly pulverized into regolith by four billion years of impact bombardment. Consequentially, the context of most Apollo highland samples[7] remains poorly understood. Exquisitely detailed measurements have been made on these rocks but we still cannot be certain about what they represent. Was there a cataclysm at 3.8 billion years ago? Currently, we are left wondering if we have sampled one, a couple, or a dozen basins.

During the Apollo missions, the astronauts did their best to sample and describe the context of representative rocks during collection, but the geological setting of most samples is still guesswork. The location of the samples returned by a robotic spacecraft will be documented to within a fraction of a millimeter. But as they are collected from regolith, their context will remain purely statistical.

By collecting hundreds of relatively small rocks (but still large enough for precision measurement) the argument is made that we will collect the desired SPA basin melt sheet through sheer statistical certainty. I suspect that the mission might well do this. But what about their context? We need to know which of the pebbles collected are from the basin melt sheet. In miniature, this situation duplicates and leaves us with exactly the same issue we currently have with the Apollo samples—which rocks represent the basins we intended to sample? With few exceptions, despite having global remote sensing data to provide context, we still do not know which (if any) impact basins are represented in the collections, which keeps our scientific understanding hobbled by degrees of uncertainty.

A simple "grab" sample from a relatively young and unmodified geological unit on the Moon could solve a major problem. A robotic spacecraft sent to the youngest lava flow on the Moon (dated relatively by crater density) could establish that flow's absolute age to high precision with a fair degree of certainty. As the age of the targeted geological unit increases, such certainty would decrease as younger events and deposits contaminate and disrupt the continuity of the older units.

The Moonrise mission proposes to sample the oldest preserved terrain on the Moon—the melt sheet floor of the SPA basin[8]. Younger units (craters, basins, and maria) are everywhere in this basin, superposed on top of the SPA melt sheet. Although pieces of the original basin floor may be preserved in places, we will not know in advance what those pieces should look like, leaving us with uncertainty over what was collected from the mission. In short, many samples will be collected, much data will be accumulated, and uncertainty will remain as to what it all means – the same knowledge gap we currently have with the Apollo samples.

Topics: Lunar Exploration, Lunar Science

Links and References

1. New Frontiers, http://www.moondaily.com/reports/Moon_Mission_In_Running_For_Next_Big_Space_Venture_999.html
2. three concepts, http://discoverynewfrontiers.nasa.gov/news/New%20Frontiers/2009/news_123009.html
3. returned samples, http://www.lpi.usra.edu/decadal/leag/AllanTreimanMoon.pdf
4. spacecraft, http://news-info.wustl.edu/news/page/normal/15260.html
5. outcrop, http://en.wikipedia.org/wiki/Outcrop
6. regolith, http://en.wikipedia.org/wiki/Regolith
7. sample context, http://www.lpi.usra.edu/meetings/LEA/whitepapers/Spudis_White_paper_SPA.pdf
8. SPA basin, http://www.psrd.hawaii.edu/July98/spa.html

Comments

2. Awesomely informative posts Paul! It's quite a coincidence (or not) that the earliest evidence for life on Earth and the "cataclysm" are both estimated to have occurred around 3.8 billion years ago. It suggests that life can occur spontaneously under the right conditions in a very short time, geologically speaking. One is tempted to invoke panspermia. The robotic probe won't be able tell the difference between a meteorite and bedrock, though, I presume. That's why I wish we could send humans to do the sample collecting. Robots are cool, and we learn a lot from them, but the results are always just so darned tantalizing! (But I guess the same could be said for the Apollo samples...) That said, it will be neat if the proposed mission is successful.

Please keep the posts flowing Paul!

Comment by Warren Platts — January 21, 2010 @ 11:49 pm

3. Warren,

Many thanks for your comments.

The robotic probe won't be able tell the difference between a meteorite and bedrock

Probably not, but this particular mission concept involves taking a random scoop of regolith (which includes rocks) and assuming that it is representative of the regional unit. Thus, your potential for success is entirely dependent upon choosing the right place to go.

Comment by Paul D. Spudis — January 22, 2010 @ 4:42 am

4. You're very welcome sir. I wonder if there are any permanently shaded spots within the useful (for Moonrise purposes) SPA Basin potential landing zone. It would be nifty if they could scratch out a little water ice along with their regolith sample!

Comment by Warren Platts — January 22, 2010 @ 9:42 pm

5. *any permanently shaded spots within the useful (for Moonrise purposes) SPA Basin potential landing zone*

As far as I can determine, the Moonrise team is focusing on landing sites at or near the center of the basin, presumably so as to provide the best geological context for sample collection. There is no permanently shadowed terrain in this region.

However, the two goals could be combined if the mission were to land near the south pole, which happens to be on the rim of the SPA basin. The problem is that the context of the samples is much less clear.

Comment by Paul D. Spudis — January 23, 2010 @ 6:17 am

6. Great post Paul. What are the weight and volume limits for samples that would be returned by Moonrise? I am wondering about both the amount that can be picked up at once and the sum total limits. Thanks and keep up the good work.

Comment by Jack Kline — March 3, 2010 @ 4:15 am

7. Jack,

What are the weight and volume limits for samples that would be returned by Moonrise?

I don't know what the current design is as I am not on the team, but it's probably on the order of 1-2 kg mass total. You don't need a great deal of sample mass to answer most scientific questions, so this keeps the sample-return mission small enough to be feasible.

Comment by Paul D. Spudis — March 3, 2010 @ 6:40 am

Beyond LEO –
Flexible Path Revisited

January 23, 2010

Teleoperated robots can emplace and build much of the lunar outpost infrastructure prior to human arrival (Astrobotic Technology Inc.)

In an interesting post at Vision Restoration[1], "Ray" tackles the desultory *Flexible Path* (FP) architecture of the Augustine committee[2], which calls for human missions to low gravity destinations and delays missions to the lunar and martian surface. The problems he finds with FP are similar to points that I've discussed in a previous post.

The principal rationale for doing *Flexible Path* rather than the current program for return to the Moon[3] is to avoid the cost of developing a new surface lander spacecraft for humans (either lunar or martian), which Augustine pronounced budget-busting for NASA. By being "flexible" and avoiding deep gravity wells, the Augustine committee saw a low cost way to send people beyond LEO. However, the Orion crew module and some type of heavy-lift booster still must be built.

Augustine committee member Jeff Greason[4] discussed the FP architecture during a recent appearance[5] on The Space Show[5]. Jeff pointed out that many people missed the principal rationale for the advancement of FP as an alternative to the existing program – that while we cannot afford the current ESAS architecture because of the requirement to do several developmental projects simultaneously (or nearly so), we might be able to afford

to do it sequentially, so that development of the Altair lunar lander[6] would only begin after we had developed and flown the Orion and its new heavy-lift launch vehicle. In his conception of FP, Jeff sees increasing space faring capability over time as robots and people visit new and more distant destinations. The FP destinations described in the Augustine report are the Lagrangian-points[7], near Earth asteroids, and martian moons Phobos and Deimos.

Ray points out that the two alternatives discussed in the Augustine report (Moon First and FP) assume a roughly $3 billion per year increase in the NASA budget. He suggests that this is unlikely, especially on a continuing basis, a supposition made even more credible by recent stories in the space press[8]. The alternative he offers to Augustine's FP takes a slightly different tack to the cost problem. Ray's solution, called *Flexible Path to the Moon*[9], shortens the destination horizon for FP and restricts it to cislunar space[10] (GEO, the Earth-Moon L-points, and lunar orbit).

With Flexible Path to the Moon, we develop routine access to all cislunar space, which adds important national security and economic dimensions to the human spaceflight program. Ray would defer not only the Altair lander but also (and this is critical) the new, proposed heavy lift vehicle called for by the Augustine report. Instead, FP to the Moon uses existing and future commercial launch vehicles for LEO access and for the subsequent build up of transfer nodes, in-space refueling of vehicles, propellant depots and other features of the Augustine FP architecture. Ray's plan further calls for "a large number" of robotic missions to the Moon and other possible destinations prior to human arrival.

I like this architecture and have advocated a very similar approach[11] that builds up space-faring capability incrementally and cumulatively—take small, affordable steps and make time and schedule the free variables. We make progress as we can with a sustainable architecture and build an infrastructure that is cumulative, inevitable and inexorable. One thing should be added to Ray's architecture: a statement of the "mission." The purpose of lunar return[12] is to learn how to use the resources of the Moon and space to create new capabilities and a sustainable human presence in space. This mission statement fits well with Ray's mission architecture. The significant level of robotic missions that he advocates in Phase 1 can be focused specifically on resource prospecting, characterization and demonstration. We can begin to produce resources using robotic missions and machines teleoperated from Earth well before the arrival of the first humans, who will then have the assets of life-support consumables, propellant, and electrical energy to draw on when they arrive.

The Flexible Path to the Moon offers the build-up of new technologies and capabilities in space by using an incremental approach that falls within existing budgetary constraints. It forgoes the building of a new heavy lift launch vehicle by creating a reusable, extensible space transportation system infrastructure using existing launch vehicles. And it focuses efforts and builds infrastructure in cislunar space, where virtually all of our assets reside. These are the stepping stones we need into the Solar System.

Topics: Lunar Exploration, Lunar Resources, Space Politics, Space Transportation, Space and Society

Links and References

1. Vision Restoration, http://restorethevision.blogspot.com/
2. Augustine committee, http://www.nasa.gov/offices/hsf/meetings/10_22_pressconference.html
3. current program, http://blogs.airspacemag.com/moon/2009/12/16/arguing-about-human-space-exploration/
4. Jeff Greason, http://www.xcor.com/bio/founders/jeff_greason.html
5. appearance on The Space Show, http://thespaceshow.com/detail.asp?q=1292
6. Altair lander, http://www.nasa.gov/mission_pages/constellation/altair/index.html
7. Lagranian points, http://en.wikipedia.org/wiki/Lagrangian_point
8. recent stories, http://www.spacenews.com/policy/100122-budget-increase-nasa-fate-ares-unclear.html
9. Flexible Path to the Moon, http://restorethevision.blogspot.com/2010/01/flexible-path-to-moon.html
10. cislunar, http://en.wikipedia.org/wiki/Cislunar_space#Geospace
11. similar approach, http://www.spaceref.com/news/viewnews.html?id=1349
12. purpose of lunar return, http://www.spudislunarresources.com/Opinion_Editorial/Why%20we%20are%20going%20to%20the%20Moon.htm

Comments

1. There has never been any significant development funds for the Altair vehicle. And funding for the Ares V and the Altair was always going to be funded sequentially– after the Ares I was developed.

As a strong opponent of the Ares I/V architecture, I would argue that its actually faster and cheaper to simultaneously fund an inline SD-HLV along with the Orion-CEV, an EDS (Earth Departure Stage) and the Altair vehicle. Why?

The Orion and Ares I programs are currently receiving $3.4 billion a year. Plus NASA's $18.7 billion budget is an increase of $1.5 billion from last years budget. If the Ares I is terminated, then a total of $2.6 billion a year is now available. Only $1.2 to $1.6 billion annually are going to required to fund the development of the Orion and space vehicle operations and integration. So that leaves at least $1 billion in extra funds.

The termination of the $3 billion a year Space Shuttle program will raise the extra funds available to $4 billion a year. That $4 billion could be used for SD-HLV development, which shouldn't cost more than $2 billion a year during the peak of its development. That leaves $2 billion a year for the development of an EDS and the Altair– vehicles only essential for lunar or beyond LEO missions. NASA has estimated that the total development cost for the EDS stage to be around $2.5 billion and the total cost of the Altair to be around $4.2 billion. That's $6.7 billion in total. But NASA should have $10 billion in extra funds over the course of just 5 years. So that's plenty of money to fund the EDS and Altair. And even if the EDS and Altair development cost doubled ($13.4 billion) and was delayed for a couple of years, NASA would still have a total of $14 billion in funds over that time period.

Of course, once the development phase ends, nearly $8 billion dollars of former development money will suddenly go into manned space flight operations for the Moon. And if the ISS program is terminated at that time then a total of $10 billion will go into lunar operations on an annual basis. And this is plenty of money for a Moon base program!

Comment by Marcel F. Williams — January 23, 2010 @ 2:57 pm

2. Marcel,

It may well be possible from a budgetary perspective to devise an Apollo-style architecture along the lines you suggest (although whether NASA is technically capable of doing it is questionable), but that isn't my point. I am arguing that such is not what we need!

I am advocating the development of an extensible, permanent infrastructure in cislunar space for the purpose of routinely accessing all of cislunar space — GEO, L-points, Moon. The whole point of going to the Moon is to use its material and energy resources to help create a true space faring infrastructure. The good part of Flexible Path develops these pieces — departure stage, in-space fueling, propellant depots and staging nodes. The bad part of FP develops new (and unnecessary) heavy-lift launch vehicles for throw-away, one-off, touch-and-go missions to NEOs and Phobos

The very last thing in the world we need in our national space program are a series of dead-end, pointless PR stunt missions and space "firsts" that leave no legacy capabilities, infrastructure or technology. In other words, FP as currently being envisioned by NASA will just be the next Apollo redux, only writ small. The problem with our space program is not the target destination — it's the way the agency does business. And if they don't figure out a way to change their business model, they'll soon be out of business.

Comment by Paul D. Spudis — January 23, 2010 @ 3:19 pm

3. Griffin and the Congress managed to turn a lunar base program into Apollo redux program.

But we do need a lunar base program since the primary purpose of the manned space program should be colonization, commercialization, and industrialization and not exploration. So I see a permanent Moon base as

the first step in that direction. Plus a continuously growing Moon base provides a desirable destination for the emerging private manned space flight industry.

Permanent structures beyond the Earth's magnetosphere are going to require a significant amount of mass shielding to protect against galactic radiation and major solar storm events. And there's no way we're going to be able to afford to launch such huge quantities of mass to a Lagrange point, plus the fuel needed to move such heavy craft through interplanetary space without developing either nuclear technology or lightsail technology.

An Asterant lightsail program could probably import thousands of tonnes of small asteroids to the Lagrange points annually for fuel, air, water, and mass shielding.

http://ntrs.nasa.gov/archive/nasa/casi.ntrs.nasa.gov/20040084672_2004088388.pdf

But I look at the Flexible path as a politician's dream program. If you want to go to an asteroid, or to the Moons of Mars, all they have to say is, well we don't have the mass shielding to do it right now with our chemical rockets— but we're working on it and maybe someday we'll have the funding for it! But, meanwhile, we can circle the Moon again like we did way back in 1968 for your entertainment! The Flexible path could well end up being missions to nowhere for decades to come while Russia, China, India, Japan and the EU may be colonizing the Moon!

Comment by Marcel F. Williams — January 23, 2010 @ 11:06 pm

4. *But we do need a lunar base program since the primary purpose of the manned space program should be colonization, commercialization, and industrialization and not exploration.*

You're preaching to the choir. Tell that to the agency, not me.

Permanent structures beyond the Earth's magnetosphere are going to require a significant amount of mass shielding to protect against galactic radiation and major solar storm events.

Not for facilities that are only sporadically and temporarily occupied. No one is advocating permanent human installations there. Propellant can be delivered to L-points via "slow boat" SEP cargo vehicles while people use "fast" chemical transport.

But I look at the Flexible path as a politician's dream program.

As currently understood by Augustine and the agency, I agree. That's why I like Ray's idea — it takes the good features of FP (incremental, infrastructure creation) and discards the bad parts (throw-away assets, PR stunts).

It wasn't just "Griffin and Congress" who turned the VSE into Apollo redux — it was the entire agency who never 1) understood or 2) embraced the VSE mission of lunar return. They dream of a human Mars mission and literally everything is subordinate to that goal. Want proof? Have a look at this:

http://www.nasaspaceflight.com/2010/01/taking-aim-phobos-nasa-flexible-path-precursor-mars/

Flexible Path to NASA is all about feeding their "Mars Forward" fantasies. The endless loop of "viewgraph engineering" studies, architecture development, and no flight missions has begun again. We are right back to where we were in 1992, after SEI died.

Comment by Paul D. Spudis — January 24, 2010 @ 5:12 am

5. I've always viewed the Flexible Path as a stealth Mars First program. The irony is that we'd probably already have permanent settlements on Mars if we had done what was logical right after the Apollo program– and established permanently manned bases on the Moon.

Plus the Augustine Flexible Path chemical rocket scenario doesn't even get us to the moons of Mars until after 2030. By 2030, the use of chemical rockets for interplanetary travel might be rendered completely obsolete by future nuclear or light sail technologies that could possibly transport humans and payloads to Mars much faster and cheaper. If President Obama chooses this scenario, NASA's manned space program could be set back for decades, IMO.

Comment by Marcel F. Williams — January 24, 2010 @ 3:04 pm

6. Interesting discussions. Genuinely wish that I could say otherwise. Unfortunately, this sounds to me like yet another proposal to further emasculate the US manned space program for the sake of private business interests.

The truth is that, although space is not cheap, it is cheap enough, compared to the federal budget, and to proposed increases in federal spending, that we can easily afford it, if we choose to do so.

It appears to me that politics must be removed as much as possible from NASA's direction in order for such a long-term program to ever accomplish anything. Get bipartisan congressional approval, maintain the required budget, without any OMB interference, and let NASA accomplish it's mission.

By 2020 there will be tested, proven commercial crew launch to LEO capabilities. Encourage NASA to include design elements into their architecture to make use of those services where advantageous. Until then, do not hold NASA hostage to the safety, reliability, functionality, and financial success of this fledgling industry.

One opinion. One voice.

Comment by Nelson Bridwell — January 24, 2010 @ 3:44 pm

7. If you want to remove politics, get a Moon base going. If one was going now, there is no way the present administration could touch it. Just listen to the clamor to preserve Shuttle and ISS. Strangely, it is these reusable, "boring" systems that only make the top fold of the front page when they blow up that have proved to be the most successful in terms of institutional staying power. It is the PR stunts that tend to get the axe. This Mission to Phobos stuff is just asking for it.

A lunar base might not make headlines everyday, but (a) like the South Pole Station and Little America, it would last as long as the USA is a going concern (i.e., it would never get technologically obsolete like Shuttle, and it would never have to be deorbited), and (b) isn't a normalized, stable space program that's ordinarily below the radar screen what we want?

But, apparently for some of these people, it's like that first landing on the Moon was their first hit of crack cocaine, and so the purpose of NASA is to chase the memory of that high and strive to regain it, even if only briefly and imperfectly. But they'll never quite get it back, and even if they come close, they will never be satisfied. The fact is that even a human landing on Mars will make less of a PR splash that Apollo 11. Why? Because of the we've-been-there-done-that factor. That second rock will never get you as high as that first rock–or so I've been told: that whole sort of mentality leaves me cold.

Anyway, Paul, I saw your terse comment at nasaspaceflight.com on the Flexible Mission to Phobos in addition to the mention above. Perhaps you could vent a little and expand on your thoughts on the matter. Your fans will appreciate it. Thanks again.

Comment by Warren Platts — January 24, 2010 @ 6:42 pm

8. Hi Warren,

apparently for some of these people, it's like that first landing on the Moon was their first hit of crack cocaine, and so the purpose of NASA is to chase the memory of that high and strive to regain it, even if only briefly and imperfectly. But they'll never quite get it back, and even if they come close, they will never be satisfied.

This is a very interesting analogy and I think that you are really on to something. I've argued in previous posts that "space firsts" and PR stunts are a poor basis for a sustainable space program. But NASA has been trying for 40 years to "recapture the Apollo excitement", even though there's no real evidence that such would lead them to the goal of their holy quest (more money). Your explanation is as good as any I've read on this. And I agree with you about the first Mars landing; not only the robot probes but 50 more years of science

fiction (and ever improving CGI effects) has made the reality of space a disappointment. But this is not a bad thing. Now, we can get down to doing business there, like a real frontier.

Comment by Paul D. Spudis — January 25, 2010 @ 4:59 am

10. I like the flexible path to the moon better than the flexible path mentioned in the Augustine Commission report. We have to get the most benefit out of the moon if we want to one day land on Mars. Does anyone know if the EELVs can launch a crew to the moon? I like the EELVs because of their proven safety record.

Comment by Rock Peterson — March 27, 2010 @ 5:36 pm

11. Rock,

There is a published report on a lunar return architecture that uses commercial launch vehicles. You can read it here:

http://www.ulalaunch.com/docs/publications/AffordableExplorationArchitecture2009.pdf

Comment by Paul D. Spudis — March 28, 2010 @ 4:58 am

Have We Forgotten What Exploration Means?

January 25, 2010

Tourists arrive at the new world (Library of Congress)

Yet again, the U.S. space program is in the slough of despond, whereby previous assumptions are questioned, the current path is discarded, the program is re-directed, and luminous enthusiasm heralds the new direction...

And then it all tapers off to nothing.

As long as we are navel-gazing during this policy hiatus, I want to examine a topic that many think is self-evident: what activities do we mean by the word "exploration?" NASA describes itself as a space *exploration* agency; we had the Vision for Space *Exploration*. The department within the agency developing the new Orion spacecraft and Ares launch vehicle is the *Exploration* Systems Mission Directorate. So clearly, the term is tightly woven into the fabric of the space program. But exactly what does exploration encompass?

Exploration can have very personal meanings, such as your own exploration of a new town, or a new and unknown field of knowledge. Here, I speak of the collective, societal exploration exemplified by our national space program. This exploration began in 1957,

when the launch of Sputnik by the Soviet Union initiated a decade-long "space race" of geopolitical dimensions with the United States. That race culminated with our first trips to the Moon. Once its primary geopolitical rationale had been served, Moon exploration was terminated. Since then, the "space program" has been astonishingly unfocused – drifting from a quest to develop a reusable spacecraft to building orbiting space stations – and despite numerous studies affirming needed direction, unfulfilled plans to send humans back to the Moon and eventually on to Mars.

When the race to the Moon began 50 years ago, space was considered just another field of exploration, similar to Earth-bound exploration of the oceans, Antarctica, and even more abstract fields such as medical research and technology development. Moreover, many used the term "frontier" when speaking about space, touching a very familiar chord in our national psyche by drawing an analogy with the westward movement in American history. What better way to motivate a nation shaped by the development of the western frontier than by enticing it with the prospect of a new (and boundless) frontier to explore? After all, we are descended from immigrants and explorers. Over time however, few recognized that there had been a shift in the definition and understanding of just what exploration represented.

Starting around the turn of the last century, while still retaining its geopolitical context, exploration became closely associated with science. Although first detectable in the 19th Century exploration of America and Africa, the tendency to use science as the rationale for geopolitical exploration reached its acme during the heroic age of polar exploration. Amundsen, Nansen, Cook, Peary, Scott and Shackleton all had personal motivations to spend years of their lives in the polar regions, but all of them cloaked their ego-driven imperatives in the mantle of "scientific research." After all, the quest for new knowledge sounds much nobler than self-gratification, global power projection or land grabbing.

Science has been part of the space program from the beginning and has served as both an activity and a rationale. The more scientists got, the more they wanted. They realized that their access to space depended upon the appropriation of enormous amounts of public money and hence, supported the non-scientific aspects of the space program (although not without some resentment). Because science occurs on the cutting edge of human knowledge, its conflation with exploration is understandable. But originally, exploration was a much broader and richer term. Which brings us back to the analogy with the westward movement in American history and the changed meaning of the word "exploration." A true frontier has explorers and scientists, but it also has miners, transportation builders, settlers and entrepreneurs. Many are perfectly satisfied to limit space access to only the former.

"Exploration without science is tourism."[1] — Statement of the American Astronomical Society on the Vision for Space Exploration, July 11, 2005

This fatuous quote accurately reflects the elitist, constricted mindset of many in the scientific community. In one fell swoop, the famous explorers of history – Marco Polo, Columbus, Balboa, Drake – are consigned to the category of "tourist." Overcoming great difficulty and hardship, these men sought new lands for many varied reasons. Exploration includes obtaining new knowledge but it does

not end there; it begins there. The quest for new lands has always been a search for new territories, resources, and riches. Historically, survival and wealth creation are stronger drivers of exploration and settlement than curiosity.

What is missing from our current program of space exploration is a firm understanding that it must generate wealth, not just consume it. Exploration is more than an experiment. The idea of space as a sanctuary for science has trapped us in an endless loop of building expendable hardware to support science experiments. Once the data are obtained, of what use is an empty booster or a used rover? We've "been there" and a pipeline of new inquiry awaits, to be facilitated by new spacecraft and new sensors designed to reach new destinations of study. Hugely expensive equipment must be developed to support science while the idea of creating transportation infrastructure or settlement is branded as "budget busting" (i.e., manned space exploration cuts into science's budget). So "exploration" lives to enable science, period.

This is our current model of space exploration. I contend that it is not exploration as historically understood and practiced. Traditionally, science (knowledge gathering) was a tool in the long process of exploration, which included surveys, mining, infrastructure creation and settlement (all advanced and protected with military assistance). This was the model of national exploration prior to the 20th Century and it is readily applicable today – if we change our business model for space. **What is needed is the incremental, cumulative build-up of space faring infrastructure that is both extensible and maintainable, a growing system whose aim is to transport us anywhere we want to go, for whatever reasons we can imagine, with whatever capabilities we may need.**

These changes do not require that an ever-increasing amount of new money be spent on space. Instead, true exploration requires only the understanding that it must contribute more to society than it consumes. And the American people have every right to expect as much in return for their years of supporting NASA.

Topics: Lunar Exploration, Space Politics, Space and Society

Link and Reference

1. Exploration without science, http://www.spaceref.com/news/viewpr.html?pid=17398

Comments

1. Space pioneering should be NASA's focus!

pioneer |paɪəˈnɪə(r)| noun - a person who is among the first to explore or settle a new country or area. - a person who is among the first to research and develop a new area of knowledge or activity

Comment by Marcel F. Williams — January 25, 2010 @ 7:08 pm

2. Paul,

Your post on the meaning of the term, "Exploration" is exactly on point. Being a NASA employee, the narrow definition is exactly how the word is interpreted at NASA (I believe), and I daresay many outside of NASA including Congress and the public have gravitated to that definition. I absolutely support the broader definition, which has definitely been lost or greatly diluted within NASA. The question in my mind is, do we re-educate everyone on the true or earlier definition of the term, or do we keep the meaning and its intent but change the word? I don't really care, pragmatically, as long as we get back to the original intent of NASA's "mission" by treating space as a true frontier to be exploited (use of local space resources, establishing infrastructure, etc) and settled, thereby creating wealth as you state and not just consuming it. Good luck to us.....

Comment by TonyL — January 25, 2010 @ 9:22 pm

**3. Well said, and right on target. Unfortunately, there is a reason that "exploration" has been redefined: 19th century-style frontier expansion is held in great contempt by today's academic, government and media elites. It is considered capitalistic, jingoistic, exploitative and unsustainable. You will be accused of wrecking the earth and then wanting to go and wreck the moon, too. The ethos of the far left is decidedly misanthropic. People will just spoil any place they go. Best to just send the robots and leave it pristine.

Comment by Bill Hensley — January 27, 2010 @ 12:06 am

**4. Dr. Spudis' article is spot on. I have speculated with friends on what America would be like today, were we to have had a lunar base since 1969. Would the population have been in the thousands? Would children have been born up there? Would the people have demanded political independence?

We always assumed that at first, such a base would have been supported by a well run government program with broad, deep popular support from the voter. We assumed eventual discovery of rich mineral deposits, although we could not say what these would have been. We read Clarke before watching Kubrick; the economic benefits of tourism seemed to go without saying, it was so obvious. Sometime near the end of the forty year speculation, we also assumed that the base, or colony, would have achieved economic profitability.

Why didn't any of this happen? I believe that the "fatuous quote" above goes a long way towards explaining why our human space flight efforts have not come to fruition. It is an incomplete explanation, even as it is pertinent.

Our domestic scene, first dominated by Eisenhower's military industrial complex, and now seemingly dominated by what might be called a social engineering complex, has effectively seen to it that people stay on the home planet. Our public education system is capable of encouraging prodigious talent, but public opinion polls would seem to indicate that our public is far less well educated than it probably should be, were we to implement off-planet exploration in a serious fashion. Our Congress is caught up in Machiavellian political shenanigans which do not serve the general welfare except in a limited sense, and cannot find in itself the ability to cooperate in achieving "a more perfect union".

While we have many blessings in America, the subject at hand is the stagnation of our space efforts, and this is partly caused by the willful discarding of our various nascent programs that Dr. Spudis touches on in his first paragraph, which seems in turn, to me, to be caused by the various "complexes" that I just mentioned.

Speaking metaphorically, Epimenides was correct: "All scientists are liars". Science will not allow a higher purpose of any sort to be created. It is a fundamental flaw in the scientific method. The fatuous quote points to this flaw. Currently, there are only two profitable activities in space: satellites and tourism; science is an expense. As for real wealth, the mineral deposits are surely out there, awaiting the eventual manufacturing ability to be realized. Exploration is simply, truthfully, and finally, not scientific at all. The scientists are experiencing the thrill of power at the controls of our space efforts, but they lack the ability to set space policy, even on a rational basis.

After forty years of trying, they have been unable to generate a profit. This is what we the people, or should I say, we the shareholders, have been asking for all along. The space program cries out for new, honest leadership.

We should go back to stay; we should build incrementally; we should not waste our efforts; we should commercialize. We have been waiting too long to purchase our Pan Am tickets to the Moon. It is indeed our right.

Comment by John Fornaro — January 27, 2010 @ 12:00 pm

5. John, there's no doubt in my mind that this country would be much richer and maybe even happier today if we had established a permanent base on the Moon right after the end of the Apollo program!

While I'm no fan of the Ares I/V architecture, I hope the Congress rejects any idea of not returning to the Moon!

The idea that developing the Altair lunar lander is exorbitantly expensive is pure mythology, IMO. In fact, what really inflates the cost of the Altair is the delay in fully funding its development. The lunar module only cost $11 billion to develop in today's dollars. Over a 7 year development time period, there's no doubt in my mind that we could fund the development of the Altair lunar lander at less than 2 billion a year.

Comment by Marcel F. Williams — January 28, 2010 @ 5:34 am

6. Hi Paul,

Great article. I have been thinking about your space rationale most of the week as I read the many stories prophesying the end of human space flight. It certainly appears to me that the VSE as you envision is gone and to be replace by a far future program that does not have sustainability as its core tenet.

When George Bush announced the VSE 6 years ago I was excited. When I later read the Commission's report to the President on US space exploration policy (of which you were a member) I was doubly excited because in those pages I saw the foundation of a feasible exploration program, not just a science program – exploration not strictly for or about science, but exploration enabled by science. I was so dammed excited I talked about that report for years with my students in a space exploration course I teach.

But I also remember other thoughts from 2004: 'Time is marching on, other initiatives have failed, this VSE is America's last chance. If it doesn't stick, it's over.' I certainly had no concrete evidence to support that thought, but it was there. And now, 6 years on, it does not appear that the VSE has stuck. Today I feel not pessimistic, but fully and rationally convinced that beyond LEO human space EXPLORATION is gone, maybe not forever, but for a very very very ....very long time.

Comment by Phil Backman — January 29, 2010 @ 11:59 am

7. Phil,

Thanks for your remarks. I will not yet despair; for one thing, Congress has yet to weigh in on NASA's future.

But beyond that, someday, somehow, some-one — a country or a corporation — will go to the Moon to learn how to use its resources. It just makes too much sense not to. It may not be the United States and it probably won't be NASA, but somebody will do it.

Comment by Paul D. Spudis — January 29, 2010 @ 12:35 pm

8. Paul –

You are right on with this and your previous post on the "flexible path" to nowhere. Going back to the Moon is so obvious. It's right there – two days away! Ready to be explored, exploited, mined etc. I can't believe it's rejected by people who should know better. Meanwhile we're throwing away the Shuttle for commercial LEO flights. What madness!

Comment by Greg — January 29, 2010 @ 9:33 pm

9. [...] in space as its #1 objective. The best arguments I have seen in that direction are the ones on Paul Spudis's blog. When you make that the rationale, the moon becomes a required destination, not an optional [...]

Pingback by Who Hung the Moon? » Blog Archive » There is No Santa Claus; Is There an Enterprise? — January 30, 2010 @ 12:58 pm

10. Paul,

This is an interesting and much needed article. The comments from your readers have also been excellent. The notion of exploration in our contemporary society does indeed need to be revisited.

Your quote, "NASA describes itself as a space exploration agency," gets to the heart of the matter. Regardless of how NASA describes itself, NASA's legal charter – The National Aeronautics and Space Act of 1958 – does not actually describe NASA as an exploration agency. Nor does The Space Act establish a space exploration agency.

I wrote a paper for the AIAA Space 2008 Conference concerning changes to the Space Act that may be required if NASA is to be a true space exploration agency. If you wish, I can send you a copy for review. The reference for the paper is AIAA-2008-7718-402: "From NASA to a National Space Exploration Administration."

Regards,

Comment by Arthur M. Hingerty — February 1, 2010 @ 1:18 am

11. Arthur,

Thanks for your comments. I would be happy to look at your paper.

Please send it to the e-mail address shown at my web site: http://www.spudislunarresources.

com/index.htm

Comment by Paul D. Spudis — February 1, 2010 @ 5:52 am

12. Right on! Every organism, organization, or system must bring in more resources and energy than it expends, in order to thrive and grow. Exploration needs to contribute to society in order to be sustained. Unfortunately, the US signed the 1967 Space Treaty that essentially prohibits anyone from making a profit from space development, and the de facto acceptance of the wording of the Moon Treaty, that the resources of space are the heritage of all mankind, currently presumes that any profits made by a private commercial enterprise on the Moon belong to the UN, and require that any such enterprise create an exact copy of their facility to give to the UN, along with all their research. Under those circumstances, no private organization would or should spend any resources in developing a Moon based facility.

A farmer doesn't plant a kernel of corn in order to only get one kernel back. They expect to get at least 3 cobs of corn on each stalk, with about 30 rows of 40 kernels each – or a return of 1200 percent. When Queen Isabella loaned the money to Columbus, she looked him in the eye and told him to bring her ships back filled with loot, or she would find him. So when his men tried to get him to turn back in the middle of the Atlantic, he really couldn't turn back empty handed. Exploration can not exist as an altruistic fantasy.

NASA, as a government agency, can not take the next step from research and exploration, to development and exploitation. And the current international treaties practically prohibit anyone else from doing so, either.

Russia, China and the US governments aren't bothered by that, either suspecting that if they do develop a facility on the Moon or an asteroid that produces a return, that their military can tell the rest of the world to pound sand, or ignoring the question entirely, for the same reason. A private commercial facility doesn't have that ability.

The non-space-faring nations are comfortable with the current situation, expecting that they don't have to do anything, and if someone else does produce a return with a Moon based facility (or any "celestial body"), the UN will make them share it.

So, what needs to happen, is that something political needs to change, so that all the nations of the world will not be left out of space development, while at the same time providing a rational property ownership regime that allows private (or public, I don't care) commercial development to reap the returns they can from investing in development.

I propose that we cede legal ownership of the Moon to the UN, with the proviso that every nation on Earth be granted 10,000 square hectares of Moon property in perpetuity, without them having to invest anything, but they can not sell their rights to it. They could lease it to anyone who would want to develop it. The Moon is big enough, each nation could have 10,000 square hectares on the near side, and 10,000 square hectares on the far side. Conflicts of property wanted by several different countries to be decided by lot.

Further, a "World Heritage Site" be declared of 1,000 square hectares around any equipment or probes currently landed on the Moon, to be undisturbed in perpetuity.

Further, the UN agency in charge of this Moon property be instructed to sell (with proceeds going to the UN) remaining property on the Moon in some sort of "Homesteading" type arrangement, at a nominal fee (something like $1,000.00 US per hectare) to any person or organization willing and capable of launching and maintaining a functioning facility on their property on the Moon, in lots not to exceed 5,000 square hectares.

This would protect the interests of the nations of the world, and provide for development of space resources by those interested in doing so.

Comment by Harmon — February 1, 2010 @ 11:31 pm

13. Harmon,

Unfortunately, the US signed the 1967 Space Treaty that essentially prohibits anyone from making a profit from space development,

Actually, it says nothing of the sort. It only says that no nation can claim sovereignty over another extraterrestrial body.

I propose that we cede legal ownership of the Moon to the UN

Yes, they've done such a bang-up job with everything that they've ever touched.

Comment by Paul D. Spudis — February 2, 2010 @ 8:17 am

14. Well, with the cat out of the bag Paul, I expect you will be writing about what President Obama's decision for the future course of NASA means. I appreciate your latest discussing what exploration means. This is one reason why have quit using space exploration as being synonymous with human spaceflight. The term is used to broadly and its meaning has been corrupted.

I just cannot help but feel the FY2011 budget proposal is a step in the wrong direction. With no market to support commercial development, commercial human spaceflight will fall flat on its face. Space tourism alone cannot generate enough revenue. I would have been encouraged if there had been a more specific plan for expanding the market in LEO, and a detailed program to develop a heavy lifter whether through commercial or NASA's design bureau.

Comment by Gary Miles — February 2, 2010 @ 10:11 am

15. This article has provoked me to do a little historical research. I highly recommend the official history of the US Geological Survey (who incidentally trained the Apollo astronauts; Harrison Schmitt was initially employed at USGS; NASA may have got the astronauts to the Moon and back safely, but when those guys were actually walking around "doing science" it was a USGS gig). The history of exploration by US federal and state agencies has always had an economic focus, ever since the Coastal Survey of 1803 was formed. The idea that NASA should only be involved in "pure" science is a brand new idea--even within NASA itself. After all, what is the purpose of the aeronautics branch of NASA if not to develop new designs that will ultimately be of some practical value.

http://pubs.usgs.gov/circ/c1050/index.htm

And as for grandiose engineering projects (like would be required to create a substantial Moon base), I refer interested readers to review the history of the Bureau of Reclamation. Because of the giant dams that Burec built, they are the largest wholesalers of freshwater in the United States, and the second largest seller of hydroelectric power in the western United States. Therefore, to say that it's not the federal government's business to, for example, develop the techniques that could make possible large scale space based solar power stations using lunar resources: well, that's an interesting philosophical standpoint, perhaps, but it certainly has no grounding in American history.

Comment by Warren Platts — February 6, 2010 @ 11:06 am

16. Science is, in general, a laudable enterprise. However, it is engineering that brings the abstract results of science into concrete realizations that can benefit everyone. The former without the latter is little better than the scholasticism of the Middle Ages--interesting for the practitioners, though susceptible to erroneous conclusions, while being of little value to those who produced the food, clothing, and shelter of the scholars.

I have wished for a more robust effort at space development for many years. In fact, I think it is imperative for anyone who does not so despise their fellow man as to be content with our extinction and the consequent loss of all human achievement when the next large asteroid or comet impacts the Earth. (In fact, it may be a very near thing if we should experience terrestrial disasters on the order of another eruption of the Yellowstone caldera.)

Fortunately, such disasters are unlikely to occur in our lifetimes or even within in a few centuries of now, but we can not know for sure. Moreover, it is certain that nothing more than possibly launching a few "time capsules" into space in a vain hope to leave some kind of legacy will be possible given the continued lack of a sustained effort to expand human habitation beyond the Earth.

With that in mind and with the intent to provide positive rewards to help motivate some effort, I am in the process of establishing a prize fund that will eventually receive the bulk of my estate as well as any donations that other interested parties may wish to make once the required legal entities have been established and the necessary registrations are completed. The fund itself will ultimately depend on growth from investment in equity index investments before the potential rewards will be of any real consequence. However, there will be plenty of time for that, especially given the current pace.

In brief, the fund will provide the following awards for the stated milestones:

1. 50% of the current value of the fund to the organization that first establishes and occupies a permanent base on the Moon.

2. 50% of the current value of the fund to the organization that first establishes and occupies a permanent base on Mars.

3. The balance of the fund to the organization that first sends a manned space mission to another solar system payable once the mission travels beyond the heliopause.

Obviously the last milestone will be a very long time into the future at the current rate, but the "miracle" of compounding should ensure that the award is a very substantial, inflation-adjusted sum commensurate with the expense of the endeavor.

The fund is the Boundless Frontier Fund and you will be able to learn more about it and track progress by visiting: http://www.boundless-frontier-fund.org/

Of course, there is nothing much to see yet and the fund has not yet been seeded. First, the lawyers must be paid and the government requirements satisfied. Still, stay tuned. Hopefully it can be at least one positive incentive in contrast to the widespread apathy (as well as some very real hostility) with respect to manned space exploration and pioneering.

Comment by Ned Nowotny, Boundless Frontier Foundation — March 11, 2010 @ 4:43 am

Vision Impaired

February 3, 2010

The administration proposes; Congress disposes

The release of the new proposed budget for NASA[1] has unleashed a blizzard of news articles and commentary. The administration proposes to terminate Constellation[2], the agency effort to design and build a new space transportation system to carry people to low Earth orbit and beyond. In its place, they plan to let contracts with several companies to provide orbital launch and spaceflight services, both as transport to ISS and to "destinations beyond LEO." This major change in the agency's business model follows in the wake of last summer's Augustine committee report[3], which concluded that NASA's "program of record[4]" to return to the Moon and beyond was inadequately funded and possibly, misdirected as well.

The Augustine "Flexible Path[5]" was an architecture designed to take people beyond LEO, but to low gravity targets: L-points[6], near-Earth asteroids, and Phobos and Deimos, the asteroid-like moons of Mars. The idea behind that concept was two-fold. First, it was a way to send people into deep space without the very high programmatic expense of developing a lunar landing spacecraft. Given that Constellation is significantly over budget, cost control is certainly an issue. The second motivation for FP was the feeling (not explicitly stated in the report, but clearly implied) that the agency plan for lunar return was largely a repeat of the Apollo experience of 40 years ago. The strength of this impression varied among the committee members, with some thinking that the chosen architecture was simply the wrong approach while others questioned the value of going to the Moon at all. The new proposed budgetary direction seems to follow the Augustine Flexible Path (FP).

I have previously discussed[7] what I perceive as the most significant problem with FP, namely, that it is activity without direction. The administration's budgetary version of this path[8] confirms this perception. Much verbiage is thrown around about multiple missions to all sorts of destinations, blazing new trails with new technology, trips to Mars that last weeks instead of months, and "people fanning out across the inner solar system, exploring the Moon, asteroids and Mars nearly simultaneously in a steady stream of firsts[9]." But nowhere in the budget documents or agency statements is there anything about the mission that we are undertaking. So we're going to an asteroid. What will we do there? Why are we going there? What benefit accrues from it?

The Vision for Space Exploration (VSE)[10] of 2004 not only laid out a clear path, but also described exactly why such a path was being taken. It is not a repeat of the Apollo experience. We go to the Moon to learn how to create a sustainable human presence in space[11]. We do this by experimenting with and learning to use the material and energy resources of the Moon to create new space faring capability. These skills enable us to build a space transportation infrastructure that allows routine access not only to the Moon, but all of cis-lunar space (where our space assets reside) and the planets beyond. All of this activity is to be accomplished under the existing budgetary envelope; as there is no deadline, we trade time for money.

Many conflate the VSE with Constellation, the agency's program to build the Ares launcher and Orion spacecraft, but they are different and distinct. The former is a strategic direction; the latter is an implementation of that direction. This is not some academic distinction; it goes to the essence of the current debate about NASA and the space program. Virtually all of the argument and debate about our future in space has been about means rather than ends[12]. Launch vehicles, spacecraft, and architectures have been grist for the mills of the space blogosphere. Beyond a vague notion that people should "move into the Solar System," the purpose and meaning of that movement has been articulated much less often. Partly that's because different people have differing notions of what those motivations should be – science, settlement, curiosity, and technical innovation all have their adherents. But if you do not clearly understand what your mission is, you are not likely to successfully implement it.

The VSE was a clear strategic direction[13]. It not only identified the path forward, but also the specific activities that would enable that path to be followed. The new budget outlines the means (new commercial launch and transport) but not the object of our space program. But more critically, it discards the clear and practical direction of the original VSE. Before the new budget, we knew exactly where we were headed and why: a return to the Moon to learn how to live and work productively on another world. Now, all we know is that at some point in the future, we will somehow go somewhere to do something. Or other.

I wrote recently[14] about a variant of the Flexible Path architecture outlined at the blog Vision Restoration[15]. I think that this approach

has a lot of merit, but suggest one critical modification: it does not have a statement of the mission. The VSE in its original guise should be stated up front and made a clear and unalterable part of the architecture. If during the course of the program the implementation somehow falls short, change the implementation, not the mission. The failure to do this in the Constellation Program led us into a blind alley of cost and schedule overruns, the Augustine committee, and now, cancellation.

This new policy will increase NASA's natural tendency to engage in organizational "Brownian motion." We are already seeing agency leaders call for new studies to determine what will be done at the (so far unspecified) new destinations. The current program looks upon itself as a transportation architecture; the activities undertaken at any given destination are irrelevant. The new "direction" outlined in the budget request is similarly focused on means (e.g., commercial launch and transport) rather than ends (e.g., What will humans do at Earth-Sun L-1?). And it will likely come down the same path, as indeed it appears to be starting to. NASA as an organization will adjust to this; after all, viewgraphs are easily changed and mission studies easily re-written. But what about the aerospace industry? They find it very difficult to pivot on a dime when the direction changes.

I've often written about how I think the VSE ought to be implemented[16] and have found the existing program of record wanting in several respects. But at least it aimed in one direction. We need a program plan that gets us beyond LEO using small, incremental, cumulative steps and the new model promises to do just that. But small, incremental steps taken in random directions yield uncertain progress.

What is your mission? It's not just the most important thing; it's the only thing. NASA forgot that during the last 6 years. Now, the White House has joined them.

Topics: Lunar Exploration, Lunar Resources, Space Politics, Space Transportation, Space and Society

Links and References

1. proposed budget, http://www.spaceref.com/news/viewpr.html?pid=17398
2. Constellation, http://www.nasa.gov/exploration/news/ESAS_report.html
3. Augustine report, http://www.nasa.gov/offices/hsf/meetings/10_22_pressconference.html
4. program of record, http://www.nasa.gov/exploration/news/ESAS_report.html
5. Flexible Path, http://www.nasa.gov/offices/hsf/meetings/10_22_pressconference.html

6. L-points, http://en.wikipedia.org/wiki/Lagrangian_point
7. previously discussed, http://blogs.airspacemag.com/moon/2009/12/16/arguing-about-human-space-exploration/
8. budgetary version, http://www.nasa.gov/pdf/420990main_FY_201_%20Budget_Overview_1_Feb_2010.pdf
9. steady stream of firsts, http://www.nasa.gov/pdf/420994main_2011_Budget_Administrator_Remarks.pdf
10. VSE, http://www.spaceref.com/news/viewpr.html?pid=13404
11. go to the Moon, http://www.washingtonpost.com/wp-dyn/content/article/2005/12/26/AR2005122600648.html
12. means rather than ends, http://www.spaceref.com/news/viewnews.html?id=1349
13. clear strategic direction, http://www.spaceref.com/news/viewsr.html?pid=19999
14. wrote recently, http://blogs.airspacemag.com/moon/2010/01/23/beyond-leo-flexible-path-revisited/
15. Vision Restoration, http://restorethevision.blogspot.com/
16. ought to be implemented, http://www.spaceref.com/news/viewnews.html?id=1349

Comments

1. Excellent article. I know you and I do not agree on the path NASA took with Constellation in the past, but now there is simply no path at all. The administration has taken to comparing this new shift to emphasizing commercial spaceflight to the rise of commercial aviation.

My question for them is what existing market is available to support human spaceflight beyond LEO? Commercial aviation was able to advance due to the available of the existing mail distribution contracts offered by the US Post Office. Furthermore, the market in LEO is limited to ISS and the FY2011 has provided no program or insight on how to expand that market.

Without clear program and mission objectives, NASA will become very vulnerable to budget cuts in the future. This means that robotic missions will also likely be cut as well. I cannot imagine that the new proposal is a win-win for anybody. Not even commercial spaceflight.

Comment by Gary Miles — February 3, 2010 @ 1:45 pm

2. Since President Obama's budget doesn't include any funding for the development of an HLV, only research in the possible development of an HLV, there really is no Flexible Path program.

Bolden clearly didn't want NASA to return to the Moon because he wanted us to go to Mars. But I believe that when he finally realized that we simply don't have the technological know how to protect astronauts from the dangers of galactic radiation and solar events, he decided to turn NASA into an R&D program to get to Mars. This decision is going to set us back a decade, IMO.

Once some other nation, or nations, start setting up permanent settlements on the Moon, then our politicians will finally realize the full political and economic consequences of our decision to abandon the Moon!

Comment by Marcel F. Williams — February 3, 2010 @ 8:39 pm

3. You nailed it Paul and very eloquently. However concerns with the New FLEX based NASA go well beyond the lack of vision and inspiration. The New Commercial NASA is laden with old NASA baggage and wasteful investments. A six billion dollar bribe to gain commercial and private space buy-in. Multi-billion dollar shuttle derived HLV (bridge to nowhere) bribe for southern politicians ensures jobs for votes continues. And the multi-billion dollar ISS boondoggle orbits endlessly. Let's not forget additional missions and spending to support "Climate Gate". Lots of smoking mirrors and amidst all the cheers and hoopla. CCDev could provide some of the focus and direction for incremental VSE related technology development. However CCDev receives mere funding crumbs and is overshadowed by the LEO based funding support Space-X appears destined to receive for ISS supply.

Comment by Doug Gard — February 4, 2010 @ 12:01 am

4. Loved the article, as usual.

If there's no declared mission other than to do research, then it's very hard to fail. A politician who announces a singular goal might then have his or her progress measured with respect to the achievement of that goal. If things don't go well, then that could be bad for the politician.

Solution: Get rid of the goal!

Your comments on Brownian motion are spot-on.

Comment by Itokawa — February 4, 2010 @ 8:56 am

5. In your previous post, you stated:

"What is needed is the incremental, cumulative build-up of space faring infrastructure that is both extensible and maintainable, a growing system whose aim is to transport us anywhere we want to go, for whatever reasons we can imagine, with whatever capabilities we may need."

The new Obama/Bolden direction appears to be moving in exactly this direction; it does so much more than Constellation did and with one additional feature, by opening the development to 'commercial' firms, it gets the nation out of the 'solely NASA' inability to provide development experience, which BTW NASA has very little of any longer at least in its existing managerial ranks.

A few years ago, Jeff Hanley Constellation Program Manager, stated that "we no longer need a Vision; we now have a program". In many respects that was where NASA went wrong.

At least for movement and transport of people in space, including cis-lunar space, developing the technologies and infrastructure that establishes the capability to "transport us anywhere we want to go, for whatever reasons we can imagine" does not require significant changes in hardware from one destination to another.

It sounds like in your latest post, you are now reversing your previous statement and saying, what we really want is a definitive program of destinations.

There are additional elements that need to be considered. One is the budget.

When the Bush Vision was established, the discussion was that the NASA budget had been stable at approximately .5% of the annual US budget for 40 years and that NASA needed to build a program that fit within this constraint. There would be no significant increase in funding.

The previous NASA Administrator figured he could raid all of the other NASA programs, including aero, science, shut down ISS, etc, in order to redirect the funds to Constellation. The Constellation management and charter also sought to 'take over' almost all of NASA's efforts beyond low earth orbit.

Yet, as you have pointed out, they selected only one mission implementation, which was an Apollo redux.

As many others have pointed out, Constellation forgot about technology development. Constellation management also chose to neglect that their vehicle designs and architecture really did not accommodate some of the missions they were pursuing. The Administrator said that Orion was going to be a Mars transit vehicle. But the spacecraft was not large enough to accommodate that size crew for a Mars length mission. It did not have the living space, facilities, or logistical/storage space for deep space missions. Constellation management conveniently changed their safety standards or neglected human ratings when it served their interests. This was something we'd learned to expect of certain of their managers on ISS, where they did exactly the same thing.

Orion wound up without the systems redundancy that should be expected of a deep space vehicle. Next thing we knew, Constellation was showing Orion capsules traveling in tandem. Two would provide more living space and redundant systems.

Another aspect that deserves consideration and that is how poorly Constellation was being implemented. It is interesting to read from prime contractor Lockheed Martin by Thursday this week was announcing that Orion would be ready to fly in 2013. L-M has been working the Orion vehicle design since 2005, 5 years, yet has not gotten through preliminary design. Preliminary design is generally 1/5 to 1/4 of the development

program. If preliminary design has taken this long, the anticipation is that getting to initial operational capability is another 10 years. Augustine said maybe the people involved will figure out what they are doing and the situation will improve so Orion might fly as early as 2017, though 2019 is a better bet. This is important because it means we have little confidence that the existing Constellation/Orion management will do better than they have done until now. I have no doubt that this is the reason that the new Obama/Bolden plan opens the program to industry. We need to seek support and leadership from a broader community, some of whom have the appropriate abilities in systems development.

I don't think that the new plan precludes what you are now saying we need, which would be a schedule of destinations and capabilities. I think the first step in the process is to recognize that Constellation had not done this, which is what you have been saying for several years. The second step, about to take place is to establish the schedule and plan as best can be done, and figure out how to go about implementing it. NASA Headquarters are saying that many of the Exploration people will simply redirect their function towards establishing the technology development program. I think if NASA is going to do this, they need to get out of the 'Mission' mode, which is exactly what you have been saying.

Maybe the NASA Administrator and POTUS are paying attention to what you have been saying?

Comment by Guest — February 5, 2010 @ 5:56 am

6. Guest,

Thank you for your thoughtful comments.

It sounds like in your latest post, you are now reversing your previous statement and saying, what we really want is a definitive program of destinations.

Hardly. My previous statement was about means; in order to go beyond LEO, we needed an incremental, cumulative program that could be paid for as we established these incremental capabilities. But increments with no direction get you nowhere. The Moon is a destination, but more importantly, it is an enabling asset. By using lunar resources, would learn how to "cut the cord" with Earth of space logistics.

Your point about the problems with Constellation are well taken and I myself have pointed out that we were focusing on the wrong things. Hanley's comment that you quote, if accurate, reflects the basic problem with it. But Constellation was an implementation of the VSE and if your implementation falls short, you should change it, not your objectives.

My criticism of the "new path" is that it has abandoned ends; it has no destination, no activities at any destination, and no obvious benefits from it, except for a "nearly simultaneous steady stream of firsts" as Bolden put it. In other words, the object of human spaceflight is to be a series of flag-and-footprint stunts, then abandonment of that destination for a new, further destination and a new "first" for flags-and-photo-ops.

My previous comments about the "go anywhere, at any time" is a statement of the ultimate objective of human spaceflight; I also said that we are many years away from such a capability. The VSE made lunar return a means to that end, by using the Moon's resources to create a cislunar space faring system.

The final problem with the new path is organizational. NASA as an entity doesn't do well without specific direction. They barely function WITH it. The "new path" is primarily a money distribution program for new widgets. As there is no overarching goal, there is nothing to tie together the technology development. It is a prescription for another 20 years of busy work, pointless studies and viewgraph engineering.

And lots of "management."

NASA Headquarters folks are saying that many of the Exploration people will simply redirect their function towards establishing the technology development program. I think if NASA is going to do this, they need to get out of the 'Mission' mode, which is exactly what you have been saying.

That is the antithesis of what I've been saying. "Understand your mission" is not the same statement as "you shouldn't have a mission."

I tried to clarify my position at these two links:

First Nail Down the Mission (Chapter 25)

Objectives Before Architectures – Strategies Before Tactics http://www.spaceref.com/news/viewnews.html?id=1349

Comment by Paul D. Spudis — February 5, 2010 @ 9:34 am

7. Paul,

You have it exactly right. Without the anchor of the VSE there is no yardstick to measure progress in space. Plans have to be anchored in a specific objective. Otherwise they just drift.

Advances in Technology are always the result of demand pulling the technology forward. That is why the English and not the Greeks invented the steam engine. The Greeks had no demand for the curiosity of steam while the English had coal mines to pump out. The technology of sailing vessels didn't develop in the 1400's until Prince Henry set a goal of reaching India. Then ship building technology saw rapid development.

The new policy is very much a repeat of the 1990's when the Space Exploration Initiative was replaced by CATS and Administrator's Goldin's plan to fill the sky with X-vehicles for SSTO.

It also recalls the space policy decision in the 1970's to build the Space Shuttle itself in the belief that "good things" will happen when the cost to space is lowered. The result was a great spacecraft that was always looking for a mission.

I hope this will turn out different but the track record shows the probably is small.

Comment by Thomas Matula — February 5, 2010 @ 1:57 pm

8. I work at KSC where we feel like we were just sucker punched in the gut. It's funny how NASA liked to brag that they were keeping the workforce informed, yet when Cabana talked right after the announcement of the cancellation of Constellation, he stated that they have been working on this for a year and a half!

Comment by sucker punched — February 6, 2010 @ 7:07 am

9. sucker punched,

Thanks for your post. I am hearing many similar stories from all around NASA about exactly this situation. Interestingly, I have heard that many senior managers at NASA were also kept in the dark, including some at HQ. It looks like the "new path" was formulated by a relatively small group inside the administration.

If the budget roll out presentation is any guide, this new paradigm has been given very little serious and considered thought — Amateur hour lives again.

Comment by Paul D. Spudis — February 6, 2010 @ 8:13 am

10. Sucker Punched,

Sad to say but it sounds like NASA has once again given me a good example to use in my Strategic Management courses of how NOT to develop and implement a strategic plan....

Comment by Thomas Matula — February 6, 2010 @ 12:45 pm

11. *But nowhere in the budget documents or agency statements is there anything about the mission that we are undertaking. So we're going to an asteroid. What will we do there? Why are we going there? What benefit accrues from it?*

That's an excellent question, and it's one that the Augustine committee didn't address very well about human space flight. But you know, you can ask the same question about the Moon. The Moon may well have value in resources, but it is hardly obvious that it is necessary to send humans there to get them, especially since the cost to get even the first ones there is $100 billion.

Comment by Doug Lassiter — February 6, 2010 @ 4:07 pm

12. Doug,

The Moon may well have value in resources, but it is hardly obvious that it is necessary to send humans there to get them

It may not be "obvious," but on reflection, anyone having significant robotic systems experience will attest that things break, get stuck, jam and quit working. You need someone there to prod it back into action. A system for lunar resource harvesting will likely have many distinct steps, processes and types of equipment. A purely robotic system may be feasible eventually, but as we are just learning how to do this, I suspect that human intervention will be necessary from time to time.

None of this should be construed to mean that I am against robotic resource prospecting, harvesting and extraction missions. As precursors, they can be invaluable in sorting out how we want to design the actual production systems.

And I do not except the agency and Augustine wisdom that human lunar return costs in excess of $100 billion. They both view spaceflight through the prism of the Apollo mindset.

Comment by Paul D. Spudis — February 6, 2010 @ 6:41 pm

14. Great article. I agree that most viewed the VSE and Constellation program as one in the same. To me, Constellation was the chosen means to achieve the VSE with the Moon selected as the first incremental step. Canceling Constellation shouldn't mean canceling the VSE.

What I find most also miss is that one of the VSE's goals was to expand the global economic sphere to the moon. It wasn't the primary goal but was always considered in any Lunar Exploration meetings I've attended. Maybe this should have been emphasized more. Obama's new budget claims to consider the promotion of jobs as priority #1 but as I see it, the VSE was doing the same by laying the groundwork for an expanded global economic sphere encompassing cis-lunar space.

I believe Dr Spudis wrote the following in his 'Objectives Before Architectures" article....

"NASA has always seen public indifference as a problem. I believe it is a great opportunity. If we are freed from the self-imposed requirement to "excite" people, we are free to carry out the more mundane task of laying the groundwork for a lasting, sustainable infrastructure in space. In other words, space ceases to be a modern gladiatorial contest and becomes instead a piece of our economic infrastructure, available for use by many different parties and not restricted to only

scientists and cold warriors. An analogy might be with railroads. Most people couldn't care less about railroads. Yes, they are aware that they exist and many people will read and know about the latest derailment disaster. But they don't think about them unless circumstance compels them to. Yet for all this, most recognize the importance of railroads to our lives, our well being, and our economy."

I think our goal should be to make our program immune from the sorts of political games that go on whenever a new administration comes in. We had a plan, now let us do our jobs.

Comment by gsfc guest — February 6, 2010 @ 9:44 pm

15. Guest,

Your comments are quite articulate and interesting. If I may, I'd like to offer the following perspective. Without an overall Vision driving us at NASA, we will be limited in the following ways. First, without a driving end goal (i.e., Vision), our selection of Robotic missions and flight Tech Demos will be based upon only the decision maker's biased idea of what is important, rather than a more thought out Vision or strategy THAT HAS BEEN VETTED AT THE HIGHEST LEVELS ABOVE NASA. Most likely, those decisions will be made for political reasons, or to balance workforce in the absence of strong goals set forth in a Vision. In fact, even when we had the VSE, NASA had a very difficult time sticking to the intent of that vision. I am not saying that the senior leadership will do bad things, only that the pressures and influences on them based upon real world issues will influence their decisions away from more "noble" goals, especially if there are none formally.

Secondly, you should realize that leaving the Earth's gravity well is pretty hard. If you haven't seen a major game-changing breakthrough in launch technology, it is not likely to improve any time soon; after all, demand has been there for tens of years. A big LOX-RP engine is good, but it is not going to cut launch cost significantly. We can spend money looking for incremental improvements, but it won't lower the cost all that much in my opinion. If the country is willing for NASA to take a little more collective risk, we can address the culture of launch preparations to reduce costs today; we don't need to wait for new technology to support a HLLV project. My point is that we can develop a heavy lift capability now. If we wait, we can get incremental improvement, but when will we know how good is good enough? To answer that, one would ask, "what's my need? What do you want to do with it?"

Thirdly, potential game changers like propellant depots have the same question of Ends. OK, I don't think it is really that hard to develop automated fluid transfer, but how big should the depot be? What is the Ops Concept? When do we say that we can do it? What does it look like? We can demo the capability at ISS in fairly short order, but the question is, what do you want the depot to do specifically? What is its mission?

Paul has consistently articulated his vision, and from my perspective, it makes total sense. I hope that we can influence NASA leaders and through them, the Administration, that a VISION, coupled with this new shift, might be the best thing for NASA and the country. I'm OK with the new approach, although I think that the Administration is expecting a bit much from Commercial ability to deliver, but I'm good to try hard to see it work. What disheartens me is the abdication of any vision that we can believe in regarding where we go from here except, "take nice pictures and plant a flag and do human "exploring" (see Paul's earlier blog).

Comment by TonyL — February 8, 2010 @ 7:02 pm

16. Salutations. Project Constellation was, and could still be our one golden opportunity for leaving earth orbit, finally, after 40+ years. The Mars zealots killed this thing. Pure and simple. They have been completely inflexible with their demands for NASA's space future: "Mars first, Mars now, plus anywhere-but-the-Moon is okay by us." I loath their contempt for the efforts of putting astronauts elsewhere. What in heaven's name were they thinking, when they set out to finish off with lunar flights?!?! Without a clear out-of-earth-orbit mission, what possible motive would there be for a heavy-lift launcher? Do they really believe that some private commercial firm is going to single-handedly deliver to NASA a huge multi-stage rocket like Aries 5? This kind of La La Land dreaming will get you nowhere! Only a government agency can deliver such a Saturn 5-capability launch system. Commercial business will at best deliver us some teeny tiny space plane, plus a minor-league rocket to get it to the ISS. But that'll be it, folks! Without an earth-escape stage to get a spacecraft out of parking orbit, you basically CAN'T go anywhere! Constellation was all slated to bring us that: a heavy-lift, heavy payload launch system; which would've had plenty of applications—sending landing vehicles to Mars included. The Zubrinites just could not see beyond their hard-headed, dead-set goals to realize that the Orion-Aries architecture was just what the doctor ordered, or would've ordered, to later advance their agenda. Don't they remember Project Gemini, from the mid-60s, which tested out the bugs in the flight systems before committing to the bigger enterprise? Intermediate goals have always aided well, the subsequent objective. I say for all of us in the space advocacy community to make a strong effort, from this moment on, of lobbying & petitioning Congress into restoring funding for Constellation. We must try to re-start this program, even if it awaits being begun again in the next year's federal budget. Let's put on a vibrant campaign. LET'S SAVE PROJECT CONSTELLATION!!!

Comment by Chris Castro — February 9, 2010 @ 10:40 am

17. Excellent article and commentary. However, I disagree with you a bit on one fundamental point that I think merits discussion: the basic direction and purpose of the path laid out for space exploration.

The VSE had it partly right: create a sustainable human presence in space. However, I prefer to take a longer view (like on the order of millions of years ;-)). The proper long-term purpose of space exploration is to move the human race OFF of the planet Earth and to other planets and star systems, insuring (or at least increasing the odds a LOT) our survival as a species. Stellar evolutionary theory clearly indicates that our solar system will not be habitable forever…we have a time frame of about 500 million to 1 billion years to find new homes within or (more likely) outside our solar system. In order to accomplish this task (particularly the element of long-duration travel outside of the solar system), we should be learning how to maintain a permanent presence in space NOW at a moderate level of consistent funding.

The issue is not "news-worthy" and "pop-culture" "firsts"; the issue is constant incremental improvements and expansion in our space travel, transport, and exploration capabilities.

The OTHER fundamental problem is that this view plays very poorly with the political folks: the people who control the purse strings, and have just about as short-term a view as you cant get: 2 to 4 years to the next election cycle.

Comment by Lewis Van Atta — February 9, 2010 @ 8:22 pm

18. Lewis,

Thanks for your comments.

I see the original Vision for Space Exploration in perfect alignment with your point. If the long-term goal of space exploration is settlement, we must use the materials and energy that we find in space to survive. Because learning how to use lunar and space resources is the central goal of the VSE, it IS the perfect near-term goal that helps create long-term capabilities.

Comment by Paul D. Spudis — February 10, 2010 @ 5:06 am

Confusing the Means and the Ends

February 13, 2010

Our new space program. "You think that you hate it now... But wait till you drive it."

The release of the proposed NASA budget and new "direction" has led to an intense "cage fight" in the blogosphere over who has the best rocket and the best architecture. Many "New Space" advocates are ecstatic[1], viewing the cancellation of the Constellation program[2] as vindication of their view that: a) this was a stupid architecture to begin with; and b) the purchase of launch services by NASA[3] is more desirable than the development of same by the agency. In the other corner, defenders of the existing program[4] and paradigm see human spaceflight as still largely an experimental activity and that by contracting for launch services, astronauts' lives will be put in danger, leading to the eventual termination[5] of America's human spaceflight program. Both sides are locked in a fierce battle over the ownership of the "how," while seemingly unconcerned as to the "why" or the "what" they are fighting for.

Once again the debate focuses on launch vehicles, the need or lack thereof for a heavy lift vehicle, and all the wonderful new technical development and leaps forward possible once NASA is freed from its responsibility to build and operate a space transportation system. I agree with the New Space people that alternative options for launch and orbit are desirable and that a flexible, extensible architecture is the way to move beyond LEO. On the other hand, I agree with the "Ares huggers" that this change will not result in the space utopia[6] its advocates promise and that an agency saddled with an unworkable approach is a ripe target for elimination.

Those cheering the new path should step back from their celebrations, take a sober look at the landscape and ask themselves, "Now what?" The "new path" has no mission. Despite what many believe or have said, Project Constellation was not same thing as the Vision for Space Exploration (VSE)[7]. Constellation was the implementation that NASA chose to carry out that mission. The VSE was both a set of destinations and a group of specific activities at those locations. The Vision's objective was to give us new spaceflight capability by learning to use the material and energy resources of space, first on the Moon and then from other objects in space.

The new policy indicates a lack of understanding of the difference between "means" and "ends" within both NASA and the current administration. When they cancelled Project Constellation, the Vision was terminated as well. And what was put in its place?

Nothing.

All of the current hand wringing and angst is focused on which rocket and spacecraft to build. But to what end? The "Flexible Path" concept came from the Augustine Commission[8]. It's main focus was to find an affordable way to move people beyond low Earth orbit. Using their concept, we would visit places beyond low Earth orbit that had very low gravity – libration points, near Earth asteroids, and the moons of Mars. The supposed advantage of such places is that they do not require a large propulsive maneuver to land on (or more accurately rendezvous with) them. Thus, the supposed enormous cost of building a landing vehicle is saved.

The "new path" called for in the budget envisions a government funded and commercially built and operated space launch system, freeing NASA from the necessity of building rockets. The agency would "invest" in new technology. Somehow, these new and wonderful approaches will lead to the spontaneous generation of a space faring infrastructure capable of taking us beyond LEO into the Solar System – anywhere and everywhere. But to do what?

NASA Administrator Charles Bolden seems to think that a return to the Moon should be ruled out because "there are already six American flags there[9]." It is hard to imagine that he believes that the purpose of space exploration is to plant a flag and move on to the next destination. Such a template will exhaust possible destinations quickly. If the goals of travel beyond LEO are more significant than that, what are they? What will people do at an asteroid? What do we get from such a trip? What capability does it create? What are we buying? Again, the "means" and "ends" argument attempts to focus on outcome.

We had a well considered and well crafted strategic direction in space – to go to the Moon and use its resources (which we now know are even more abundant and accessible than we thought) to create a new transportation system[10] that will reduce costs and increase access to cislunar space. That mission was not just the proposal of the former President; it was endorsed by two different Congresses (in 2005 and 2008), under the leadership of different parties, and both times, by huge bipartisan majorities. The Vision for Space Exploration is our national space policy and will be until the Congress passes a new authorization bill, changing the mission and goals of the space program.

Currently, the proposed budget casts aside this hard-won, bipartisan policy and puts nothing in its place. This new policy is striking in that, rather than serving America's national security, economic and scientific interests, it undermines them. The "new path" was apparently put together by a very small group of people, without significant debate or input from outside sources. Whatever the circumstances of its genesis, it is poorly conceived; if it were well considered, we would know exactly where we were going, what we would do there, and what benefits would accrue from these voyages. The idea expressed by some in the blogosphere, that we will now be able to "go everywhere and do everything" is ludicrously naïve. Given the past performance of this agency (or any agency) given no direction, random motion is a much more likely outcome, at $20 billion per year.

If the current architecture is broken or unaffordable, fix it or change it. If getting NASA out of the rocket-building business is the right way to go, do that. But don't discard our strategic direction. The space program can survive a change in the business model of its implementing agency; it won't survive fecklessness and a complete lack of direction.

Topics: Lunar Exploration, Lunar Resources, Space Politics, Space Transportation, Space and Society

Links and References

1. ecstatic, http://spacefrontier.org/2010/02/12/the-battle-for-a-new-space-age-begins/
2. Constellation program, http://www.nasa.gov/mission_pages/constellation/main/
3. purchase of launch services, http://washingtontimes.com/news/2010/feb/12/obamas-brave-reboot-for-nasa/
4. defenders, http://www.washingtonpost.com/wp-dyn/content/article/2010/02/11/AR2010021103484.html
5. termination, http://online.wsj.com/article/SB10001424052748703382904575059263418508030.html
6. space utopia, http://blogs.airspacemag.com/moon/2010/02/03/vision-impaired/
7. VSE, http://www.spaceref.com/news/viewpr.html?pid=13404
8. Augustine, http://www.nasa.gov/offices/hsf/meetings/10_22_pressconference.html
9. six American flags, http://www.denverpost.com/headlines/ci_14385263
10. create a new transportation system, http://www.spaceref.com/news/viewnews.html?id=1376

Comments

1. A Flexible Path requires vehicles to travel beyond LEO. There is nothing in the Obama budget that funds the development of such vehicles. Sure there's money to test concepts, but no money to develop anything. So there is no Flexible Path.

And I still reject the notion that NASA should not have its own vehicles to access LEO and beyond. Why not! It doesn't make any sense for the Russians and Chinese to have a government presence in space and the US to have none.

Whose going to protect the interest of private US companies in the New Frontier? China and Russia?

There's no way the US military is going to allow other nations to dominate the New Frontier. Pushing NASA out of the manned space flight pretty much guarantees that the US military will fill in the US government vacuum. And for the US military, NASA's tiny manned space flight budget will look like chump change! The US military space program is already larger than NASA's budget. And with NASA out of the manned spaceflight business, I predict the military's space budget is only going to get a lot larger.

Comment by Marcel F. Williams — February 14, 2010 @ 11:15 pm

2. I just found this article on the cost of developing and operating a lunar base. Its very interesting:

http://csis.org/publication/costs-international-lunar-base

Comment by Marcel F. Williams — February 15, 2010 @ 3:48 am

3. Excellent article once again, Paul. You are the clearest thinker on these subjects.

Comment by Itokawa — February 15, 2010 @ 8:54 am

4. Paul, I can only half agree with you on this article, and your past musings on the new NASA budget. I totally agree that there hasn't been nearly enough discussion of objectives and the "why" of space exploration. However, I feel you have artificially married the VSE with Constellation.

The following quotes from the article show what I mean perfectly. In one paragraph you state:

"Despite what many believe or have said, Project Constellation was not same thing as the Vision for Space Exploration (VSE). Constellation was the implementation that NASA chose to carry out that mission."

Yes, the VSE is a set of objectives and Constellation is implementation. However, the *very next paragraph* you state the following:

"When they cancelled Project Constellation, the Vision was terminated as well."

What? You kill the architecture and the objectives disappear too? How did that happen?

However noble and correct the objectives may be, a flawed and unsustainable implementation will get us nowhere. The ends do not justify the means.

Furthermore, as it is just a set of objectives, there is absolutely no reason why the VSE cannot be implemented with a better architecture. If

you are so fond of the VSE, fight for that set of objectives under the new budget, by all means. But don't cry for an architecture that was so over-budget and over-time that NASA was actively slashing other programs just to keep it afloat. It would not have been worth the price.

Comment by Andrew — February 16, 2010 @ 10:55 am

5. Andrew,

The VSE goals were not terminated accidentally by ending Project Constellation — they were eliminated deliberately, as an act of policy. I refer you to the statements of the NASA administrator to the effect that there was no need to go back to the Moon as "it already has six American flags on it." The slides issued with the proposed budget discuss the Augustine committee's Flexible Path destinations, including Lagrange Points, NEO's and the moons of Mars. FP specifically eliminated the lunar surface because of the "expense" of developing a landing spacecraft. Conspicuously absent from all the FP discussion is any mention of purpose or objectives once we get to whatever destination is chosen, hence, this post.

Furthermore, as it is just a set of objectives, there is absolutely no reason why the VSE cannot be implemented with a better architecture

Agreed, which is exactly what I have been advocating since I started writing this blog. The problem is the current leadership of NASA, which has been, at best, lukewarm in its support of the goals of the VSE.

Comment by Paul D. Spudis — February 16, 2010 @ 11:34 am

6. Thank you so, so much, Dr Spudis. President Obama's proposals really looks to SLAUGHTERING any ambitious NASA human spaceflight. It really opens the door for cutting the glorious, vital centers in our world leading, education energizing national space heritage, to the DESTRUCTION OF THE GREAT CENTERS of 'Cape Kennedy' (KSC), Houston Mission Control, Johnson Space Center, etc. When our youth are slipping in education (and inspiration) and China is preparing to surge in technology, science & engineering, it seems insane of the Administration to bring such a proposal. If the President leaves our space program floundering, and with the space shuttle's end, the way would be cleared for an oppressive, military government to lay siege to the world spotlight with CHINA'S IMMINENT FLIGHTS TO THE MOON and a likely landing thereafter. What kind of inspiration to our youth, our education and our world will it be to have the flags of an oppressive, military government unfurled across the moon? Lets tell Washington now.....Stop the Slaughter!!!! Restore & rebuild the Vision!!!!

Comment by J Hansen — February 16, 2010

@ 8:50 pm

7. Paul, you have highlighted much of what I feared would come to pass if Constellation program was cancelled with no alternate vision of architecture put forth by administration. Furthermore, judging from the reactions of upper NASA management and members of Congress the decision to cancel was made within a very small group of people as you pointed out.

As someone who has actively supported President Obama in his election campaign, I find this lack of open transparency within NASA and lack of communications between President's administration and members of Congress, especially those whose district will be directly affected, regarding NASA's future a direct contradiction to the promises Obama made during his campaign. What is even more disturbing is that the President himself has not spoken a word about these major policy changes that have been proposed.

Again, as you pointed out, the VSE and Constellation program had significant bipartisan support. So why go looking for a fight that wasn't there? Especially in the light of so many other equally important issues facing our nations on which Congress and the public are sharply divided. NASA should have been a great opportunity for President Obama to display the spirit of bipartisanship and garner some support from crucial electoral states like Florida, Ohio, and Texas.

You and I disagree over the execution of VSE strategy. You seem favor an incremental approach through ISRU development of lunar resources to build a space-based launch and transportation infrastructure that can be developed within existing budget. You agree there needs to be some sort of heavy lift system or in-space fueling system to support transport of heavy upmass equipment and supplies.

Whereas, I support an overall large scale program for beyond LEO exploration. The high visibility of such a program serves both the parochial interests of Congressional legislators and creating economic activity while providing a format for space development. Granted this may not be the most efficient way to develop spaceflight and would waste greater resources. But this visibility and large economic impact is how the space shuttle program and the ISS were able to survive multiple attempts to cut these programs. The valuable lesson that the ISS has provided, and Mir space station before that, was establishing a permanent outpost in LEO created a new market for human spaceflight that thus far the Russians have been able to capitalize. This new market has created new interests in developing commercial spaceflight here in the US and resulted in greater commercial investments by the private sector. Thus, establishing permanent facilities in a new, unin-

habited area with no market can create new markets that will generate revenue to support greater development efforts. As long as the Constellation programs objectives could have been refined to focus greater resources on lunar development and establishing a permanent facility, then the goals of the VSE could have been achieved.

So the question becomes, outside of ISS, what market exist that can support spaceflight development to LEO and beyond LEO based on what the administration has proposed so far? People talk about potential markets, but as one friend of mine pointed out that potential markets are equivalent to inside straights. Most investors are not going put major money up on those kind of odds.

Finally, I am a big supporter of SpaceX and want very badly for the Falcon 9 to have a successful launch. However, the new FY2011 proposal has create a schism within me. In light of SpaceX initial launch failures with Falcon 1, there is a strong chance that the Falcon 9 could fail. How will such failure impact commercial space development? What will a successful Falcon 9 launch mean for NASA? Should Falcon 9 fail, such result could hamstring further commercial spaceflight development for many years. The government funds for commercial spaceflight could dry up or worse be eliminated altogether. If Falcon 9 succeeds, such success could damage NASA and its overall mission to extending human spaceflight beyond LEO leaving a void where no markets exist to serve commercial spaceflight thus hurting commercial development. Under the new proposal, success or failure could be a catch-22.

Comment by Gary Miles — February 17, 2010 @ 2:08 pm

8. Hi Gary,

So the question becomes, outside of ISS, what market exist that can support spaceflight development to LEO and beyond LEO based on what the administration has proposed so far?

This problem has vexed the space business from the beginning. It's a chicken-or-egg situation: you need a market for space products and services but you don't yet have any products or services to offer. How do markets emerge from that?

The VSE was a way for government (NASA) to take on risk by answering the questions: Can the Moon and cislunar space be industrialized? Can lunar resources be used to make a product that has commercial value? We suspect that the answer to these questions is "yes" but we don't know it. That's why it's perfect for NASA to tackle.

But they never understood this. Or if they did understand, they did not want to find the answers. So here we are.....

Comment by Paul D. Spudis — February 17, 2010 @ 2:48 pm

9. Paul, you have provided the perfect icon for Obama's space program, the 'truckster'. So, I would like to nominate the Talking Head's song "Road to Nowhere" as the theme song.

Comment by John G. — February 18, 2010 @ 12:54 pm

10. Paul,

I completely agree with your original blog on this topic, and moreover, especially concur with the comments of Gary Miles in his fifth paragraph. For some two decades, I have argued that manned space exploration beyond LEO must be large scale, if not immense, and the emphasis must not be on optimization of efficiency. Rather, such a program must become a matter of national policy, not just space policy, let alone selection of an architecture or its implementation.

Unless and until it can be shown that human space exploration delivers broader tangible and relatively near-term socioeconomic benefits to the country, I do not believe there will ever be wide public support for human exploration of the Moon or Mars. Such a program most seek to address fundamental societal issues such as increasing unemployment and underemployment in the face of globalization and productivity improvements from technological advances. And such benefits can not wait for decades following achievement of the exploration objectives.

I call this the Parakeet Paradox. I once asked a room of space enthusiasts how many owned a parakeet or some such talking bird. To my amazement, nobody raised their hand. Yet there may well be more parakeet owners in this country than space enthusiasts. (An uncle has such a bird and is definitely not a space enthusiast.) Benefits of a large-scale exploration and settlement must appeal to non-space enthusiasts, extending beyond the usual advantages held dear by space enthusiasts such as exploration excitement, science, education, and technology spin-offs as valid as all these may be. They must substantively accrue during the prolonged development phase of such a program.

I still believe it was the benighted, over bloated Space Exploration Initiative of the first President Bush that elicited the greater excitement in non-aerospace industry than did its more recent avatars. Government space advocates who are constantly subjected to the parochial political interests cited by Gary Miles, and space entrepreneurs who often still refuse to recognize the iron law of their own philosophy: are both needed, even if each side may hold its nose at the thought of the other. But is the non-aerospace segment of business that needs to be attracted.

Alas, I do not think the present atmosphere of political partisanship and economic doldrums is not conducive to examining this alternative, from which space program objectives would be derived, with any seriousness. Thus, short of national prestige or military necessity I doubt there will be serious human exploration of the Moon or Mars in my lifetime.

Comment by Robert E. Becker — February 18, 2010 @ 4:30 pm

11. Robert,

Thanks for your thoughtful comments.

Benefits of a large-scale exploration and settlement must appeal to non-space enthusiasts, extending beyond the usual advantages held dear by space enthusiasts such as exploration excitement, science, education, and technology spin-offs as valid as all these may be.

I am in full agreement with this, but only to a point. I think that moderate agreement from the public is all you really need, if you can keep the rate of expenditure below a certain level. My claim is that the VSE does that; it assumes that the NASA budget will stay at its current level plus inflation indefinitely. Thus, the problem becomes one of crafting an architectural approach that allows you to go to the Moon and establish a foothold there under such a fiscal environment.

Unlike Augustine, I think this is possible, if approached correctly. The key is to understand your mission — in this case, to create a reusable, extensible cislunar transportation system based on the use of lunar manufactured propellant. Build up lunar surface capability with robotic presence, extend human missions beyond LEO incrementally, and end with a lunar outpost that serves as a logistics depot for the cislunar transportation system. A clear mission, an incremental path, an extensible and flexible result.

So of course, they don't see it.

I doubt there will be serious human exploration of the Moon or Mars in my lifetime.

I gave up on that a decade ago. Now, I'm just trying to ensure that we're heading in the right direction so that my grandchildren may live to see it.

Comment by Paul D. Spudis — February 19, 2010 @ 4:07 am

12. Good day Paul,

I have read a number of your articles over the years, but certainly not everything, and so I apologize if the question I ask here has already been answered.

I recently read an article, linked through NASAWATCH, that again tackles the view that space is about science, and robots do it much better than people. If we look strictly at the science component of the space frontier I certainly can only agree: the majority of our scientific voyages to the Planets, and moons, and asteroids have been accomplished vicariously through our machines. Now I have said it before; I very much agree with you that the focus of our human space endeavors should be the incremental establishment of a space infrastructure that utilizes (if possible) natural assets that already exist in space – thus giving us the economical means to move even farther out into the solar system. The NASAWATCH article, however, caused me to reflect (again!) on what I believe is the fundamental question, and also the most difficult to answer, related to human space flight: Why humans in space?

Although in principle no reason exists preventing space resources from being used to support unmanned space vehicles, how do you defend your human spaceflight rational if no rational can be identified for placing humans in space in the first place? (I am approaching the question from a particular angle: the space resource infrastructure is only necessary if your aim is to move humans out in the solar system. I don't think such an established capability would serve any purpose for those living on the Earth's surface.) The proponents for 'robots only' in space seem to be interested solely in science, where as supporters of human spaceflight seem to approach the idea from an emotional side that is not easily defended by pure logic.

All the best,

Comment by Phil Backman — February 19, 2010 @ 11:00 am

13. Hi Phil,

Why humans in space?

There are lots of different answers to this question and which one you latch onto depends on your perspective. For the short term (say, the next hundred years), we need people in space to do all the different things that we want to do there — robotic spacecraft are quite accomplished, but there are simply some things that they cannot fully do, like service and maintain complex machines or conduct field science and observations. If we want to process space resources to create new capability, much of the "grunt work" can be automated, but I suspect that those machines will need to be tended by human operators and service technicians.

In the long term (millennia hence), there is a broader objective. The ultimate goal of human spaceflight is to create sustainable human communities off-planet. This is the rationale given in the Augustine Committee report (although they stated it and left it there, without carrying it to its logical conclusion). But in brief, the story of life

on Earth is the story of extinction. Extinction could come from external causes, such as asteroid impacts, from internal disasters, such as a super-volcano, or even a self-inflicted disaster, such as global nuclear war. But there is simply no doubt that at some point, human life on Earth will be erased. It's simply a question of when, not if. By settling other bodies in the Solar System, we preserve human culture, that and our genes being the only thing truly lasting in human existence.

Comment by Paul D. Spudis — February 19, 2010 @ 11:14 am

14. NASA astronauts need to be more than just explorers. They need to be pioneers!

We need a base on the Moon to see if humans can adjust to a 1/6 microgravity environment over months and even years on the lunar surface. We need to know how easy or how hard it is to use regolith to protect shelters from radiation. We need to know how efficiently we can extract oxygen from lunar regolith for air and rocket fuel. We need to know if water can really be harvested from the lunar regolith. We simply need to know how well humans can live off the land on the moon because that's what pioneers do. And this important information tells privateers what the possibilities so that they know what the investment possibilities are and how to go about safely exploiting these resources.

A government base also gives privateers a place of refuge just in case they have difficulty setting up their own private facilities on the lunar surface.

Its a shame that the public vs. private ideological battle that has already paralyzed the economic growth of this country now threatens one of the best government agencies ever created. I can't think of a government agency that has created more wealth for America and the world, per dollar, than NASA!

Comment by Marcel F. Williams — February 19, 2010 @ 10:50 pm

17. I believe I've simmered down, at least a little bit, since I made the fevered remarks in Comment #6.

But we all know China is rising fast and they have definitely made the moon a major target with the upcoming rover & sample return mission.

Overhaul, yes. But not this strange destruction of the written Vision for Space Exploration goals. Call your congressmen today, the committees, and the President, to stop this sudden abandonment of the Moon as our national goal.

Comment by J Hansen — February 23, 2010 @ 1:32 pm

A Lunar Visionary

February 23, 2010

Klaus Heiss and President Reagan

My good friend Klaus Heiss[1] is resting in the hospital after recently suffering a stoke. Klaus is not widely known or familiar to many in the space community, but over the years, he has had a major impact on our national space program – a major player in both the Shuttle program and in helping to promote a return to the Moon. His work is thoughtful and visionary and deserves a wide readership.

Klaus comes from the Tyrol of Austria, near the Italian border. His academic background is in economics, specifically the economics of space transportation, but he also has considerable expertise in physics and engineering. In the late 1960s and 70s he worked on the rationale for the Space Shuttle[2], calling for the development of a fully reusable, liquid-fueled version. For a variety of reasons, such a path was not chosen and the Shuttle failed to live up to his expectations as an economical method for launching payloads into space. In the 1980s, Klaus participated in a variety of studies for the Strategic Defense Initiative, including satellite interceptors.

In the wake of the loss of Shuttle Columbia in 2003, many were concerned about the future direction of NASA. Why do we have a national space program? What should we be striving for – pure exploration, space applications, or some combination of the two? Is there some goal or objective that creatively combines these two threads of spaceflight and is it attainable on reasonable timescales and budgets?

Klaus had long held an interest and fascination for the Moon. In the spring of 2003, he visited President Bush and Vice-President Cheney and presented to them his ideas about a return to the Moon. In his view, making the establishment a lunar base as the next NASA goal offered the country two principal advantages. First, the Moon is a strategic destination that is in reach within a decade or so without a substantial increase in the agency budget. Second, he recognized both the scientific and cultural value of establishing a human community on another world. The Moon is close and possesses the resources necessary to permit its permanent habitation. Once there, we can use the Moon as both an observing platform and a natural laboratory for scientific and engineering research.

Klaus was tasked by the White House to develop his ideas in a more detailed manner. He spent the next few months mapping a pathway from where we were to where he thought we should be. His report, *Columbia: A Permanent Lunar Base*[3], cogently summarizes all the various threads of lunar return, including not only an architecture for launch and transportation (based on Shuttle-derived vehicle components) but also the spectrum of surface activities that we will undertake there. This document was presented to the NASA Office of Spaceflight in December 2003.

I first met Klaus at a meeting on the future of the American space program convened by Buzz Aldrin in Washington DC in that same month. At the time, there was a widespread rumor that at Kitty Hawk, during a ceremony celebrating the 100th anniversary of the Wright brothers first powered airplane flight, President Bush would announce a major new direction for the American space program.

For years I had been advocating a return to the Moon mostly from the "bottom up" as I worked my way up the NASA chain of command from the field centers to Headquarters, telling anyone who would listen about the advantages of lunar return. I was unaware that while I was pursuing the Moon from the bottom up, Klaus had been doing the same thing from the top down. He knew of my work and told me that a major decision would be forthcoming from the White House and not to be discouraged by the lack of an announcement at Kitty Hawk – that I would be "pleased" with the new direction. President Bush announced the Vision for Space Exploration about a month later, in January 2004.

As Klaus and I got to know each other over the next few years, we found that we saw space issues in a similar way. Both of us were frustrated at the attempts by some in NASA to thwart the Vision, in some cases by slow-rolling it and in others by more deliberate action. Yet at the same time, we both worked closely with and attempted to help our allies within the agency, a group of smart, dedicated people who were trying to do the right thing and implement the Vision, even when it wasn't the desire of their immediate superiors. For his leadership in fostering the Vision for Space Exploration[4], he was awarded the NASA Distinguished Public Service Medal in January, 2008.

Klaus was one of the first to see that the ESAS architecture[5] (NASA's chosen implementation path for the Vision) was unaffordable and unsustainable. He specifically outlined a path[6] by which the goals of permanent presence might yet be implemented, but his counsel and pleas fell on deaf ears. Now, six years and $8 billion later, we see a new proposed budget that terminates the program designed to take us beyond low Earth orbit. Despite the positive spin from many in the space community, a return to the Moon is as far away now as it was six years ago before the Vision. In fact, it's actually farther away as much of the tooling for Shuttle parts that could enable the quick and relatively inexpensive fabrication of a moderately heavy-lift launch vehicle for the VSE are being mothballed or destroyed.

Klaus's work on the activities of lunar settlement[7] will stand as a lasting contribution to the literature of space travel. We have too few clear thinkers in this business and we were indeed fortunate to have his informed and authoritative voice to help guide our journey out into the Solar System. We need others who can clearly see the importance of a sustained lunar return[8] to step forward and pick up where Klaus left off.

A growing catalogue of the resources of the Moon continues to spill out. Much of this knowledge was gained through the vision and efforts of Klaus Heiss. I am grateful for his contributions and his friendship. Klaus will be with us in spirit as news of the Moon and its resources is presented at the annual Lunar and Planetary Science Conference next week in Houston.

Topics: Lunar Resources, Space Politics, Space and Society

Links and References

1. Klaus Heiss, http://www.highfrontier.org/Highfrontier/main/Contact/Meet%20the%20Staff.htm
2. rationale for the Space Shuttle, http://www.pmview.com/spaceodysseytwo/spacelvs/sld037.htm
3. Columbia: A Permanent Lunar Base, http://www.spudislunarresources.com/Links/klaus/ColumbiaLunarBase.pdf
4. Vision for Space Exploration, http://www.spaceref.com/news/viewpr.html?pid=13404
5. ESAS, http://www.nasa.gov/exploration/news/ESAS_report.html
6. outlined a path, http://www.spudislunarresources.com/Links/klaus/EqualBudget.pdf
7. activities of lunar settlement, http://www.spudislunarresources.com/Links/klaus/JamestownPaper.pdf
8. sustained lunar return, http://www.spudislunarresources.com/Links/klaus/JamestownSlides.pdf

Comment

1. I regret to note that Klaus passed away on July 24, 2010. As a long time supporter of lunar return, I know that he would tell us all to carry on the struggle. May he rest in peace.

Comment by Paul D. Spudis — July 28, 2010 @ 8:01 am

Talismanic Thinking

February 27, 2010

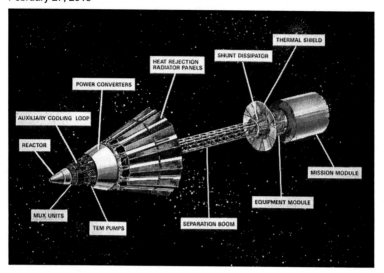

The SP-100 space nuclear reactor

Wild claims are being tossed about regarding the future U.S. space program. Recipes for success are touted and e-mailed around – concepts based more on wishful thinking than on solid science and engineering. My friend Rand Simberg refers to those who would replicate anew the means we devised to go to the Moon several decades ago, as having an "Apollo cargo cult[1]" mentality (i.e., Pacific islanders waiting for parachutes to once again drop wondrous things in crates from planes, as they did during World War II). A counterpart to the so-called "Apollo cargo cult" also exists in the space community and they rely on their own talismanic thinking – a belief in some technique or item that allows us to go farther and longer in space, with incredible new capabilities. The talisman takes different forms for different groups, but in all cases, they ward off the evil spirits of physical and bureaucratic reality.

Early in the history of the Vision for Space Exploration, talismanic thinking was apparent with Project Prometheus[2]. This was a program to develop an advanced space nuclear reactor for missions to the outer Solar System – where the Sun's rays are too weak to provide enough energy to power systems. Used anywhere, such a capability enables activities to take place in a power-rich environment, making many necessary and routine operations

easier, safer and more efficient. Former NASA Administrator Sean O'Keefe was enamored of Prometheus, so much so that he often unintentionally overstated its capabilities. For him, Prometheus was a talisman – a unique capability that enabled the otherwise unobtainable.

NASA Administrator Charles Bolden fancies his own talisman – the technology to "go to Mars in days and weeks, rather than months[3]." Bolden is probably referring to VASIMR[4], the plasma rocket engine designed and undergoing testing by former astronaut Franklin Chang-Diaz. In principle, a VASIMR-powered vehicle could go to Mars on non-minimum energy trajectories, thereby cutting transit time between planets to a fraction of that required for a chemical rocket.

VASIMR is an interesting concept and some form of it will be very useful when we are ready to travel to the outer planets. However, one aspect about it that I have not heard mentioned by Bolden is the low mass, high power system needed to run it. The only known systems approaching the necessary power density needed are nuclear reactors. Which brings us back to Project Prometheus, a joint NASA-Department of Energy (DoE) effort.

Prometheus was canceled in the FY2006 budget. It was deemed too complex and too costly for its proposed use, the Jupiter Icy

Moons Orbiter[5]. This was a robotic spacecraft designed to tour the Jupiter system and obtain data on its satellites during multiple flybys. Note well: this power system was thought to be both too complex and expensive for a robotic mission. A similar system for human missions – which involves many more systems, power requirements, and propulsion – would be even more complex and expensive. Tack on international participation and – well, you get the picture.

So where does this leave VASIMR? Chang-Diaz notes[6] that nuclear reactors can be launched empty and then assembled and fueled in space, presumably by human astronauts. Thus, there are no safety considerations associated with its launch. The problem is that the pieces of this reactor don't exist and aren't even being thought about being built. For decades the DoE community has talked about a space reactor of the 100 to 1000 kW class; a VASIMR-powered Mars vehicle would need a 10 megawatt reactor[6]. Billions of dollars went into the SP-100 program in the 1980s and 1990s and still the reactors needed to power VASIMR exist only in the mind's eye of some space dreamers. The United States Navy has been building and operating nuclear reactors for over 60 years, so one would think that building a space reactor would be achievable, but practice has proven otherwise.

VASIMR is Bolden's talisman, the magic beans that will grow a stalk that we can climb to Mars. Such a rocket engine would be a technological breakthrough promising capabilities well beyond our current reach. But for now, a Mars craft using VASIMR is imaginary. Reality will not come about by spending massive amounts of money on general technology investment. When VASIMR is finally built, it will be because it is needed for a specific application or mission. Once again, the ends will drive the means, not the other way around.

Talismanic thinking is common in much of the current discussion about the new path for NASA. Other talismans include cheap access to low Earth orbit, commercial transport replacing Orion, and an "exciting space goal" to engage the public. These new dogmas (all of them means, not ends[7]) clearly illustrate that there is no strategic thinking or thoughtful leadership guiding America's space program. Those at the top need to know where they are going and understand why; the fact that they currently do not bodes ill for the future of our country.

NASA Administrator Charles Bolden recently said[8] that "he is trying to find middle ground between groups "radically" in favor of keeping the Constellation program and others lobbying for reliance on commercial space entities." But he is still confusing the means with the ends. We should re-affirm that our mission is to use the resources of the Moon to build a transportation infrastructure whereby all can travel to wherever they choose as often as they want. Our direction in space goes through the Moon or we go nowhere.

Topics: Lunar Resources, Space Politics, Space Transportation, Space and Society

Links and References

1. Apollo cargo cult, http://www.popularmechanics.com/science/air_space/4345250.html
2. Project Prometheus, http://en.wikipedia.org/wiki/Project_Prometheus
3. go to Mars in days and weeks, http://www.washingtonpost.com/wp-dyn/content/article/2010/02/24/AR2010022403699.html
4. VASIMR, http://en.wikipedia.org/wiki/Variable_Specific_Impulse_Magnetoplasma_Rocket
5. JIMO, http://en.wikipedia.org/wiki/Jupiter_Icy_Moons_Orbiter
6. Chang-Diaz notes, http://www.adastrarocket.com/SciAm2000.pdf
7. means not ends, http://blogs.airspacemag.com/moon/2010/02/13/confusing-the-means-and-the-ends/
8. Bolden said, http://www.spaceflightnow.com/news/n1002/24congress/

Comments

1. Of course it's important for NASA to have goals, but having the means to meet those goals is essential. VASIMR (I think you misspelled it) may indeed be a talisman for some, but something as basic as cheaper access to LEO will be essential to any practical and affordable exploration program, and pays dividends for many other programs. There are in fact a number of basic technologies that are enablers for all kinds of exploration activities, and it is legitimate for NASA to pursue them as technology development programs.

There is a danger to having no goals, of course: wasting money and time on technologies that never get used. There is also danger a danger to the mentality of "the mission is everything": You may wind up building a throwaway system that doesn't leave us any closer to being a true space-faring civilization.

Comment by Bill Hensley — February 27, 2010 @ 4:09 pm

2. Bill,

First, thank you for the spelling correction — I have fixed this in the post.

Second, I am not against technology development and I don't think we're fundamentally in disagreement. The point I have been trying to make since this new budget came out is that in the agency's past, two kinds of technology development have occurred — those in support of specific mission goals and those part of a general technical development effort. The former includes things like fuel cells and inertial guidance; the latter includes any number of widgets and incremental improvements in existing systems. If you have a specific mission goal, you'll get all the technology development you need. If you don't have such a goal, you'll get a lot of random development, some useful, some not. But technical development alone is not why we have a space program. People expect accomplishment. My concern about the new direction is that a perfectly logical and rational strategic direction was discarded, without any real thought or rationale. I think that was a major mistake.

Comment by Paul D. Spudis — February 27, 2010 @ 4:35 pm

3. Perhaps we are not that far apart, as you say. I, too, am uncomfortable with the lack of a clear plan into which all of the technology development pieces can fit. VSE was good, but Constellation became narrowly focused on flags and footprints (at ruinous expense).

Comment by Bill Hensley — February 27, 2010 @ 9:49 pm

4. Unfortunately, General Bolden doesn't seem to understand the incredible value of a lunar base and the idea of reducing cost through the exploitation of extraterrestrial resources. Could you imagine what a game changer it would be if lunar base astronauts were manufacturing oxygen on the Moon for air and as a major component of water? And what if we really could extract significant amounts of water from the lunar regolith!

As far as interplanetary game changing technology is concerned, light sails are the key to opening up the solar system, IMO. Even if you deployed a simple 20 tonne 2 kilometer in diameter Mylar light sails at a Lagrange point, each one could potentially bring back a 50 to 100 tonne NEO asteroid to a Lagrange point every year. Just ten of these reusable vehicles could bring back 500 to 1000 tonnes of small NEO asteroids to a Lagrange point every year!

Relative to such resources from Earth, small asteroids would be an extremely cheap source of hydrogen and oxygen fuel for space

depots which could dramatically lower cost for traveling to the Moon. Asteroids could also provide mass shielding for Lagrange point space stations.

Comment by Marcel F. Williams — February 28, 2010 @ 12:38 am

9. OK, Dr. Spudis, I think your point is quite well taken by now. I think it would be interesting (in your future posts) to start exploring more positive and constructive angles re: the budget…especially since your problem seems to be not so much with the budget itself, but rather with Bolden's Mars-ish unfocused ramblings and the lack of any lunar ISRU or development planning.

I'd be interested in hearing how you think the new framework can work with your ideas, rather than against them–even if you feel it's unlikely that it will actually pan out that way. For example, as I understand it the budget includes a $1 billion increase in funding for robotic precursor missions over the next few years, at least some of which are likely to go to the lunar surface. In light of the recent LRO data and north pole ice etc., what sorts of missions become possible with this extra slice of the budget (pried from the corpse of Ares I)? Also, Robert Bigelow was quoted last month as saying that his company had already blueprinted soft-landing some of his conjoined BA-330 modules on the lunar surface, in order to rent out the lab space. This obviously isn't going to happen anytime soon, but if private sector initiatives are going to play a part in determining how things move forward in space exploration, how do you see this potentially playing out, from LEO across cislunar space to the lunar surface? What are your recommendations for the private sector, and how could NASA work with or guide them, ideally, to lead towards lunar development? You spent most of this post bashing VASIMR-to-Mars wishful thinking, but Ad Astra's shorter-term plans for the rocket have focused on providing slower, cheaper cargo transport to the moon. How much of a difference do you see this making?

In short, I think there are plenty of things to think about in the wake of the budget besides just criticizing it (although criticisms are also, of course, fair game), and I'd enjoy reading any discussions of these issues that you'd like to post. Since a detailed strategy hasn't emerged from NASA yet–and won't for a while–it seems like now would be a good time to voice as many positive ideas as possible.

Comment by Jared — March 2, 2010 @ 2:46 pm

10. Jared,

In short, I think there are plenty of things to think about in the wake of the budget besides just criticizing it (although criticisms are also, of course, fair game), and I'd enjoy reading any discussions of these issues that you'd like to post

I suggest that you can start by going back in this blog and reading any and all of my previous 50-odd posts, where I lay out exactly why we are going to the Moon, what it's value is, and how it can help us create a sustainable human presence in space.

Comment by Paul D. Spudis — March 2, 2010 @ 4:15 pm

11. "*Our direction in space goes through the Moon or we go nowhere. Once again, the ends will drive the means, not the other way around*"

Awesome Paul you are in tune to reality check. As an engineer I know full well where the random Garver "flex-path" will lead. We are already seeing that upfront flex will become the "lobby path". From there over the next few decades who knows. Bolden is looking more and more like a Garver/Obama puppet a mere figure head. It is now apparent to me who is really calling the shots at NASA. And that is very troubling.

The moon offers the resources to develop free markets and to define the technologies required to harvest those resources and move outward. Putting the technologies in place prior to defining the markets will not guarantee that those technologies will match what the market demands.

For me its not Constellation vs. commercial. It about applying some common sense project management practices and accountability in place to ensure a successful lasting transition to commercial development.

Comment by Doug Gard — March 3, 2010 @ 1:36 am

13. Dr. Spudis, What is your explanation why the Augustine commission, NASA refuse to mention or acknowledge space nuclear technology and at the same time proportion funds for research in this area?

Doesn't this smell like irreconcilable differences between public policy and R&D.

Wouldn't a clear public commitment signal a willingness taxpayer funds are well spent on advancement in space technology?

What is your position on Nuclear thermal rocket propulsion and power ?

Comment by Bruce Behrhorst — March 24, 2010 @ 4:24 pm

14. Bruce,

What is your explanation why the Augustine commission, NASA refuse to mention or acknowledge space nuclear technology and at the same time proportion funds for research in this area?

I don't know what their thinking was, but I can hazard a guess that it's largely because of the high perceived cost of space nuclear power. This high cost is caused not only by technical issues, but almost more significantly by the enormous bureaucratic/legal overhead that would have to be paid on the launch of any nuclear reactor.

I have no "position" on space nuclear power/propulsion. If it makes good architectural sense (i.e., both technically and fiscally), then it should be developed and used. If it doesn't, we should seek alternatives to it. I look upon it as a means, not an end.

Comment by Paul D. Spudis — March 24, 2010 @ 4:31 pm

15. Wouldn't the 'enormous' bureaucratic/legal overhead be of the gov't own making. How could past launchings of RTG (Radio-isotope thermoelectric generator) missions be any different in launch logistics other than launch capability of for example, on an Ares HLV?

Dr. Spudis, what I don't understand is why everyone is ignoring the obvious elephant in the space capsule. Propulsion and power is the name of the game in space. This superior alternative to chemical is proven technology and has been sitting on the shelves of gov't labs for 50 yrs. Is waiting for a politically correct alternate method of power and propulsion in space any inspiration ?

Comment by Bruce Behrhorst — March 24, 2010 @ 5:08 pm

Ice At the North Pole of the Moon

March 1, 2010

 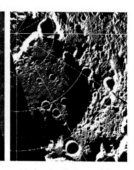

SAR mosaic   CPR mosaic   Clementine hires mosaic

Floor of Peary
73 km diameter
88.6° N, 33° E

Radar mosaic of the floor of the north polar crater Peary, showing many craters with elevated CPR inside, but not outside, their rims. This material is probably water ice.

Last year, India's Chandrayaan-1 lunar orbiter[1] spent eight months mapping the surface of the Moon. I had the honor of being the Principal Investigator of an experiment on that mission, the Mini-SAR imaging radar[2]. The purpose of this experiment is to map and characterize the deposits within permanently dark areas of the poles. These dark areas are extremely cold and it has been hypothesized that volatile material, including water ice, may be present in quantity here. Our radar team has just finished the first round of analysis of data returned by the Mini-SAR for the north pole and results will soon be published in the technical journal, Geophysical Research Letters.

Mini-SAR[2] is a lightweight, low power imaging radar. It uses the polarization properties[3] of reflected radio waves to characterize the lunar surface composition and physical state. Mini-SAR transmits pulses of left-circularly polarized radar. Typically, reflection from planetary surfaces reverses the transmitted polarization, so that Mini-SAR radar echoes from the Moon are right circularly polarized. The ratio of received power in the same sense transmitted (left circular) to the opposite sense (right circular) is called the circular polarization ratio (CPR). Most of the Moon has low CPR (about 0.3), meaning that a reversal of polarization is the norm, but some specific areas have high CPR (greater than 1.0). These include

very rough, rocky surfaces (such as a young, fresh crater) and ice, which is transparent to radio energy. In this latter case, the radar penetrates the ice and is scattered and reflected multiple times by inclusions and flaws in the ice, resulting in the reflection of many same sense polarization echoes, leading to higher CPR values than normal. High values of CPR are not uniquely diagnostic[4] of either surface roughness or ice; we must take into account the geological setting of the high CPR signal to interpret its cause.

Many craters near the poles of the Moon have interiors that are in permanent shadow from the Sun. These areas are very cold and water ice is stable permanently there. Fresh craters show high degrees of surface roughness (high CPR) both inside and outside the crater rim, caused by sharp rocks and block fields that are distributed over the entire crater area. However, Mini-SAR found craters near the north pole[5] that have high CPR values inside, but not outside their rims. This relation suggests that the high CPR is not caused by roughness, but by some material that is restricted within the interiors of these craters. It is not geologically reasonable to expect rough, fresh surfaces to be present inside a crater rim but absent outside of it. The craters that show this enhancement are all permanently cold and dark, where ice is stable. We thus interpret this high CPR to mean that water ice is present in these craters.

Over forty small (2-15 km diameter) craters near the north pole of the Moon are found to contain this elevated CPR material. The total mount of ice present at the pole depends on how thick it is; to see this elevated CPR effect, the ice must have a thickness on the order of tens of wavelengths of the radar used. Our radar wavelength is 12.6 cm, therefore we think that the ice must be at least two meters thick and relatively pure. At such a thickness, more than 600 million metric tones of water ice are present in this area. Such an amount is comparable to the quantity estimated from the 1998 Lunar Prospector (LP) mission's neutron spectrometer data[6] (several hundred million metric tones). The LP neutron spectrometer only sees to depths of about one-half meter, while we penetrate at least a couple of meters, so the neutron data would underestimate the total quantity of water ice present.

The emerging picture from many experiments on several different lunar missions indicates that the creation, migration, deposition and retention of vast amounts of water are occurring on the Moon. Such an astounding result was totally unexpected by most lunar scientists, including myself. The emerging picture is consistent with earlier studies from the 1994 Clementine mission[7] and subsequent Lunar Prospector as well as the more recent reports of the presence of water-bearing minerals at high latitudes (Moon Mineralogy Mapper[8]), the detection of water vapor in the LCROSS[9] impact plume (a few percent water content at its target site), and a variety of new supporting measurements, such as the discovery of unexpectedly cold polar temperatures by the DIVINER experiment[10] on NASA's Lunar Reconnaissance Orbiter[11] (as cold as 25 degrees above absolute zero, colder than the estimated surface temperature of Pluto). The Moon experiences complex geological processes that were wholly unexpected before the recent results.

The quantity of water present at the lunar poles is significant; at the north pole alone, the 600 million metric tons of water there – turned into rocket fuel – is enough to launch the equivalent of one Space Shuttle (735 mT of propellant) per day for over 2000 years. The discoveries we are now making show that the Moon is an even more interesting and attractive scientific and operational destination than we had previously thought. The Moon is the key to sustainable human presence in space[12]. Its resources enable us to create a reusable, sustainable transportation system, one that can routinely access not only the Moon, but all points of cislunar space. Once established, such a system can be used to go forward into the Solar System.

Topics: Lunar Exploration, Lunar Resources, Lunar Science, Space Transportation

Links and References

1. Chandrayaan-1, http://www.isro.org/Chandrayaan/htmls/home.htm
2. Mini-SAR, http://www.isro.org/chandrayaan/htmls/minisar_nasa.htm
3. polarization properties, http://www.nasa.gov/mission_pages/Mini-RF/science/index.html
4. CPR not uniquely diagnostic, http://www.thespacereview.com/article/740/1
5. craters near the north pole, http://www.nasa.gov/mission_pages/Mini-RF/multimedia/feature_ice_like_deposits.html
6. neutron spectrometer data, http://lunar.arc.nasa.gov/results/neures.htm
7. Clementine mission, http://www.sciencemag.org/cgi/content/full/274/5292/1495
8. Moon Mineralogy Mapper, http://m3.jpl.nasa.gov/NEWS/
9. LCROSS, http://www.nasa.gov/mission_pages/LCROSS/main/prelim_water_results.html
10. DIVINER, http://www.diviner.ucla.edu/blog/?p=232
11. LRO, http://lunar.gsfc.nasa.gov/
12. key to sustainable presence, http://www.spaceref.com/news/viewnews.html?id=1376

Comments

1. *The emerging picture from many experiments on several different lunar missions indicates that the creation, migration, deposition and retention of vast amounts of water are occurring on the Moon.*"

One question: Where is this water coming from?

Comment by Steve — March 1, 2010 @ 11:15 pm

2. The Moon is so close that you can almost touch it. We could live there, work there, and grow the American economy there for less than 0.5% of our total Federal budget. I wonder what future historians will say about a country that had a chance to own the rest of the solar system through its own technological know how for such a relatively tiny investment, but turned down the deal.

Its almost as if Thomas Jefferson decided not to purchase the Louisiana territories from France. What a mistake that would have been for the future of our country.

Napoleon Bonaparte, upon completion of the agreement, stated, "This accession of territory affirms forever the power of the United States, and I have given England a maritime rival who sooner or later will humble her pride."

Comment by Marcel F. Williams — March 2, 2010 @ 12:45 am

3. Steve,

Where is this water coming from?

We're not totally certain. There are a variety of possible external water sources, including the impact of comets and water-bearing asteroids, cosmic dust particles, and the reduction of surface oxides by solar wind protons (hydrogen). Apparently, some or all of these various processes are continually adding water to the Moon's surface. Much of it is destroyed or lost to space by a variety of processes, but if it gets to the cold polar areas, it gets trapped and is stable there, essentially forever. If it sounds like a very slow process, it is, but over geological timescales of billions of years, appreciable amounts of water can accumulate.

Comment by Paul D. Spudis — March 2, 2010 @ 5:31 am

4. Paul – Any notion of what the carbon or nitrogen abundances in the cold traps might be? Is there a known/plausible lower bound?

Comment by Jay Manifold — March 2, 2010 @ 11:40 am

5. A good question to ask is how would we go about mining these resources at the lunar poles. And how could you deliver these resources to lunar bases or colonies not located near the poles? And how would you deliver these resources to a Lagrange point or lunar orbit?

Comment by Marcel F. Williams — March 2, 2010 @ 2:02 pm

6. Jay,

Any notion of what the carbon or nitrogen abundances in the cold traps might be? Is there a known/plausible lower bound?

The actual species present in the polar cold traps cannot be determined from radar data. But the spectral results from LCROSS suggest a variety of cometary volatiles are present, including ammonia, CO_2 and simple organics. If they are present in cometary abundance, they would represent less than 5% of the total amount of water ice.

Comment by Paul D. Spudis — March 2, 2010 @ 4:18 pm

7. Marcel,

All very good questions indeed. I'm not sure that I have all the answers yet, but one thing I would say right now is that I see no reason NOT to locate a lunar outpost or base directly AT one of the lunar poles. Not only is water ice available there, but you also have near-permanent sunlight on certain mountain peaks, which both provides abundant, clean constant solar energy as well as a near-constant, benign thermal environment (about -50° C, +/- 10° C)

Whether the south or north pole is the optimum location remains to be determined.

Comment by Paul D. Spudis — March 2, 2010 @ 4:21 pm

8. Maybe hydrogen powered cannons like those proposed by the Quicklaunch company could be used to distribute solid polar materials to particular areas of the Moon during that area's lunar night (just make sure your base is not located there:-)) while also being able to deliver payloads to lunar orbit and the Langrange points.

In fact, such as system might also be industrially useful on the surface of Mars, Mercury, Callisto, and the larger asteroids.

http://quicklaunchinc.com/

Comment by Marcel F. Williams — March 2, 2010 @ 7:56 pm

9. Paul!

Kudos on your latest results that confirm the previous findings. This is science at its best!

Comment by Warren Platts — March 5, 2010 @ 8:23 pm

10. Warren,

Many thanks! Getting the radar built and flown was an accomplishment by a lot of people from a variety of places around the country and in India, all of whom share in this discovery. It was and is my honor to be associated with them.

Comment by Paul D. Spudis — March 7, 2010 @ 11:44 am

Stuck in Transit – Unchaining Ourselves From the Rocket Equation

March 11, 2010

The Moon is the key resource needed to open up the frontier of space

Last fall, after much anticipation, the Augustine Committee presented us with their assessment of the future of space exploration[1]. Its basic conclusion was that at currently envisioned budgets, the Program of Record (a.k.a. ESAS[2], Project Constellation) would not get us back to the Moon before many decades had passed, if then. This meme has been picked up by many in the space community to the point where is it now cliché to claim that we don't have enough money to do anything in space. Hence, the direction proposed in the new budget[3] takes NASA out of the space transportation business entirely, freeing up their budget to focus on technology development, and contracting with commercial providers to create access to low Earth orbit (LEO) and the International Space Station (ISS).

How are costs estimated for space systems? The costing exercise for the Augustine Committee was done by The Aerospace Corporation[4], a non-profit science and engineering company run for the U.S. Air Force. Their costing procedures (described briefly on page 82 of the committee report[1]) includes estimating the time and level of effort it takes to develop a system, informed by data from past projects. The vast bulk of this costing effort deals with launch vehicles and systems.

Looking over cost estimates is a strange experience. Almost anyone can immediately see inflated levels of costing for things they know about, but are uncertain for other items. Bob Zubrin wrote a stinging rejection[5] of the Aerospace Corporation's costing just before the Augustine Committee released their report. He noted in particular that the estimates included several years of increasing ground operations costs[6], even while nothing was being launched. Of course, if you pull together a ground crew, you have to pay them to keep them around, even during slack times. But his point is a good one; why should it cost more than Shuttle does now to support a launch system that requires an order of magnitude less preparation than the highly complex Shuttle Orbiter?

Using these estimates of the cost of the existing architecture, the Augustine Committee concluded that it was unaffordable. What did they do then? Rather than fix the problems with the ESAS architecture[2], they discarded the entire Vision for Space Exploration[7] and came up with the so-called "Flexible Path[8]" (FP). Although cloaked in platitudes about how technology development will give us options to go to many destinations beyond LEO, the real motivation for this idea is revealed by the committee's words on "public engagement" (e.g., "It (FP) would provide the public and other stakeholders with a series of interesting "firsts" to keep them engaged and supportive." – Augustine report[1], p. 15). Thus, the goal of FP is to create Apollo-like spectacles for public consumption, rather than creating steps toward increased space faring capability.

We can wait and hope for the proposed technology development program to provide us with magic beans[9], or we can begin that process now by returning to the Moon with robots and humans to learn how to harvest and use its material and energy resources. Creating a sustainable system of space faring[10] that can take us anywhere we want to go would be a real accomplishment. By gaining this knowledge and expertise, mankind will be free to choose many space goals, thereby achieving "at will" space destination capability.

Jeff Greason, President and co-founder of XCOR Aerospace[11] and a member of the Augustine Committee, recently spoke at the annual Goddard Memorial Symposium. He asserted that for the near future, we have no path to move people beyond low Earth orbit because the options the Augustine Committee looked at cost more than the United States can afford or is willing to spend. His principal message[12] to Symposium attendees was to "deal with it."

According to the Augustine Committee, "The cost of exploration is dominated by the costs of launch to low-Earth orbit and of in-space systems." This outlook is one reason why so much of the costing focus was on building Ares V, the super-heavy lift (188 mT) launcher designed for human Mars missions. For such a mission with chemical propulsion (the only technology currently available) you need about one million pounds in LEO, of which more than 70% is propellant. Going to Mars is expensive because you must lift all of that fuel out of the deep gravity well of Earth. Even with the economies of scale provided by

a super heavy lift rocket, it still costs tens of billions of dollars to mount such a mission.

Making propellant on the Moon completely changes these numbers[13], yet use of lunar resources is discussed in only a few brief paragraphs of the Augustine report. We now know (as the committee did then) that water is present at the lunar poles[14] in significant quantity and that its use to make rocket propellant can create a transportation system that could routinely access all of cislunar space. This should be the objective of lunar return: to create a space "transcontinental railroad" through the use of lunar resources. Once established, we can go to the planets with relative ease.

Is any of this possible under the existing budget? Not if we dissipate our money with pointless and unfocused technology development. Of the many advantages of the Moon, one of the biggest is that it is close enough that preliminary work can be done by robots on the lunar surface – controlled and remotely operated from Earth. By emplacing robotic assets on the Moon before human arrival, we can begin to survey, process and store water for use well in advance of human arrival. Sending robotic assets in advance of people allows us to start creating capability now, without a major increase in budget. It simply requires a sense of clear objectives; we have the technology to work this problem now.

Simply put, our space objectives need to be – arrive, survive and thrive. To do that, the goal must be stated, mapped out and achieved before setting out to the next destination. A sustainable, expandable transportation system in space can be devised by using the resources we find in space. We will learn how (and if) we can do this on our Moon. Once we don't have to haul everything with us from the Earth, costs become lower. When you don't have to use 90% of your travel budget just to get out of town, a lot more people can take the trip. Before you know it, you have a space-based economy.

The nation has important strategic and economic interests in cislunar space and it is entirely appropriate for the federal government to develop a sustainable and extensible cislunar transportation system[15]. NASA needs to lead and point the way so that the private sector (not just aerospace companies) can invest in and develop the yet unknown technologies that will improve our lives here on Earth as we move out to explore and ultimately settle the new territory of space.

The Moon is a classroom, a test bed and a supply depot. By using its resources, humanity can create the capability to live, work and travel in and beyond cislunar space. As a nation, we cannot and must not pass on this enterprise.

Topics: Lunar Resources, Space Politics, Space Transportation, Space and Society

Links and References

1. Augustine committee, http://www.nasa.gov/pdf/396093main_HSF_Cmte_FinalReport.pdf
2. ESAS, http://www.nasa.gov/exploration/news/ESAS_report.html
3. new budget direction, http://www.nasa.gov/news/budget/index.html
4. Aerospace Corporation, http://www.aero.org/
5. stinging rejection, http://www.marssociety.org/portal/AugustineNumbers/
6. increasing costs, http://www.marssociety.org/Augustine-Cost_Comparisons_8_12.xls
7. Vision for Space Exploration, http://www.spaceref.com/news/viewpr.html?pid=13404
8. Flexible Path, http://blogs.airspacemag.com/moon/2010/02/03/vision-impaired/
9. magic beans, http://blogs.airspacemag.com/moon/2010/02/27/talismanic-thinking/
10. sustainable system, http://www.spaceref.com/news/viewnews.html?id=1349
11. XCOR Aerospace, http://www.xcor.com/
12. principal message, http://tinyurl.com/2cpsyoc
13. changes these numbers, http://www.spaceref.com/news/viewnews.html?id=1334
14. water is present, http://blogs.airspacemag.com/moon/2010/03/01/ice-at-the-north-pole-of-the-moon/
15. sustainable and extensible, http://www.spaceref.com/news/viewnews.html?id=1376

Comments

1. The cost of Constellation was horrendous. no doubt it was possible (in theory anyway) to restructure it to bring the costs down. But there was a second problem. Time. Even with unlimited budgets (according to Augustine) it would be impossible to bring Ares I on line in any reasonable time frame. Taken together these are fatal flaws.

On the other hand Atlas V exists. Delta IV exists. Falcon 9 is on the launch pad now. There is no practical option but to go with the vehicles that exist to reduce the gap as much as possible and restore HSF.

And relying on EELVs is not a bad thing. Higher flight rates means lower costs. Lower costs in turn allow higher flight rates. A virtuous circle. Even more important than establishing lunar ISRU is getting the cost of access to LEO down. Getting that cost to LEO down enables everything else

Comment by Frediiiie — March 12, 2010 @ 1:53 am

2. Paul, this was so close to being a good

article, it's a shame you wrote it with such obvious anger and bias. The personal attacks were completely unnecessary. The misrepresentations did not help your argument.

Orbital propellant transfer and storage, and in-situ resource utilization are both listed as technologies to be developed and demonstrated under the FY11 budget – they are flagship programs. Under the budget of previous years they were assigned no money at all. So it baffles me why you trying to imply the opposite.

Robotic precursors to the Moon, both orbiting and lander/rovers, are funded under the FY11 budget. They have received little to no budget allocation under previous budgets. Again, I don't understand why you seem to be implying the opposite.

I think if you just stop being angry and actually look at what is going on, and get involved, you'll see NASA is better off now than it was under Griffin.

Comment by Trent Waddington — March 12, 2010 @ 4:39 am

3. *The personal attacks were completely unnecessary*

What personal attacks?

Orbital propellant transfer and storage, and in-situ resource utilization are both listed as technologies to be developed and demonstrated under the FY11 budget – they are flagship programs....Robotic precursors to the Moon, both orbiting and lander/rovers, are funded under the FY11 budget

Simply put, although the budget outlines these as items in the "technology development" lines, I do not believe that we'll ever get anything from it. The history of the agency shows that they are very good at spending money on "technology development" but less proficient at producing any flight hardware from it.

you'll see NASA is better off now than it was under Griffin.

Although I have had my differences with both Mike Griffin and his chosen architectural implementation of the VSE, at least he was aiming in the right direction. The "new path" aims nowhere.

Comment by Paul D. Spudis — March 12, 2010 @ 6:12 am

4. Paul, the personal attacks are obvious,

I'm not going to play the game of pointing them out to you. If you're going to advocate the use of technology like propellant depots and ISRU, I simply can't understand why you would be opposed to a technology development program that focuses on that technology... The alternative is to focus on transport systems which don't have that technology, which is what ESAS was all about... it really does just seem like cognitive dissonance.

Comment by Trent Waddington — March 12, 2010 @ 6:28 am

5. If NASA was doing something innovative, something different than has been done before by building the Constellation, I might see that as a valid investment into the future. About the only thing the Ares I was doing that was different from the past is to try and develop a man-rated solid fuel rocket capable of independently (instead of jointly with liquid-fueled rockets like the Shuttle program) achieving orbital velocities. From just about every rational study on the topic, that sounds almost an insane proposition anyway to even consider such a concept in the sense that abort options on a solid fuel rocket are embarrassingly few as once the rocket is lit off, it won't stop (barring some incredibly expensive and wasteful from a payload standpoint fire suppression system). A liquid-fueled rocket at least can be turned off as an abort option.

Constellation also was not focused on reducing the cost for access getting into space. At least the Space Shuttle was nominally oriented in that direction... at least in the 1970's. What was the primary selling point for the Shuttle program during the Nixon administration was that it could provide a much lower cost per flight than using the Saturn I or similar vehicle. Everybody forgets now, but the Shuttle was considered the one super cheap way for getting into orbit. If it had gone through a couple dozen iteration in an engineering design cycle, it might have actually achieved that sort of reputation as well.

One engineering mantra or philosophy is that any item can be produced cheaper, sooner, or more reliable. Unfortunately you must choose only two of those options for whatever you want to make or build. Apollo choose to build something sooner and more reliable, at tremendous expense. Constellation and many of the proposals based on the experience from the Apollo project were based on the same philosophy, with the retirement of the Shuttle looming all that harder on the time pressure to get something done now.

SpaceX and some of the other newer spaceflight companies are trying an approach to build something cheaper and reliable. This is a very different philosophy and something that takes patience to get it to work correctly. I've seen critics complain about how long it is taking SpaceX to get the Falcon 9 to launch, thinking that this private company is trying to follow an Apollo crash program of trying to reach the market first and damn the costs of getting there. If the Falcon 9 doesn't launch for another year.... so what? SpaceX already has a proven system with the Falcon 1, so they can at least be making some money and staying profitable.

I look at Armadillo Aerospace as an example of what really should be happening, and that company is definitely taking the slow but steady progress approach toward spaceflight, where fuel costs are a significant concern and budget item. In spite of being one of the original Ansari X-Prize participants, they still don't necessarily have a genuine sub-orbital vehicle, but they are getting closer to that goal every day and are making a profit from some of the technologies they have already developed in trying to reach that goal.

It still gets back to doing something new and innovative. Constellation was bad because the Ares I vehicle was essentially duplicating existing vehicles in terms of function if not design, and did so at an incredible premium over those other vehicles. While there was innovation with Constellation, it hat nothing to do with lowering the cost of getting into space, as that wasn't even a design goal. The real purpose of Constellation was to act as an engineering/technology jobs program for the districts and states where the factories and design bureaus were located at. That has nothing to do with making it cheaper or easier to get into space.

Comment by Robert Horning — March 12, 2010 @ 6:56 am

6. Great article Paul. None of the EELVs (or Falcon 9) are remotely close to providing human transportation to the space station. And, I work for one of the companies that bid the EGLS (Exploration Ground Launch Services) contract for KSC that was recently cancelled. Launch operations costs were approximately 25% of the costs for the SSP. As Griffin has pointed out, the Augustine Commission used real estimates for Constellation and phony ones for the commercial alternatives. More important, both the Augustine Commission

and the Administration are ignoring significant safety concerns.

Comment by Rocketman — March 12, 2010 @ 7:45 am

7. *Paul, the personal attacks are obvious, I'm not going to play the game of pointing them out to you*

They are not obvious to me. You made this charge. Prove it.

Comment by Paul D. Spudis — March 12, 2010 @ 8:04 am

8. Robert,

It still gets back to doing something new and innovative.

It also involves understanding your mission and designing to accomplish it. The "new direction" has no mission, but spends on technology in general (or so they claim). Giving NASA money to New Space companies is merely trading one group of aerospace contractors for another.

Comment by Paul D. Spudis — March 12, 2010 @ 8:07 am

9. The "Exploration Technology and Demonstrations" program includes an in-orbit propellant transfer and storage technology demonstration, probably started in 2011. It looks like more of a demonstration than a technology development effort.

That program also has ISRU work, including extracting water and oxygen from lunar ice. A lunar ISRU technology demonstrator/robotic precursor is also to be considered.

The robotic precursor line most likely starts in 2011 with a lunar surface robot using telerobotics. Part of its job would be to validate the availability of resources for extraction. One of the potential subsequent missions is to land on the lunar surface and transform lunar materials to fuel.

There are other potential robotic precursor destinations listed. I can't say I object to those, although I'd prioritize the Moon for this sort of work. However, it looks like the Moon is fairly high in the robotic precursor queue already.

The robotic precursor work also includes a new line of affordable "scouts". Although the budget's description of this line doesn't call out any specific destinations, I have to think that the Moon is front and center there considering the private lunar surface competition.

Comment by red — March 12, 2010 @ 8:08 am

11. A great article. What gets me riled about all this is that the Augustine Commission stated, I believe, that unless Constellation receives an additional $3 billion a year for the next 5 years to meet the dateline as originally outlined in VSE, the plan was unworkable, as is. Don't tell me that after squandering 100's of billions of dollars on a "stimulus package", "omnibus" bill, bank bailouts, Wall Streets bailouts, auto bailouts, giving billions to shady groups such as ACORN, that Obama did not feel it was necessary to give NASA an extra $15 billion to complete Constellation. Where is NASA's bailout? And then he insults everybody's intelligence by stating that private industry will be used to transport personnel to ISS and into LEO by giving them $6 billion to kickstart a program. Who is he fooling? Even Burt Rutan says that Obama's plan is stupid. And private industry might take 10 years, if not longer to devise a system that is viable, if they even ever get one off the ground. The Augustine Commission was a smoke screen right at the outset to cancel the entire project. Obama even stated in his 2008 campaign platform that he intended to "postpone" Constellation for 5 years to help pay for his education reforms. I don't understand why there can't be both. Private and government projects. Obama instead decides to scrap 5 years and $10 billion worth of research and development and fork everything over to the private sector, essentially starting from scratch. If this decision doesn't prove that Obama is the new Walter Mondale, I don't know what is. Hopefully, Congress will reverse Obama's decision and if not, that a new US President with more common sense gets elected in 2012 and reverses the "new" policy.

Comment by Paul Mense — March 12, 2010 @ 11:27 am

12. Paul,

the Augustine Commission stated, I believe, that unless Constellation receives an additional $3 billion a year for the next 5 years to meet the dateline as originally outlined in VSE, the plan was unworkable, as is.

The committee did indeed claim this. But note carefully the qualifier: "the plan was unworkable, **as is.**" (emphasis added) That is a different statement than saying "the VSE cannot be implemented" under existing budgets, although the Augustine committee jumped to that conclusion. The more I dig into the costing numbers, the more questionable the assumption that we cannot go anywhere beyond LEO becomes. Stay tuned for more on this soon.

Comment by Paul D. Spudis — March 12, 2010 @ 11:44 am

13. Paul,

I don't know if you were in the audience on Wednesday, but I was and of course am very familiar with Jeff's thinking and the Augustine report.

In no way did Jeff or the report say that the VISION was unexecutable. They said that CONSTELLATION was unexecutable. They are different. In fact, Jeff repeatedly referred to the Aldridge Commission, on which you served, as supporting evidence for the technology plan and for commercialization.

Jeff remains a strong advocate for the Vision and, as some of your commenters stated, ALL of the technologies you discussed above.

You need to talk to Jeff or me before writing another piece like this.

Comment by Jim Muncy — March 12, 2010 @ 1:34 pm

14. Jim,

Thank you for your offer to educate me on what's in the Augustine report, but I can read. I am also aware that Constellation and the VSE are different, a point I have made repeatedly in these very pages.

The Augustine report states "Human exploration beyond low Earth orbit is not viable under the FY 2010 budget guideline." (Executive Summary, p. 17). But the committee largely ignored Constellation alterations and alternatives that are executable under the existing budget, including one using a Shuttle-derived HLV (side-mount) and a different one using commercial EELVs.

The biggest problem I have with the new proposed NASA budget is that it discards the solid strategic direction of the VSE — lunar surface (to learn how to use its resources to create sustainable human presence), and then beyond (including Mars and other destinations). In contrast to your statements, the VSE has been discarded; the Augustine committee left in "lunar orbit," but eliminated the lunar surface. Orbital lunar missions have no value because they cannot access and harvest resources. Instead, the "new path" proposes to spend billions on "technology development," with no specific applications,

missions or spacecraft envisioned. My point is that the history of this agency with such programs is not encouraging; NASA is now and always has been a mission-oriented entity. A few years of unfocused technology money down the drain and the entire program will become a ripe target for cancellation. In contrast, technology development in pursuit of a clear mission goal and set of objectives and activities will give you all the new technology you can handle.

I know that many in the space community hated the Constellation architecture with a passion and are taking great pleasure dancing on its grave. I myself had issues with some of it. But the Vision is a good direction, one that made logical programmatic sense and contributed to national needs. It enables us to extend our reach in space incrementally and cumulatively. All I see coming out of this new path is a lot of widget-making and the exchange of one batch of aerospace contractors for a new and different batch.

Thank you for your offer to check and approve my posts before publication in the future, but I'll pass on taking you up on that, if you don't mind.

Comment by Paul D. Spudis — March 12, 2010 @ 2:20 pm

15. re: Rocketman "None of the EELVs (or Falcon 9) are remotely close to providing human transportation to the space station." Atlas & Delta are real vehicles, whose reliability and costs are known. The path for upgrading them to be man-rated is fairly well understood, and ULA is probably the most qualified company around to depend on for launching valuable cargo. If we had gone with Atlas/Delta instead of Ares 1, crew would have been launching within a year of the shuttle ending.

Instead we have Ares 1, which has not lived up to any part of Simple, Safe, Soon. Choosing a new technology (SRB only) to launch crew to space was ballsy, and there have been too many design compromises for it to launch a crew of six to LEO — and we still don't know if it will truly be safe at all points of flight.

NASA is not a manufacturing company, and they are not in the business of being a business. We need NASA to oversee our commercial space programs, and to create the new technologies that will decrease the cost of access to space, and to move us beyond earth orbit. Constellation was a deviation in their focus, and their inherent high operating costs were going to freeze out any private use of their launchers.

Without frequent launches, you can't increase reliability and drive down cost. Atlas & Delta can do that, and Falcon 9 is on a path to do it too. No more eggs-in-one-basket. We need a true space industry.

Comment by Coastal Ron — March 12, 2010 @ 2:25 pm

16. *why should it cost more than Shuttle does now to support a launch system that requires an order of magnitude less preparation than the highly complex Shuttle Orbiter?*

Why do you assume Ares/Orion was dramatically less complex or cheaper to launch prep then Shuttle? That's certainly not what I saw in it – nor what GAO saw. Building a new Orion (which has about all the system complexity of the orbiter) for each flight, should reasonably cost more than tearing down and inspecting/servicing the orbiters.

Also given you'd need most of the overhead (R&D costs, facilities, maintenance), for far fewer flights, you would expect far higher cost per flights with Orion then with shuttle. Which GAO also projected.

Wings and a tail aren't big cost drivers.

Going to Mars is expensive because you must lift all of that fuel out of the deep gravity well of Earth. Even with the economies of scale provided by a super heavy lift rocket, it still costs tens of billions of dollars to mount such a mission. Making propellant on the Moon completely changes these numbers, ...

Not necessarily. After all you need a Earth to LEO launcher for anything you do in space, and most of the launch costs are fixed costs divided by the number of flights. So flying 20 times as often to fuel your Mars ship doesn't dramatically increase your Earth launch costs. Since a Lunar to space tanker dev costs would rival a Earth to LEO launcher, and then you need to develop the ISRU fuel mining/refining systems and add in their operating cost, studies not infrequently find its cheaper to buy and launch bulk supplies from Earth, rather then use Lunar resources in space.

Comment by Kelly Starks — March 12, 2010 @ 2:30 pm

17. *Orbital propellant transfer and storage, and in-situ resource utilization are both listed as technologies to be developed and demonstrated under the FY11 budget – they are*

flagship programs....*Robotic precursors to the Moon, both orbiting and lander/rovers, are funded under the FY11 budget... Simply put, although the budget outlines these as items in the "technology development" lines, I do not believe that we'll ever get anything from it. The history of the agency shows that they are very good at spending money on "technology development" but less proficient at producing any flight hardware from it.*

Sad but true. One of the main reasons stated for the VSE program was Washington saw that unless there was a milestone that required NASA to do something with their cool new tech – they never advanced it past endless studies and research. ...and now the new budget eliminates any delivery date, and just proposes endless studies.

Comment by Kelly Starks — March 12, 2010 @ 2:31 pm

18. Kelly,

Why do you assume Ares/Orion was dramatically less complex or cheaper to launch prep then Shuttle?

Because a lot of the fixed costs of the Shuttle system are spent in the OPF and the orbiter requires many man-hours of work for servicing, refurbishment and re-launch preparation. I am assuming that stacking big pieces in the VAB will cost less because it will take less time and personnel.

studies not infrequently find its cheaper to buy and launch bulk supplies from Earth, rather then use Lunar resources in space.

Yeah, I know that you can make a study to show anything you want. I base my comments on Gordon Woodcock's work, who has done the most thorough study of this that I am aware of. He concluded that for a chemical-rocket Mars mission, making the propellant on the Moon saved considerable money, even after building that lunar infrastructure. To which I add that the real benefit of lunar propellant production is the creation of routine access to all of cislunar space, where all of our space assets reside. My claim is that this completely changes the paradigm of spaceflight, from one-off, use-and-discard satellites, to maintainable and extensible distributed systems.

Comment by Paul D. Spudis — March 12, 2010 @ 2:39 pm

19. *If NASA was doing something innovative, something different than has been done before by building the Constellation, I might*

see that as a valid investment into the future. Constellation also was not focused on reducing the cost for access getting into space. At least the Space Shuttle was nominally oriented in that direction... at least in the 1970's. One engineering mantra or philosophy is that any item can be produced cheaper, sooner, or more reliable. Unfortunately you must choose only two of those options for whatever you want to make or build. Apollo choose to build something sooner and more reliable, at tremendous expense. Constellation and many of the proposals based on the experience from the Apollo project were based on the same philosophy, Comment by Robert Horning

Yes, Constellation was (in inflation adjusted dollars) FAR more expensive then Apollo, was to take twice as long to develop, and no be better in any significant way.

A good example is contrasting this with the DC-X of the mid '90's, which was projected to cost 3 years and $5 billion to develop into a production craft, and with orbital refueling could have taken gone to the lunar surface and back. A dramatic advance in technology and capacity at a reasonable cost. In contrast Orion / Ares-1 alone were to take 10-15 years and over $50 billion dollars to develop. Even comparing it to the $37 billion shuttle program makes the Constellation program a staggeringly high cost, slow to develop, crude system. Hardly something that would be a showpiece for the US or NASA.

Comment by Kelly Starks — March 12, 2010 @ 2:46 pm

20. *Because a lot of the fixed costs of the Shuttle system are spent in the OPF and the orbiter requires many man-hours of work for servicing, refurbishment and re-launch preparation.*

And funding a OPF to service and relaunch a orbiter is going to cost less then funding a factory to build a new Orion for each flight?

I am assuming that stacking big pieces in the VAB will cost less because it will take less time and personnel.

I don't see stacking a STS and stacking a Ares/Orion say would be THAT much different, especially enough to cover the sunk costs and per flight costs per Ares/Orion flight. (Ares-V + cargo would obviously be at least as big a stacking issue as a STS.)

studies not infrequently find its cheaper to buy and launch bulk supplies from Earth, rath- er then use Lunar resources in space. Yeah, I know that you can make a study to show anything you want. I base my comments on Gordon Woodcock's work

So does that mean Gordon used more valid assumptions and logic – or you just picked the study that gave you the answers you wanted? ;)

I add that the real benefit of lunar propellant production is the creation of routine access to all of cislunar space, where all of our space assets reside. My claim is that this completely changes the paradigm of spaceflight, from one-off, use-and-discard satellites, to maintainable and extensible distributed systems

I'd debate that. If you don't have cheap routine Earth to LEO transport, a lunar fuel supply won't help you (and likely isn't build able anyway). If you have cheap routine access to LEO – LEO fuel really isn't helpful.

Comment by Kelly Starks — March 12, 2010 @ 2:57 pm

21. *If you have cheap routine access to LEO – LEO fuel really isn't helpful.*

Yes, except many satellites are not in LEO, but in higher orbits that take fuel to reach. The community has been fixated on "cheap access to LEO" for years, but it's already "cheap enough" in the sense that we launch viable commercial assets now. What does not exist is a way to reach and service those higher orbit assets (MEO, HEO, and GEO) that have both commercial and national security satellites.

If you want "cheap access to LEO" we already know how to do that: outsource it to India and Russia.

Comment by Paul D. Spudis — March 12, 2010 @ 3:10 pm

22. *"If you want "cheap access to LEO" we already know how to do that: outsource it to India and Russia."*

Russia isn't so cheap anymore – $45M a seat soon? But for arguments sake, let's say that any U.S. launcher was going to be 10-20% more/seat than Russia. The multiplier effect of keeping those dollars here in the U.S. makes the overall cost far cheaper to do here. The funds flow thru our economy, and we build a stronger and robust space transportation system. I don't mind using the services and capabilities of other countries, but we need to keep a strong commercial presence here too.

NASA is not a manufacturing and operations entity, and has no incentives to lower the cost of access to space. Put them in charge of overseeing the emerging commercial crew launch programs (more than two), and give them back the charter to develop the next generation of technologies that we will need to leave LEO.

Regarding chemical fuel spacecraft to Mars, I don't think anyone can estimate the cost of operating a fueling station on the Moon, so any cost savings over alternatives is fictitious. Chemical fuel may be the best choice for Earth-Moon, but we need to start using more efficient propulsion to get us beyond.

I look forward to mining the moon someday, but I also see a strong commercial launch program as key to doing that. Without competition, we are not going to be able to afford to start any industry in space, and we the taxpayer cannot afford to shoulder the entire space program.

Comment by Coastal Ron — March 12, 2010 @ 6:31 pm

23. *"Thus, the goal of FP is to create Apollo-like spectacles for public consumption, rather than creating steps toward increased space faring capability."*

Exactly!

The Augustine Commission also distorted the development and operational cost for alternative architectures. Unlike for the Ares 1/V architecture, for the Sidemount/DIRECT alternatives, they required an extension of the shuttle program for an additional 5 years ($15 billion) plus they extended the ISS for an additional 5 years ($10 billion). So they added $25 billion in extra cost to the Sidemount/DIRECT alternatives to make it seem like these were nearly as expensive as the Ares I/V architecture. They also claimed that the operational cost of the Sidemount/DIRECT alternatives would be higher because they require three launches. Well, they only require three launches if you ban NASA from LEO and force them to use private commercial space craft. Only two launches would be required for the Sidemount/DIRECT architecture for a trip to the Moon– with over 80 to 90 tonnes transported into lunar orbit for a manned two launch scenario.

But there are all kinds of easy ways to reduce the developmental and operational cost of a lunar architecture. The simplest way, IMO, would be to use the Altair LH2/LOX descent stage as both an ascent and descent stage.

It would still have the same 17 tonne cargo capacity as an unmanned vehicle but as a manned ascent/descent vehicle it could operate as a people shuttle with about 1.5 tonnes for the pressurized transport module on top of the Altair and another 1.5 tonnes for passengers, pressure suits, and a few hundred kilograms of extra payload.

Of course, an Altair people shuttle could land several tonnes of payload with its passengers if an Altair cargo vehicle was used to previously place a solar or nuclear powered LOX manufacturing unit on the lunar surface. Then an Altair people shuttle would only have to carry the extra LH2 for future ascent down to the lunar surface. This could allow an Altair people shuttle to carry at least 10 tonnes of payload down to the lunar surface per mission.

Of course, a lunar LOX factory also means that we would only have to ship nine times less mass (LH2) to the lunar surface to produce water for a lunar base. A LOX factory on the Moon just makes too much sense to me! Plus I instinctively love when humans try to live off the land!

Comment by Marcel F. Williams — March 13, 2010 @ 4:24 pm

24. It seems that there is a difference of opinion as to what is important in the space discussion. Some feel that the ultimate goal is to send humans to Mars, but to what end? Many don't penetrate deeper than that, and so focus on simply putting boots down on the ground and planting flags and "exploring". Bolden himself refers to this when he says that it's ok that other nations go to the Moon, since the first 12 sets of footprints are already there and they are American (i.e., what's left there to do?). There are others who say that the important thing is to learn how to live, expand our sphere, create wealth, and eventually thrive off-planet. It seems like this latter thinking has more intrinsic appeal for those of us interested in space. Some of those advocates point to the resources of the Moon as the means to really dramatically reduce launch cost because you launch less "stuff". This also has some intrinsic value, as it makes sense that eventually, if we are going to evolve off-planet, sooner or later we must be able to use local resources (or wait for science fiction physics breakthroughs), so why not tackle that issue now and build that industry from infancy.

Following that line, now the question is, what role does NASA play in this "vision"? If it does have the primary role, then it should be defined and coded into its "mission". This was mostly the thinking with the VSE, but now there is nothing apparent in NASA's strategic mission. Further food for thought, there are those that say that the government is too costly and shouldn't be establishing this new infrastructure, and that it should be in the realm of the private sector. I don't think that anyone argues in principle that eventually the private sector would evolve, but the pertinent question should be, what is the government role now? Let's engage about that, since it is the driver (or should be) for most everything in the human space flight arena that NASA does. There is a lot of talk about Ares, Orion, EELV's, Direct, but all these launch vehicles are simply means. Frankly, the cost for launch is going to be high for the foreseeable future due to the fact that the physics is hard. Think about it; let's say that the cost of launching a kg can be reduced by even 25%. It is still a lot of money, and that cost will always stifle the drive to space. Look at the EELV progression from its start to now. Both Delta 4 and Atlas 5 are commercial endeavors that are still almost exclusively paid for by the government payloads they fly. Without the government, their business collapses. Why should it be any different with human payloads?

OK, so what if we use ISRU to lower the cost. Does ISRU help get it lower? If one propagates the concept past the initial development and learning cost, eventually there will be the long term payoff that reduces the cost below 25%, because you bring less and less with you. Hence, why not start that now and codify that scope in NASA's mission? But the point is to SET THE MISSION INTO NASA'S SCOPE and let it drive the decisions that NASA makes. Otherwise, you will get choices by NASA managers like sending the Project M Robonaut to the Moon on a robotic lander for a few hours of hand waving to the camera and not much other benefit for the cost of $500M – $700M. I believe that this was precisely the point that Paul was making.

Comment by flash — March 13, 2010 @ 5:52 pm

25. *The Moon is a classroom, a test bed and a supply depot. By using its resources, humanity can create the capability to live, work and travel in and beyond cislunar space. As a nation, we cannot and must not pass on this enterprise."*

YES!!!!! The moon's resources will define the "free markets" and drive the technology needed to develop those markets from LEO to cislunar and beyond. A commercial based VSE is still the right path. Flex has it backwards, its stalls the USA access to those resources for decades. Meanwhile China enjoys access and claim to the moons resources.

Comment by Doug Gard — March 14, 2010 @ 1:15 am

26. Paul

You and I have crossed verbal swords over the use of lunar hydrogen before and I won't riposte on that substantive point in your article; except to say that we seem to have a nice collection of nest eggs...

If I may, I would like to add three aspects to the Flex Path vs Moon First debate.

(1) The hoary chestnut of ownership. Whilst a FlexPath tele-robotic survey and ISRU experiment program will cause little international upset; a full blown VSE Armstrong Base exploiting these resources unilaterally would have been, shall we say, provocative. The current Administration is clearly of a multi-lateralist mind. Especially so when it comes to space. The potential for a lunar base, ISRU and all that you would like, will only be enabled WHEN the time is right. And it's not right at this moment. So rather than wait for the "New Space Order" the interim proposal lays down the technological foundations to ensure that America has a lead in such future activities.

(2) Space Faring resources. Whilst the Moon has abundant resources they are nevertheless at the bottom of a Well. With a Lunar catapult, tethers or even a Moon stalk... the future for the Moon as an Industrial resource is assured: IN THE LONG TERM. But, in the long term, there are many 'cheaper' and more valuable targets out there: providing we have the deep space infrastructure to go and get them. The Flex Path recognises that the Asteroids are the ultimate key to the future incorporation of the Solar System into Humanity's economic sphere. It is long term thinking but a correct conclusion IMHO. Again the new direction proposed lays down the technological foundations to ensure that America has a lead in such future activities. Whilst an Armstrong Base would be useful – as a tourist destination – with advances in tele-robotics and automation: Lunar Industries will need little, if any, human intervention. Compare this with a mobile ISS (International Space Ship) evolv-

ing over time into a deep space vessel with long term 'ISRU' (closed loop life support), radiation shielding, nuclear power and a halfway decent set of engines. The could call it "The Aldrin Cycler!" (3) panem et circenses Or: "No buck rogers, no bucks!"

"Thus, the goal of FP is to create Apollo-like spectacles for public consumption, rather than creating steps toward increased space faring capability."

This is where the Aldrin Cycler beats an Armstrong Base hands down. If the POR had come to fruition in the distant future and had survived the ennui: "but we did this back in the '60's dude" and had started to place real hardware on the surface, the public experience (apart from us space cadets) would have been ISS redux. I.e. a select group of civil servants performing routine micro-managed maintenance tasks. I would contend that your taxpayers would not have stood for it! Indeed I would imagine that after a couple of initial landings the architecture would have been abandoned as unaffordable. In contrast to building a monolithic architecture capable of doing one thing (very expensively) we have, in the Flex Path a sequence of demonstrations of increasing capability whilst developing those self same capabilities!

As to the fine print of the Augustine Report. This was a fig leaf to protect professional reputations that ought not to have been protected. But "History will out!" as they say!

David G. Lermit

Comment by brobof — March 14, 2010 @ 8:07 am

27. @Trent Waddington

You are seeing things that are not there. Paul Spudis made no 'personal attacks' on either Jeff Greason or any member of the Augustine committee. There not one shred of character assassination in his post.

Thanks for the article Paul.

Comment by Gary Miles — March 14, 2010 @ 3:52 pm

28. A Moon base will change everything. Once its established, the privateers and the settlers will quickly follow since reusable orbital transfer vehicles and lunar landers are a lot cheaper to develop and maintain that Earth to orbit vehicles.

Robots are the best way to exploit the asteroids– not humans. Just send a light sail out and grab a 10 to 100 tonne NEO asteroid a bring it back to a Lagrange point. Then you can produce all the oxygen and hydrogen that you want for your fuel depots. An aluminumized kite is far simpler technology than any nuclear rocket.

http://alglobus.net/NASAwork/papers/AsterAnts/paper.html

Comment by Marcel F. Williams — March 14, 2010 @ 5:38 pm

30. All space exploration is not worth much, if ultimately, space is like the center of the earth- not very accessible. It could be conceivable by some that astronomy is all that is needed in terms of space exploration; the argument "robots are cheaper" sometimes veers in this direction. It's widely understood that what is needed regarding exploration and exploitation is Cheaper Access to Space.

Our current situation is similar to desiring the internet and only having a few billion dollar computers. Until rather recently many people accepted the dogma that because space is so expensive only the government can do it. This idea still colors much of the current approach to space exploration, though this concept seems to be fading and probably within a few decades be such an odd idea that will be hard to grasp by the next generation.

Obama, whether he is actually aware of it or not, is currently putting the finishing nails in the coffin of socialism- though like all vampires, one should have no doubt that in the future, this vampire can re-emerge. The idea that "because space is so expensive, only the government can do it" is concept of the workable socialist paradigm. I believe it's axiomatic that socialism is incapable of "Opening the Space Frontier". I also think that some socialist "know" this or at least "feel" this. Btw, the only real "value" of socialism is stagnation- other than this possibly useful "product" it is mostly parasitical- it depends upon "change" and/or innovation from outside it's social system.

Opening up the space frontier will certainly involve change- it will be as some Chinese guy said "interesting times". Opening up the space frontier will radically alter this world. And if you abhor the "fast pace of modern times", you going to wish for the calmer periods of the 20th century.

Lunar water is all about Cheaper Access to Space. Lunar water isn't primarily about having water to drink or wash things, rather it's about having the ability to make cheap rocket fuel in space. Cheap rocket fuel is relative to current costs.

Today rocket fuel on earth costs a few dollars a kg, and in low earth orbit, it currently can't be bought, but if it could be, rocket fuel would be in the range of few thousand dollar per kg. In High earth orbit or lunar orbit, the rocket fuel would be about twice the cost as it would be in LEO. And on the lunar surface rocket fuel would be about 4 to 5 times as much in LEO.

So example of cheap lunar rocket fuel on the lunar surface would be in the range of hundreds dollar per kg- or about 1/10th the current costs. But the most significant aspect of lower price rocket fuel on the lunar surface is not the rocket fuel on the lunar surface, but being able to bring the lunar rocket fuel to lunar orbit. In other words if it was a matter of choosing between having 1/2 the price of rocket fuel in lunar orbit [and/or high earth orbit] or having rocket fuel at 1/10th it's current price on the Lunar surface, having rocket fuel at half the current price in orbit is more significant than having 1/10th the price on lunar surface.

Of course, it's not actually a choice of one or the other, but rather having the price be 1/10th the price on lunar surface enables the price to be 1/2 the current price in lunar orbit.

Comment by gbaikie — March 17, 2010 @ 11:01 pm

31. Can anyone comment about the cost effectiveness of rocket fuel from water ice at the lunar poles to LEO compared to rocket fuel from water ice from an NEO? Getting to the lunar surface would require new craft to drop into and out of the lunar gravity well whereas there is no significant gravity well when landing on a NEO. However the Moon is always close by whereas the few NEOs containing water ice orbit around the Sun.

Also, if the strategy were to move an entire NEO to a Lagrange point (using sails), then does going from LEO to an L point and back significantly affect the economics?

Comment by JohnHunt — March 26, 2010 @ 1:58 pm

32. John,

Can anyone comment about the cost effectiveness of rocket fuel from water ice at the lunar poles to LEO compared to rocket fuel from water ice from an NEO?

Well, as it's my blog, let me be the first to comment on it.

Each destination has specific advantages. For the Moon, it is close, accessible and we know what's there and what form it's in. Lunar water is present as ice, which is relatively easy to harvest and process (as opposed to chemically bound water). The NEOs have low surface gravity.

For disadvantages, the Moon has a 2.2 km/s escape velocity, which is an energy penalty for launching product into space. For NEOs, we don't know what form their water is in or how much a given asteroid contains, but for asteroids close to the Sun, it is very likely to be chemically bound water, meaning that significant energy is required to break those bonds during processing. Also, we cannot remotely teleoperate processing robots at NEO's because of the (minutes-long) time-lag in communication; we could do so on the Moon (3 light-second time delay.)

In short, each location both offers features and has issues. I prefer going to the Moon first to learn how to extract and use resources because it is the destination that offers the possibility of resource utilization with the least amount of unknowns. After we demonstrate that we can do ISRU on the Moon, it becomes much easier to do so on other objects.

Comment by Paul D. Spudis — March 26, 2010 @ 2:14 pm

Value for Cost:
The Determinate Path

March 24, 2010

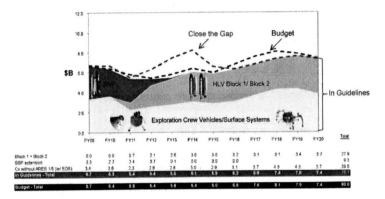

In contrast to the claims of the Augustine committee report, use of existing launch assets and infrastructure permit us to return to the Moon within the projected budget guidelines

The report of the Augustine committee[1] analyzes America's space program through a very narrow prism. Much of their report argues that the existing program of record[2] (more specifically, the Ares I and V launch system) is not affordable, a fact already apparent to most observers. Thus, the committee advocates human missions to destinations without deep gravity wells, eliminating the need to develop a lander spacecraft. The pertinent question about the original intent of the Vision for Space Exploration is not considered or addressed: Is a program to return to the Moon and create a permanent space faring infrastructure possible within the projected budgetary envelope?

So once again we have a space policy gap, a technical base mired in uncertainty and a citizenry that believes its once-heralded space agency has faltered – that NASA simply cannot send people on missions beyond low Earth orbit (LEO) because the government no longer has the interest or the money to support such efforts.

The administration's budget proposal[3] has directed NASA to concentrate on developing leap-ahead technology that will enable human exploration at some non-specific point in the future – "when we are ready" so as to excite and engage the public. As the shutdown of the Space Shuttle quickly approaches, people are surprised to realize that seats must be purchased on a Russian Soyuz to reach the Inter-

national Space Station that we've been building for twenty years. And we must continue to do this until commercial space transportation (seeded with NASA money) develops hardware certified for human flight. It is impossible to escape the sinking feeling that something just isn't right about this state of affairs.

The cost estimates of the Augustine committee report made by the Aerospace Corporation[4] perforce assumed certain developmental and operational scenarios, some of which could be seriously questioned. One of the most interesting points about the Augustine costing exercise was that it focused almost entirely on launch vehicles and not deep space systems. No scenario examined by the committee looked at the leveraging possibilities provided through the use of off-planet resources, apparently because they believed that the technology was too far out in concept and too far off into the future. "When we are ready" apparently didn't apply here.

In the committee's view of the future, human missions beyond LEO will remain staged and launched from the bottom of Earth's gravity well, along with all the required consumables for multi-week and multi-month duration missions of their Flexible Path[5] exploration option. For any significant effort, the enormous mass is launched using either existing expendable vehicles or a new, to-be-developed heavy lift rocket (for which their cost models estimate many billions of dollars). Is it any wonder that the committee's principal conclusion was that human spaceflight beyond LEO[1] is "unaffordable?"

The cost estimates the Augustine committee relied upon are unrealistic, bloated and artificially high. Although there is always uncertainty in estimating program costs, a system can be developed within the existing budgetary framework that gives us a heavy lift vehicle and human lunar return on reasonable timescales. The key is to keep new developmental costs low by maximizing the use of existing assets. A launch vehicle derived from the existing Shuttle system[6] can be developed from existing pieces while retaining virtually unmodified the existing VAB and pad infrastructure at the Cape. A single Shuttle side-mount can put 80 mT into LEO; two launches can mount a lunar mission. As shown in the accompanying illustration, this launch vehicle solution fits within the projected budgetary envelope of NASA, the very same budget profile assumed by the Augustine costing exercise.

Digging into the details of the cost numbers of alternatives suggests that Augustine assumed both significant developmental costs associated with the new launch system, along with relying on inflated recurring costs (as has been pointed out previously by others[7]). But more significantly, they also assumed we would continue indefinitely the current paradigm of launching everything we need in space directly from the surface of the Earth. Again, no consideration was given to using space-based consumables – specifically, LOX-LH2 propellant made from lunar water – to improve access and reduce total costs for missions beyond LEO. And yes, the committee was thoroughly briefed on the benefits of the Moon's extensive and accessible water[8]; they were apprised of the dynamically unfolding knowledge involving the enormous quantity of lunar water (both current knowledge and findings soon to appear in print) before, during and after their hearings. They simply ignored its significant impact and consigned the discussion of space resources to a few paragraphs describing technology development.

An affordable return to the Moon is possible under the existing budgetary profile. Such an affordable architecture based on Shuttle-derived assets was understood and in hand prior to the Exploration Systems Architecture Study. But the better became the enemy of the good enough. The needed incremental build up of cislunar spaceflight capabilities through the use of lunar resources[9] was discarded and a system was designed based on the assumption that mega-sorties and Earth-staged missions to Mars were the future of human spaceflight beyond LEO.

To unlock our Earth-bound budgetary shackles, we must employ the keys at hand. We must use existing flight assets to the maximum extent possible. We must use robotic missions and teleoperations to emplace surface infrastructure on the Moon prior to human arrival. We must begin water production from lunar resources as soon as possible using these robotic assets, with an eye toward integrating lunar propellant production into a reusable and extensible cislunar space transportation system.

To go forward, a determinate path to the Moon and beyond is required. The Moon is where humans will learn how to arrive, survive and thrive in space. To live and expand into the universe, we must take bold action to identify, access and harness those resources that are within easy reach. In order for America to remain a leader in space technology development – charting the way for human space flight – we need an objective to design and cost out, not just a budget to spend.

Topics: Lunar Resources, Space Politics, Space Transportation, Space and Society

Links and References

1. Augustine report, http://www.nasa.gov/offices/hsf/meetings/10_22_pressconference.html
2. program of record, http://www.nasa.gov/exploration/news/ESAS_report.html
3. budget proposal, http://www.nasa.gov/news/budget/index.html
4. Aerospace Corporation, http://www.aero.org/
5. Flexible Path, http://blogs.airspacemag.com/moon/2009/12/16/arguing-about-human-space-exploration/
6. existing Shuttle system, http://www.spudislunarresources.com/Links/klaus/EqualBudget.pdf
7. pointed out previously, http://www.marssociety.org/portal/AugustineNumbers/
8. Moon's water, http://science.nasa.gov/headlines/y2010/18mar_moonwater.htm?list1327682
9. use of lunar resources, http://www.thespacereview.com/article/1590/1

Comments

1. There is also the water in vapour form that ISRO's CHACE instrument found. This has been reported in Planetary and Space Science Journal. This is really exciting because it means that the moons atmosphere may contain more matter in it than previously thought.

Comment by lvs — March 24, 2010 @ 10:40 am

2. The scenarios painted here sound like science fiction. But it is exciting to think that this could be the reality tomorrow!

Comment by Ivs — March 24, 2010 @ 1:09 pm

4. Flex has one main purpose. And that is to fill the pockets of a select few "new space" companies fielding backwards splash down capsule cold war era technology. Flex is about easy quick money funneled to commercial companies to field big dumb boosters, even dumber splash down capsules and deep space projects that simply look-but-don't touch. Nothing concrete beyond flying to LEO ISS until 2020-2030. Then maybe think about landing somewhere. Easy money at the expense of our national manned space program.

Certainly Flex is no worse than Griffin's bumbled attempt to fulfill VSE through the ill fated Constellation program. So why would we want to be restricted to choose only among these two flawed programs. We need a rational, worthy, goal oriented program that embraces commercial development, while forwarding our national interest and searching for resources to spur new space based markets. Those resources are to be found on the moon, not LEO. This overstating of landing based programs demonstrates how slanted and one sided the Augustine panels finding were. Due to the vagueness of the Flex plan it now appears NASA's future rest with congress...and that is a long shot.

I must commend the likes of Rutan for steering clear of this subsidized quick easy money fray. Rutan has chosen to remain free market and to field worthy innovative advanced technology. Between the "New Space-X NASA" and Rutan my money is on Rutan and XCOR. Gods speed Burt!

Comment by Doug Gard — March 24, 2010 @ 8:56 pm

5. I agree with Paul's goals regarding the moon, but I'm a little fuzzy on your budget numbers.

On NASA TV last week, a Shuttle official (the Program Manager?) was talking about the costs associated with extending shuttle flights. He said that the operations overhead was $200M/month, regardless how many flights they flew. For an SDHLV, I would imagine it would be less, but still a lot. For that kind of money, we can build & launch a lot of 5 meter wide cargo on existing launchers (Atlas, Delta, Falcon) until we start seeing a true need for heavy lift.

I'm not against heavy lift. I tend to look at the need for heavy lift from a supply/demand standpoint. Until you get to the point that you would need too many 5 meter wide payloads to accomplish your goal, you don't need a 10 meter wide heavy lifter. It would be nice to have one, but from a mission & economic standpoint (that pesky deficit), I think we can get the moon program going earlier without one.

Comment by Coastal Ron — March 24, 2010 @ 9:43 pm

6. The Sidemount is probably the cheapest and the fastest HLV we could develop. And there's no doubt in my mind that we could get to the Moon before the end of the decade using directly shuttle derived technologies.

Unfortunately, NASA never put any effort in promoting Constellation as a Moon base program. And Bolden seems hostile to the whole idea of a Moon base.

Most people just don't seem to realize that once there is a permanent base on the Moon where people are mining the regolith for oxygen and water and other resources that our cultural universe will change forever! It will be the beginning of a new level of economic growth and will probably insure the long term survival of our species. I also think it would be culturally inspirational for our youth to see that they have an exciting future with new worlds to visit, colonize, and to explore!

Comment by Marcel F. Williams — March 24, 2010 @ 10:16 pm

7. Ron,

the costs associated with extending shuttle flights. He said that the operations overhead was $200M/month, regardless how many flights they flew. For an SDHLV, I would imagine it would be less, but still a lot.

A Shuttle-derived heavy lift system would be significantly lower in cost than the Shuttle proper, largely because much of the high fixed costs of flying the Shuttle are spent in orbiter refurbishment and preparation, especially the time-consuming Thermal Protection System (heat tile) overhauls. This work requires a large "marching army" of the Orbiter Processing Facility. A Shuttle HLLV involves stacking the large pieces in the VAB, a much less labor-intensive set of operations.

Comment by Paul D. Spudis — March 25, 2010 @ 4:11 am

8. Paul, as I explained in private correspondence, you can't plan a mission around technologies that have not been demonstrated. If you watch the proceedings you'll see Jeff Greason making the case for in-space propellant transfer.. which is why the smaller heavy lift vehicles were even considered.. he goes on to say that propellant storage is not mature and cannot be considered, even though he would very much like to explore infrastructure which includes it. ISRU of lunar ice is even less developed.

To me, your position at this point in time should be advocating robotic rovers to characterize the lunar ice and pushing ISRU and Propellant transfer/storage to be the focus of the initial Flagship Technology Demonstration missions. Developing a human beyond-LEO transportation system, now, will be a transportation system that does *not* utilize this technology and there will be no motivation to switch over to using the new technology if and when it becomes available.

Comment by Trent Waddington — March 25, 2010 @ 5:00 am

9. Trent,

Your reading comprehension skills need some work. I do not propose planning a mission around lunar surface ISRU; determining how to do ISRU **is the mission** of lunar return. But we cannot do that mission unless we are on the Moon, which the Augustine Flexible Path discards (and they do discard it, regardless of their subsequent claims to have retained it as an "option"). What this post focuses on is the claim in the Augustine report that Project Constellation is "unaffordable." Perhaps the current configuration of it is, but they did not seriously consider options to change Constellation so that it could be affordable. My other point is that money given to NASA for technology development in the absence of any specific strategic direction and destination will produce nothing and that will make the program a ripe target for cancellation in the future.

I have already advocated in this blog new robotic missions to the lunar surface to prospect for resources and demonstrate ISRU techniques. Again, such missions have little value unless they prepare for future human lunar return.

Comment by Paul D. Spudis — March 25, 2010 @ 5:57 am

10. One possible problem in the side-mount budget chart is that the exploration and Shuttle accounts in the NASA budget are smaller than the ones in this chart, even if those were the budget guidelines. That funding is probably going to other things (ISS, Earth observations, Aeronautics, general space technology) that are probably going to get higher priority.

Another is the opportunity cost. If this budget chart is implemented, I think one sacrifice would have to be the commercial spaceflight line. That leaves us without ISS crew support and with less capable commercial cargo support. Side-mount could step in, but those ISS missions and capabilities could be expensive, and could have safety issues if side-mount supports crew (which block II could). Possibly more important are the non-ISS opportunities presented by these potential services. For COTS cargo, examples might be DragonLab, Taurus II filling the otherwise lost Delta II niche, and the Falcon 9 attempt at reducing launch costs somewhat. Crew support could have even bigger "side benefits". Could commercial crew be fit in the budget?

I'm not sure if it's counted in the chart's "surface systems", but you might also lose items in the new NASA budget like the robotic precursors, demos for things like accurate landing, ISRU, propellant depots, automated docking, inflatable habitats, and closed-loop life support, a U.S. version of RD-180 (which would be useful beyond exploration or HLV), and so on. I guess the question becomes can we fit enough of this sort of thing into that side-mount chart to make the exploration effort cost-effective and worthwhile?

Comment by red — March 25, 2010 @ 6:47 am

13. red,

but you might also lose items in the new NASA budget like the robotic precursors, demos for things like accurate landing, ISRU, propellant depots, automated docking, inflatable habitats, and closed-loop life support, a U.S. version of RD-180 (which would be useful beyond exploration or HLV), and so on.

One cannot lose what one never had. All that you list here are in the new budget as "technology development" projects. One of my points is that historically, this agency has never produced any flight hardware from a technology development line. It's the other way around — they develop technology by designing and building flight hardware. The new budget proposal trades in a spacecraft development program (irrespective of its fiscal and technical issues) for Powerpoint promises.

Comment by Paul D. Spudis — March 25, 2010 @ 7:40 am

14. Suppose there was a mission to mine and store 1 ton of lunar water. How much would such a mission cost? Could 500 million dollar prize be enough for such a task? Or perhaps, one would first need more information about where precisely on the moon that one could mine a ton lunar water. How many missions would this require? Could one mission have a reasonable chance of successfully doing this? Perhaps this type of mission could also be a prize- say something added to Goggle X-prize, if one finds minable water then you get in addition 50 million. And if there was a known location where lunar water could be mined, could the prize to mine it be significantly less than 500 million- say 300 million? And once ton of water is mined then you could have missions which used the water.

One way to use the water would having facility that made fuel which can be used to power larger machinery which mined larger amounts of water. Other uses could be used to power lunar vehicles fly around the general area doing further exploration- for more water or other types of resources.

With such an approach it's possible that before a NASA manned mission the Moon one already has commercial lunar water mining, NASA may at that point modify it's manned mission or perhaps even bypass the Moon and go to Mars- and use lunar water which been processed into rocket fuel for this Mars manned mission.

Comment by gbaikie — March 25, 2010 @ 8:26 am

15. *perhaps, one would first need more information about where precisely on the moon that one could mine a ton lunar water. How many missions would this require? Could one mission have a reasonable chance of successfully doing this?*

Based on what we now understand about the Moon, probably we could. What is needed is a carefully thought out plan to characterize and demonstrate ISRU on the lunar surface first with a series of robotic precursor missions, building up knowledge and capability with time, and ultimately supporting and provisioning human presence on the Moon. Such an architecture is possible within the proposed budget envelope.

Comment by Paul D. Spudis — March 25, 2010 @ 8:44 am

16. Think of all the machinery needed to mine materials on the moon, together with the cost of getting them there. Instead, think of a "Base Camp" in LEO, where vehicles can be docked together for journeys beyond. Energy to get to LEO is about four times energy needed to go from LEO to moon or Mars. (delta v squared). Propellant transfer may not be needed, only relatively short times of propellant storage.

Comment by O Glenn Smith — March 25, 2010 @ 9:21 am

17. With our experience operating rovers on Mars, I think this points the way for how we should exploit water resources on the Moon.

We already struggle to keep a small contingent of humans aboard the ISS in LEO, and supporting humans on the Moon will require a large supply infrastructure. Until that happens (i.e. commercial cargo + crew), we should pursue landing robot exploration and mining equipment.

There is a lot of work we can get done from afar, especially with the low communication latency (seconds to/from the Moon versus minutes to/from Mars). We can overcome the lack of human maintenance by use of modular design and sending a constant stream of additional equipment. Without the need of an ascent system, we can put large pieces of equipment on the surface. The Altair lander ascent module weighed 20K lbs, which would give you a huge vehicle to drive around the Moon. I'm all for human exploration, but this to me is low hanging fruit.

All of this can be done using existing launchers, which postpones the need for funding an HLV until it's needed for human exploration needs.

Comment by Coastal Ron — March 25, 2010 @ 12:20 pm

18. *Think of all the machinery needed to mine materials on the moon, together with the cost of getting them there.*

The quantity and sizes of robotic machines needed to begin resource processing are not large. A small rover (MER-sized), dirt-moving equipment (drag line), and a processing center (about the size of a large office desk) can produce several metric tons of water per lunar day. The required infrastructure investment can start small and build up both size and capacity with time.

Comment by Paul D. Spudis — March 25, 2010 @ 12:34 pm

19. Ron & Paul, Sidemount or DIRECT operational cost could be further reduced if the SRBs were expendable and no longer had to be refurbished. This might also add several tonnes of additional payload per launch which might also lower cost.

http://www.spaceref.com/news/viewnews.html?id=1177 http://ghostnasa.blogspot.com/

Comment by Marcel F. Williams — March 25, 2010 @ 4:19 pm

20. Hi Paul,

Thanks again for an interesting and rational article. Always gets me thinking. Just wondering about your thoughts on the following hypothetical situation: in comparison to the moon's surface, if for less delta V (from LEO) we could reach a low gravity object in a Helio-centric orbit – one that contained useful compounds such as water – would this make the moon less attractive as our resource supply for cis-lunar space? Just wondering this scenario would give support to the flexible path idea.

Comment by Phil Backman — March 26, 2010 @ 1:26 pm

21. Phil,

if for less delta V (from LEO) we could reach a low gravity object in a Helio-centric orbit – one that contained useful compounds such as water – would this make the moon less attractive as our resource supply for cis-lunar space?

If delta-v were the only consideration, perhaps. But a big advantage of the Moon is that it is both close and accessible. Its closeness means that round-trip light-travel time is < 3 seconds, meaning that near real-time telerobotic control from Earth is possible; we can do much of the resource prospecting and processing by robot machines. Such telerobotic control is not possible for NEOs because they are usually much more distant (delays of many light-minutes). By accessible, I mean that we can go to and come from the Moon pretty much at will; there are always abort options that permit return to Earth within a few days at most. Some NEOs have very few windows in which they can be accessed or from which you can return to Earth easily.

Comment by Paul D. Spudis — March 26, 2010 @ 1:51 pm

22. *if for less delta V (from LEO) we could reach a low gravity object in a Helio-centric orbit – one that contained useful compounds such as water – would this make the moon less attractive as our resource supply for cis-lunar space?*

No, it would make more attractive.

The problem with NEO's is time. But there is a lot of water and other resources in NEOs. If you need large quantities of resources in cis-lunar space then NEOs will be the place to get them. In other words if need somewhere in the order of hundreds of thousands of tonnes of water- and at a price of somewhere around $10 kg [$10,000 per ton or $1 billion per 100,000 ton].

I would say by the time you mining NEOs, you will be on the verge or in process of powering the world with electrical power shipped from Space.

The problem is we don't need a lot of stuff in cis lunar in near future- less than a hundred tons to thousands of tonnes within a decade or so. Once you at the point of having a demand for thousands of tons per year, then NEOs will be an obvious place to get it cheaper [which makes getting to the Moon even cheaper].

A way to look at it. Put a thousand tons of water in a barrel on best rock you can find. And put thousand of water in barrel on the Moon. Which is worth more money?

One aspect is the water on the rock has to moved to cis lunar to have a market and the one on Moon doesn't need to be move off the Moon to have market. In other words lunar water has a market on the lunar surface and cis space, and water on the rock has it's market in cis space [though by time you are actually mining NEOs the market will be in cis lunar, LEO and maybe Mars and the lunar surface {and or cyclers going between Earth and Mars or other places and etc}].

Comment by gbaikie — March 26, 2010 @ 7:19 pm

23. Paul: "All that you list here are in the new budget as "technology development" projects. One of my points is that historically, this agency has never produced any flight hardware from a technology development line. It's the other way around — they develop technology by designing and building flight hardware. The new budget proposal trades in a spacecraft development program (irrespective of its fiscal and technical issues) for Pow-

erpoint promises."

I think the 2011 budget line for things like inflatable or light-weight habitats, closed-loop life support, ISRU, fuel depots, and other items is mainly a technology demonstration line. These missions are intended to be real flight hardware in space. I'd compare them to DS-1 or EO-1: real space missions whose main purpose is to demonstrate new technologies in space so later "production" missions can comfortably use these technologies.

The robotic precursors (the line of larger robot precursors and the line of smaller "robotic scouts") are also flight hardware.

The RD-180 equivalent development is different. It has an implied mission focus (improved equivalent to the RD-180 on the Atlas V and ultimately HLVs that is made in the USA) that I suspect may flounder as a tech project unless NASA gets DoD to contribute budget and mission focus – but the 2011 budget implies that NASA is looking for that DoD focus.

I wouldn't argue that NASA will be successful with all of these technology flight demonstrations, since there will undoubtedly be failures. There is also a lot of non-flight technology research and development in the budget of the sort I think you're critical of. I also think there could well be a case for shifting the balance somewhat so NASA is doing more "production" rather than "demonstration" flight hardware, since there are probably cases where we don't need new technology. However, I'm a lot more optimistic that the technology demonstrations will be valuable, so I wouldn't want to shift the balance too far in the other direction, either. I think there's a balance where we could make progress on developing actual beyond-LEO missions while at the same time doing robotic precursors, technology demonstrations, and maybe some tech R&D ... but that balance probably wouldn't satisfy either "camp".

Comment by red — March 27, 2010 @ 11:36 am

24. We need a Moon base. And we've needed a permanently manned lunar facility since the end of the Apollo program.

So we need to develop the architecture to establish that Moon base. But no break through technologies are really required to do this. And the simpler and faster we do it, the better. And once this new space architecture is developed, everything that we attempt to do– beyond the Moon– will be a lot easier and a lot cheaper and a lot faster to develop.

But first we need a Moon base!

Comment by Marcel F. Williams — March 27, 2010 @ 5:55 pm

25. red,

I think there's a balance where we could make progress on developing actual beyond-LEO missions while at the same time doing robotic precursors, technology demonstrations, and maybe some tech R&D ... but that balance probably wouldn't satisfy either "camp".

This may be right, but it's not my point. My contention is that NASA as an entity has shown itself unable to wisely spend technology money and then produce flight hardware from it. Such a condition is independent of your hypothetical program balance.

Comment by Paul D. Spudis — March 28, 2010 @ 5:05 am

26. EML-1 and EML-2 lie at the crossroads of cis-lunar space and the NEOs. A depot and/or transfer station and/or warehouse at EML-1 or EML-2 would be immensely valuable.

24/7 global lunar access without concerns over orbital inclination or launch windows, easy on-ramps to favorable trajectories to beyond cis-lunar space, and the ability to accumulate supplies, equipment and propellant delivered from LEO by "slow boat" — high delta t, low delta v trajectories.

These "slow boat" routes include ion propulsion and ~100 day single impulse ballistic trajectories. See Jeffery S. Parker's paper, here:

http://ccar.colorado.edu/nag/papers/AAS%2006-132.pdf

It's all about the gravity . . .

Comment by Bill White — March 28, 2010 @ 12:53 pm

NASA Lost Its Way

April 2, 2010

From the film 2001: A Space Odyssey. *Not in my lifetime. And probably, not in yours either.*

As we survey the wreckage and ruin of yet another NASA "return to the Moon" program, the inevitable "what went wrong?" arguments play out. We're in a much different place today than we were when Apollo 11 reached the Moon (and each year there are fewer of us alive who witnessed it). To some of us, this is not a new movie – we've been to this show many times before. Although some aspects of the experience convey a startling sense of déjà vu, in other respects, this time it was a very different event. While one could rightly blame previous unsuccessful efforts on politics, this time another culprit brought us to the tipping point.

Our efforts to return people to the Moon after Apollo came from a sense that such a move was inevitable. This feeling largely came from Wernher von Braun's vision[1] of our future in space. von Braun wrote a series of magazine articles and books[2] in the early 1950s that outlined a sequence of steps that would lead us into space. They were so logical that despite the out-of-sequence Apollo lunar landings, NASA returned to this template after we "jumped ahead" of his vision sequence. The von Braun architecture began with a rocket that could routinely go to and from Earth to low Earth orbit, followed by the construction of a space station, the building of a transfer vehicle, lunar landings, and finally a manned mission to Mars. This "stepping stone" sequence was to give us both routine access to space as well as move humanity out into the Solar System.

In order to answer the challenge posed to America by the Soviet Union in space, von Braun's vision was altered when President

John F. Kennedy called for Americans to go directly to the Moon. Afterward, NASA tried to pick up von Braun's original sequence (shuttle, station, Moon tug, Mars) but by then, the logical appeal of his architecture had faded. The Moon landing ignited passions about space, popularizing and expanding study of science and engineering. But each failed "vision" has seen our country retreat from space exploration, fall behind in engineering, and our dreams of moving into space have faded away. The logical sequence of manned exploration of the Solar System had stalled; the program fought for its very existence by promising new rockets, new space stations, and landings on the planets, "sometime" in the future.

I have argued previously[3] that you must understand your mission before you make decisions on how to accomplish it – the objective of your trip may well have relevance to decisions on launch vehicles and architectures. The Vision for Space Exploration (VSE) proposed by President George W. Bush in 2004[4] was exciting not for its chosen destinations ("Moon, Mars and Beyond") but because it clearly and cogently articulated the mission (the purpose) for space exploration and the need for lunar return – to use the resources of the Moon[4] to create a sustainable human presence there. Once we learned how to do that on the Moon, the entire Solar System was the objective.

The VSE laid out a path forward that would change the paradigm of spaceflight, from one-off missions where everything we needed must be brought from Earth, to one where fuel and other consumables are extracted from what we find in space, thus creating an extensible, reusable space faring infrastructure that conquers the budget busting limitations imposed by our residence inside Earth's gravity well. Numerous articles[5] since the announcement of the VSE expounded on these goals and objectives[6]. They were widely discussed[7] and disseminated to the public[8]. The legislative branch responded to President Bush's proposed mission twice, with strong, bipartisan endorsements of the VSE in 2005 (Republican majority) and in 2008 (Democratic majority).

Although the purpose of the VSE was clear to the White House and the Congress, it became increasingly clear over time that NASA was having difficulty understanding the mission. They eventually embarked on a multi-year study to define exactly why they had been tasked to go to the Moon and to understand what they might do once they got there. The mission to understand their mission involved lots of meetings, workshops and conferences, whereby all the "stakeholders" had an opportunity to give their input. All this "input" was distilled into a series of documents[9] containing six themes and 181 different specific objectives.

No one at NASA could state the mission of the VSE in a single sentence.

Recently, former Administrator Michael Griffin was interviewed[10] about the "new path" for NASA. Among the questions about the demise of the Constellation program, he was asked:

Ars: What was the imperative for developing the Moon? Was it because it was felt its resources could be used for longer-term exploration?

Griffin: Well, the Moon is interesting in itself. And the United States bypasses the Moon at its peril, because other nations—as they develop space capability—will not bypass it. So, the Moon is interesting in and of itself. Secondarily, the experience of learning to live and work off-planet will be valuable... it may not be essential, it may be possible to go to Mars without learning how to utilize the Moon. But, as I say, it is not advisable. So, the experience of learning how to live on another planet only three days from home, I think, is enormously valuable, before we set out on a voyage where our astronauts will be seven or eight months from home.

I read this answer in stunned amazement. The former Administrator of the agency charged with executing the VSE omitted the principal reason for going to the Moon: to use its resources to create new space faring capability. The interviewer seemed more informed about the reason for a lunar return than the former Administrator; he even teed up the answer within his question!

This lack of understanding of the mission didn't just emanate from the top. A recent quote from Jeff Volosin[11]:

"We really never had a compelling reason to send humans back to the Moon .... More than that, we really, really don't have a compelling reason to set up a permanent presence on the lunar surface – we really don't."

Really Jeff?

Jeff Volosin worked at NASA in the Exploration Systems Mission Directorate. He was in charge of collating and synthesizing the results of the agency's efforts to articulate the reasons we were returning to the Moon (the one with six themes with 181 different specific objectives).

Lest you think that such was then and now is different, the current NASA Administrator, Charles Bolden, was recently asked[12] what he thought about China going to the Moon. His response:

"There are six national flags on the surface of the moon today. All six of them are American flags. That's not going to change."

In other words, because an American planted a flag on the Moon 40 years ago, there is no possible reason for the United States to want to go there, nor to be concerned about another country doing so. Plus ça change, plus c'est la même chose.

You may well ask, what is the purpose of wallowing in this sordid saga of utter cluelessness? First, it is important to understand that while the reasons for the VSE were clearly understood by some[13] (and that includes many dedicated, smart, hard-working people within NASA), many more either never understood it or refused to accept it and could not explain it to those who needed to know. So their selection of a doomed architecture may well have been inevitable.

Second, this experience offers food for thought to those who think the "new path" for NASA[14] (as laid out in the proposed budget[15]) will somehow magically transform the agency into a fount of technical and scientific excellence. NASA couldn't understand why we were going back to the Moon, which confused their reasoning about "how" to get back under the existing budgetary envelope. So why should anyone believe that with the "new path," NASA will be able to go to Mars and beyond?

Finally, unless and until scientists and engineers jointly embrace the objective of making human reach into the Solar System permanent and affordable, our country and its space program will continue to diminish. Robotic missions are important but their true value lies in enabling sustained human exploration and settlement of space. With the increasing evidence of vast lunar resources[16], a logical sequence of stepping stones into the Solar System is more relevant than ever. It is the way back to what was once the promise of NASA.

Topics: Lunar Exploration, Lunar Resources, Space Politics, Space Transportation, Space and Society

Links and References

1. von Braun vision, http://history.msfc.nasa.gov/vonbraun/disney_article.html
2. series of magazine articles, http://home.flash.net/~aajiv/bd/colliers.html
3. argued previously, http://blogs.airspacemag.com/moon/2010/02/13/confusing-the-means-and-the-ends/
4. VSE, http://www.spaceref.com/news/viewpr.html?pid=13404
5. numerous articles, http://www.spudislunarresources.com/Links.htm
6. these goals and objectives, http://www.spaceref.com/news/viewsr.html?pid=19999
7. discussed, http://www.spaceref.com/news/viewnews.html?id=1116
8. disseminated to the public, http://www.spudislunarresources.com/Opinion_Editorial/SVM.htm
9. series of documents, http://www.nasa.gov/exploration/home/why_moon.html
10. Griffin interview, http://tinyurl.com/yensdox
11. recent quote from Jeff Volosin, http://www.spaceref.com/news/viewnews.html?id=1378
12. Bolden was recently asked, http://www.denverpost.com/headlines/ci_14385263
13. understood by some, http://www.spaceref.com/news/viewnews.html?id=1334
14. new path for NASA, http://blogs.airspacemag.com/moon/2010/02/03/vision-impaired/
15. proposed budget, http://www.nasa.gov/news/budget/index.html
16. vast lunar resources, http://tinyurl.com/3xgkff3

Comments

1. Thanks for another excellent article. I love the simple logic that von Braun had for our space program. And I still think its viable today. Using a space station as a way station for missions beyond LEO is still a good idea. Unfortunately, the ISS is in a great orbit for Russian launches but not for American launches.

Of course, if we attempted to use the shuttle to deploy a simple space station in an orbit more appropriate for US manned launches, maybe using Bigelow technology, there would probably be heavy criticism that we already

have a space station.

Still there is the question about why the tax payers should spend $100 billion over the next 5 years for a program that doesn't build anything or go anywhere.

I was reading some old Moon base research papers by you and other scientist and thought to myself that its too bad the rest of the public are not aware of these fascinating and exciting visions of the future.

And that was the core problem with the Constellation program, IMO. It was never promoted as a Moon base program. In fact, it appears that NASA was afraid to promote it as such which caused a lot of confusion. A permanent and continuously growing lunar facility would absolutely capture the public imagination, IMO.

But instead the Moon was treated as a place that we really didn't want to go to because we really wanted to go to Mars. Ironically, I don't think we really can get to Mars without a Moon base.

Comment by Marcel F. Williams — April 2, 2010 @ 6:49 pm

2. The White House and Congress need to take some of the blame for the saga of utter institutional cluelessness. The VSE vision voice by President Bush was great: but there was no follow-through to make sure the VSE did not get perverted into a crypto-Mars program. Heads needed to roll–and they still do: they can start with Bolden.

I don't know why it's so hard to pick competent NASA administrators.

Paul, we've got to get you in there somehow as the next NASA administrator. I know the job won't be any fun, but you've got your duty to your country to think about. Please take the job when it is offered.

And us Constant Readers can help out by creating buzz and starting a letter writing campaign. I think when Mitt Romney gets elected in 2012, he can pick the right person, but only if he's made aware of who the right person is. As space activists, that's where we need to focus for the next couple of years.

Comment by Warren Platts — April 2, 2010 @ 7:33 pm

3. "Using the resources of the Moon" causes people to think about Helium-3 for fusion reactors that don't exist yet. It will take a very big marketing effort (more than just this blog) to get people thinking about the resources of the Moon as raw materials for propellant processing plants that don't exist yet. A great way to

break away from that would be a robotic precursor mission to demonstrate that the technology is ready.

Comment by Trent Waddington — April 2, 2010 @ 8:21 pm

7. I'm not sure why Helium-3 keeps being mentioned as a power source since we don't have any fusion reactors and we don't even know what the capital cost of a fusion reactors will be. Fission reactor fuel is extremely cheap, but it still has high capital cost.

You don't need Helium-3 to provide Earth with power from the moon. Solar energy, or even nuclear fission energy, could be beamed back to Earth from the Moon via invisible infrared lasers. The beam spread from the Moon to the Earth should be less than 2 kilometers in diameter. Infrared photovoltaic energy farms on Earth could then operate 24 hours a day instead of only during the daytime.

Comment by Marcel F. Williams — April 3, 2010 @ 5:38 am

8. By the way, check out this March 31st video of what astrophysicist Neil deGrasse Tyson has to say about the NASA budget and the Obama plan. Its very interesting.

http://newpapyrusmagazine.blogspot.com/2010/04/how-much-would-you-pay-for-universe.html

Comment by Marcel F. Williams — April 3, 2010 @ 5:42 am

9. Marcel,

I'm not sure why Helium-3 keeps being mentioned as a power source

It keeps coming up largely because of its energetic and continuous advocacy by Jack Schmitt, who is adamant that it is "the answer" to Earth's energy problems. It has never been a part of any architecture or strategy for lunar return that I have advanced or promoted. I want to go after the easy stuff first (water) to bootstrap a sustainable human presence in cislunar space. Once that is established, we can look at other possibilities.

Comment by Paul D. Spudis — April 3, 2010 @ 8:52 am

10. Yes, there is some sort of disconnect mechanism in the brains of NASA higher-ups when asked about purposes for spaceflight. The answer ALWAYS is that space is interesting, space offers scope for great science, space will generate friendlier and more productive contacts among all the peoples of Earth.

Which butters no parsnips. Yet this is as much as NASA leaders will allow themselves

to say, it is what US presidents and their science advisors (mostly) echo, it is the language of Congressmen and US government publications, this is the discourse of the academic/think tank world, it is most passionate rhetoric of the space policy world.

I'm not sure why. And to be honest, it seems to puzzle other people as well, to judge by their books, Joan Johnson-Freese for example (SPACE, THE DORMANT FRONTIER), or W.D. Kay (CAN DEMOCRACIES FLY IN SPACE?, DEFINING NASA).

My suspicion is that at the highest levels of the American government (and the Russian and Chinese and Swiss and Sumatran governments) civilian space programs simply aren't seen as having much value at all — they're a lure for recruiting clever young engineers into the aerospace job market, and a convenient cover for the military and intelligence space programs that Really Matter. Which is a partial explanation for civilian space programs proceed at such a glacial pace and with such lukewarm endorsement, but obviously not a complete one.

Comment by mike shupp — April 3, 2010 @ 6:02 pm

11. Hell, I'd be satisfied if we were simply extracting oxygen from the regolith, 89% of the molecular component of water:-) Humans have been melting rocks for thousands of years. Extracting hydrogen, carbon, and nitrogen from the polar areas would be icing on the cake!

But it's going to be difficult to go to the moon with a president and administrator that doesn't want to go there. But its also going to be difficult for Congress to fund NASA at $20 billion a year, $100 billion over the next 5 years, and not build something!

So I do think there could be immediate funding for a heavy lift vehicle and perhaps an Earth Departure Stage. And with those things already funded, developing a lunar lander and lunar base modules might look much more tenable in the near future.

Comment by Marcel F. Williams — April 3, 2010 @ 9:55 pm

12. It's just my opinion, but I think you'll see space travel, including to the moon, become privatized/commercialized. The real objective is no longer science; the objective is to make money rather than spend it. You'll see commercialization of the moon (resorts and advertising visible from the earth) before you'll see the government shelling out any further vast amounts of money for moon or space exploration.

The moon will belong to whoever can get there first, build those resorts, and establish tourism. Keeping the moon a pristine laboratory where the scientists can play and space vehicles are produced and launched...I really don't think that's going to happen. Man keeps nothing pristine and his greed is rampant. Exploitation of the moon and space is where man is headed, just as man has exploited the planet. Science now is an afterthought. We can't afford it.

Comment by tasha — April 4, 2010 @ 11:41 am

15. tasha,

You'll see commercialization of the moon (resorts and advertising visible from the earth) before you'll see the government shelling out any further vast amounts of money for moon or space exploration.

There are considerations for government involvement on the Moon other than science. I have argued that routine access to cislunar space, enabled by developing and producing propellant from lunar materials, is reason enough for a federal government space program. All of our satellite assets reside in cislunar space. This includes not only national strategic assets, but maintenance and protection of commercial assets too. Science is only something done to serve national interests, not an end unto itself on the Moon.

Comment by Paul D. Spudis — April 4, 2010 @ 3:47 pm

17. *to use its resources to create new space faring capability*

Amazing how clear and concise that is. Water on the moon...water on the moon! Go get it! Get it? Use the proven resources at hand, ULA, Atlas V, Delta, Bigelow keep it commercial based, low cost, innovative in application not hardware development and most of all sustainable. We do not need to flex around LEO for 2-3 decades. The mission is clear, concise, focused and worthy. Get the water first then go for the H-3. The rest will follow.

Comment by Doug Gard — April 4, 2010 @ 5:59 pm

19. Paul,

A robotic LH_2/LOX factory on the Moon would be the commercial, profit-center solution we've been looking for to could get government (NASA/President/Congress) out of the way in the road to space permanently. Who wouldn't want to own the gateway "filling station" to the rest of the solar system?

I wrote up a description of what it might take

to build such an "icebase". It's at

http://www.bobcarver.net/icebase/icebase.pdf

Comment by Bob Carver — April 4, 2010 @ 8:13 pm

20. Paul,

I am saddened, but not all that surprised. I remember talking to many folks who recognized something major needed to be accomplished before administrations changed or it would all be reset to zero. That is why I feel if another administration, or even this, propose going to Moon as a space goal it will only last as long as they are in office. If the program is not able to be implemented in that time frame, or be at the point is also ready to go as was the case with Apollo, it won't happen.

That is also why the only practical solution I see will be creating an organization dedicated to lunar exploration and development. Its the model that worked well with the major dam projects in the U.S., TVA, BPA, etc., as well as with the building of the Alaska Railroad. A federally, or internationally, chartered Lunar Development Corporation, with a clear mission statement in its charter and focused on bringing government and commercial resources together is the best bet for moving forward on both lunar exploration and economic development.

Comment by Thomas Matula — April 4, 2010 @ 9:51 pm

22. The problem with Constellation isn't that the budget was too small; the problem was that NASA is simply incapable of operating efficiently. For all its flaws, Russia still manages to operate a very effective space program for a fraction of what NASA spends.

NASA has become an enormously bloated government agency with appallingly bad managers (two shuttles destroyed as a result of management incompetence should be proof enough of that, but the handling of Constellation is the icing on the cake). The organization is so horribly dysfunctional that hammering some "single sentence" mission statement into its collective skull won't help a whit. They're simply not capable of replicating — much less bettering — the great achievements of yore.

I think the president's emphasis on private industry is a subtle recognition of NASA's hopeless deficiencies.

I say close the book on NASA's management. The place is broke and I don't think it can be fixed. Time to try a new approach. One that might actually succeed.

NASA Lost Its Way

Comment by Matt — April 5, 2010 @ 12:26 am

23. All that commercial/ private aerospace companies will be able to deliver will be Mini Space Shuttles. The Space Shuttle on steroids! It'll be 1981 all over again. Privatizing spaceflight—100 Percent—will be a wholesale disaster, as far as any real future space exploration is concerned; because then we will never leave LEO! If the rocket lacks an earth-escape stage, then you basically CAN'T go anywhere! Hence, once you are in LEO, some 200 or 300 miles up, that is the finish of your "journey": just going around in circles, in a free-fall state around Earth. What about getting NASA out of LEO, and into deep space?? You CAN'T launch a Moon or Mars expedition on board the Venture Star! There's simply NO payload space. And this piece-by-little-piece approach, used for the ISS, will just not work, when it comes to manned deep space operations; going to planetary destinations. President Obama has proven to be ignorant, naive & gullible, to be thinking in terms that commercial firms are going to jump-start deep spaceflight. Believe you me: THEY WON'T!!! Thanks to that misguided segment of the space interest community who are dead-set against renewed Lunar exploration, now NASA is going to be trapped in low earth orbit for another twenty years!! This was the genesis of "Flexible Path". The "Anywhere-but-the-Moon" fanatics, desiring to wrecking ball any future lunar flights. They finally found a President dumb enough to go along with their anti-Moon agenda! If Project Constellation actually stays dead, then America will henceforth get very inferior spacecraft, that will only be adequate for LEO flights. This move to kill the would've-been new Lunar program, stands as the most damaging-to-the-long-run decisions ever made by a U.S. President, with regard to spaceflight. The nation's space capabilities will dwindle to nothing, under the Obama administration.

Comment by Chris Castro — April 5, 2010 @ 3:07 am

24. Tom,

That is also why the only practical solution I see will be creating an organization dedicated to lunar exploration and development. Its the model that worked well with the major dam projects in the U.S., TVA, BPA, etc., as well as with the building of the Alaska Railroad.

I think that your idea for a development corporation has a lot of merit, but is premature. In the examples you cite above, no one doubted that a dam could be built and operated or a railroad could be constructed across a wilderness. But one of the reasons we're having such difficulty getting extraterrestrial development started is that no one has yet shown that it is possible. Yes, the chemistry and physics say that it is, but it has not yet been demonstrated on the Moon or in space.

I speak to a lot of people in the space business, both NASA and industry. When I talk about using lunar resources, I get a lot of blank stares. It is a topic beyond their comprehension because it's never been done. For NASA, this means that they are deathly afraid of it — if something has never been done before, there is no guarantee that it can be done. Thus, ISRU is a mission risk and they've been trained to minimize risk. In fact, they have become totally risk-averse, a complete change of culture from their glory days during the Cold War, when technical people were willing to try almost anything at least once.

My point is that the VSE was an opportunity for NASA to return to its roots as a technical pioneer. The "mission" was not to industrialize the Moon — it was to determine if the Moon could be industrialized. Your Lunar Development Corporation is the logical step to take after we have shown that we can arrive at the Moon, survive using its resources, and thrive by producing a product for export and profit to cislunar space.

Comment by Paul D. Spudis — April 5, 2010 @ 4:21 am

25. This is a great article. In fact, it seems like one of the clearest I've read on this topic since the NASA budget came out in February. I fully support the notion of fostering commercial human spaceflight, recasting our primary goal as Mars (I am thinking that the moon would be an ideal training ground for a Mars trip anyway), and cooperating with other nations. What concerns me is the lack of a focused goal. I am only an avid observer of all this, but there seem to be many good ideas floating around for a platform for deep space exploration. I would bet many of them would work. Or we could combine ideas. But we need to take a stand and declare a "before this decade is out" kind of goal.

Comment by Matt Collister — April 5, 2010 @ 11:20 am

26. You are right on target Mr. Spudis, your last line summing it up so well – "Finally, unless and until scientists and engineers jointly embrace the objective of making human reach into the Solar System permanent and affordable, our country and its space program will continue to diminish."

Yet it is not easy to have engineers and scientists focus on the "affordable", that "-ility". At

207

the R&D level, very new and innovative ideas, at the birth of a capability, there is usually no focus at all on how a technology may mature to be widespread, how it might become afford-able in a widespread sense. Later, once value has been shown, the ability to do something that could not be done before, and if also use-ful, other organizations run with a capability to mature it, to add in the qualities like affordable, reliable (and safer), producible, more produc-tive, etc.

In this sense the groups who work R&D must by definition embrace different parts of the world before them than those who em-brace matters of cost and such. It's this as-pect, in the move to breaking out into healthy R&D vs. commercial arms in NASA that holds hope in the strategic layout of the budget as proposed.

So perhaps the embrace that is needed is of engineers and scientists in NASA (and their contractors) to develop capability that adds value (such as beyond Earth orbit transport, for using resources in-situ, for refueling in space, etc). It is for the commercial sector to embrace the capabilities that already exist to get to LEO and make such capability (existing technology) more affordable, reliable (safer), and produc-tive (more flights per year, more people per year to orbit, more tonnage per year, etc along a growth curve expanding the market, revenue and overall access to new comers / customers.

So as new capabilities – "if the Moon could be industrialized" for example – are shown, then the cycle again changes, moving govern-ment capability ever further out, and the com-mercial role into maturation and growth of the previous frontier.

Comment by Edgar Zapata — April 5, 2010 @ 11:49 am

27. I have to disagree with the basic prem-ise. Rereading the 2004 Presidential Com-mission Report it reads much like Obama's plan. There is no clear goal or vision on why going back to the moon or even Mars is desir-able. You have to get some major return for the investment and while I believe it is there (resources, energy (He3 etc)) I have not seen anything beyond a vague "enhance our se-curity and foster economic development" ra-tionale. Colonizing Mars may be a wonderful fantasy but its a long way off and we are still far from understanding how not to destabilize our ecosphere much less create a new one. VSE pushed commercial transport to LEO just like Obama's plan.."… the Commission believes that NASA should procure all of its low-Earth orbit launch services competitively on the com-mercial market."

To summarize my assessment, NASA did not understand the VSE vision, there *wasn't* any real vision and it was not mentioned by Bush beyond his SOU address. Until there us a vision for manned space which provides us with real tangible returns and value, we are like to see this stumbling along on inadequate budgets continue. We spend more in a month in Iraq than we do for the entire space program in a year so its easy to see where our priorities are and until we provide a vision that is worth a higher priority, it will remain in the state it is.

Comment by Jamie Ross — April 5, 2010 @ 1:26 pm

28. I too was stunned when Charlie Bolden made the remark about how there were al-ready six American flags on the Moon. Those six flags were standard issue American flags, bought through government supply or at the local Sears, depending on the story you read. Either way, they were made of nylon. After 40 years of being on the Moon, the extreme radia-tion from the Sun and galactic cosmic rays, the extreme diurnal temperature variations experi-enced on the lunar surface (from approx. -230 to +230 °F), and the constant micrometeoroid bombardment, have very likely made those ny-lon flags unrecognizable. The colors are likely completely faded, and I'm willing to bet the ma-terial has lost much of its integrity.

Comment by JohnG — April 5, 2010 @ 1:35 pm

29. Jamie,

There is no clear goal or vision on why go-ing back to the moon or even Mars is desirable

I suggest that you re-read the VSE an-nouncement speech here:

http://www.spaceref.com/news/viewpr. html?pid=13404

Note in particular paragraphs 16 and 17 (beginning "Our third goal is to return to the Moon,....")

Now have a look at the simultaneously re-leased document that describes the three key elements of national space policy (to address national scientific, security, and economic in-terests):

http://georgewbush-whitehouse.archives. gov/space/renewed_spirit.html

If that isn't clear enough, I direct you to Presidential Science Advisor John Marburg-er's exegesis of the VSE here:

http://www.spaceref.com/news/viewsr. html?pid=19999

The "mission" on the Moon was clearly stated in the original announcements and elaborated on by Marburger (twice!) in public, just in case NASA had not gotten the message (they still didn't get it).

Finally, have a look at my presentation that charts the words of the VSE founding documents and how the agency changed its meaning and emphasis:

http://www.spudislunarresources.com/Papers/The%20Vision%20and%20the%20Mission.pdf

The direction given to NASA — as well as an explanation for what we hoped to accomplish — was as clear as any Presidential space policy direction we have ever had.

Comment by Paul D. Spudis — April 5, 2010 @ 1:46 pm

30. I've always been in favor of placing NASA into a US Department of Extraterrestrial Resources. NASA could easily be a profitable TVA type of corporation, IMO. Of course the rest of the aerospace industry would hate that! And the Congress has done everything to make NASA more of a burden to the tax payers by banning the shuttle from launch commercial satellites, for instance.

If every shuttle launch were allowed to carry a commercial satellite, that would save the tax payers at least $500 million a year. If they reserved 8 seats on the shuttle for wealthy tourist, that could save the tax payers at least another $200 million per flight at perhaps another $1 billion a year. Of course, if the payload area were used to carry perhaps 45 wealthy tourist into space per launch at perhaps $20 million per tourist, then the shuttle would make a $450 million profit per launch.

Comment by Marcel F. Williams — April 5, 2010 @ 3:13 pm

31. As Paul Spudis has pointed out, we need an integrated, re-usable Earth to orbit and Earth orbit to Moon transport system that is financially sustainable. One of the ways to make it sustainable is to use lunar resources for rocket propellants, at least to start up a base operation. I suspect that a lunar mining base including propellant extraction could be set up with as little as 500 tons of equipment on the lunar surface. This would be no more than about 25 payloads of 20 tons each delivered to the lunar surface using unmanned but re-usable cargo landers, in addition to crew landers. All of these payloads would need to be delivered within 3-4 years to sustain a practical effort to "launch" or "boot" a mining base. The mining base could then support a science base, and provide "LUNOX" for operations in cis-lunar space. A successful base could then

evolve into a permanent base and eventually a colony. It might take several decades for us to develop lunar based fabrication techniques to allow expansion of a base into a colony using primarily lunar materials.

Until the actual quantities of usable lunar volatile resources are known, we do not know how long we could use them for rocket propellants like hydrogen. Oxygen is far more abundant on the Moon and is the heavier propellant component by far, so that it would be a very critical propellant component to be brought up to lunar orbit than hydrogen. Lunar hydrogen could be used primarily for lunar ascent in the near term.

The publicly quoted financially sustainable rates of two Ares 5 launches per year show that a successful lunar base boot will NOT occur using an expendable HLV such as the Ares 5, combined with expendable cargo and crew landers, since it would take 20 years to launch this many cargo payloads and thermal cycling would render most of them useless within 5 years without human maintenance. A minimum set of payloads have to be landed within a few years to safely support a crew and then to allow the crew to establish propellant extraction. This underscores the importance of having a re-usable HLV design, and not rushing to build a quickie HLV based on the Shuttle solids and ET simply due to shuttle-related jobs. If you were a NASA employee, would you rather be building expendable booster parts or re-usable lunar lander or Mars lander or hab parts. If the transport costs can be brought low enough, we can do both Moon and Mars.

Comment by John K. Strickland Jr. — April 5, 2010 @ 9:25 pm

33. The fundamental error in Mr. Spudis' article is that it misinterprets the objective of the Apollo program. As John Kennedy said clearly:

"If we are to win the battle that is going on around the world between freedom and tyranny, if we are to win the battle for men's minds, the [Soviet Union's] dramatic achievements in space which occurred in recent weeks should have made clear to us all...the impact of this adventure on the minds of men everywhere who are attempting to make a determination of which road they should take.... We go into space because whatever mankind must undertake, free men must fully share.... I believe this Nation should commit itself to achieving the goal, before this decade is out, of landing a man on the Moon and returning him safely to earth."

President John F. Kennedy, May 25, 1961[1]

Apollo was funded because it was a symbolic substitute for a perilous race to build nuclear arms. Space Station was funded only when the Russians joined, to provide Russian rocket scientists with something to do other than build missiles for Iran, and to serve as a catalyst for international trust and cooperation. Yet Bush, not understanding this, wanted ISS cancelled in 2010.

This is not the Sixties. China is as capitalist as we are, and the nonaligned countries are looking to them because of their willingness to invest huge sums in local projects, not their space program. To expect US taxpayers would put up $150 billion just because we space enthusiasts want to run around on the moon is simply unrealistic. If Constellation went forward in this taxes-are-a-sin political climate, the money would have to be borrowed from China, and would hurt US interests

John Strickland is right on target. The problem with Constellation is that it costs far too much to be sustainable. The cost of human spaceflight must be reduced to WELL UNDER $10 million per seat to LEO before it can even BEGIN to be considered practical. This requires fully reusable launch vehicles, and is a lot harder than a $150 billion lunar joyride with obsolete solid-fueled throw-away rockets, but it would be a practical long-term goal.

Comment by Dan Woodard, MD — April 6, 2010 @ 6:09 pm

34. This would be no more than about 25 payloads of 20 tons each delivered to the lunar surface using unmanned but re-usable cargo landers, in addition to crew landers. All of these payloads would need to be delivered within 3-4 years to sustain a practical effort to "launch" or "boot" a mining base.

That's kind of what I was thinking; an Earth-based analogue might be a land-based oil rig. They have comparable numbers of "modules" that must be able to be carried by ordinary tractor-trailers: ~20 tons each. To do it all in 3-4 years is a fast launch schedule though. The ULA "Affordable Architecture" paper says that 2 modules per year could be done for $7 billion USD per year (which is comparable to Paul's estimate in the blog post before this one—see the first figure). Assuming 13 years, that would cost maybe ~ $100 billion USD, and still not break NASA's budget. In other words, for the cost of an ISS, we could have some serious lunar ISRU capability. The money has to be spent on something, so we might as well spend in on something that will give us back a practical benefit.

Comment by Warren Platts — April 6, 2010 @ 11:56 pm

35. Dan,

The fundamental error in Mr. Spudis' article is that it misinterprets the objective of the Apollo program

Just curious — have you ever actually read anything I have written? It's NASA who are continuously attempting to repeat Apollo and that is a mistake, as I have written on this blog repeatedly over the past year. It is the current Administrator who talks about "planting flags in an endless series of space 'firsts'", not me.

You also missed the point I made that the reason to go to the Moon is not because "we space enthusiasts want to run around on the moon." It is to use the resources of the Moon to create new space faring capabilities. That is a mission objective as far removed from the rationale for Apollo as I can imagine.

And I have written on the meaning of Apollo here:

http://www.spudislunarresources.com/Opinion_Editorial/Apollo_30_op-ed.htm

Yet Bush, not understanding this, wanted ISS cancelled in 2010.

Wrong again. The VSE policy specifically instructed NASA to complete construction of the ISS by 2010. Our participation in research there was to be finished in or around 2016.

Comment by Paul D. Spudis — April 7, 2010 @ 5:03 am

36. John Kennedy also frequently said that space is the new ocean. So he fully understood that this was more than just posturing with the Soviet Union this was about the pioneering of a New Frontier!

Comment by Marcel F. Williams — April 7, 2010 @ 2:59 pm

37. Polls show that 7% of those wealthy enough to pay $20 million to fly into space would do so. That's a population of over 7000 people world wide.

If just 10% of that population paid to travel into space every year then the cost of space travel would fall dramatically since space rockets would have to be mass produced instead of kraft produced in order to keep up with the demand. And then would probably only increase the demand for space tourist. The energy cost for traveling into space are relatively tiny. You could add even more demand by setting up an international space lotto for the average Janes and Joes that want to travel into space.

Comment by Marcel F. Williams — April 7, 2010 @ 3:09 pm

38. The new budget proposal is much closer

to Von Braun's vision. I have no doubt it will get us further into space and sooner than continuing the path with Constellation. And we will go back to the moon as well.

Comment by Jack Vaughn — April 8, 2010 @ 3:25 pm

39. Jack,

The new budget proposal is much closer to Von Braun's vision. I have no doubt it will get us further into space and sooner than continuing the path with Constellation.

You imply that the choice must be either Constellation or the "new budget proposal." An alternative possibility is that neither will do what you claim. I'm glad that you "have no doubt." Some of us do have them.

Comment by Paul D. Spudis — April 8, 2010 @ 3:31 pm

40. The new budget proposal reduces our investment in manned space travel from $8.4 billion a year down to only $4.1 billion a year. And it pretty much keeps us at LEO for the next decade or two!

If its not Obama's intention to gut our Federal manned space program– he sure is doing a good imitation of it!

Comment by Marcel F. Williams — April 8, 2010 @ 5:42 pm

42. Good article, though I disagree somewhat with the motivation for returning to the Moon. Lunar resources will become very important, but the principal reason for going back to the Moon initially, then on to Mars, will be science, at least for the foreseeable future. Whether new space faring capability is useful depends on what we do with it. Apollo barely scratched the surface of lunar science, and the Mars landers and orbiters have revealed a world with a very complex geologic history that has raised myriad questions for scientific investigation that can only realistically be answered with a long-term human presence on the planet. There is plenty of useful science to be done.

The question, as you touched on in your last paragraph, is how do we do it in a way that is "permanent and affordable", which translates immediately to "How much will it cost"? Apollo died because it simply cost too much to throw away a Saturn-V and Apollo CSM and LM to land 2 men on the surface of the Moon for a few days. All of NASA's attempts to restart our manned lunar exploration program have amounted to little more than trying to redo the Apollo Program (with all new technology), and they have suffered from the same fatal flaw that killed Apollo: a reliance on expendable

Heavy Lift Launch Vehicles (HLLVs). The expendable HLLV paradigm is dead. It is time to stop beating on that dead horse and give it a decent burial.

Fortunately, there is an alternative (which I have been promoting on the internet, off and on, for the last 8 or 9 years). The alternative is to develop a fully-Reusable, 2-stage, Vertical TakeOff and Landing (VTOL) Launch Vehicle (RLV). We have had the technology to do that for some 40 years. Unfortunately, NASA fixated on the notion, back in the mid-60s, that Reusable Launch Vehicles Must Have Wings. Thus, they proposed a design for the Shuttle, a 2-stage, fully-reusable, VTHL design with fly-back booster, that would have cost much more to develop than the limit imposed by the Office of Management and Budget. As a result, NASA was forced to enlist the aid of the Air Force, which demanded an Orbiter with a very large cross-range (which has never been used), as their price for participation, resulting in the very complex, partly reusable design we have now. When the Challenger disaster showed that the Shuttle was not, and never would be, the reliable, low-cost launcher that NASA wanted, they embarked on two successive programs to replace it: NASP and X-33. Both were single stage, horizontal landing designs, and both failed.

A 2-stage, VTOL design has many advantages, but, as far as I know, the concept has never been seriously considered by NASA. A 2-stage design is much easier than a single-stage design, since it has much less challenging dry mass requirements. It is also much simpler, structurally and aerodynamically, and will, therefore, cost much less to develop. But the principal advantage is that, if we can refuel the Orbiter stage at an LEO propellant depot, the Orbiter can then do the TLI/TMI burns to send our manned lunar and Mars spacecraft on their way, thus dispensing with the need for an HLLV (which I define as >150,000 lb. payload) entirely. The Orbiter would be sized for that capability, and the Booster stage would be sized to get the Orbiter, and it's attached lunar/Mars payload, into LEO, arriving dry. Once in LEO, the Orbiter would dock with the propellant depot, nose to nose with the tank set, then the entire assembly would be rotated around it's common centerline to settle the propellants in their tanks, and the propellants would be transferred from one to the other with differential gas pressure.

Not only do we eliminate the need for an HLLV but, once we have the capability of refueling on the Moon with lunar LOX, the Orbiter stage can fly out to the Moon (after refueling in LEO) land on the Moon, refuel, and return directly to LEO with aerobraking. And we can do all of that without expending a pound of hardware. That gives us the capability of sending a much larger expedition to the Moon than Apollo could for only a few percent of the cost. Once that happens, lunar tourism becomes a serious possibility.

Eventually, we will be able to do the same for Mars. A complicating factor is that, initially at least, we will be manufacturing our Earth-return propellants from CO2 in Mars' atmosphere and a small amount of seed LH2 brought from Earth to produce LOX/Methane propellants (the Mars Direct scenario). There is plenty of water on Mars, but, initially, we will not know where and in what quantities and what impurities may be present. (Brine might be a better description.) It is also much more difficult to store LH2 on Mars than Methane. But CO2 is available everywhere – just open a valve and start pumping.

So we might have to develop a separate Earth Return Vehicle (ERV) to make use of LOX/Methane propellants, but there may be a way around that. The RL-10 has been run on methane, with some modifications, and Pratt & Whitney is now working on the CECE (Common Extensible Cryogenic Engine) which will run on both LH2 and Methane. The Methane turbopump has one less stage than the LH2 turbopump, but that could be handled by placing a bypass valve around the first stage of the turbopump. If all other differences could be handled by simply designing for the worst case, then we could dispense with a separate ERV and just use the Orbiter. It would fly out to Mars using LOX/LH2 and land, then switch the engines to LOX/Methane mode. When the Earth-return launch window opened, it would fly back to Earth using LOX/Methane. The Earth return Delta-V is much less than the Earth-to-LEO Delta-V, plus it will not be carrying a substantial payload, so it could do that with a single stage.

The 2-stage, VTOL RLV gives us the capability to establish fully-reusable transportation systems between the Earth and the Moon and Mars. Being a medium lift design (approx. 80,000 lb. lift capacity), it would have many other uses as well. It could launch communications satellites (two at a time like the Shuttle used to do), space station modules, various kinds of scientific satellites and planetary probes, as well as establish a robust LEO tourism industry carrying approx. 50 passengers per flight for less than $1 million per ticket – eventually much less. Those other uses would help support the fixed costs of the manned lunar/Mars transportation system and immunize that program from cancellation by shut-ting down the launch vehicle production line, as happened with Apollo. It can be flying by the end of the decade for a lot less than what NASA wants to spend on Constellation – to take a lot longer to do much less.

Comment by Dick Morris — April 8, 2010 @ 10:41 pm

43. Dick,

I want to comment on two of your points.

There is plenty of useful science to be done.

I do not disagree with this statement, but science alone is an insufficient rationale for human spaceflight. The real goals are to create a permanent, extensible space faring system. I think that the key to this is to learn to convert what we find in space into what we need, i.e., learn to use off-planet resources of materials and energy. I focus on the Moon because it has both, in forms that we already know how to manipulate.

Apollo died because it simply cost too much to throw away a Saturn-V and Apollo CSM and LM to land 2 men on the surface of the Moon for a few days.

No, the Apollo program ended because it had achieved its mission objective: to beat the Soviets to the Moon. That's why I think we need a more compelling reason to go into space than pulling off stunt missions. Learning how to create sustainable presence off-planet and a system that can routinely access cislunar space (where all our satellite assets reside) seems like a good rationale to me.

Comment by Paul D. Spudis — April 9, 2010 @ 8:51 am

44. Apollo ended because it accomplished it's goal of beating the Soviets to the Moon – and it was too expensive to keep it going just for science. Science is a good reason to go back to the Moon, and, especially, on to Mars, but we will need major cost reductions to make it happen. Once we have a low cost launch vehicle, we will be able to go back to the Moon, and developing lunar resources, particularly liquid oxygen, will be critical to supporting the transportation system.

Once we have that capability, it would make sense to set up a propellant depot in LLO to transfer lunar LOX from vehicles returning to Earth to vehicles arriving from Earth. That would make a major improvement in the economics of the transportation system. It might then make economic sense to export LOX to other destinations, possibly even LEO.

The metals produced as a byproduct of LOX manufacture might also be exported at some point, but it should be noted that the RLV, which will be absolutely necessary to get to that point, will, itself, be pretty stiff competition for lunar "outsourcing". How the economics will work out is difficult to predict at this point. Lunar resources will be used wherever it makes economic sense to do so, and I'm sure it will be a major industry eventually.

I completely agree that creating a space-faring civilization is the long term goal, I just have a slightly different perspective on the strategy: Once we get the cost down, science will be a good and sufficient reason for expanding the frontier, and everything else will follow. But other strategies are certainly possible, and I will support whatever works.

Comment by Dick Morris — April 9, 2010 @ 1:29 pm

45. Apollo ended because Nixon and a Democratic Congress didn't want to use the heavy lift architecture to build a lunar base. Instead, Nixon decided to invest in a– game changing– architecture (the Space Shuttle) that would allow America to cheaply and conveniently access LEO :-)

Be careful what you wish for!

Comment by Marcel F. Williams — April 9, 2010 @ 8:33 pm

46. *"Be careful what you wish for!"*

Indeed. But are you suggesting that we forego a fully-reusable design now because of past blunders that ruined the Shuttle design (and caused NASP and X-33 to be such dismal failures)?

Building a fully-reusable launch vehicle would have been the right thing to do at that time. The Nixon administration and Congress didn't proceed with a lunar base, as NASA proposed with the Apollo Applications Program, because of cost. The administration actually cancelled the last two Apollo flights, even though the hardware had already been built. It was not considered worthwhile to assemble the vehicles on the pad and launch them, so a lunar base was completely out of the question.

Unfortunately, we allowed the Air Force and Office of Management and Budget to drive the Shuttle design, then NASA downsized the vehicle as much as possible during preliminary design in order to maximize the performance. The result was the very complex, partly reusable, extremely fragile and unreliable vehicle we have now. Our manned space program has never recovered from those catastrophic blunders.

We have another opportunity now to do it right. If we don't, I can easily see us beating our heads against the brick wall (expendable launchers) for another 10 or 20 years.

Comment by Dick Morris — April 12, 2010 @ 12:48 pm

48. Just what has NASA being doing for 40 years? The united States, the most powerful nation on the planet, has no idea of where it is going, paralysed by political bundling and interference. When China puts a man on the moon, only then will the US wake up. Obviously deep space exploration is difficult, but why does it take another 25 years just to put a manned flight in Mars orbit? How many hundreds of billions of dollars has the US spent on Iraq, which could have been spent elsewhere and NASA is promised a lousy $6 billion over five years. Well no doubt that will just about pay for the "bubbly" at the senior management meeting over that time. Who would have thought it, that the US, would end up like this: utterly rudderless.

Comment by Peter Hall — April 15, 2010 @ 5:27 pm

49. Paul- One point of clarification for the record – You did describe the role I played a number of years ago for NASA – but – you forgot to mention that I resigned from this position at NASA almost 3 years ago… and – based on what you read about my perspective today – I would say that being at the front end of trying to establish the value of lunar missions helped shape my current opinions – and led to my resignations…. Ad Astra Per Aspera!!!

Comment by JeffSpaceGeek — April 17, 2010 @ 8:21 pm

50. All Constellation would've required was a 3 billion dollars per year commitment from the U.S. Government. Now President Obama says that the project was "underfunded & unsustainable". Now he proposes to instead spend way more than this sum on NASA per year, solely to do advanced propulsion research!! WTF!! Hello, Mr. President?!?! If a major project is struggling for funding, what you're supposed to do then is ACTUALLY FUND IT. And furthermore: WE ALREADY HAVE A VIABLE HEAVY-LIFT ROCKET DESIGN & PLAN: it was called the Aries 5 rocket. Why don't we just go ahead with this heavy-lift launcher? Or can it be that the Obama administration just hates the initial, slated destination? And they think that if they terminate Constellation and start all over in 2015, that they'll get to build a new NEW heavy lift rocket solely for their Guinness Book of World Records highest altitude spectacular asteroid mission. By the way, all this asteroid mission bull dung that Mr. Obama is speaking of: it serves only the purpose of bypassing & avoiding the Moon! Yeah, right....on 2025

NASA will have huge & elaborate bases on an asteroid far, far off in interplanetary space—but at the same time, the Moon will be completely empty & devoid of any human activity. Does anybody out there, besides me, see something very flawed & wrong with this scenario?? Why the freak do we have to bypass the Moon?? What is wrong with these Anti-Moon zealots?!

Comment by Chris Castro — April 18, 2010 @ 3:26 am

51. Jeff the Geek,

I would say that being at the front end of trying to establish the value of lunar missions helped shape my current opinions – and led to my resignations

Given the bang up job you guys did while taking taxpayer dollars at ESMD, I wish you a similar level of success in your new advocacy.

Comment by Paul D. Spudis — April 18, 2010 @ 5:09 am

52. Chris,

All Constellation would've required was a 3 billion dollars per year commitment from the U.S. Government.

That was a $3 B increase in the existing levels of funding for Exploration ($ 5-6 B per year) and it comes from the Augustine report. But that group spent all their time "proving" that the Ares boosters were too costly, not on looking at options to make Constellation affordable.

NASA gets plenty of money — they just don't know how to spend it wisely.

Comment by Paul D. Spudis — April 18, 2010 @ 5:20 am

53. They don't want to spend it wisely. Maintaining employment levels at their research centers is their principal motivation, so they always try to do everything the hard way, as with NASP and X-33, in order to "justify" a great deal of technology development work. As a former Space Station official put it "We've GOT to push the technology – it's our mandate."

Even that, however, doesn't fully explain the Ares cost estimates I've seen. To think that it should cost 4 or 5 times as much to develop a largely expendable launcher as it does to develop a new commercial aircraft model, which is much more complex, is simply insane.

Frankly, I don't understand why anybody is still talking about the Ares launchers. The expendable HLLV paradigm has killed every proposal for manned lunar and planetary exploration since the end of Apollo, and now Constel-

lation appears to be going down in flames. On April 8 I outlined a proposal for using a 2-stage, VTOL RLV, plus a propellant depot in LEO, which would enable us to do manned lunar and planetary exploration for a small fraction of the cost, and give us a much more useful launch vehicle as well. I wish someone would explain why it is not obvious to everybody that that is the way to go.

Comment by Dick Morris — April 26, 2010 @ 12:48 pm

54. Dick,

The expendable HLLV paradigm has killed every proposal for manned lunar and planetary exploration since the end of Apollo, and now Constellation appears to be going down in flames.

I do not agree — I think that programs to go beyond LEO have floundered because of a failure to articulate what we would do there and why it is important. Arguments about rockets, destinations and architectures are secondary to this fundamental requirement. If you don't know what you're doing and why you're going, it doesn't matter what other plans you come up with.

Comment by Paul D. Spudis — April 26, 2010 @ 1:47 pm

55. And if you DO know what you're doing and why you're going, it STILL doesn't matter if it costs too much to make it worthwhile. The cost of Earth-to-orbit launch is the fundamental driver of the cost of everything we do in space, and with expendable launchers it will simply cost too much to set up and maintain a lunar resource utilization capability to justify any application of such resources that I've ever heard of.

I believe that my proposal for a 2-stage, VTOL RLV, plus an LEO propellant depot, is critical for establishing a manned lunar exploration and development capability. Based on my 50+ years of study and observation, I cannot imagine that we are going to do it any other way. Goals are indeed paramount, but, as they say, the devil is in the details, and how we propose to accomplish those goals will determine whether we will be given the chance to do it.

Comment by Dick Morris — April 26, 2010 @ 8:29 pm

56. Dick,

Goals are indeed paramount

As we agree on this, we can leave it at that.

Comment by Paul D. Spudis — April 27, 2010 @ 4:52 am

To Do The Heavy Lifting

April 14, 2010

To heavy lift or not to heavy lift - that is the question.

A recent talking points memo by the Office of Science and Technology Policy (OSTP)[1] seeks to clarify some aspects of the new direction in regards to the cancelled Project Constellation[2]. Touted by some as "compromise," it asserts[3] that NASA will develop and build a new "Orion lite" crew vehicle whose primary mission will be to serve as an escape pod for the crew of the International Space Station (ISS). And more interestingly, the policy "Begins major work on building a new heavy lift rocket sooner, with a commitment to decide in 2015 on the specific heavy-lift rocket that will take us deeper into space[3]."

I'm confused. If a heavy lift launch vehicle (HLLV) is not needed for future human missions beyond LEO, why are we spending billions of dollars researching aspects of it in order to make a design decision five years hence? If a heavy lift launch vehicle is needed for such missions, why are we waiting five years to make that decision when we have the parts and workforce needed to make the vehicle now?

Are you confused?

Let's break this down a bit. What exactly is a heavy lift launch vehicle and what is its significance to spaceflight and more specifically, to sending people beyond low Earth orbit (LEO)? The Space Shuttle orbiter carries a bit more than 24 metric tonnes (mT; 52,800 lbs.) to orbit. That would seem to qualify as heavy lift. But the old Saturn V could launch 118 mT (260,000 lbs.) to low Earth orbit. That capability allowed us to launch the complete Apollo spacecraft (three modules) and its Earth departure stage (S-IVB) in one fell swoop. The cancelled Ares V launch vehicle[4] was to have carried up to 188 mT to LEO (over 400,000 lbs). These indeed are heavy lift launchers.

Why do we need heavy lift? Thoughts among space engineers on this topic fall into two broad categories – those who think that heavy lift of the Saturn V variety (~100 mT to LEO) is the *sine qua non* for human missions beyond LEO, and those who think we can develop an incremental approach that uses the smaller vehicles currently available, such as the medium-lift class Atlas V[5] (21 mT to LEO) and Delta IV-Heavy[6] (25.8 mT to LEO) launch-

215

ers. The basic philosophical difference centers around what can be done in space versus what has to be done on Earth.

The Saturn V was built so that a single, self-contained lunar mission could be launched on one vehicle. This meant that the crew could focus their efforts solely on their lunar voyage and not have to concern themselves with assembling a larger, more complex vehicle in Earth orbit. In modern terms of human flight beyond LEO, the big advantage of heavy lift is that you can use fewer launches to get the same amount of equipment and material in space, thus reducing the chance that a single launch failure might cause a mission abort. Propellant can "boil off" in space, especially the very cold (cryogenic) liquid oxygen and hydrogen that fuel the most powerful chemical rockets. It is also assumed that assembling a few large pieces in space is simpler than assembling many more smaller ones. But over time, we've learned quite a bit about assembly in space from construction of the ISS.

Critics of new heavy lift vehicle development point to the high costs associated with new vehicle development. They also point out that there are new techniques and technologies that can help us venture into the Solar System with existing launch vehicles. One of those ideas are depots where propellant is collected[7] and used to fuel empty vehicles for journeys beyond LEO. A detailed architecture that features propellant depots[8] and that does not require a new heavy lift vehicle has already been published. Many groups strongly advocate this approach.

Which brings us back to the new OSTP document[3]. This new document indicates that work will proceed on development of a heavy lift launch vehicle, with a decision on what vehicle to build coming in 2015 (**note well**: not *building* a vehicle, but *making a decision* on what vehicle to build). How will our decision on heavy lift be more informed in five years than it is now?

In five years, all the Shuttle manufacturing and assembly infrastructure, including tooling dies, assembly jigs, milling facilities and solid-rocket booster production lines will have been disassembled, mothballed or discarded. Some of this shutdown has already begun. The skilled workforce that now builds and operates the Shuttle launch system will dissipate (people have families to support and cannot stand around waiting for a decision five years – maybe never – down the road). There will

be no experience base to build, assemble and launch a heavy-lift vehicle within NASA or industry.

I have written previously[9] that a Shuttle-derived heavy lift vehicle **can be built now**, with existing piece parts and minimal changes to the assembly and launch infrastructure at Michoud and the Cape. Such a vehicle can launch 80 mT to LEO; two launches can send a human mission to the Moon. And it is completely affordable[10], fitting into the existing run-out budget and available for use within a few years. Currently, robotic missions are discovering and analyzing the vast resources of the Moon. More robotic missions are needed[11] to begin the processing of lunar resources in preparation for human return and expansion beyond LEO. These are all things NASA can afford to do now.

This re-invention of NASA, as trickled out by the administration, has been eagerly seized upon by many in the "New Space" community, as their long sought, free-market (but government funded – for now) opening at building a commercial rocket industry. Some of us (who also believe in free markets) see different motives for this new direction and are particularly concerned that the new "flexible path (FP)[12]" doesn't have any specified destination or mission.

The fundamental fecklessness of the new direction is exposed in this new OSTP document[3]; we will build a Crew Return Vehicle for ISS that is not needed (if we can get there on the Soyuz, we can certainly return on it) and we will conduct "research" on heavy lift technologies that are already well understood.

Maybe it's not so confusing after all.

Topics: Lunar Exploration, Space Politics, Space Transportation, Space and Society

Links and References

1. OSTP, http://www.whitehouse.gov/administration/eop/ostp
2. Constellation, http://www.nasa.gov/mission_pages/constellation/main/
3. it asserts, http://www.whitehouse.gov/sites/default/files/microsites/ostp/ostp-space-conf-factsheet.pdf
4. Ares V, http://www.nasa.gov/mission_pages/constellation/ares/aresV/index.html
5. Atlas V, http://en.wikipedia.org/wiki/Atlas_V
6. Delta IV-Heavy, http://en.wikipedia.org/wiki/Delta_IV
7. depots, http://www.lpi.usra.edu/meetings/leag2007/presentations/20071003.bienhoff.pdf
8. detailed architecture, http://www.ulalaunch.com/docs/publications/AffordableExplorationArchitecture2009.pdf
9. written previously, http://blogs.airspacemag.com/moon/2010/03/24/value-for-cost-the-determinate-path/
10. see figure in chapter XX, http://blogs.airspacemag.com/moon/files/2010/03/SDHLLV-costs.jpg
11. robotic missions, http://www.spudislunarresources.com/

Papers/Spudis_VSE%20mission.pdf
12. flexible path, http://blogs.airspacemag.com/
moon/2009/12/16/arguing-about-human-space-exploration/

Comments

2. I agree that both decisions (use Orion as a return vehicle only and wait 5 years to decide what kind of HLV to build) make no sense (at least without more information). If the Orion is used as an emergency return vehicle only, then it cannot be cycled through every 6 months as is done now with the Soyuz. If it cannot be cycled, how long could one be left in place before being disposed of and without even using it for a crew descent! This is a very poor direction for a jobs program. Instead, put the same people to work designing lunar and Mars habs, depots, space tugs, etc. right away.

I agree that the minimum throw weight for a "true" HLV is about 75 tons. The question is, right now – how soon do we need an HLV? If we wanted to continue using part of the shuttle technology (as a jobs program) it would make sense to develop and use a true shuttle-derived (side-mount) HLV as soon as possible to maintain the work force.

Unfortunately, it has cost about 5 billion a year to maintain the ability to launch shuttles, but without the orbiters to maintain, the annual cost would be much less. However, if we want to be able to afford to support real, functioning lunar and Mars bases, a shuttle derived HLV would still be far too expensive for the launch rate needed. For that we need a re-usable HLV (first stage at least), and that is why we might open a design competition for 1 year to create at least 2 designs for such a vehicle. Within another year, we could choose 1 or the other, thus advancing the HLV decision date by 3 years.

It is also not a HLV versus a propellant depot situation. It is both and we need both. For certain scientific payloads, we need the total mass and volume of a true HLV. For real space base operations, we need the low cost and high payload capacity of an HLV (combined with the propellant depots to allow payloads and stages to be launched dry). We also need the depot to allow launches of station-bound payloads with a space tug to deliver them to the station as a replacement for the shuttles capacity.

Comment by John K. Strickland Jr. — April 14, 2010 @ 3:48 pm

3. about the "new" new-plan...

"Begins major work on building a new heavy lift rocket sooner, with a commitment to decide in 2015 on the specific heavy-lift rocket that will take us deeper into space."

develop the hardware for lunar missions needs 8-10 years, so, "decide [the HLV] in 2015" means NO lunar landings before 2023-2025

"Restructures Constellation and directs NASA to develop the Orion crew capsule effort in order to provide stand-by emergency escape capabilities for the Space Station – thereby reducing our reliance on foreign providers."

an Orion built ONLY to serve as ISS "rescue capsule" is very expensive and useless... it's much cheaper to (just) dock a third Soyuz to the ISS

more discussions and proposals on the Space Summit Facebook Group:

http://www.facebook.com/group.php?v=wall&gid=356261201268
Comment by one — April 14, 2010 @ 3:58 pm

4. Well said, Paul. Spending money on a space program just to keep people employed is hardly frontier-enabling.

Comment by James Pura — April 14, 2010 @ 4:17 pm

5. This "new and improved" human spaceflight policy has nothing to do with spaceflight, and has everything to do with politics. Go to Florida (a swing state in recent elections), promise money and jobs, and get your party reelected. And while you're there make sure you spend more time raising money for your party than you do talking about space. There was as much ink used in the OSTP announcement on the amount of money that would be sent to Florida, and the amount of jobs it would "create" in Florida as there was about human spaceflight. So placate the Florida delegation with jobs and money, placate the Alabama delegation with heavy-lift technology and four program offices, placate the Texas delegation with an 'Orion-lite', extended ISS operations, and flag-ship technologies, and placate the Colorado delegation with jobs to design and build the Orion-lite, and move on. When mild-mannered Neil Armstrong signs on to a letter of disapproval for the President's new direction for human spaceflight, you know something is terribly wrong. Why waste money

building an Orion life boat that will never get used? I know, I know, the jobs. Why not just build a 'real' Orion that can carry crew to ISS on an EELV, as an insurance policy to the ill-advised gamble on commercial carriers. Why not build a 'real' Shuttle-C starting in 2011, not 2015. This new recipe for human spaceflight was made with rotten eggs. It stinks.

Comment by JohnG — April 14, 2010 @ 4:40 pm

6. The current Constellation program was not going to be a very good jobs program, since it was only a trickle of work, and did not really build anything substantial for heavy lift until the end of this decade. Except for design engineers, there wasn't going to be a lot of real hardware to test or process.

I think the decision in 2015 will based on which of the two camps (HLLV or current lifter) is able to win their argument. As of now, there are no payloads planned that cannot be launched by our current Atlas/Delta/Falcon Heavy families, so I think the onus will be on the HLLV community to identify specific programs, products and needs that can only be satisfied by such a vehicle.

Of course the elephant in the room regarding an HLLV engine program is what is the fuel? If it's LOX/LH$_2$, then maybe a Jupiter/Ares is possible, but if it's RP-1/LOX (like the RD-180), then we're talking Saturn V redux. Maybe I'm reading too much into this?

Comment by Coastal Ron — April 14, 2010 @ 5:18 pm

7. We should not simply throw away the successful and reliable technology from the Space Shuttle era. We should use it to immediately build a heavy lift vehicle, either a side-mount or an inline vehicle.

Making a decision to build a particular heavy lift vehicle 'sometime' in 2015 would be almost at the end of President Obama's second term (if he gets a second term). Which pretty much means that any decision on a heavy lift vehicle could be easily canceled by the next administration.

And why spend billions studying global warming and then build a 'hydrocarbon' HLV which would increase global warming? A Space Shuttle derived HLV would be a 'green' vehicle especially if the hydrogen is derived from the electrolysis of water using either nuclear or hydroelectric power. And nuclear and renewable energy are supposed to be the future energy resources for our country and the rest of the world.

I will not vote for any Senator or member of the House of Representatives that supports this pitiful space policy.

Comment by Marcel F. Williams — April 14, 2010 @ 6:07 pm

8. After reading one of your links: http://www.ulalaunch.com/docs/publications/AffordableExplorationArchitecture2009.pdf

I found it very encouraging and I will make some comments about.

This is exactly the type of competition that lunar water miner and rocket making would face. Though obviously such a plan would enable lunar water mining. The lunar water miner and lunar rocket fuel maker will be able to deliver rocket fuel to L-2 cheaper than from Earth. But such a simplified system of deliver rocket fuel will lower their capital cost and will already establish a market for rocket fuel in space. Not to mention the bonus of having all those empty tanks all over the place.

In one part of paper: "Each LEO to L2 tanker transit delivers 29 mT of propellant to L2 where the propellant is consolidated into a single depot. The spent stages perform a small disposal maneuver leaving L2."

That's a business opportunity- sure it makes no sense for them to keep those tanks- but to a different party they are gold. And they are worth more than gold just as scrap- though as tanks for future use they could have more value. So rather then just toss them the rocket companies should sell them cheap to a "space junk dealer". Of course that space junker dealer is also a potential customer for the launch company. If the major launch companies actually do this they going to make a killing- it will make Bill Gates will look like a poor boy.

And when they eventually lose the competition to supply rocket fuel to L-2 to the Lunatics, then their main problem will be where to their spend all the money from all the profit they are making.

I seriously doubt these dinosaurs will actually grab this opportunity- they are more likely to drop the bone for it's reflection in the stream.

And this part in paper: "The ACES/Altair is loaded or topped from the L2 depot just prior to lunar descent. This includes the loading of the Ascender propellant tanks which are used during the terminal hover/landing phase. Fully loaded, it can deliver a combined mass of vehicles (such as the ascender), cargo and unused propellants greater than 40t to the lunar surface."

So, that's 2 or almost 3 times more payload to the lunar surface as compared to what the Saturn V did. All from earth launcher which about 1/5 of Saturn V lift capability. Sweet.

One thing I had a slight problem with was it was envisioning a pretty big NASA lunar operation. And I am slightly worried about NASA getting bogged down on the Moon- as is Zubrin's worry [or terror]. I think as long as we get commercial lunar mining this won't be a problem at all- they will be a lot of political pressure from Lunatics for a manned Mars [more profit]. And of course this plan actually makes going to Mars possible- and colonies on Mars, doable.

Now on to reading the other links.

Comment by gbaikie — April 14, 2010 @ 6:36 pm

9. Oh, one more thing, if you start fuel depots now, as in yesterday, you will get the heavier lifts sooner. Simply because what should drive larger boosters is they are built to meet the growing demand [and save in costs due to economy of scale]. Larger booster do not create more demand. You would think that after the Saturn V, people would realize this.

Comment by gbaikie — April 14, 2010 @ 7:08 pm

10. In response to gbaikie, while I enthusiastically agree with the idea of fuel depots, I disagree with your point that fuel depots get you heavy lift sooner. If anything, fuel depots make the need for heavy lift less, since they allow the current generation of lifters to launch and supply modular spacecraft in orbit.

In response to Marcel F. Williams, the argument for any HLLV would be easier to solve if someone could point to a specific need for HLLV. Can you name a funded program that needs an HLLV? Until Atlas/Delta/Falcon 9 Heavy are keeping us from launching individual pieces because the payload won't fit on them, then we don't need HLLV. Atlas/Delta/Falcon can all launch the same mass as the Space Shuttle (if not more), so what is it that you want to launch that needs an HLLV?

Comment by Coastal Ron — April 14, 2010 @ 10:54 pm

11. *This new document indicates that work will proceed on development of a heavy lift launch vehicle, with a decision on what vehicle to build coming in 2015 ... How will our decision on heavy lift be more informed in five years than it is now?*

After 5 years, they will have results from the propellant depot and autonomous rendezvous and docking flagship technology demonstration missions, as well as the EVA assembly and servicing technology demonstrations. Success in those demonstrations would tend to reduce our need for a really big HLV.

They will also have results from more robotic precursor missions to guide decisions.

By 2015 NASA will also have a better idea how commercial crew works out. NASA will have Atlas V, Delta IV, Falcon 9, Taurus II, and perhaps other choices for their HLV "starting point", all with track records to inform their decisions. NASA will also have had time to work our commercial and international partnerships.

They don't have any missions for the HLV yet, so they aren't in a hurry to make one.

In five years, all the Shuttle ... infrastructure ... will have been disassembled, mothballed or discarded. ... There will be no experience base to build, assemble and launch a heavy-lift vehicle within NASA or industry.

The type of heavy lift work that's described in the 2011 budget document sounds like it's for heavy lift based on EELVs or something similar. It doesn't look like Shuttle infrastructure is needed for the sort of heavy lift they have in mind. If they go with something like the Phase I EELV HLV to start with, and maybe move on to Phase II at some point, they should be able to share costs with the EELV infrastructure that's going to be there anyway. The same would be true if it's based on other rockets that are already used for other reasons. NASA would have to pay for all costs of a Shuttle-based HLV.

Some of the HLV work looks like it's for making a U.S. RD-180 class engine. They probably don't want an HLV with lots of Russian engines. That may be another reason why they don't just start working on EELV Phase I or II. If they decide they don't need an HLV at all, the U.S. engine may be useful for other rockets like Atlas V anyway (if they succeed in making it cheap like the RD-180).

we will build a Crew Return Vehicle for ISS

that is not needed (if we can get there on the Soyuz, we can certainly return on it)

They may be looking ahead for the time when U.S. commercial crew ISS transportation is ready. If they can support crew return with Orion Lite, they don't need to burden the commercial crew transport vendors with emergency crew return requirements for the astronauts brought to the ISS by commercial crew. This could make it easier for commercial crew to succeed technically, and could also make it easier for commercial crew business to succeed (if they have a reusable system that doesn't need to sit at the ISS for months).

Comment by red — April 14, 2010 @ 11:28 pm

12. red,

After 5 years, they will have results from the propellant depot and autonomous rendezvous and docking flagship technology demonstration missions

Sorry — I'm just not buying any of this. Five years is barely enough time for them to complete the preliminary studies for all that, let alone the actual flying of those missions. My evidence for this? The last six years of the agency's implementation of the Vision.

You've nicely summarized their storyline, though. Thanks.

Comment by Paul D. Spudis — April 15, 2010 @ 3:10 am

13. @Coastal Ron

So what would I use an HLV for? Well, let's use the SD-HLV (Sidemount Shuttle) as an example.

With a single SD-HLV launch with an EDS stage, we should be able to land at least 10 tonnes of payloads that can be at least 5 meters diameters to the lunar surface. So you can start telerobotically assembling the habitat modules for your lunar base right away. Just three launches a year could place two large habitat modules plus one node module on the lunar surface for a continuously growing lunar base. You'd probably need at least another cargo launch every year for lunar transport vehicles, LOX processing machines, power plants, etc. plus supplies to the lunar surface.

It would probably require two SD-HLV launches per manned mission to the lunar surface. If there are two manned missions per year, that would be 4 SD-HLV launches. So that would require about 8 SD-HLV per year.

So the SD-HLV under a lunar base program would be heavily used with only two SD-HLV launches actually having humans in them.

You can also instantly launch huge space stations weighing nearly 100 tonnes into Earth orbit. Skylab weighed about 77 tonnes. Such, instant, space stations could be used as way stations for manned beyond LEO missions. Companies like Bigelow have talked about using HLVs to launch super large inflatable space stations for tourist.

Such heavy lift capability might also allow us to finally launch the first large artificial gravity space stations into orbit with perhaps just three or four launches to assemble it.

I'd also like to use a SD-HLV to deploy a couple of light sails at a Lagrange point that I would utilize to grab small 50 to 100 tonne NEO asteroids (the Asterant concept) and transport them back to a Lagrange point for oxygen and water extraction and space station mass shielding.

So an SD-HLV could be heavily used, IMO, while making things a lot easier.

Comment by Marcel F. Williams — April 15, 2010 @ 5:03 am

14. First, the Augustine commission concluded that a launcher capable of lifting more than 25 tons but less than 75 tons would be needed. The range reflects the uncertainty of the weight and volume of the biggest piece of the transportation architecture chosen.

The first stage requires a large hydrocarbon engine. The United States does not know how to build a cost-effective large LOX hydrocarbon engine. The lead time for developing the engine, before the launcher design is chosen is three years at a minimum.

United Launch Alliance now purchases high-performance Russian engines. It does so because US industry cannot produce engines with competitive performance at their price. And the reason for the new development program is to give America that capability again.

Comment by Lee Valentine — April 15, 2010 @ 5:04 am

15. Lee,

the Augustine commission concluded that a launcher capable of lifting more than 25 tons but less than 75 tons would be needed...... The first stage requires a large hydrocarbon engine.

Just because that report makes an assertion does not mean that the assertion is either true or required. Their cost estimates are inflated and completely unbelievable. The "Flexible Path" architecture is a plan to do nothing for 20 years while we pretend we are going to Mars.

There is no reason to suppose any particular launch mass to LEO is "needed" — it all depends on what your mission is and what architecture you choose to implement it. My point is that if we need a heavy lift vehicle, we have the capability to make one right now. Yet the "new path" chooses to "study" the issue for five years and then (maybe) make a decision. To me, it sounds like the usual agency Brownian motion.

Comment by Paul D. Spudis — April 15, 2010 @ 5:44 am

16. In general, you are correct about the nonsense that this Administration is putting forward.

One correction, however, the payload that the Orbiter carries is between 35000 and 60000 lb, depending on the orbit. But the Orbiter itself is most of the payload which goes into orbit on every Shuttle launch. That is about 250000 lb which goes into orbit carrying astronauts, their life support, robotic arms, fuel and engines for orbital maneuvering, and wings and a thermal protection system for the ride home. So, swap the Orbiter out and replace it with a 200000 lb payload. With other improvements the existing Shuttle system could carry nearly the same as the Saturn V.

But you are 100% correct, this nation is about to flush away the Shuttle system, just as we did 40 years ago with Saturn. Absolute stupidity. You could shut down Orbiter and save most of the cost of the Shuttle rocket system and for minimal cost, and before 2015, you would have a HLV.

What would it launch ? Besides fuel and depots, BTW people have been talking since before the Shuttles first flew about flying the Shuttle ETs into orbit to scavenge their left-over fuel, the ISS modules and systems are designed for long duration spaceflight. You can easily produce what Aldrin calls the XM vehicle, which can be used to carry people to lunar orbit, onto Mars trajectories, or to asteroids, and it is done with vehicles, modules and systems that we already have. This is a logical follow on, at minimal cost because you continue to use the existing people, to the ISS that

is flying today. With these systems, the only totally new vehicle that needs to be invested in are the lunar or planetary landers.

The wastefulness, sheer stupidity and political recklessness of the latest Obama plan, if it proceeds, is treasonous.

Comment by BAL — April 15, 2010 @ 7:51 am

17. Dear Dr. Spudis: I think what you see as a bug is intended to be a feature.

You wrote:

In five years, all the Shuttle manufacturing and assembly infrastructure... will have been disassembled, mothballed or discarded... The skilled workforce that now builds and operates the Shuttle launch system will dissipate... There will be no experience base to build, assemble and launch a heavy-lift vehicle within NASA or industry.

Well, we didn't have that experience base when we built the Saturn V, now did we?

I think this is a political decision, but one made for technical reasons. One of the major causes of the collapse of Constellation was the jobs/contracts issue — that not only did the systems have to be designed to preserve jobs in the traditional space industry states, but they had to be designed to preserve the very, very specific jobs embodied in the current assembly plants, design shops, refurbishment centers, etc., etc. built for Shuttle. Re-use of these facilities was initially proposed as saving money but quickly became a shackle on NASA's legs.

The HLV delay is designed to do a single thing: burn away the detritus of the accumulated thicket of NASA contracts, organizations, bureaucracy, etc. associated with Shuttle component procurement. If Shuttle parts are really, truly useful for our HLV plans, then five years is not too long a time to de-mothball equipment and knowledge (unlike, say, Saturn Vs, which are completely irrecoverable). But enforcing this pause — and this budgetary starvation plan — will remove the incentive to build inefficient vehicles with inefficient procurement simply because it is the path of least institutional resistance.

A final thought: if 2015 is too late to start work on an HLV, what of the plans to start Ares V work in 2017?

Comment by Catfish N. Cod — April 15, 2010 @ 9:18 am

221

18. Cat,

I think what you see as a bug is intended to be a feature.

Did I say it was a "bug"? I am well aware that this is being done deliberately.

In contrast to your assessment, re-establishing a new HLLV manufacturing capability is not enabled or even facilitated by destroying the existing Shuttle-derived capability. We will simply spend more and take longer to accomplish less.

A final thought: if 2015 is too late to start work on an HLV, what of the plans to start Ares V work in 2017?

If you had read any of the posts I've made here over the last year or so, you would see that I've always had issues with the current Program of Record. That does not mean the Vision should be abandoned, as the new path does.

Comment by Paul D. Spudis — April 15, 2010 @ 9:44 am

19. Well Paul....just heard the President abandoned the Moon, given that we've already been there! I don't understand why he thinks like that: such an ignorant, particular statement deeply challenges the purported integrity of his other general statements. You, and others, have such a sound plan for developing our space infrastructure, a plan that seems obviously to support the President's concerns that we need to do things differently. I simply do not accept that Mars is in our future if the Moon is not.

Thank you for all your hard work, and I'm still hopeful your voice will eventually rise above the noise.

Comment by Phil Backman — April 15, 2010 @ 3:47 pm

20. Well the president just said that we've already been to the Moon so there's no reason to go back :-) Unbelievable!

Hopefully, Congress will have a different opinion on the value of our closest neighbor in space!

Comment by Marcel F. Williams — April 15, 2010 @ 3:57 pm

21. "In response to gbaikie, while I enthusiastically agree with the idea of fuel depots, I disagree with your point that fuel depots get you heavy lift sooner."

Oh, I guess I should have said heavier lifters which are sustainable, sooner.

In others words it's one of those "everyone knows" that if increase the amount rocket launches one will get lower cost per lb. that is given in free markets [even in highly restrained free markets]. And much of reason given for really big launchers is the idea that it "should be" cheaper. As it is cheaper to make a super tanker or it's cheaper to make a high rise building- in terms of amount crude oil delivered somewhere or the unit cost of renting/leasing sq ft of office space. But of course this assumes there is enough of a market for it-people needing enough crude oil and people needing enough office space.

And some think that big booster would generate a market- cheaper unit cost will mean more demand. And of course they have the notion that you NEED these booster for going to the Moon or Mars. And of course if you had colonies on Moon or Mars that would increase the need for more rocket launches. Building a large rocket is sort of like building a bridge to nowhere- any merit of such idea, is it would generate commerce to that area [that the bridge was going to]. Though the obvious question is how much commerce would it generate [assuming Time is also considered] and what does it actually cost to build.

Rather than merely being cheaper, fuel depots are actually more important because the create a market in space. Having market will drive down the cost further [and almost as important bring money "into the area"- or bring people in- involve the world, so to speak. Fuel depots will start mining water in space- and other things. Fuel depot will *start* vast amounts of money to be spent in space. Which is not saying NASA budget will increase [though it will] but rather is talking about real and large quantities of money.

[If you want space to be the exclusive domain of government, the last thing you should want to do is encourage markets in space.]

So, if have say doubling or tripling of the current space industry- say instead a hundred or so billion a year to say 300 billion per year, then Hvy lift [when one is considering potential future growth] could become desirable- it could actually save money in terms of costs.

Comment by gbaikie — April 15, 2010 @ 8:23 pm

22. I agree with Marcel, especially with what LCROSS uncovered at the lunar south pole.

Canceling the return-to-the-moon-first is just plain dumb. Write your representatives and senators!

Comment by Robin Chew — April 15, 2010 @ 9:04 pm

23. Instead of funding an HLV, we would be better off spending the money developing a lunar lander.

Comment by Warren Platts — April 16, 2010 @ 1:18 am

24. After the Presidents speech yesterday, I think we need to rename this blog to 'The Once and Future Mars'

Comment by Darth Vader — April 16, 2010 @ 8:43 am

25. Darth,

After the Presidents speech yesterday, I think we need to rename this blog to 'The Once and Future Mars'

If you actually think that the new policy will result in a human mission to Mars — in 2035 or ever — you haven't been paying attention to the history of NASA.

Comment by Paul D. Spudis — April 16, 2010 @ 9:12 am

26. *Sorry — I'm just not buying any of this. Five years is barely enough time for them to complete the preliminary studies for all that, let alone the actual flying of those missions. My evidence for this? The last six years of the agency's implementation of the Vision.*

That may very well be true. If, after 5 years, we don't have enough information from robotic precursors, technology demonstrations, commercial development, and partner agreements to tell us what sort of HLV we actually need, then the HLV decision might get delayed by a corresponding amount. However, I imagine that whatever approach NASA takes, it will be subject to similar potential delays.

Comment by red — April 16, 2010 @ 8:18 pm

27. Note to Darth – ... Could also rename the blog "The Once and Future Asteroid". (Note to Paul Spudis – as well as some people who have stated they are "confused" by the new direction.) This is a paradigm shift. Comparing what we want to do in the future to the history of NASA's getting us there in the past... just doesn't make as much sense as it used to, if we all realize the players are going to be comprised of so many MORE corporations and cultures than the one(albeit, exemplary) of the National Aeronautics and Space

Administration. It would be like saying "That's not how the Army Air Corps does things."

Comment by Claudia — April 17, 2010 @ 12:32 am

28. P.S. My Daddy was an inventor who worked in the *private sector*, so I guess that colors my perspective — literally, I had the pleasure of seeing color T.V. before most of my peers in the early days, thanks to his inventiveness. And I saw men walk on the moon, real time.

Comment by Claudia — April 17, 2010 @ 12:35 am

29. Claudia,

This is a paradigm shift.

Really? Let's see — we're canceling the Constellation contracts but we will let new contracts with New Space corporations to provide transport to ISS. We're going to Mars in 30 years) because it is a "first" that will "excite and engage the public." Presumably, that includes live TV of the first steps and flag planting. Oh, and all this is to be accompanied by huge amounts of consensus management, extensive paper study, and bureaucratic overhead.

Yeah — as Rick Tumlinson would say, "Welcome to the Revolution!!"

Comment by Paul D. Spudis — April 17, 2010 @ 5:53 am

30. A small nit-pick, the President did not say IN 2015. I see about 7 out of 10 reports on the President's speech make this exact same mistake.

The President said a design decision would be make NO LATER than 2015, not IN 2015. That is a huge distinction. If people can not see this and make that distinction they are being disingenuous.

Just like he said he had to have that employment plan on his desk no later than august 15th, he did not say it had to on his desk ON the 15th.

For me that that means they will have to settle on materials, composite tanks? et cetera .. and how the new engine will shape up, will it be a knock off of the 180 with a simpler turbo pump, or something closer to the F1 of Saturn V fame? Elon Musk has talked about a bigger upgrade to the Merlin pushing it closer to the F1, has he found an ear in the President to help fund development of such an engine?

Even if Santa Claus dropped a HLV on a pad tomorrow, we would not have anything to launch. Five years to test systems and turn the ISS into a development park, find out what works, and finalize a launch vehicle, if we need it, starting sometime approaching 2015.

Comment by Vladislaw — April 17, 2010 @ 3:38 pm

31. Vlad,

A small nit-pick, the President did not say IN 2015….. The President said a design decision would be make NO LATER than 2015, not IN 2015. That is a huge distinction.

How does a "small nit-pick" equal a "huge distinction"?

Anyway, I do not agree that it is a distinction at all. Give NASA a direction like "make a decision by 2015" and on December 31, 2014, you will still have one hundred senior managers in a meeting, showing Powerpoint slides to each other, arguing whether the deadline begins at midnight that evening or COB the first federal work day after January 1st.

Comment by Paul D. Spudis — April 17, 2010 @ 5:47 pm

32. I thought I was nitpicking because I was not arguing the premise of your arguments, either for or against. Only that there is a distinction in WHEN, theoretically a decision could be made and construction start. Although NASA could, through some miracle come out next week and say, "we are going to brush the dust off the 1984 SD side mount plans and start building in 2011", I highly doubt it will happen. The President said "And we will finalize a rocket design no later than 2015 and then begin to build it"

It is just a question of how serious the Nation and NASA is for a HLV.

Comment by Vladislaw — April 18, 2010 @ 12:56 pm

33.   "Give NASA a direction like "make a decision by 2015" and on December 31, 2014, you will still have one hundred senior managers in a meeting, showing Powerpoint slides to each other, arguing whether the deadline begins at midnight that evening or COB the first federal work day after January 1st"

Is it honestly that bad at NASA? No matter who is President, no matter who is Administrator, no matter the congressional mandate? They truly are that frozen? If so .. I am glad we are moving to commercial.

Comment by Vladislaw — April 18, 2010 @ 1:00 pm

34. Vlad,

Although NASA could, through some miracle come out next week and say, "we are going

to brush the dust off the 1984 SD side mount plans and start building in 2011", I highly doubt it will happen.

People within the agency have been looking at side-mount HLLV plans for the last six years. One could be built and flown by 2014. And completely within the run-out budget envelope.

I am glad we are moving to commercial.

We're not "moving" to commercial — we're only talking about moving to commercial. In the same manner as we're "going" to Mars.

Comment by Paul D. Spudis — April 18, 2010 @ 1:16 pm

35. Just a nit. I'm burned out trying to explain that no ones proposing a big move to commercial — if anything the same commercial firms will now get far less then they do now with shuttle and station.

Ignoring that. Folks often talk about the importance of researching and developing on orbit fuel transfer systems, and doing demonstration missions. Am I missing something, or do people not know we've been using such systems for decades? I mean the Russian Progress has been doing automated docking and refueling of space stations since '78, and currently still is the prime tanker refueling the ISS.

Does 30 years of operational experience not count as enough demonstration for this technology?

Comment by Kelly Starks — April 19, 2010 @ 1:13 pm

36. Paul, I am sure you saw this, would this be something fairly easy to mitigate against or would this pose any serious problems?

"Craters Around Lunar Poles Could Be Electrified?" http://tinyurl.com/34v4dd4

Can static charges be harnessed in any form as some kind of supplementary power for a base or powering equipment/experiments? That is another whole area for examination. The moon gives up her mysteries slowly but has so many to offer as you have written about so often.

Comment by Vladislaw — April 19, 2010 @ 1:32 pm

37. Ok, I'm dense. It just occurred to me the central theme of Obama's space plans. Cut out the Russians.

Commercial Crew transport, really is just to get Boeing and L/M to field a space taxi, so they don't have to depend on the '60's era Russian Soyuz. HLV technology research.

Boiled down to develop a new US LOx/Kero rocket engine to compete with the '70's era Russian RD-180's. On orbit fuel transfer demonstration missions, means show folks we could refuel the ISS without the '70's era Russian progress freighter/tankers that do it now.

Comment by Kelly Starks — April 19, 2010 @ 1:57 pm

38. Vlad,

Can static charges be harnessed in any form as some kind of supplementary power for a base or powering equipment/experiments?

We don't really know very much about static charging of the lunar surface, but the seven-year survival of all the ALSEP experiment packages and a couple of the Surveyor spacecraft indicate that the surface charging, if it exists at all, is very weak. It is not a good source for power. However, the near-permanent sunlit areas near both poles are potential candidate for solar arrays to provide electricity for surface operations.

Comment by Paul D. Spudis — April 19, 2010 @ 2:18 pm

39. Kelly,

the Russian Progress has been doing automated docking and refueling of space stations since '78, and currently still is the prime tanker refueling the ISS.

I think that the technical issues people are concerned about with in-space fueling revolves around cryogenic propellant transfer and management. The Soyuz spacecraft uses storables.

Comment by Paul D. Spudis — April 19, 2010 @ 2:21 pm

40. I have a map of Shackleton that I had photocopied and marked in a base a long time ago showing the different light densities. Always thought it would happen before this.

Did you ever hear about the proposal that Stone Aerospace put out. The one where you have to make your fuel there for the return trip. You don't take it with you.

http://www.stoneaerospace.com/news-/pictures/ShackletonCraterExpeditionV32.pdf

He was talking about a real daring plan, wonder how far off something like that is.

Comment by Vladislaw — April 19, 2010 @ 3:19 pm

41. *I think that the technical issues people are concerned about with in-space fueling revolves around cryogenic propellant transfer and management. The Soyuz spacecraft uses storables.*

That's a pretty trivial issue to be concerned about? Moving fluids around in zero G and auto docking the fueling connector are issues (long known, but at least challenging.) the cryo part just means more insulation.

Comment by Kelly Starks — April 20, 2010 @ 8:29 am

42. *That's a pretty trivial issue to be concerned about?*

Tell them, not me.

Comment by Paul D. Spudis — April 20, 2010 @ 9:13 am

43. *Tell them, not me.*

;) Sounds like a excuse to go back and study non issues for a few years, rather than do anything productive.

Comment by Kelly Starks — April 20, 2010 @ 12:32 pm

44. *Sounds like a excuse to go back and study non issues for a few years, rather then do anything productive.*

You have just described the essence of the "new direction"!

Comment by Paul D. Spudis — April 20, 2010 @ 2:36 pm

45. Sadly agree. More sadly – or stunned/disgusted – that so many space advocates don't see this as the trap it is.

Comment by Kelly Starks — April 20, 2010 @ 3:25 pm

46. *stunned/disgusted – that so many space advocates don't see this as the trap it is.*

Agreed. Denial and delusion.

Comment by Paul D. Spudis — April 20, 2010 @ 4:26 pm

47. Flexible Path stinks!! This whole Obama-space plan reeks!! We would've been well on the way to building & cargo-rating the Aries 5 rocket—were it not for the anemic funding, that Constellation got since the 2004 inception. The Moon SHOULD be dealt with first, ahead & before any asteroid. It is a FAR better destination, for base-building & resource-utilization. The Aries 5 should be properly funded, and made into America's next heavy-lift work-horse. We DO NOT need to waste & squander another five years, just on researching a possible new design.

Comment by Chris Castro — May 4, 2010 @ 12:18 am

"We've been there before. Buzz has been there."

April 16, 2010

President Obama and Apollo 11 astronaut Buzz Aldrin walk toward Air Force One Thursday, April 15, 2010, for a day trip to Florida. (AP)

During a carefully staged appearance at Kennedy Space Center yesterday, President Barack Obama rolled out his plans for the U. S. space program. Although there weren't many surprises (the White House Office of Science and Technology, under the direction of John P. Holdren, had released a fact sheet[1] days earlier outlining details), one startling part of the speech was that we are abandoning the Moon as a goal[2]. Though hinted at in several statements by people around the President, including NASA Administrator Charles Bolden[3] and Apollo 11 Astronaut Buzz Aldrin[4], a path away from human return to the Moon is now officially the direction of Obama's space policy.

Given the topic of this blog, it shouldn't surprise many of you to learn that people are calling and writing me, asking for my reaction to the new policy. Although it wasn't much of a surprise, it is disappointing to me, but not for the reasons you might suspect.

The speech detailed aspects of the administration's new space budget[5], which will eliminate Project Constellation[6], contract with commercial entities for human transport to LEO, and spend money for development of new technology so as to "revolutionize" our access and capabilities in space. The Moon was finally mentioned near the end of the speech and I felt it would be fitting to use the President's own words[2] as the title for this post, and

then give my views of the Moon's place in the template of space exploration.

I've heard the "been there" line many times since 2004 when President George W. Bush announced the Vision for Space Exploration[7], so hearing it one more time was not a particularly jarring experience. But stop for a moment to consider exactly what President Obama said. Lunar return critics give many reasons to NOT go to the Moon: they think that it's scientifically uninteresting, it doesn't contain what we need, it will turn into a money sink (preventing voyages to many other destinations in space – perhaps number one on their list), that there are more pressing needs here on Earth, and I'm sure others that I haven't yet heard. But this new space policy rationale is unique and carries with it different and significant implications for our nation's exploration of space.

We have now added a new requirement for U.S. space missions – we must go to a place never before visited by humans. Of course, some will argue that such a concept is implicit in the word "exploration" but until recently, exploration encompassed a much wider concept[8] where exploration was followed by exploitation and settlement by many people from many walks of life using many different skills toward a myriad of goals. I wonder if supporters of this new space policy have stopped to consider the implications of the "not been there" requirement. The new meaning of exploration contains within it the seeds of its own termination: after you've touched the surface, planted a flag, and collected some rocks or deployed an instrument, that destination is "done." Or does such a formulation apply only to the Moon?

One of the biggest criticisms hurled at Project Constellation is that it is largely a grandiose repeat of the Apollo explorations of the Moon undertaken over 40 years ago. Certainly, as had been outlined by NASA, lunar return consisted of sortie missions that landed crews all over the Moon to do local field exploration. Such a mission template is indeed Apollo writ large. But that is not and was never the intent of lunar return under the Vision for Space Exploration which is now under assault. Constellation was largely NASA's rocket development program, while the Vision for Space Exploration was strategic direction outlining a sustainable lunar return, whereby we would bootstrap our way "beyond" by learning how to use the resources of the Moon and other bodies.

So let me respond to the President's new plan by reminding the readers of this column why the Moon is our goal[9] and of its significance and value to space exploration.

It's close. Unlike virtually all other destinations in space beyond low Earth orbit, the Moon is near in time (a few days) and energy (a few hundreds of meters per second.) In addition to its proximity, because the Moon orbits the Earth, it is the most accessible target beyond LEO, having nearly continuous windows for arrival and departure. This routine accessibility is in contrast to all of the planets and asteroids, which orbit the Sun and have narrow, irregular windows of access that depend on their alignment with respect to the Earth. The closeness and accessibility of the Moon permit modes of operation not possible with other space destinations, such as a near real-time (less than 3 seconds) communication link. Robotic machines can be teleoperated directly from Earth, permitting hard, dangerous manual labor on the Moon to be done by machines controlled by humans either on the Moon or from Earth. The closeness of the Moon also permits easy and continuous abort capability, certainly something we do not want to take advantage of, but comforting to know is handy until we have more robust and reliable space subsystems. If you don't believe this is important, ask the crew of Apollo 13[10].

It's interesting. The Moon offers scientific value that is unique within the family of objects in the Solar System. The Moon has no atmosphere or global magnetic field so plasmas and streams of energetic particles impinge directly on its surface, embedding themselves onto the lunar dust grains. Thus, the Moon contains a detailed record of the Sun's output through geological time (over at least the last 4 billion years). The value of such a record is that the Sun is the principal driver of Earth's climate and by recovering that detailed record (unavailable anywhere on the Earth), it can help us understand the details of solar output, both its cycles and singular events, throughout the history of the Solar System. Additionally, because of the Moon's ancient surface and proximity to the Earth, it retains a record of the impact bombardment history of both bodies. We now know that the collision of large bodies has drastic effects on the geological and biological evolution of the Earth and occur at quasi-regular intervals. Because our very survival depends on understanding the nature and history of these events as a basis for the prediction of future events, the record on the lunar surface is critical to our understanding. A radio telescope on the far side of the Moon can "see" into deep space from the only platform in the Solar System that is permanently free from Earth's radio noise. The Moon is a unique, rich and valuable scientific asset[11].

It's useful. In my opinion, this is the most important and pressing argument for making the Moon our first destination beyond LEO. Because of the detailed exploration of the Moon undertaken during the last 20 years, we have a very different understanding of its properties than we did immediately following Apollo. Specifically, the Moon has accessible and immediately usable resources of both energy and materials[12] in its polar regions, something about which we were almost completely ignorant only a few years ago. For energy, both poles offer benign surface temperatures and near-permanent sunlight, as the lunar spin axis obliquity is nearly perpendicular to the plane of Earth's orbit around the Sun. This relation solves one of the most difficult issues of lunar habitation – the 14-day long lunar night, which challenges the design of thermal and power systems. In addition, once thought to be a barren desert, we have recently found that the Moon contains abundant and accessible deposits of water, in a variety of forms and concentrations. There is enough water on the Moon[13] to bootstrap a permanent, sustained human presence there. Water is the most important substance to find and use in space; not only does it support human life by its consumption and provision of breathable oxygen, in its form as cryogenic liquid oxygen and hydrogen, it is the most powerful chemical rocket propellant known. A transportation system that can routinely access the lunar surface to refuel, can also access all of cislunar space, where all of our national strategic and commercial (and much of our scientific) assets reside (many satellites reside above LEO and are inaccessible for repair). Such a system would truly and fundamentally change the paradigm of spaceflight and can be realized through the mining and processing of the water ice deposits near the poles of the Moon. Space exploration should be a driving force in our economy not merely a playground for scientists or a venue for public entertainment.

Given the real and potential benefits of lunar return, the question is no longer "Why the Moon?" but "Why bypass the Moon?" I'm glad that "Buzz has been there" but that fact is irrelevant to either the value or the desirability of lunar return. By proposing to eliminate the Moon as a destination, the President has fundamentally altered the societal value of the space program in a significant and qualitatively different way.

If our new space program is to be made into a simple instrument of public spectacle ("cheap thrills[14]" and "colossal feats[15]," as variously reported by news columnists) with each new mission requiring a "series of 'firsts' to engage and excite the public[16]", it will no longer have any more real long term benefit to our national security and wealth than did the bread and circus shows that heralded the demise of ancient Rome. Yes, there were and will be some exciting spectacles. And when such events are finished, people turn away and go home – none the wiser, none the richer, and none the better off. We won't be staying at any destination long enough to fully characterize it and use what it has to offer.

Is this the kind of space exploration we want?

The seeds of the termination of our national space program were planted yesterday in Florida.

Topics: Lunar Exploration, Lunar Resources, Lunar Science, Space Politics, Space Transportation, Space and Society

Links and References

1. a fact sheet, http://www.whitehouse.gov/sites/default/files/microsites/ostp/ostp-space-conf-factsheet.pdf
2. abandoning the Moon, http://www.nasa.gov/news/media/trans/obama_ksc_trans.html
3. Charles Bolden, http://www.denverpost.com/headlines/ci_14385263
4. Buzz Aldrin, http://www.usatoday.com/news/opinion/forum/2010-04-15-column15_ST2_N.htm
5. new space budget, http://www.nasa.gov/news/budget/index.html
6. Project Constellation, http://www.nasa.gov/exploration/news/ESAS_report.html
7. Bush announced, http://www.spaceref.com/news/viewpr.html?pid=13404
8. exploration encompassed, http://blogs.airspacemag.com/moon/2010/01/25/have-we-forgotten-what-exploration-means/
9. why the Moon, http://www.washingtonpost.com/wp-dyn/content/article/2005/12/26/AR2005122600648.html
10. Apollo 13, http://en.wikipedia.org/wiki/Apollo_13
11. scientific asset, http://www.nap.edu/catalog.php?record_id=11954
12. resources of energy and materials, http://www.spudislunarresources.com/Opinion_Editorial/status_quo.pdf
13. enough water on the Moon, http://blogs.airspacemag.com/moon/2010/03/01/ice-at-the-north-pole-of-the-moon/
14. cheap thrills, http://www.pasadenastarnews.com/ci_14893500
15. colossal feats, http://www.philly.com/philly/news/nation_world/20100416_Obama_s_space_plan_sees__leap__into_future.html
16. series of "firsts," http://www.nasa.gov/pdf/384767main_SUMMARY%20REPORT%20-%20FINAL.pdf

Comments

2. I haven't had time to write about this, although I feel as if your article sums up my point exactly. While I would love to go to Mars (yes, I love Mars!), it's just simply too far away right now to be of any real use. Unless VASMIR can be perfected, taking a 6 month journey one way to Mars isn't practical due to the amount of food, water and supplies one will need en route, not to mention the radiation, microgravity, etc. that can drastically affect health. Hopefully the private sector won't follow NASA's lead and head towards the Moon first, then Mars as establishing a permanent colony is in our species best interests.

Comment by Darnell Clayton — April 16, 2010 @ 1:32 pm

3. Paul – At first glance, the President's remarks sounded internally self-inconsistent with a theme of firsts in exploration and pushing the boundaries, then followed towards the end with this remark: "Our goal is the capacity for people to work and learn and operate and live safely beyond the Earth for extended periods of time, ultimately in ways that are more sustainable and even indefinite."

I fully expect that when plans are developed involving both the commercial and public sectors to realize this "goal" (actually more properly cast as a vision), the Moon must be considered as a critical element to guarantee sustainability. Incorporating the Moon as a key element will occur after Obama is after office anyway, so I

"We've been there before. Buzz has been there."

fully expect this oversight to be rectified in the not-too-distant future; perhaps as early as this year as the details are worked.

Comment by Joe Williams — April 16, 2010 @ 1:50 pm

4. Joe,

"Our goal is the capacity for people to work and learn and operate and live safely beyond the Earth for extended periods of time, ultimately in ways that are more sustainable and even indefinite."

Yes, I heard this too, but I do not believe that he believes this. Otherwise, why the dismissive — "We've been there" in regard to lunar return? Note well: he could have simply ignored the Moon completely, but instead went out of his way to repudiate it as a goal.

Nope – I think there is another agenda at work here.

I fully expect this oversight to be rectified in the not-too-distant future; perhaps as early as this year as the details are worked.

I doubt that. In a year, we'll still be on a funding CR for NASA, the various Congressmen will be wrestling their district pork, and we'll all be "another year older and deeper in debt."

By the way, many thanks for the very nice call-out to me on your blog the other day.

Comment by Paul D. Spudis — April 16, 2010 @ 2:01 pm

5. Darnell,

Unless VASMIR can be perfected…

You should be aware that a VASIMR-powered human Mars craft requires a 10-15 megawatt nuclear reactor to power it, something not only well beyond existing technology but given the aversion to nuclear power in space by this country, something not likely to be achieved on anything less than a multi-decadal timescale. See Franklin Chang-Diaz's Scientific American article:

http://www.adastrarocket.com/SciAm2000.pdf

Comment by Paul D. Spudis — April 16, 2010 @ 2:05 pm

6. I absolutely agree with you. And I posted the President's statements on my blog last night:

http://newpapyrusmagazine.blogspot.com/2010/04/how-congress-should-respond-to-obamas.html

But we still have a Congress! And scientist like yourself still have a voice! In fact, one of my biggest complaints is that not enough scientist and NASA engineers have been talking about lunar bases, lunar colonization and the long term social and economic value of the Moon.

Comment by Marcel F. Williams — April 16, 2010 @ 3:02 pm

7. Marcel,

I am largely in complete agreement with all of your recommendations, except for the building of a new space station using inflatables. The existing ISS could support lunar return, especially if we develop an SEP cislunar tug, as Dennis Wingo advocates. I would rather use the existing station and pay the delta-v penalty than spend more money to build one that is optimized for lunar return — the better is the enemy of the good enough.

And I am talking to members of Congress, staffers, and other people in Washington and industry. I agree with you that we have an obligation to explain the benefits of lunar return to them.

Thanks for reading and for your support!

Comment by Paul D. Spudis — April 16, 2010 @ 3:20 pm

8. Thanks, Paul, for taking the time to outline, once again, the reasons why it's folly to ignore the deep-water port (and Rosetta Stone for the Solar System) situated so conveniently close-by.

All for the sake of popinjay politics.

It's depressingly familiar, as though we were somehow back in 1972 watching the retreat from the Moon, completely unaware how far the reality would fall so short of the selling points and even our worst expectations. I wouldn't have believed it then, if someone had told me, and it makes me sick just thinking what the next 40 years of spinning our wheels might look like.

Like standing on the docks, crying "bon voyage" as the Titanic slips off on its maiden voyage.

The hubris is just stunning. The True Belief of those who want to trust this administration is sincere, that the costs of scaling back aren't unfathomable.

A last question, if I may. Let's pretend, since it seems so fashionable, that it's 1972, and you knew what the next 40 years would certainly be, in detail. Ignoring the paradoxes, etc., seriously… what course would you take?

Comment by Joel Raupe — April 16, 2010 @ 5:36 pm

9. Joel,

it's 1972, and you knew what the next 40 years would certainly be, in detail...... what course would you take?

An interesting question. Clearly, although the Apollo system was technically superb, it was financially unsustainable. So we needed a cheaper way to get back and forth to orbit with people.

I would have developed a Shuttle, but a fully reusable, liquid booster one. It would be used only for crew transport to and from LEO. I would have kept Saturn V for cargo launch only; the Shuttle development budgets would have supported at least one cargo launch per year. I would have built a space station, largely to be used as a transportation node, out of Skylab-like modules. The next vehicle we need is a lunar "tug" or OTV, based at station and used for routine transport to and from Earth-Moon L-1. At the L-point, I would station a human-tended (not permanently occupied) staging node to transfer to and from the lunar surface. The goal of lunar return would have been focused on manufacturing: first oxygen production, then bulk materials (for shielding, aerobrakes for OTV return, structures, etc.) and finally, hydrogen. Materials would first be used to build up and establish the lunar outpost, but the goal is to export products to L-1 and cislunar. For a timeline, I think we would have had Shuttle/station by the late 70's-early 80's, OTV and L-1 node by the mid-80's, and lunar outpost by the late 80's-early 90's. Routine cislunar access and construction of satellite distributed systems would have begun by the mid-90's. By now, we would be on Mars.

Fun to speculate about, but I am under no illusions that this is all imagined with hindsight, which is always 20-20.

Comment by Paul D. Spudis — April 16, 2010 @ 6:09 pm

10. The way I interpret the "We've been there before" comment is more in the light what can we afford to do, and what is NASA's true mission

Sure we will go back to the Moon, and there is more to explore. But remember that we're still exploring the Earth, so exploration will go on for centuries. We're not going to hold off on other destinations because we haven't finished exploring the Moon. I see exploration as part of what we need to do to start our exploitation of the Moon. I think NASA will be involved with this part (most likely robotic missions), but I see commercial companies as building and doing everything. Some of it may be under a NASA contract (like lunar fuel & water depots), and as industry becomes more established, this will decrease.

The job that we demand from NASA regarding exploration is to go to places and do things that no company can do. When the early explorers found someplace new, they may have created an outpost or started a food supply (i.e. goat islands), but they kept pushing on to the next new place. Others followed, and they were the ones to start the towns and create the local industries. That's where we are today with the Moon. Armstrong, Aldrin and others found that the Moon wasn't made of cheese, and that the lander didn't sink into a puddle of dust. Subsequent missions showed us hints of how we'll be able to move and work on the Moon. The explorers have done the hard work, and now the followers can start doing theirs.

For NASA the Explorer, the next goal should be to reach out beyond the Moon. This will entail a whole host of new knowledge & technology – this is where the affordability issue comes in. Some of this new knowledge & technology will be useful for Moon exploration & exploitation, but it is absolutely critical if man ever wants to explore the harsh environment of our solar system.

This is what I saw in the message of "We've been there before".

Comment by Coastal Ron — April 16, 2010 @ 6:51 pm

11. The Moon is clearly not the central immediate goal as it was. However, I think it's been given the status of "one goal among many, and not the first" rather than being crossed off the list entirely. The transcript has the following statement:

"I understand that some believe that we should attempt a return to the surface of the Moon first, as previously planned. But the simple fact is, we have been there before. There is a lot more space to explore, and a lot more to learn when we do."

I suppose you could read this 2 ways. Obama could be saying the Moon isn't going to be the first place we go to. He could also be saying that we aren't going to the Moon at all, since we've already been there before. I'm inclined to give it the first interpretation because the 2011 NASA budget has a Planetary Science budget still having the Lunar Quest line. It also has the following phrases:

"NASA is taking a new approach to this long-term goal; by laying the ground work that will enable humans to safely reach multiple potential destinations, including the Moon, asteroids, Lagrange points, and Mars and its environs."

"These missions will help us determine the next step for crews beyond low Earth orbit, answering such questions as ... Do the resources at the lunar poles have the potential for crew utilization?"

"ESMD will also develop multi-mission operational concepts for future human space flight campaigns to such targets as the Moon, asteroids, Martian moons, and Mars itself…"

"NASA will fund research in a variety of ISRU activities aimed at using lunar, asteroidal, and Martian materials to produce oxygen and extract water from ice reservoirs. A flight experiment to demonstrate lunar resource prospecting, characterization, and extraction will be considered for testing on a future Flagship Technology Demonstration or robotic precursor exploration mission."

"NASA will work with industry and academia to develop advanced spacesuits to improve the ability of astronauts to assemble and service in-space systems, and to explore the surfaces of the Moon, Mars, and small bodies."

"NASA will begin funding at least two dedicated precursor missions in 2011. One will likely be a lunar mission to demonstrate teleoperation capability from Earth and potentially from the International Space Station, including the ability to transmit near-live video to Earth. This will also result in investigations for validating the availability of resources for extraction. NASA will provide opportunities to participate in the payloads and observation teams, and potentially portions of the spacecraft, through open competition."

"Potential missions may include: … Landing a facility to test processing technologies for transforming lunar or asteroid materials for fuel could eventually allow astronauts to partially "live off the land.""

"Additionally, a new portfolio of explorer scouts will execute small, rapid turn-around, highly competitive missions to exploration destinations. Generally budgeted at between $100M and $200M lifecycle cost…" [I'm having trouble picturing missions from $100M to $200M going to, say, Mars, so I think the Moon is a prime candidate for this scout program run by NASA Ames.]

Comment by red — April 16, 2010 @ 8:50 pm

12. Paul, I think you're exactly right! Ending the Apollo program was one thing. But decommissioning the Saturn V was a tragedy!

All we needed was one of those classic two staged winged people shuttles to go with it and we'd have a much different and much better space program today. And, I agree, we'd already have a Moon bases, or bases, and a Mars base.

My favorite two stage shuttle concept was the Bono and Gatland 1969 Perseus concept which used a reusable SSTO LOX/LH2 plug nozzle booster to launch a space plane glider situated on top of the booster. The booster would return to Earth immediately after launch, landing vertically like the Delta Clipper; the space plane would return to Earth after the completion of its orbital mission.

A reusable SSTO booster would have been the ultimate game changing technology. And I guess if you could refuel it in orbit, it would also be the ultimate reusable orbital transfer vehicle.

It's a shame there hasn't been any significant funding for plug nozzle technology since the termination of the X-33.

Comment by Marcel F. Williams — April 16, 2010 @ 9:17 pm

13. It is at times like these we can realize that, no matter how wonderful the Apollo program was and no matter how much it achieved so early in the age of spaceflight, that it ultimately taught the American public and politicians the wrong lesson. Apollo taught us that the way to do space was to make a huge spectacle of it: have a huge rocket, send up superheroes and have them leave flags and footprints. Then, when the public is no longer interested, close up shop.

So yesterday, BHO mentioned going to an asteroid. Great! How I'd love that! But then the depression set in: NASA would only go to an asteroid just once or twice just to say that we'd achieved a new first in space. Take some pretty pictures, onto the next destination. There was no mention of platinum group materials nor of space industrialization that could be achieved with NEOs. No, just a way stop on the way to Mars, that's all. Impressive firsts is what it's all about.

Apollo taught us the wrong lessons. By contrast, the Soviets lost and now they launch Proton 12x per year and launch Soyuz 12x per year without much fanfare. Their first stage booster engines are awesome machines mastered 4 decades ago that the U.S. will now spend billions to replicate. The Russians launch Progress several times per year to ISS and deliver crews regularly too.

Maybe it would have been better had we lost the race to the Moon!

Just a side note on Proton launching 12x per year… Proton is no small rocket: it is in the same class as the heavier Atlas V and Delta IV variants. Imagine if we could launch 6 Atlas Vs and 6 Delta IV Heavies per year. Each one of those flights could deliver, say, 3 metric tons to the North Pole of the Moon. Imagine delivering 36 metric tons of robotic equipment to the Moon's North Pole each year. Now that would be a space program!

Comment by Itokawa — April 16, 2010 @ 9:29 pm

14. What saddens me and frustrates me about these debates about alternative destinations is that it often seems to result in ignoring some of the best aspects of the new plan.

I think that the moon is great, as well as asteroids, the moons of Mars and Mars itself. What I think is even greater is developing a broadly useful space transportation system that could be used to go wherever we ultimately decide to go.

This general space infrastructure includes much cheaper access to LEO. It includes fuel depots to increase the capacity of current and future rockets. It includes things like an orbital transfer vehicle and expandable structures. It includes autonomous rendezvous and docking. It includes other technology development (e.g. life support, radiation shielding, etc). All of these things will make space so much more doable in the long run, and they are all in the new plan!!! (The only item that wasn't mentioned in the plan was an OTV, but it's a very reasonable option.)

If Obama doesn't have a destination/deadline or the "right" destination, then so much people predict the downfall of our space program. I, on the contrary, think that this new plan finally puts the space program on the right track for the long term. It's the development of these enabling technologies that excites me.

It's the involvement of the commercial access to space that is also exciting for the future development of space. Once the commercial people really get involved, there is the long-term potential for an explosion of space activities. I would guess that the moon would be of interest to them. I know that Bigelow is already interested in landing his modules on the moon.

These are exciting times! Let's support this new direction of technology development and sustainability in space.

Comment by Stefan — April 16, 2010 @ 10:52 pm

15. I forgot to mention in my prior post that there is also the very practical issue of not being able to afford the development of a lunar lander (or any lander) at this time.

By making space activities more affordable, we'll ultimately be more able to go wherever we want. We can let the long-term destinations take care of themselves. As everyone keeps mentioning, Obama won't be in office by the time we're ready to really pick our destination.

One final thought:

This is something that I never see mentioned anywhere. There is also the option to team up with other countries and have them develop elements of an exploration program that we can't afford. For instance, Europe and/or Russia could pay for and develop a lunar lander and/or deep space vehicle. This would make ambitious exploration goals more doable. Any comments on that idea? Or does our national pride say that we have to do everything alone?

Comment by Stefan — April 16, 2010 @ 10:59 pm

16. red and Coastal Ron,

Obama could be saying the Moon isn't going to be the first place we go to. He could also be saying that we aren't going to the Moon at all, since we've already been there before. I'm inclined to give it the first interpretation

I understand that the statement as written is subject to individual interpretation. I read it in the second sense, largely on the grounds that people in the agency (with few exceptions) never really understood the reasons for going to the Moon to begin with and as the President gets his technical input from them (and Buzz), he doesn't know the value of lunar return either. So I wrote this column to lay out those reasons yet again.

I am not one of those who treat political speeches as holy writ, designed for exegesis and commentary. I think of them as straightforward statements of intent and belief. I see the preservation of the Lunar Quest program as indicative of nothing in particular, except to let Ames keep its small mission work (LADEE) and to bureaucratic inertia in general. As for the alleged new ESMD robotic missions to the Moon, I'll believe it when I see it.

Comment by Paul D. Spudis — April 17, 2010 @ 5:11 am

17. Stefan,

I, on the contrary, think that this new plan finally puts the space program on the right track for the long term. It's the development of these enabling technologies that excites me.

I am not against technology development. My point on this (made in several past columns) is that you can get technology development in two ways: through developing and flying real missions to specified destinations or through a random R&D "investment" program. The historical record shows that NASA is good at the former and abysmal at the latter.

there is also the very practical issue of not being able to afford the development of a lunar

"We've been there before. Buzz has been there."

lander (or any lander) at this time.

Yes, I know this is an article of faith among many, especially fans of the Augustine report — I do not agree with them and I question both their costing methodology and results (on that issue, see this very interesting article in the current AvWeek:

http://tinyurl.com/Obama-explains

By the time we would need a lander (the end of this decade), we could have one ready; in contrast to widespread opinion, a lunar return program is affordable under the projected run-out budget. Just because Constellation was unaffordable does not mean that lunar return is — it all depends on the program architecture.

Comment by Paul D. Spudis — April 17, 2010 @ 5:23 am

18. Itokawa,

no matter how wonderful the Apollo program was and no matter how much it achieved so early in the age of spaceflight, that it ultimately taught the American public and politicians the wrong lesson

Only because we insist on taking the wrong lesson from it. Apollo was never about space faring — it was a battle in the Cold War. It just happened to occur in space.

Comment by Paul D. Spudis — April 17, 2010 @ 5:33 am

19. Paul, I agree with all the points in your article about why the Moon is interesting and why we should revisit it. But that's not necessarily the reason why it should be the next destination for human exploration by NASA.

I think there were far better ways that the President could have outlined the new NASA vision, but I do agree with the new direction. I would like to suggest the following descriptions:

NASA Human Exploration = Visiting new places beyond Earth, while perfecting the methods and technologies that let us do it safely. Apollo 11 – 17, including Apollo 13, epitomize this.

NASA Human Exploitation (for lack of a better word) = Re-visiting a prior destination, and expanding our knowledge of that place. Apollo did some of this on the Moon, and next it should be robotic explorers to help us refine our future human plans. Some of this may be geology related, but I think ultimately NASA will use the Moon as a testing ground for exploration technologies needed for human expansion beyond Earth's orbit. I also think that NASA will do this by contracting with commercial firms to provide the transportation and logistics.

Commercial Exploration = Utilizing the knowledge accumulated by NASA Exploration/Exploitation to identify future commercial areas of interest. I see this as an extension of current business R&D funding, and these companies will be relying on the pathfinder technologies that NASA has funded directly or indirectly.

Commercial Exploitation = Creating profitable enterprises in space. In order to have a robust commercial space industry on Earth, there must be profitable materials, products or knowledge that we can extract from space. NASA's Exploration/Exploitation activities are crucial for companies to understand what the risks/rewards will be if they want to generate revenue from space activities. The industries generated from these activities are the ultimate payoff from the knowledge NASA has passed down from the prior years of exploration & exploitation.

NASA is planning robotic explorers for the Moon, and this will help us understand the next steps for exploitation. But we already know how to get to/from the moon, and we already understand the basic risks/rewards of going there, so while going back will be exciting in it's own way, it's not really new.

However, if we want to venture beyond the Moon and Earth's orbit, that will require a new set of solutions and technologies, all of which trickle down to our growing commercial space industries. We haven't been there yet, and NASA is the only organization that can do it.

Last thought. I think part of the reason this is not crystal clear to all of us, is that we don't have a robust commercial space industry yet. I think in their own way, Obama/Bolden are trying to encourage the nascent space companies that are out there, but maybe they're doing it in too coy of a way.

Comment by Coastal Ron — April 17, 2010 @ 1:06 pm

20. Paul, may I suggest that you write another article entitled, "Why I Am Against Space Exploration"?

Of course, you are not against space exploration with respect to the manner in which *you* interpret that term. However, space exploration to the politicians means an Apollo-like program with a sequence of new firsts, flags and footprints of superheroes, gigantic rockets, big spectacles, on to the next destination. That's what the politician wants.

You and I, on the other hand, do not want space exploration in that sense. What we want is space development, space industrialization. You would say, using your definition of the term space exploration, that space exploration is necessary for space industrialization. But we don't want *exploration*, per se, we want *prospecting*. Of course it is good to know the history of the Moon and you are a scientist for whom this knowledge is of keen interest. But your articles show that what you deem more important, long term, is space development and industrialization, not space exploration as the politicians define it.

We now have that term *exploration* everywhere: Vision for Space Exploration (VSE), Exploration Systems Architecture Study (ESAS), Exploration Systems Mission Directorate (ESMD) and so forth. It's time to come out against space exploration.

I, for one, am against space exploration. I am for space development and industrialization.

Comment by Itokawa — April 17, 2010 @ 1:19 pm

21. While I'm at it, let me say that I'm against "Human Space Flight", too.

Amongst us girls, am I against human space flight as we might define it? No, of course not: no one will be more thrilled to see regular people visiting space hotels.

But Human Space Flight to the politicians and to the leadership of NASA who serve them means a focus first on getting humans to space and later, if ever, figuring out what to do with them once they're there. As you've pointed out many times, that's backwards: we should decide on goals and then decide what combination of humans and/or robots best achieves that goal.

So this apparent Luddite is now coming out against Space Exploration and Human Space Flight, because those terms I don't control and the interpretation of those terms by the politicians is completely corrupt.

Comment by Itokawa — April 17, 2010 @ 1:22 pm

22. @Stefan

Private enterprise isn't into the charity business. Their in business to make money. And any American manned spaceflight company is going to have to compete with other private and sometimes not so private manned spaceflight companies. This pretty much means that they're going to have to compete with the mighty Russian Energia company for cost, safety, and reliability!

Maybe a company like Space X somehow knows a lot more about flying into space than NASA or the Russians. But I don't think so! And its very likely that Space X will go through the same growing pains that NASA and Russia went through. So just one fatal accident aboard

a Space X vessel would probably send 99.9% of customers to Energia– with a brand new meaning in the English language for the term– Space X.

Its going to take a long time for American private companies to get off the ground and prove that they can deliver people into space as safely and reliable as NASA or Energia can. So obviously, you don't completely shut down the government's ability to fly into space until the private companies have proven that they can fly humans safely and reliably to and from orbit.

And that's what President Obama doesn't seem to understand! He's not turning over Space Shuttle operations to private industry. He's completely shutting down the government's ability to fly into space– with absolutely no replacement vehicle. And that would be extremely foolish!

Secondly, why should the corporations be the only ones able to fly into space. Why shouldn't an agency owned by the American people be able to fly into space too since they're the once who invented America's manned space program in the first place!

I strongly support NASA giving private companies money to develop private access to orbit. However, that $1.2 billion that NASA is giving to private companies is only a tiny portion of the $20 billion dollar a year NASA budget. So I'm a lot more concerned about the president's policy wasting the other $19 billion dollars, nearly $300 billion over the next 15 years with only the ISS and a single visit to an asteroid to show for it.

A permanent Moon base would be a much better investment for that $300 billion, IMO, for both public and private enterprise. And when we finally build that base, the privateers, tourist, and the settlers would quickly follow and America will be a much richer country with a much more exciting future!

Comment by Marcel F. Williams — April 17, 2010 @ 4:37 pm

23. Itokawa,

*may I suggest that you write another article entitled, "Why I Am Against Space Exploration"? Of course, you are not against space exploration with respect to the manner in which *you* interpret that term.*

You may suggest anything you want, but I am **not** against space exploration. I am against stupid space tricks.

I have already directly addressed the issue of what exploration means and entails — now and then — here (Chapter 45):

"We've been there before. Buzz has been there."

moon/2010/01/25/have-we-forgotten-what-ex-
ploration-means/

I refuse to submit to someone's re-definition
of the perfectly good word "exploration" that has
had a given and clear understanding for well
over 500 years.

Comment by Paul D. Spudis — April 17,
2010 @ 5:28 pm

24. Ron,

*if we want to venture beyond the Moon and
Earth's orbit, that will require a new set of solutions and technologies, all of which trickle down
to our growing commercial space industries.
We haven't been there yet, and NASA is the
only organization that can do it.*

Absolutely and one of these technical areas
is In Situ Resource Utilization, which should be
NASA's mission in a lunar return program. The
agency's job is not to industrialize the Moon — it
is to determine if the Moon can be industrialized.

Oh yes — to learn how to do lunar ISRU, you
need to be on the surface of the Moon for some
time with some degree of capability.

Ooops! I guess the Moon is in the critical
path.

Comment by Paul D. Spudis — April 17,
2010 @ 5:36 pm

25. One of the most important things we
could discover from a permanent base on the
Moon is whether humans and other animals
can actually remain healthy and reproduce under a 1/6 gravity environment.

We already know that microgravity environments are inherently deleterious to human
health. If men and women are stationed on the
lunar surface, we'll also find out if a hypogravity environment is deleterious to human health.

If a 1/6 hypogravity environment is not harmful to the human body then private industry is
probably going to be a lot more enthusiastic
about investing big money into lunar colonization. It will also probably make the colonization
of Mars look more attractive.

Scientist might also wonder just how low
gravity has to be before it becomes deleterious
to human health. Could human colonist remain
healthy on Callisto or even on Ceres?

But if it turns out that the low lunar gravity
is deleterious to humans permanently stationed
there then we'll quickly discover something
about the environmental limitations of the human body. But even this problem could probably be solved by undergoing high levels of arti-

ficial gravity for a few hours a day with portable
centrifuges.

But how humans and other animals respond
to a 1/6 lunar gravity for long periods of time
(years) will probably be one of the most important discoveries in the history of our species.
And once a permanent base is established, it
won't take too long to find out!

Comment by Marcel F. Williams — April 17,
2010 @ 10:28 pm

27. Paul, ISRU would be nice, and it's part
of the exploitation that should happen. I also
agree, as I mentioned in an earlier post, that
some of it should be funded by NASA. But exploitation is not exploration.

Part of the reason for discourse on this issue
may be in the interpretation of the timeline of
when exploration & exploitation occur, and even
how quickly. I see robotic missions to the Moon
as precursors to eventual manned exploitation
missions. But I see quite a few robotic missions
needed before we decide what task & equipment will be needed for manned mission. What
are we going to exploit, where is it, what equipment do we need, how do we make it work in
low gravity and a vacuum, how do we establish
a logistical supply line? This could take a decade or more to put in place.

In the meantime, and in parallel, NASA will
be working on exploration beyond the Moon,
and all of the technology being developed will
contribute to the Moon exploitation missions.
The Moon is not needed for fuel, but it will eventually be the cheapest place to find it. But I think
that will be at least 20 years.

To accomplish all of this, we will need the
ability to launch many missions per year. This in
itself will spur a more robust commercial sector,
but unless costs can be contained or reduced
over time, it will be hard to afford everything at
once. Without exploitation opportunities, commercial companies won't share the risk. And
without non-government money, we will be limited in pace of activity. There's no magic formula
for this, but if you don't start the process (i.e.
The New Plan), then it won't happen.

Comment by Coastal Ron — April 18, 2010
@ 1:41 am

28. In response to comments from Marcel F.
Williams:

"Maybe a company like Space X somehow
knows a lot more about flying into space than
NASA or the Russians". Why do you say this?
They have never boasted about being better
than anyone. You're making a false comparison.

"So just one fatal accident aboard a Space

235

X vessel would probably send 99.9% of customers to Energia…". This didn't happen with the two launch failures Soyuz had, and it didn't happen after we lost Challenger, or later when we lost Columbia. Astronauts are smart people, and one of the first things I'm sure their contract says is "You Could Die!". Commercial space will have to prove itself, but I see NASA as being one of the many overseers of their services. Keep in mind also that the CCDev winners are part of an effort to "spread the wealth" for creating human-rated vehicles.

"Its going to take a long time for American private companies to get off the ground and prove that they can deliver people into space as safely and reliable as NASA or Energia can…." Hence NASA providing some financial incentives to hurry this along.

"you don't completely shut down the government's ability to fly into space until the private companies have proven that they can fly humans safely and reliably to and from orbit.". Talk to your buddy Bush about this, although I agree with his timing (after finishing the ISS). Other than supporting the ISS, what would you use the Shuttle for? We're talking taxpayer money here – pretend like you only have a limited budget…

"Why shouldn't an agency owned by the American people be able to fly into space too…". This is one of the funnier arguments I hear. Look around the U.S. government, and you see will plenty of examples of how they will do this – they will buy what they need! And like all other forms of transportation, the government controls the infrastructure & licensing, and allows companies to operate. All they need to do is buy launch services, or whole rockets, or whatever they want. This is so obvious…

"However, that $1.2 billion that NASA is giving to private companies… more concerned about the president's policy wasting the other $19 billion dollars…". So you think we should outsource NASA, or shut it down? I don't understand what your point is… :-(

"A permanent Moon base would be a much better investment …" How do you magically invent the technology to do this? How do you afford it? Constellation was only going to provide a few trips of temporary visits, with no permanent infrastructure being left behind. That was going to take 15-20 years, and $50B+. Don't you think this is a big leap, even from Constellation? I think a more incremental plan would be more helpful…

Comment by Coastal Ron — April 18, 2010 @ 2:26 am

29. I viewed Obama's speech, and I got to tell you: IT WAS JUST AWFUL!! He totally trivialized & excluded the Moon as a destination. He totally drew the battle lines, along making it seem like a stark choice between the Moon or Mars—a complete falsehood! We as a nation CAN do both! We merely do the Moon first, because it's closer and more viable for resource exploitation sooner. We of the space interest community need to articulate, right now more than ever, the firm need of a Lunar Return as the nation's prime intermediate goal. President Obama basically sold out to the Planetary Society and the "Anywhere but the Moon" lobby, in putting his Presidential weight on the flimsy "We've-been-there-already" jazz. HOW COME THAT ARGUMENT NEVER APPLIES TO LOW EARTH ORBIT—which is scarcely even a real destination, other than the same old space stations that we and the Russians have been placing up there since the 1970's…?? Mr. Obama pandered to the Mars extremists—those people who are dead-set against any further manned journeys to the Moon. (Lately, I try to call them the Anti-Moon people or the "Anywhere-but-the-Moon" people, to be more conciliatory. I darn well know, that there are plenty of Mars enthusiasts who actually do NOT oppose further ventures on the Moon. These Mars moderates are precisely the kind of persons that we need to keep an open dialogue with; with our continuing campaign to save Constellation. We need to together, lobby Congress into keeping the Aries 5 heavy-lift rocket—which ALREADY is a viable plan & design—plus all the other elements of the Constellation program: the fully developed, Lunar mode Orion spacecraft, the Altair-class Lander, and the Aries 1 rocket to launch the astronaut crew separate from the massive trans-lunar cargo. Are you all with me on this?? Regardless of what your favorite destination is??) THE MOON FIRST!! Before & ahead of any asteroid!!

Comment by Chris Castro — April 18, 2010 @ 2:52 am

30. By the way: Low Earth Orbit….HAVEN'T WE BEEN THERE ALREADY?!?! This should be a protest sign message from us Moon enthusiasts, who only want to do what is pragmatic, prudent, & most practical.

Comment by Chris Castro — April 18, 2010 @ 2:56 am

32. Ron,

To accomplish all of this, we will need the ability to launch many missions per year.

My point is that by going to the Moon to learn how to use the resources we find there, your statement becomes obsolete — we still need "many launches" per year, but not from the

Earth. As long as we are confined to the existing template of spaceflight that requires launching everything we need from the bottom of the deepest gravity well in the inner Solar System, we will be mass- and power-limited in space and therefore, capability limited. Our mission in going to the Moon is to change those rules.

a more robust commercial sector

We all want this; I sure do. But the idea that the proposed "new path" will give it to us is a statement of faith, not fact. It is the triumph of hope over experience.

You can discount everything I say about the value of lunar return and the capabilities it will create, but you cannot gainsay the history of NASA as an organization. They will take the $20 billion per year and happily spend it. And we will get nothing for it.

You may not believe me now. But you will.

Comment by Paul D. Spudis — April 18, 2010 @ 5:47 am

33. @Coastal Ron

Space X and its advocates are constantly touting how more successful and efficient they're going to be relative to NASA, yet they've never flown one individual into space. They need to be quiet and simply start putting people into space and bringing them back safely.

Space X advocates are some of the harshest critics of NASA safety. Space travel is risky business. NASA has made it look so easy that any fatal accident is considered a catastrophe. Yet they somehow think that a company with no manned spaceflight experience is going to improve safety.

Sorry, but I never voted for Bush and was against terminating the shuttle and also against the Ares I/V architecture. Obama is the President and he had plenty of time to make sure the shuttle program continued. But he clearly doesn't want it to continue it.

I don't consider a $20 billion a year budget a limited budget– especially if you're not going to build anything or go any place. And I'd use the space shuttle to prepare for manned beyond LEO missions until a heavy lift vehicle is finally built.

NASA ships are already built by private vendors. But private manned spaceflight companies needs to avoid the panacea of NASA contracts and focus on launching commercial satellites and space tourist. There's simply not enough manned space flight traffic from NASA to support more than one company– unless you're advocating a monopoly for just one company.

You can read my blog on the subject if you'd like to know how I'd spend that $300 billion in NASA expenditures over the next 15 years:

http://newpapyrusmagazine.blogspot.com/2010/04/how-congress-should-respond-to-obamas.html

Comment by Marcel F. Williams — April 18, 2010 @ 6:17 am

34. I agree with you 100% about why the Moon needs to be our next destination. That said, I'm not worried about the new policy simply because the Mars proposal is so far away (mid-2030s) and upcoming exploration of the Moon is not slowing down at all, with some 20+ missions over the next eight years or so. Mars only has about half that, and whenever one of them misses a launch window it gets set back by another 2.5 years.

I'm a big supporter of the near-Earth asteroid mission, and what I actually think will happen is that the new program will invest heavily in attaining launch capability again, the asteroid mission and any other related technologies, and then after about 5+ years preparations for the asteroid mission will be well underway, but the increasing presence of missions on the Moon by other countries and even private industry will cause the US to think again about whether it really wants to go all the way to Mars when exploration of the Moon has already begun. The Mars mission will be put on the shelf and the US will decide to help out with the international effort and the Moon will become our next destination, along with the odd near-Earth asteroid mission.

Comment by Mithridates — April 18, 2010 @ 6:30 am

35. *You think so, huh? Well consider this. The historical record shows that when NASA is given a pot of money and no specific direction except to "do good innovation," we usually get nothing — no flight hardware, no useful product, no missions. Just Powerpoint charts.*

Now stay with me for just a few seconds longer — If they implement the "new path" as I have outlined above, what do you think is the likely greeting they will get in Congress and in OMB after spending tens of billions with nothing to show and announce that they are now ready to go to Mars?

Paul, you are so right! If you give federal workers hundreds of billions of dollars of tax payer money to build absolutely nothing, I guarantee that they'll do an excellent job at that:-) And so will their private vendors!

Comment by Marcel F. Williams — April 18, 2010 @ 6:30 am

36. Mithridates,

upcoming exploration of the Moon is not slowing down at all, with some 20+ missions over the next eight years or so.

There's an interesting prediction. What missions are those? Where are they described?

Or maybe you meant missions by countries other than the United States....

Comment by Paul D. Spudis — April 18, 2010 @ 7:48 am

37. Yes, I meant missions by other countries plus the United States, which thankfully is still sending robotic missions to the Moon even if manned exploration is shelved for the time being.

http://en.wikipedia.org/wiki/List_of_current_and_future_lunar_missions#Under_development

Comment by Mithridates — April 18, 2010 @ 8:46 am

38. Mithridates,

The vast bulk of those missions will never fly (e.g., the ILN (International Lunar Network) nodes are unworkable within the budget envelope projected for it). The only missions on that list that may actually occur are GRAIL, LADEE, Chandrayaan-2 and Change'E 2. All the others are vaporware.

Comment by Paul D. Spudis — April 18, 2010 @ 10:11 am

39. I should have mentioned that the missions there should be seen in comparison with the proposed Mars missions over the same time period, only about half the number and again with a lot of vaporware missions as you mentioned. IOW, missions to the Moon are relatively more frequent and also much easier than those to Mars.

http://en.wikipedia.org/wiki/Exploration_of_Mars#Timeline_of_Mars_exploration

Comment by Mithridates — April 18, 2010 @ 11:33 am

40. *Paul D. Spudis said "My point is that by going to the Moon to learn how to use the resources we find there, your statement becomes obsolete — we still need "many launches" per year, but not from the Earth."*

I don't disagree with the goal Paul (Moon exploitation), but I think you're discounting the logistics part of building any place of work. My background is in logistics and moving new products out of engineering and into full production – manufacturing. When I look at a task, I start imagining what needs to be put in place to sup-port everything.

The most popular example of ISRU is H2O, which can be used for water and fuel. H2O is important, but its a consumable. Everything else that you need to extract it must be shipped to the Moon. We don't have self-assembling factories yet, so regardless of how much local material is at hand, you can't build a computer or drill bit out of moon dust.

For every machine you send, you must also send the spare parts – lots of them. When a machine breaks, you can either discard it and grab a replacement, or you can fix it. Both of those options involve landing tonnes of cargo on the Moon on a regular basis. And we haven't even started talking about the logistics of add-ing people into the equation. Workers, mainte-nance, staff, transportation, logistics & support, food service... plus rotating people out on a regular basis.

What is the ratio of Earth lift mass to Moon landed mass?

Any level of complexity on the Moon sud-denly becomes a lot of work. There are lots of historic examples of this (railroads, dams, ISS, etc.). You can experience this at home too – take a 3 month RV trip with your family, and count the number of times you stop for sup-plies, visit any type of business, or maybe need something fixed. In the U.S. it's so ubiquitous that we forget about it.

I do see exploitation happening, but it can only happen as fast as we can ship the peo-ple and equipment to the Moon. And also re-member that the 1st generation of any product does not last very long, so you have to plan for continuous replacement and improvement (i.e. more stuff shipped).

All of this activity supports the commercial space industry. More launches supports more competition, and helps to innovate and keep costs in check. I'm not saying it will be cheap, because you are right about us being in a deep gravity well. But I think we can use the ISS ex-perience as a guide post to what it will take to expand our presence in space, and the lesson I get is that we're going to need lots of upmass capability.

Comment by Coastal Ron — April 18, 2010 @ 12:43 pm

41. Mithridates

IOW, missions to the Moon are relatively more frequent and also much easier than those to Mars.

The series of hypothetical future missions to the Moon listed on Wikipedia outnumbers a se-ries of hypothetical future missions to Mars and that proves what exactly?

If you look at the actual number of missions in the last 20 years, Mars has had (17) versus the Moon (7). This shows where the exploration emphasis has been.

Comment by Paul D. Spudis — April 18, 2010 @ 1:02 pm

42. Ron,

More launches supports more competition, and helps to innovate and keep costs in check

I don't disagree with this in theory or as a goal, but there is no evidence that the "new direction" will produce this result. I also do not for a moment minimize the difficulty of what I am proposing — you are correct, it is a long-term, difficult task. But we are going to spend ca. $ 20 B per year on NASA in any event and my contention is that they should do something useful, not useless. Doing technology "development" and paper studies for a Mars mission that is 25 years into the future is not the former, but the latter.

Comment by Paul D. Spudis — April 18, 2010 @ 1:06 pm

43. The past 20 years when the Moon was completely ignored isn't really relevant to the current situation. Since 2007 there has been a total of one mission to Mars, and five to eight (depending on how you count) to the Moon. Why? Because countries like China and India are now capable of contributing there while Mars is still more or less out of their league. At the same time even Russia and the US have failed to launch probes to Mars in 2009. And in the next decade or so as well we should expect to see missions at a similar rate, perhaps two or three Moon missions for every one to Mars.

Comment by Mithridates — April 18, 2010 @ 2:17 pm

45. *The past 20 years when the Moon was completely ignored isn't really relevant to the current situation*

On the contrary, it is very relevant. There has been no Mars mission since 2007 only because JPL and its MSL mission cost overruns have sucked all the money out of the Science Mission Directorate, not because NASA has lost interest in Mars robotic missions.

The agency's Mars obsession has kept them from designing and implementing a rational lunar program and from understanding why lunar return is important and what our mission there is. Thus, they dissipated the political will to do attainable lunar return and exchanged it for pie-in-the-sky, "sometime in the future" man on Mars dreams.

This continues to the present day and is clearly manifested in the new announced space policy.

Comment by Paul D. Spudis — April 18, 2010 @ 2:35 pm

46. Well, I agree there. I meant that it isn't relevant if one were to take the number of missions over the past 20 years to each destination and try to use that as a gauge for the difficulty of each destination (i.e. "See! There were more missions to Mars over 20 years because it's an easier target!").

Comment by Mithridates — April 18, 2010 @ 5:32 pm

47. Paul, I think there are a lot actionable items they are doing, that are the building blocks for future systems:

1. The COTS program (from the prior administration) takes two companies that don't have experience in cargo or ISS docking, and teaches them how to do it the NASA way. With shared risk (lack of supplies for NASA, lack of payment for the companies), these inexperienced cargo companies get to acquire a lot of real experience, and will position them for future launch opportunities. For SpaceX, the COTS program provides a paying customer to gain experience with their Falcon 9 & Dragon systems – something they could not afford on their own. They will leverage this into crew capability quicker because of the COTS program. Time will tell if they both are up to the task, but the tasks on the milestone schedule are part of reducing some of that risk.

2. CCDev, which although small, is spread out to technologies that can be applied fairly quickly if NASA decides to fund a commercial crew capsule. Based on the plans for Orion, I think they will. The payoff for this contract is not clear yet, but I think they are starting to connect the dots for further investment in commercial crew. More on this will happen after the budget is approved.

3. The ISS. We shouldn't overlook the elephant in space – a $100B, fully functioning space station that has a crew & logistics supply line in place through 2015. In the world of space Supply (launchers) and Demand (places to go), this gives us demand for lots of launches. Without a clear future market, companies find it hard to invest long term. The ISS provides a potential market for services through 2020, and hopefully beyond.

I will admit that these are small steps, but they are in the right direction. Based on the new budget, and the announced plans, I think we will see a clearer commercial plan emerge over the next two years. Obama/Bolden have deviated only slightly from their proposed budget (Orion CRV), so I think they are serious about what they said.

There really isn't too much NASA can do until their new budget is approved, so I think there is a lot more to come. For the talk about Mars, I would call them "forward looking statements", not actionable plans. They are concentrating on building the building blocks, but the public needs to be weaned from "A Place" that we're going to in 15 years. What a lot of us see, is that they are building the infrastructure for "Lots Of Places".

Comment by Coastal Ron — April 18, 2010 @ 9:00 pm

49. Ron,

For the talk about Mars, I would call them "forward looking statements", not actionable plans. They are concentrating on building the building blocks, but the public needs to be weaned from "A Place" that we're going to in 15 years

They have been making these "forward looking statements" for 30 years and we are no closer to the capability we need to go to Mars than we were since then. The public doesn't need to be "weaned" from anything because they don't follow it closely one way or the other — all they want is a space program that 1) seems to be doing something important; 2) doesn't kill people too often; and 3) doesn't cost too much.

I find it amusing that supporters of the new path have unlimited faith in technical research and the agency's capabilities now that we've exchanged a goal for which near-term measurable accomplishment is possible (lunar return) for one in which the payoff (Mars landing) comes 30 years down the road.

Comment by Paul D. Spudis — April 19, 2010 @ 3:54 am

51. So it now appears that the Augustine Commission was just a tool for his science advisers to shut the Federal manned program down with an easy solution, "Just let private industry do it!" And then we hear the president arrogantly saying that we really don't need to return to the Moon. I'm pretty sure the president spent– at least an hour– deciding that one:-)

And that's really a shame when you think about how John Kennedy's enthusiasm for space captured our imagination and our hopes for a better tomorrow.

Of course, we don't need to return to the Moon– if you're not really interested in America's space program!!! If President Obama, who I voted for, was interested in NASA, he wouldn't just make one short speech about space and then rush out to attend a fund raiser. And he

would have had that town hall meeting he promised initially. But he didn't!

If the president's really not interested in our space program then Congress needs to step up and use its– power of the purse– to take charge of our extremely valuable space assets. And I think that's exactly what they're going to do!

Comment by Marcel F. Williams — April 19, 2010 @ 4:05 pm

56. I often speak to students and teachers about the human and robotic exploration of space, and I always talk to the "been there, done that" attitude when it comes to our return to the Moon. Here's how it usually goes: Though a small planetary body, the Moon has almost as much surface area as North and South America (about 90%). We've landed at only six places on the Moon, all relatively near to each other. If you take the distances between the landing sites and put them on a map of the United States, this is what you get. Start at Houston, the heart and soul of human spaceflight, and use that as the Apollo 16 landing site. The farthest west we've been would be near El Paso, Tx with Apollo 12 (Apollo 14 would be near the Midland/Odessa area). The farthest north we've been would be Apollo 15 near Salina, Kansas. The farthest East would be Apollo 17 near Memphis, Tennessee. Apollo 11 would be somewhere in the woods and swamps near the Arkansas/Louisiana border. Then ask yourself, if those were the only places we had been to in North and South America, what would we know about? The Appalachian Mountains? Rocky Mountains? Mount St. Helens, Mt. Rainier, or any of the other volcanoes all the way down to South America associated with the 'Pacific Ring of Fire'? The hugely complex Amazon Basin? How much would we know about plate tectonics? The answers of course are nada, zero, zilch. So, when it comes to the Moon, 'been there'-sure, 'done that'-no way. We've only begun to understand the importance of the Moon, and to bypass it makes no sense what so ever.

Comment by John G. — April 20, 2010 @ 1:05 pm

58. If we utilize one of the space cannon concepts on the lunar surface, we could probably toss thousands of tonnes of lunar material, including water, off the Moon to a Lagrange point every year even cheaper. That would turn the space depot idea from a marginally attractive concept to a revolutionary concept that would make it almost as easy to travel to the Moon as it is to get into orbit.

Propellant depots are not marginally attractive: they are mandatory. Moreover, it is counterproductive to send unrefined bulk material from the lunar surface to orbit. A factory in orbit will have every problem that a factory on the Moon's surface will have, and then a bunch more as well. Not to mention the mass of the factory itself. A factory built on the Moon from lunar material doesn't need to be launched anywhere by anything. Then there's the problem with waste disposal and accidental collisions that an orbital manufacturing facility would entail. The Kessler syndrome is getting bad enough as it is. Let's try and keep it confined to LEO.

Comment by Warren Platts — April 20, 2010 @ 5:54 pm

59. One of the biggest criticisms hurled at Project Constellation is that it is largely a grandiose repeat of the Apollo explorations of the Moon undertaken over 40 years ago. Certainly, as had been outlined by NASA, lunar return consisted of sortie missions that landed crews all over the Moon to do local field exploration. Such a mission template is indeed Apollo writ large.

Let's not kid ourselves: Constellation had little to do with the Moon: it was a Mars program. Cripes, they named their rockets "Ares". There's no better proof than that. President Obama did us all a big favor by canceling that dinosaur. But then he took what could have been a grand opportunity, and blew it. Every poll I've seen shows that Americans prefer a Moon program over a Mars program. Yet the vocal minority of Mars advocates managed to get to the President's ear, and so now we get Constellation redux. Instead of Apollo on Steroids, we get Apollo on Meth: a toothless, skinny version of Constellation.

The sad part is that there was a viable alternative: the affordable, commercially based lunar architecture devised at United Launch Alliance by Frank Zegler, Bernard Kutter and many others. These guys are the top aerospace engineers at the top aerospace company in the world. Their proposal could have got us to the Moon within a decade using today's technology. It didn't require HLV's or fancy robots that don't exist. It was a lean 'n' mean Viking-like architecture that could have got us there in small boats by breaking up the voyage into small parts, just as the Vikings broke up the voyage to North America with stops at Iceland and Greenland.

Moreover, the timing of the ULA proposal was perfect: it came out at about the same time as when Augustine was wrapping up. The ULA proposal could have been the perfect antidote to Constellation. But that would have entailed, in effect, following in the footsteps of President Bush's VSE. Presumably, President Obama himself is dismayed at how he's been snookered into following President Bush's footsteps

on issue after issue. So he probably figures that if he can find a relatively harmless issue where he can take a different path, he's going to take it if he can. Start with medical marijuana dispensaries; now America's space program.

Unfortunately, it is the Mars advocates like Buzz Aldrin, who have convinced the President that taking the different path in this case is actually the correct path. No other lobby has done more to destroy the US space program. Bypassing the Moon for the sake of Mars gets us neither.

But look on the bright side, Paul. The Obama plan contains seeds of destruction all right: but they are the seeds of its own destruction rather than of the US space program as a whole. Politically, this will be a net loss for the President. On the one hand, he's managed to alienate the majority of those who care about space at all–Marsnauts are a minority compared to MoonFirst!ers by at least 2 to 1. On the other hand, he's alienating his base who don't give a hoot about space. I was watching the speech at the local truck stop, and the waitress said it best right before she changed the channel: "He's talkin' 'bout the International Space Station and satellites in space when there's still people starvin' on this planet!" (So we watched Dr. Phil instead.) Then there was the USA Today political cartoon the other day, where a guy is complaining about how the national debt "is astronomical"–"but we're not going to cut the space program!" This goes to show that inspiration from space is the last thing most people want, and it's certainly not something anybody needs. The next election is going to be close. This may well be the straw that breaks the camel's back. Then it will be back to square one with a clean slate. The VSE will still be there, and so will at least one sane architecture for bringing the vision to fruition.

It's time to start recouping some practical economic and strategic and practical scientific returns from our human spaceflight program. Only the Moon offers at least a chance of that happening.

(And let it not be said that there is more important "pure" planetary science to be done on Mars. Mars is interesting in itself, but it is unique, and thus doesn't have much to say about the Earth, or the Solar System as a whole. The Moon on the other hand, has a history inextricably tied to Earth, and it preserves a record of what has happened to the Solar System and the Sun for billions of years. From the perspective of "pure" planetary science, the Moon offers more scientific bang for less bucks.)

Comment by Warren Platts — April 20, 2010 @ 6:54 pm

60. Mandatory means that you can't conduct space travel without them. Since we don't have any space depots, obviously they're not mandatory.

Unrefined material has immediate value since it can instantly provide the hundreds and even thousands of tonnes of mass shielding that are going to be required for permanent space stations located at a Lagrange point. Being able to extract oxygen from lunar material and from asteroids is going to be essential if we are to dramatically lower the cost of space travel.

Comment by Marcel F. Williams — April 20, 2010 @ 9:05 pm

61. Warren,

I think that your analysis is spot on. One slight departure — I do not think it will be either easy or cheap to re-establish the industrial capabilities that are being destroyed now. So it's not simply a matter of putting the VSE "on hold" for a time and then picking up where we left off sometime in the future.

We are discarding an existing spaceflight capability (for all its faults) for the promise of a new and better future capability ("The better is the enemy of the good enough") and exchanging a near destination against which measurable progress is being made for future, distant destinations which are so far off in time and space that we cannot judge whether progress toward them is being made or not.

I suspect that suits some just fine.

Comment by Paul D. Spudis — April 21, 2010 @ 8:18 am

62. @ Marcel: We don't currently need propellant depots because we never leave LEO. Yes, we didn't have depots during Apollo; but that's why we had Saturn V–it all had to go up at once. So unless we have Ares V or an equivalent, depots are pretty much mandatory. I know you like HLV for some reason, but if our goal truly is to go to the Moon and establish an infrastructure capable of producing hundreds to thousands of tons of propellant per year, I think we should be leery of building heavy lift as a matter of strategic politics. By starving the beast of heavy lift capability, then Mars is no longer a viable option (never mind that it never was anyway). By taking heavy lift off the table, we take Mars off the table. Yes, 1,000 cubic meter space stations that can be lofted with one launch would be nice–but as Paul says, the better is the enemy of the good enough. All we're asking for is 10 or 15 years without distractions so we can get the Moon Base going good. Then we can talk about Mars again; by then the lunar infrastructure will make a Mars flight much more

doable. For right now, we would be better off spending the money building a lander instead of an HLV. Of course, all this will have to wait for the next regime change.

@ Paul: There is one industrial capability that is not being discarded, and that's the EELV family of launch vehicles, thanks to the Air Force. What is your take on the United Launch Alliance affordable, commercially based, lunar architecture developed by Zegler, Kutter, Barr, et al. Is this architecture "good enough"? Can they get the job done for $7 billion USD per year?

Comment by Warren Platts — April 21, 2010 @ 7:44 pm

63. A few comments:

(1) The ISS – fully functioning that has a crew & logistics supply line in place….

Sorry, not really. They can probably keep enough supplies coming up for a few years to keep a crew of at least 3 and maybe even 6, via Progress and ATV; HTV is not on the same schedule. There is at present NO return capacity so science experiments, samples, and anything larger than what fits in an astronaut's pocket won't be coming back once Shuttle stops. We are already hearing that the lack of return capacity is already restricting the number of interested experimenters from places like the NIH.

(2) the VSE was the product of a year-long study by NASA, the Cabinet Secretaries, and the White House. Both houses of Congress voted overwhelmingly to authorize it; the agency has been trying to implement it for 6 years.

Yes the VSE was developed by extended planning and review. Constellation was NOT the VSE.

The people testifying before Congress were the Constellation/Exploration people and Dr Griffin. They always said it was on schedule, on budget...It was always on until they would move the schedule further out, or reduce the requirements... Everyone was happy if they were getting truthful answers.

(3) Those of us outside NASA look at the Constellation program in general, and the Ares 1 in particular, and shake our heads in wonder that such colossal failures as these could have come out of NASA. Ares 1 represents a triumph of ego over physics and engineering; the fact that it got as far as it did indicates that something is seriously wrong inside of NASA. Large bureaucracies are supposed to be self-correcting, avoiding serious mistakes such as Ares 1 — NASA didn't do that, and any president in Obama's position that let such an agency continue making mistakes without investigating

why and correcting them would be shirking their duty.

You are right on this. I was bewildered to see Constellation failing to make the necessary milestones starting almost from the start more than 3 years ago. By 2 years ago it was well known that they could not close the architecture. The defined Orion could not be carried by an Ares 1. I watched from the sidelines. Where was the senior NASA management. They should have gotten on top of this situation and corrected it years ago; they had to let Augustine point out the fallacies ??? Is the senior NASA management worth anything ?

Who are these people ? They should be summarily dismissed or worse. Not only were they wasting our money ($10 billion) and time (4+ years). They've now created the near term GAP, and they may very well have killed US HSF for the foreseeable future.

(4) I do not think it will be either easy or cheap to re-establish the industrial capabilities that are being destroyed now. We are discarding an existing spaceflight capability (for all its faults) for the promise…

This is the REAL IMMEDIATE issue. Make no mistake about it, you will need heavy lift capability and a Shuttle derive HLV is your best, and relatively inexpensive hope, for the foreseeable future. If you really want to test out the Buzz XM Cruiser, you need to be able to launch ISS-sized or larger modules. If you want in-flight refueling, you'll need similar sized and heavier mass payloads. A Shuttle derived heavy lift is a relatively inexpensive way to do this. It could be done now with relatively few dollars and in a relatively short amount of time, by people who have already begun to lose their jobs.

Instead, Obama chose a purely political decision to support Colorado with an Orion-lite that there should be no need for. Does not make any sense at all.

The last one, the Obama-Orion save, is really troubling. There is no rationale for that vehicle, except as a jobs program for Colorado. This shows that Obama cannot be trusted. No telling whether any of his words last week meant anything at all.

Comment by JerryL — April 21, 2010 @ 11:05 pm

64. Warren,

What is your take on the United Launch Alliance affordable, commercially based, lunar architecture developed by Zegler, Kutter, Barr, et al. Is this architecture "good enough"? Can they get the job done for $7 billion USD per year?

Possibly it is. What is needed is a thorough

and objective technical review (Augustine was not that) and re-build of the entire architecture, not cancellation of the program wholesale. First, define the mission objectives of lunar return (to learn the skills we need to live and work off-planet, including especially new enabling activities that have never been done, like resource utilization.) Second, design the architecture you need to implement this mission; if you need Shuttle-derived heavy lift, build it — if you don't need it, use EELV. This decision should be driven strictly by the objectives to be met and the projected cost envelope (you need to know both — ESAS was clueless about the former).

The issue that I have with the current new direction is that it discards the overarching strategic direction of the VSE. How we implement it is of less (although not zero) concern to me.

As for the "$7 B per year" cost envelope, it has always been my belief that an exploration program can be built that fits into any cost profile. The important thing is to keep the strategic horizon in sight. If we get less money than we expect, we stretch the program out longer. But we make continuous, steady progress towards the goal, which is sustainable and permanent human presence beyond LEO.

Comment by Paul D. Spudis — April 22, 2010 @ 4:50 am

65. The article by Paul D. Spudis, gives a cogent analysis of the importance, inevitability and irrepressibility of the Moon's role in space exploration. A pleasure to tweet it in the hope of a wider audience.

I would add that the moon is a natural launch platform, with one-sixth our gravity and yet stable enough for our gradual amassing of paraphernalia, prior to any launch into Space beyond. Bases are inevitable, as is our colonizing of our Moon. Man's history and innate tendency has been to explore and colonize uninhabited places, and this has always followed a logical progression in terms of effort and distance.

Buzz, with tremendous respect for you and your place in history (and you remain a hero to me and many) I nevertheless would plead that you have 'been there', but we, lesser mortals have not – and we're eager.

While President Obama's emphasis on better vehicles than those planned for Constellation, is to be applauded, sidelining the Moon and in the meantime the Shuttles (man's fleet of smoothly-performing, workaday cargo spaceships), could prove a false economy. Falling behind the rest of the world may be irreversible for the US, as Britain experienced with manned Space after Blue Streak et al. If The traditional home of Space Travel, the USA, does relinquishes its hard-earned place at the forefront of manned space exploration (easily lost), by not returning to our moon to finish the job Kennedy started, and establish man's foothold on our immediate heavenly body, then the BRIC countries, Brazil, Russia, India & China will step in to fill the vacuum (pun half-intended). Matters can be reversed at present – it's OK to err in the wrong direction, but less so to persist with it.

Comment by Vinay (@vinluce) — April 23, 2010 @ 10:05 pm

66. *Possibly it is. What is needed is a thorough and objective technical review (Augustine was not that) and re-build of the entire architecture, not cancellation of the program wholesale.*

Arguably, there has been such a technical review and rebuild of the entire architecture, the EELV based architecture by Frank Zegler and his colleagues at ULA. It requires no shuttle derived heavy lift, and relies entirely on existing assets, or on gradual evolutionary modifications of such assets; e.g., the ACES-41 EDS is descended from Centaur; cryogenic propellant depots and lunar landers are then derived from the ACES-41. There are no "clean sheet" designs, so reliability should be good. The dual thrust axis landers (DTAL) land horizontally after the fashion of the landers depicted in the old TV show "Space 1999". These are superior to the Altair design in at least 20 different respects; there's not enough room here to go into it. It can deposit 15 tons onto the lunar surface, and it has a 5 m diameter cargo bay. I don't know if it could carry the 250 kilo Watt solar arrays you all contemplated in your Going Beyond the Status Quo paper; however the "DTAL-R's 5 m cargo hold and lunar performance are capable of supporting all of NASA's planned lunar surface systems, including hard shell and inflatable habitats, crewed rovers, ATHLETE, in-situ resource plants, lunar telescopes, or large drilling rigs." A reusable SSTO version of the DTAL can deliver 7 tons of propellant to an L2 depot. A tanker version of the DTAL based on the ACES-71 would be able to deliver even more.

They claim they can get us to the Moon permanently within a decade after the starting gun goes off, with two manned missions per year, all within a $7 billion USD budget envelope. What more do we need? I'm surprised this potential architecture isn't talked about more. I've certainly never ran into even one article, blog post, or discussion forum rant that can point out even one showstopper in this design.

Comment by Warren Platts — April 26, 2010 @ 12:00 am

67. Warren,

They claim they can get us to the Moon permanently within a decade after the starting gun goes off, with two manned missions per year, all within a $7 billion USD budget envelope. What more do we need?

What's needed is a clear understanding of what our mission is on the Moon. I have written on this previously (Chapter 25):

http://blogs.airspacemag.com/moon/2009/06/25/first-nail-down-the-mission/

Clearly, no one currently in authority at NASA understands the purpose of lunar return and they and others who similarly don't understand it are the ones informing the decision makers.

Which architecture you choose is irrelevant if you don't understand what you're doing and why you're doing it.

Comment by Paul D. Spudis — April 26, 2010 @ 4:48 am

68. I think there's a growing consensus in Congress for immediately funding a heavy lift vehicle. That in itself would be a game changer from the Constellation program where a HLV vehicle was not scheduled to be developed until after the Ares I was completed.

But we really don't need a super duper HLV for the Moon or even for Mars, IMO. A simple shuttle derived HLV, either sidemount or inline, that can transport at least 100 tonnes into orbit (with an EDS stage) should suffice.

So instead of a final presidential decision on an HLV in 2015, we could have an HLV ready to go for beyond LEO missions by 2015.

Comment by Marcel F. Williams — April 27, 2010 @ 3:35 am

69. I am SO sick of that stupid argument being brought up, whenever there is talk of a Lunar Return! If we can never ever go back to where we've been, then there will NEVER be bases nor resource-utilization! Mr. Obama has just turned deep space exploration into nothing more than a Guinness Book of World Records farthest-distance stunt! People, going farther in space DOESN'T mean doing better in space! The destination does matter. After that first "Flags & Footprints" docking mission with an asteroid…just what in heaven's name, happens next?? Remember, we can NEVER ever visit a planetoid more than once, according to our supreme leader, Mr. Obama.

Comment by Chris Castro — April 28, 2010 @ 1:17 am

The Four Flavors of Lunar Water

May 2, 2010

Earth over the north pole of the Moon as seen from Clementine, 1994

The Moon is constantly bombarded by the solid debris of the Solar System. Comets, asteroids and interplanetary dust, all containing varying amounts of water, have pounded the lunar surface for billions of years. Yet until recently, the Moon was considered to be barren and bone-dry. Rock and soil samples returned by the Apollo missions lacked any hydrous mineral phases or water-bearing weathering products. Since water is not stable on the Moon under ordinary conditions, what happens to it?

New studies of lunar samples, along with results from several missions in recent years, have given us a revolutionary new picture of water on the Moon. Study of volcanic glass

from the Apollo 15 landing site[1] in 2008 demonstrated that tiny amounts of water (about 50 parts per million) are present in the interiors of these glasses, suggesting that the lunar mantle (whence they came) contains about ten times this amount. This was a startling result, considering the extreme dryness of other lunar samples.

Because the Moon's spin axis is nearly perpendicular (1.5° from vertical) to the ecliptic plane, the Sun is always on the horizon at the poles, keeping the floors of deep craters in permanent shadow. These dark areas only receive heat from the interior of the Moon and are extremely cold; recent measurements by the DIVINER instrument[2] on the Lunar Reconnaissance Orbiter (LRO)[3] spacecraft indicate temperatures as cold as 25-35° C above absolute zero. Water molecules are trapped by the cold as soon as they find their way into these craters. Over the more than 4.5 billion years of lunar history, significant amounts of water could accumulate in many of these crater "cold traps" at the Moon's poles.

The first hint of water ice in these polar cold traps[4] came from a radio experiment aboard the 1994 Clementine[5] mapping mission orbiting the Moon. The polarization characteristics of echoes from the south pole were consistent with the presence of ice in the crater Shackleton. Four years later, the Lunar Prospector (LP)[6] spacecraft carried an instrument designed to measure the amount and energy of neutrons given off the Moon's surface. Hydrogen absorbs neutrons, so when LP investigators saw a decrease in the flux of medium-energy neutrons near the lunar poles, they concluded that excess amounts of hydrogen were present there. Although this observation is consistent with the presence of polar ice, neutron data alone do not tell us what form the hydrogen is in, and it was alternatively postulated that this enhancement was caused by excess solar wind hydrogen.

The Moon Mineralogy Mapper[7] (M3) instrument on the 2008-09 Indian Chandrayaan-1 mission collected reflectance spectra for most of the Moon. It found both water (H_2O) and hydroxyl (OH) molecules[8], present either as a monolayer on lunar dust grains or bound into the mineral structures in surface materials, poleward of about 65° latitude at both poles. Moreover, the abundance of this surface water varies with time, being present in greater quantity in both local early morning and late evening and it increases in abundance with increasing latitude. These results were verified by observations from the Cassini[9] and EP-OXI[10] spacecraft during separate flybys of the Moon. The new observations indicate significant quantities of water moving towards areas with lower mean surface temperatures and increasing in abundance with latitude. Taken all together, the results mean that water is being deposited (e.g., by comet impact) and/or created (e.g., by reduction of metal oxides in the surface by solar wind protons) and then transported to the poles. By this process, significant quantities of water ice could accumulate at the poles over geological time.

Last October, the companion satellite to LRO, LCROSS[11], slammed the upper stage of its launch vehicle into the Moon's south pole and observed the ejected material. Results show that both water vapor and ice particles[12] were ejected from the LCROSS impact crater; initial analyses indicate that water is present at about the 5-10 wt.% level. The LCROSS impact site exhibits no anomalous radar behavior, suggesting that such an amount of water ice cannot be detected by radar. However, the results do indicate that significant amounts of lunar polar water may be present even in the absence of specific radar evidence for it. Spectra from this impact event show evidence for other volatile substances, including ammonia and simple carbon compounds. The presence of such material may indicate a cometary source for these volatile materials.

Both poles were covered by radar images from the Mini-SAR[13] instrument on Chandrayaan-1. Much of the north polar region displays backscattering properties typical for the ordinary Moon, but one group of craters in the region show elevated polarization enhancements in their interiors, but not in deposits exterior to their rims. Almost all of these anomalous craters are in permanent sun shadow and correlate with proposed locations of ice modeled on the basis of the Lunar Prospector neutron data. These relations suggest that the interiors of these craters contain nearly pure water ice[14], with approximately 600 million metric tonnes of ice present in over 40 small craters within 10 degrees of the pole. The south polar region shows similar relations, except that it has fewer anomalous craters than the north pole. Small areas of polarization enhancement are found in some craters, notably Shoemaker, Haworth and Faustini; these areas might be deposits of water ice.

So water on the Moon is present in large quantity in at least four different "flavors." Water was in the deep lunar interior 3.3 billion years ago, at concentration levels of a few hundred parts per million. This water would have been released during the eruption of lunar magma and could have made its way into the polar cold traps. Water is either being made or being deposited nearly continuously by impact all over the Moon. Most of this water is subsequently lost to space (e.g., by sputtering, ionization or thermal escape) but some is retained on the Moon. Any water arriving at a cold trap near the pole will be captured. Water, once in the polar areas, is stable as ice in the permanent darkness or where sublimation is prevented when buried by a thin layer of soil. Significant quantities of water may accumulate there; the LCROSS results suggest several to tens of weight percent water ice may exist in the polar soils. Finally, some of this migrating water apparently collects at rates high enough so that significant soil cannot mix with it during normal impact bombardment, as shown by the presence of relatively "pure" water ice deposits in selected lunar craters imaged by radar.

A significant amount of water at the poles of the Moon is present, with many billions of metric tonnes at each pole (detailed estimates of the water reserves are in progress). Such an amount is more than enough to support both permanent, sustainable human presence[15] on the Moon and for export to cislunar space. Water is useful as rocket fuel and energy storage (hydrogen and oxygen are the two most powerful chemical propellants known) and for life support (water and oxygen) in space. These new discoveries fundamentally alter our understanding of the Moon's processes and history and highlight both it's scientific value and utilization potential. The Moon is on the critical path to human expansion into the Solar System.

Addendum. In the comments below, reader Pradeep Mohandas reminds me about the findings of the Moon Impact Probe[16], released from Chandrayaan-1, which discovered water vapor in very small concentrations in the space just above the Moon during its descent to the south pole. This exospheric water (i.e., water in extremely small concentrations) may be related to the time-variable water seen in the spectral data from M3, Cassini, and EP-OXI — in other words, it may represent water molecules in motion, migrating toward the poles. Work on the nature and processes of the lunar hydrosphere continues and I will keep you up to date on the latest research results on this new and exciting subtopic of lunar science.

Topics: Lunar Exploration, Lunar Resources, Lunar Science, Space Transportation

Links and References

1. volcanic glass, http://www.newscientist.com/article/mg19926644.200-now-the-moon-reveals-its-water.html
2. DIVINER, http://diviner.ucla.edu/
3. LRO, http://lunar.gsfc.nasa.gov/
4. water ice, http://www.thespacereview.com/article/740/1
5. Clementine, http://en.wikipedia.org/wiki/Clementine_%28spacecraft%29
6. Lunar Prospector, http://en.wikipedia.org/wiki/Lunar_Prospector
7. Moon Mineralogy Mapper, http://m3.jpl.nasa.gov/
8. water and hydroxyl, http://www.nasa.gov/topics/moonmars/features/moon20090924.html
9. Cassini, http://www.nasa.gov/cassini
10. EPOXI, http://epoxi.umd.edu/1mission/status.shtml
11. LCROSS, http://lcross.arc.nasa.gov/
12. water vapor and ice particles, http://www.nasa.gov/mission_pages/LCROSS/main/prelim_water_results.html
13. Mini-SAR, http://www.nasa.gov/mission_pages/Mini-RF/main/index.html
14. craters contain nearly pure water ice, http://www.nasa.gov/mission_pages/Mini-RF/multimedia/feature_ice_like_deposits.html
15. more than enough, http://blogs.airspacemag.com/moon/2009/10/04/space-exploration-sets-sail-on-lunar-water/
16. Moon Impact Probe, http://tinyurl.com/3yxt9r3

Comments

2. It's not exactly a surprise to hear that there is some form of water on the Moon, but is there enough to support a lunar colony. 600 million tons of water isn't really all that much, especially since it would be very hard to extract. I want Man to get out into space. Does knowing that there is ice at the Moon's poles give us a sufficient reason to construct rockets and go there? Does it give us a good reason to establish a base or build a lunar colony? These things need to be done.

Comment by Ken St. Andre — May 2, 2010 @ 10:15 am

3. Ken,

It's not exactly a surprise to hear that there is some form of water on the Moon

It was to about 95% of working lunar scientists. We've been debating and fighting about this since 1996.

600 million tons of water isn't really all that much, especially since it would be very hard to extract... Does knowing that there is ice at the Moon's poles give us a sufficient reason to construct rockets and go there?

It's a considerable quantity. If all of it were converted to LOX-LH$_2$ fuel, it is enough to launch the equivalent of a Space Shuttle, every day for over 2200 years. And if the radar data are any indication, it won't be that hard to extract — just collect it with a bulldozer, separate the rock and soil debris and store it as block ice in a polar cold crater.

And the answer to your second question is "Yes."

Comment by Paul D. Spudis — May 2, 2010 @ 10:57 am

4. So, what about the water vapour traced by the Moon Impact Probe in the lunar atmosphere which published article says the amount of water found in the atmosphere is 100 times than previously thought.

How would that fit into the lunar hydrology model?

Comment by Pradeep Mohandas — May 2, 2010 @ 1:20 pm

6. The existence of water on the Moon and asteroids is indeed a remarkable discovery. But the question by the previous poster wasn't really answered. Yes, if 600 million tons of water were converted into LOX-LH$_2$, you'd have a lot of propulsion. But let's extend that answer. On the Earth, there are 300 million •cubic miles• of water. (I'll leave it to someone else to work out the tonnage. But it's a LOT!) Now, if we converted all that into LOX-LH2, it would completely offset the higher gravitational force of the Earth compared to the Moon, and we could probably put the entire human race into space!

I'll also suggest that it is vastly easier to extract water on the Earth compared to extracting water on the Moon. You don't even need bulldozers! A bucket will do.

I'm not trying to be glib here, but just pointing out that explanations that rely on simplistic conversion resource estimates into capability aren't necessarily convincing, especially when those capabilities aren't unique.

Assuming that water on the Moon is intrinsically useful, a fair question might be the relative expense of refining water on the Moon compared to just taking it there. Putting a bulldozer on the Moon isn't exactly cheap, especially one that has to move a huge amount of regolith to get a substantial amount of water. Putting a separation plant there might be kind of pricey as well.

Comment by Heinrich Monroe — May 2, 2010 @ 1:49 pm

7. Heinrich,

I do not mean to minimize the difficulties, but you are assuming an initial level of effort that implies more capability than is needed, at least initially. If the polar water is present in the quantities we estimate, a very small initial operation can harvest sufficient water to supply first the needs of the outpost, then to refuel the lunar lander, and finally make enough for export. A system that can collect and process a few metric tonnes per month is small and deliverable in one robotic landing. The machines needed can be teleoperated from Earth, something not possible for resource processing on asteroids or Mars.

People tend to have the impression that resource processing on the Moon requires a high level of industrial capability. Such is not

the case; we start off small, use remote operations to build up an initial capability, and when people arrive, a resource processing infrastructure is already up and operating. It is then extended until needed capacity is reached.

Alternatively, we could just "study the technology" for ten years while doing nothing.

Comment by Paul D. Spudis — May 2, 2010 @ 2:56 pm

8. Extracting the oxygen alone from the lunar regolith would substantially reduce the cost of space travel within cis-lunar space and the cost of supplying space stations with oxygen for air and as a component of water.

Lunar sources of water, of course, would reduce those cost significantly more. Supplying space depots with oxygen and hydrogen from the Moon would be substantially cheaper than supply those resources from Earth. Lunar regolith and lunar water could also be used for mass shielding space stations and manned interplanetary vehicles from galactic radiation.

The addition of carbon and nitrogen resources from the lunar poles could, in theory, make a future industrialized lunar colony completely independent of the Earth's resources.

So it is obvious that the first nation, or nations, to establish a permanent presence on the surface of the Moon will be the first to reap the economic benefits of utilizing lunar resources for reducing the cost of manned and unmanned space travel with the added benefit of economic gain from lunar tourism thanks to the reduced cost of traveling to the Moon.

Comment by Marcel F. Williams — May 2, 2010 @ 3:12 pm

9. @Heinrich Monroe

Its probably going to cost $50,000 to $100,000 per kilogram to send oxygen or water to the Moon from Earth. So I think its pretty obvious that a lunar base could manufacture oxygen from lunar regolith substantially cheaper than $50,000 per kilogram. And I think the same would be true if we extracted water from the lunar poles.

Comment by Marcel F. Williams — May 2, 2010 @ 3:49 pm

10. Pradeep,

So, what about the water vapour traced by the Moon Impact Probe

My presumption is that this water is the same water that M3 is detecting, only in motion — on its migration to the cold traps. But there is still much about the lunar "hydrosphere" that we don't fully understand.

Thanks for reminding me about this observation. I have added an addendum to the article about it.

Comment by Paul D. Spudis — May 2, 2010 @ 5:50 pm

11. It's a good thing Obama thinks the Moon is a been there done that destination..

Yes that was sarcasm!

Comment by Johnny — May 2, 2010 @ 9:24 pm

12. Hi Paul,

Many thanks for a very informative article, it is a subject I have been following for a long time. One question though. What does cislunar space mean in this context? Cis according to my Latin dictionary means "on the near side of".

Comment by Brian Sheen, Roseland Observatory — May 3, 2010 @ 4:09 am

13. Brian,

Cislunar means the volume of space between Earth and Moon. It derives by analogy from the Latin "Cisalpine Gaul" (i.e., Gaul before the Alps, or northern Italy) and "Transalpine Gaul" (Gaul beyond the Alps, or southern France).

Comment by Paul D. Spudis — May 3, 2010 @ 4:14 am

14. Dear Dr. Spudis,

you write that *"significant quantities of water moving towards areas with lower mean surface temperatures and increasing in abundance with latitude"* on the Moon.

I would like to ask if such areas of lower temperature could be made artificially. For example, could cold spots be produced by simply building big sunlight reflecting umbrella constructions?

If "the abundance of this surface water varies with time, being present in greater quantity in both local early morning and late evening and it increases in abundance with increasing latitude" then would this water molecules also move into those human-made cold traps and thus be gathered?

Comment by Hans-Peter Dollhopf — May 3, 2010 @ 7:14 am

15. Hans-Peter,

if such areas of lower temperature could be made artificially?

In principle, yes. But there are two problems. First, the temperatures would be lowered, but the diffuse thermal IR from surrounding, non-shaded areas would significantly raise the temperatures in the man-made trap. There is no way that cold on the order of the polar cold traps could be achieved. Second, even if we could make these artificial cold traps, the rates of water formation and deposition are so slow, we would never accumulate significant amounts of it. The polar cold traps have a lot of water because they have been there accumulating it for billions of years.

Comment by Paul D. Spudis — May 3, 2010 @ 7:39 am

16. As it seems to be the case that the Administration has nixed the Moon as a destination for *human* crews in the foreseeable future, we might consider how the Moon can become the domain of tele-operated robots. How much could be done without a single human hand on location? For example, might it be feasible to develop lunar water production tele-robotically and to then transport it, store it and crack it at an EML1 depot, all without a human hand touching it?

The future of robotics is tremendously exciting!

Comment by Itokawa — May 3, 2010 @ 8:56 am

17. What about ice being in caves on the Moon. Could there be ice in caves in equatorial regions?

If water migrates does follow any kind of route- would the lunar topography have any effect upon it's movement.

If you were to measure the temperature say 10' below the lunar regolith, would there be a significant difference between polar areas and equatorial?

Could it be that the geological formation under particular polar craters is a significant factor in determining whether the dark crater are cold enough to collect water?

Comment by gbaikie — May 3, 2010 @ 9:32 am

18. gbaikie,

Could there be ice in caves in equatorial regions?

Possibly. However, as we still don't fully understand how the water is made/deposited and how it migrates, we cannot estimate the rate at which such water might accumulate in favorable, non-polar localities. My suspicion is that it will not accumulate as much water as the poles simply because the net flux of water available for capture is much less.

The temperature of the regolith below the diurnal layer (about the upper meter or so) is pretty constant at about 250 K, as measured at the Apollo 15 and 17 landing sites.

Comment by Paul D. Spudis — May 3, 2010 @ 10:10 am

19. Paul, do you have references for the escape vs diffusion rates of water molecules? I kinda sorta had the impression that escape requires photodissociation by solar UV with a half life of a few weeks, while brownian motion by ballistics hops of individual molecules (mean free path is hundreds of km?) delivers them to the polar cold traps within a few tens of hours. In that case, I'd expect most H2O molecules to be cold trapped before they can get nailed by a UV photon and broken up.

I don't have easy access to an analysis, maybe you can track one down or do it yourself?

Comment by Doug Jones — May 3, 2010 @ 1:17 pm

20. Doug,

Many have modelled the transport and retention of water in the polar cold traps. I suggest the following references (all *Journal of Geophysical Research*):

J. Arnold (1979) JGR 84, 5659. L. Lanzerotti and W. Brown (1981) JGR 86, 3949. R. Hodges (2002) JGR 107, 10.1029/2000JE001491 D. Crider and R. Vondrak (2003) JGR 108, 10.1029/2002JE002030

None of these studies (all based on impeccable physical analysis) predicted what has been directly observed on the Moon in the past few months. No moral drawn, but as an observational rather than theoretical scientist, I suspect that while what you are asking for can be calculated, those results would be of little value without confirming data from the real Moon.

Comment by Paul D. Spudis — May 3, 2010 @ 1:51 pm

21. The line "..Spectra from this impact event show evidence for other volatile substances, including ammonia and simple carbon compounds.." interested me. So estimates are that hydrocarbon sludge (I.E. heavy crude oil by terrestrial standards) is as common as water in NEO comet cores. Could there be oil as well as water collecting at the poles?

Comment by Kelly Starks — May 3, 2010 @ 2:22 pm

22. Kelly,

So estimates are that hydrocarbon sludge (I.E. heavy crude oil by terrestrial standards) is as common as water in NEO comet cores.

I didn't mean to imply that the simple organics found in the LCROSS spectra are anywhere near as abundant as water. Things other than water are present in very minor amounts. In comets, water ice is about 95% of the total mass, with carbon dioxide being the next most abundant component. More complex organics are in the 1-2% range.

We're still awaiting formal publication of the LCROSS results; I expect them to mirror cometary abundance, but we'll just have to wait and see.

Comment by Paul D. Spudis — May 3, 2010 @ 3:09 pm

23. There is a fifth kind of water -- the tears of lunar scientists once they heard of the current plan for human spaceflight.

Comment by Ben — May 3, 2010 @ 3:32 pm

24. Ok, and yes I suppose non NEO impacts would dominate the moon, and I've only heard of the NEO comet cores having that ratio.

Would certainly be useful though if rather then just finding water you could electrolyze only LOx/LH, you could actually find usable reserves of hydrocarbon fuel. ;)

Comment by Kelly Starks — May 4, 2010 @ 10:32 am

25. I found some discussion of mean free paths and ballistic hops here-

http://www.geoffreylandis.com/moonair.html

Comment by Doug Jones — May 4, 2010 @ 7:20 pm

26. "Ok, and yes I suppose non NEO impacts would dominate the moon, and I've only heard of the NEO comet cores having that ratio."

NEO is Near Earth Object. Comet aren't NEOs. Most NEO do have a high water content- if you consider say 10% of it's total mass to be water. Some have as much as 25%. The ones with highest water content are thought that they could be dead comets. Water is very abundant in this universe. It is comprised of the most abundant [known] element- Hydrogen. Which is about 75% of total mass of the [known] universe.

Oxygen is also quite abundant- for example about 40% of the total mass of the surface of the Moon is oxygen. About the same goes for the Earth's surface- and all other terrestrial planets or bodies in the solar system.

"Would certainly be useful though if rather then just finding water you could electrolyze only LOx/LH, you could actually find usable reserves of hydrocarbon fuel."

Hydrocarbon fuel is useless as a fuel, if you don't have oxidizer [such as Oxygen- O2]. Oxygen is highly reactive, and unless you are on earth which has 20% of it's atmosphere O2, you need to find some kind of oxidizer or separate Oxygen from some compound with oxygen in it. Water is one such compound. And compared to other types of oxidized compounds it requires less energy- such as SiO2- Silicon Dioxide is a very common material found on the Moon or Earth- sand generally is mostly SiO2 as are most rocks. The next most abundance material other than Silicon Dioxide is Aluminum oxide:

Here's a chart: http://mistupid.com/geology/earthcrust.htm

To make aluminum it is also "split" using electricity:

"Because aluminum smelting involves passing an electric current through a molten electrolyte, it requires large amounts of electrical energy. On average, production of 2 lb (1 kg) of aluminum requires 15 kilowatt-hours (kWh) of energy. The cost of electricity represents about one-third of the cost of smelting aluminum."   http://www.madehow.com/Volume-5/Aluminum.html

Water requires about twice the amount energy rather 3 times as much as compared to water. Of course the aluminum might be useful for things- though so is the hydrogen useful for various things.

Comment by gbaikie — May 4, 2010 @ 9:15 pm

27. I propose an Operation Highjump initiative to emplace bases on the Moon, and commit to expanded scientific inquiry, which includes even the sortie missions to investigate a variety of landing sites. The outpost missions could be either intermittent or semi-permanent in nature: most bases in Antarctica are permanently staffed, but temporary forays to outlaying ground has always been part of the equation, thus, both types of expeditions will be needed for this effort. (Once base modules have been emplaced, sortie missions sent elsewhere would have back-up support, in the event of problems arising.) Yes, people, we SHOULD return to the Moon, first! We do NOT need a Book of World Records one-time stunt mission to an asteroid, now! Such a "farthest distance record" category spectacle will NOT lead to bases NOR resource utilization! Flexible Path is wholesale bunk & illusion!! Project Constellation could be America's Operation Highjump quest for deep space. We deal with the Moon first, because it is the closest & easiest to reach. Remember that Lunar surface conditions are near-identical to the vacuum of interplanetary space. Keeping men alive there for ANY length of time, prepares us technologically for reaching any other far-deep space destination. Captain George Dufek, sir, we are ready to fly to the Pole!

Comment by Chris Castro — May 5, 2010 @ 1:10 am

28. *NEO is Near Earth Object. Comet aren't NEOs*

There are comet cores, or extinct comets that get stuck orbiting in NEO orbits. They boil off most of the volatiles and get down to rocky ores, water, and oil sludge – as you mentioned

The ones with highest water content are thought that they could be dead comets. "Would certainly be useful though if rather then just finding water you could electrolocize only LOx/LH, you could actually find usable reserves of hydrocarbon fuel." Hydrocarbon fuel is useless as a fuel, if you don't have oxidizer [such as Oxygen- O2]

But a slight amount of hydrogen can be used to mine Oxygen out of the lunar soil. "burn" the soil with the hydrogen. Electrolyze out the Ox from the resulting steam, reuse the hydrogen to recover more Oxygen.

Kerosene works much better in space ships then Hydrogen. Hydrogen works great in rocket engines — not so good in the overall launch vehicle.

Comment by Kelly Starks — May 5, 2010 @ 8:53 am

29. *I propose an Operation Highjump initiative to emplace bases on the Moon, and commit to expanded scientific inquiry, which includes even the sortie missions to investigate a variety of landing sites. The outpost missions could be either intermittent or semi-permanent in nature: most bases in Antarctica are permanently staffed*

Now there's a return to the moon plan I could get excited by. ;)

We do NOT need a Book of World Records one-time stunt mission to an asteroid, now! Such a "farthest distance record" category spectacle will NOT lead to bases NOR resource utilization! Flexible Path is wholesale bunk & illusion!!

Oh hell yeah!

Comment by Kelly Starks — May 5, 2010 @ 8:57 am

30. *Kerosene works much better in space ships then Hydrogen. Hydrogen works great in rocket engines — not so good in the overall launch vehicle.*

I would agree that kerosene works better getting off earth than compared to hydrogen. The reason is that kerosene is a denser fuel-more energy per gallon. Getting off of Earth is a daunting task. Getting off the Moon isn't. So if you need a lot of power within a short period of time, kerosene is good stuff. And also, suppose you needed a lot energy but could burn it for say 4 times the time as the time you need to get into earth orbit- say, 20 mins instead of 5 mins? This means one doesn't need as large of plumbing as you need for the Shuttle's main engines but can deliver as much power. So if you need a lot of Delta-v, say to get to Mars fairly quick, you could use Hydrogen and it would work as well or better than Kerosene.

In other words burning at perigee of a highly elliptical orbit/escape trajectory one has more time than compared to trying reach orbit from earth surface.

Comment by gbaikie — May 7, 2010 @ 12:12 am

38. @ Kelly Starks

Thanks for the analysis on my commentary. One thing: I always cringe when I hear this talk about NASA having to do MORE earth-orbital assembly work & further extravagant activity in LEO. To me, the light in my eyes flickers when a concept like Constellation emerges WHICH GETS US OUT OF LOW EARTH ORBIT. To me, the romance comes in our astronauts finally breaking orbit, and going off into the dark void. That moment when Apollo 8, in December of 1968, heard the directive: "You are go for TLI." THAT equivalent moment alone, for OUR generation, will be the most historically amazing ever! Can you imagine, that fine day, when a crewed spacecraft heads off to go SOMEWHERE, for the first time in half a century! The name of the game, to me, is actually going someplace! So I get very leery & consternated of all this talk of: "Let's not go anyplace now, until we've done more testing in LEO first." It's like an invitation for us to waste more decades trapped in LEO. (Obama's Plan does a real good job at this. It keeps us marooned doing endless circles round the Earth for the next 15 or 20 years.)

Look, Constellation does have that one earth-orbit rendezvous at the start of the space journey to contend with, true. But after that, it needs NO further LEO assembly work; and following that docking of those two spacecrafts, its earth-escape stage lights up, and the complement heads off to the Moon. This maneuver, can be replicated, with some variation, when the time comes to send unmanned modules & cargo craft over to the Red Planet. I say, that we support the Constellation effort, because it is the best & most realistic proposal, to get NASA out of LEO, and onto deep space—where the action REALLY should be.

Comment by Chris Castro — May 11, 2010 @ 2:56 am

Using the Earth to Study the Moon

May 15, 2010

Fig. 1 SP cone and flow, a very rough, fresh volcanic feature in northern Arizona. Radar image courtesy of L. Carter, Smithsonian Institution

Last week, the Science Team[1] of the Mini-RF imaging radar[2] experiment aboard the Lunar Reconnaissance Orbiter (LRO)[3] mission, met in Flagstaff, Arizona. We were there to conduct field studies of some interesting lunar analogs that occur in this area. Scientists study the planets through a variety of means, including images, remote-sensing, and sample return. One technique involves studying the processes and deposits of the Earth as a guide or analog to understand similar features on the Moon and other bodies. Analogs have been studied since the beginning of the space program and have been essential to unraveling the complex histories of rocky objects in the Solar System.

The team gathered early Wednesday morning north of Flagstaff. Our field guides pictured the three areas we would spend the day visiting, along with geologically similar features found on the Moon. Our technique used airborne radar[4] images of our targets: The SP cone and lava flow[5], Sunset Crater National Monument[6] and Meteor Crater[7]. Each site offers specific features that one can observe

and walk across, using it as a guide toward understanding the same processes that have shaped our Moon. Our field trip illuminated the radar data in a "real world" environment, assisting us as we continue to explore and map with our instrument now orbiting the Moon.

The SP cone and flow is one of the most remarkable volcanic features in the region, with a beautifully symmetrical cinder cone and an extremely rough, blocky lava flow (Fig. 1). As viewed from the ground, the lava flow is blocky and extremely rough at the scale of the L-band radar wavelength (about 25 cm, or almost a foot). Steep flow fronts of blocky lava lie directly upon a smooth plateau of flat-lying sedimentary rocks. These remarkable flow fronts can be up to 50 m high (over 150 feet) and their rubbly, rugged fronts provide a spectacular contrast to the featureless plain upon which they rest. In the radar image, the lava flow is extremely bright, indicating high radar returns and its circular polarization ratio (CPR), one measure of its surface roughness at wavelength scales, is very high.

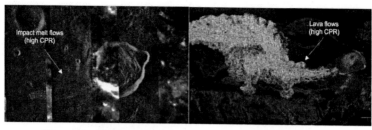

Moon - Gerasimovich D   Earth - SP lava flow

Fig. 2 The lunar crater Gerasimovich D, showing an outflow of impact melt rock with high CPR. This relation is similar to the high CPR seen in the SP lava flow

The relations seen at the SP flow indicate the very high CPR features on the Moon could likewise represent very rough, block-rich surfaces. An example of such is the unusual flow of shock impact melt (not volcanic lava, although quite similar in terms of its physical properties) seen emanating from the far side crater Gerasimovich D (22°S, 122°W, 26 km diameter; Fig. 2). Both of these lobate flows (volcanic lava on the Earth, impact melt on the Moon) show high CPR, indicating the surface of the flow on the Moon probably has similar properties to the SP flow north of Flagstaff. One exception is that the mean block size on the Moon may be smaller, as the Mini-RF S-band radar has a shorter wavelength (12.6 cm or about 5 inches) than the longer wavelength AIRSAR L-band image (23 cm wavelength) of SP crater.

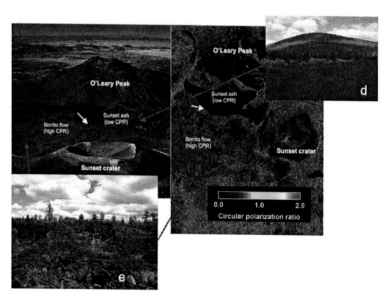

Fig. 3 Sunset crater lava flow (high CPR) and ash deposits (low CPR). Radar image courtesy of L. Carter, Smithsonian Institution

Fifteen miles away from the SP flow, an instructive set of geologic relations are seen at Sunset Crater National Monument (Fig. 3). At this feature, the extremely rough lava surface of the Bonito flow is in direct contact with smooth, ash mantled hills of the same age. This contact is shown by the sharp boundary between high CPR lava and the extremely low CPR ash-covered hills in the radar image. Such a relation is also evident on the Moon, where regional dark mantle deposits of lunar volcanic ash (such as the Sulpicius Gallus dark mantle[8] on the rim of Mare Serenitatis) show low CPR, exactly as does its terrestrial counterpart. Once again, the Earth example allows us to better interpret our remote-sensing data for the Moon.

Fig. 4 Meteor crater, showing blocky, rough exterior rim deposits, wall outcrop, and fine-grained floor materials. AIRSAR radar image

The Moon is covered with millions of impact craters and we were anxious to visit and compare the radar data of Meteor Crater[7], the world's first proven impact structure, with surface conditions within and near the crater rim to better understand the surface of the Moon. The rugged, blocky ejecta of rocks thrown out of the crater is evident by the radar bright halo surrounding Meteor crater (Fig. 4). On the ground, this is manifested by abundant boulders of rock, strewn about the outer rim of the crater. The crater interior is filled with ancient lake bed sediments. This fine-grained material results in lower radar echoes for the floor of Meteor crater than for its rocky walls and rim. Similar features are found in certain lunar craters where fine-grained material, moved downhill by gravity, partly fills the crater interiors. The afternoon's hike down into and across the floor of Meteor crater gave all of us a better appreciation for the surface topography and conditions on the Moon. The climb back up to the rim, capped off our long day of field work.

Many of the geological features seen in radar images of the Earth are also seen in the radar images from the Moon. As we continue to map the Moon with the Mini-RF radar, the sometimes puzzling relations seen in the lunar data are understood better by comparison with Earth analogs. Our entire team acquired valuable insight into how the Moon works and what the surface is like from our day in the field. For a geologist,

there is simply no substitute for directly observed field data to fully comprehend the complex history and processes of the Moon.

Equally interesting and important will be the insight and knowledge gained when we sample the Moon in more detail. The Moon has been described as a "dead planet" because compared to the Earth, which has rapid, dynamic processes of erosion, the Moon remains unchanged for millions of years. However, for its ability to retain the ancient historical record of the Earth-Moon system, advantage goes to the Moon. The multi-billion year records of impact and solar wind embedded in the lunar surface awaits our recovery, and will tell us about both the past and possible future of our home planet.

Topics: Lunar Exploration, Lunar Science

Links and References

1. Science team, http://www.nasa.gov/mission_pages/Mini-RF/team/index.html
2. Mini-RF, http://www.nasa.gov/mission_pages/Mini-RF/main/index.html
3. LRO, http://lunar.gsfc.nasa.gov/
4. airborne radar, http://airsar.jpl.nasa.gov/
5. SP cone and flow, http://geology.com/volcano/sp-crater.shtml
6. Sunset Crater, http://www.nps.gov/sucr/index.htm
7. Meteor crater, http://www.meteorcrater.com/
8. Sulpicius Gallus dark mantle, http://adsabs.harvard.edu/full/1980LPI....11..418H

It's the Space Economy, Stupid!

May 21, 2010

It is now 40 years since the United States first landed men on the moon. Do you think the space program has brought enough benefits to this country to justify its costs, or don't you think so?

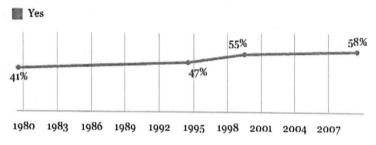

■ Yes

41% 47% 55% 58%

1980 1983 1986 1989 1992 1995 1998 2001 2004 2007

1979 poll conducted by NBC News/Associated Press

GALLUP POLL

Flatlined -- in every sense of the word (Gallup Polls[1])

Those of us in favor of human lunar return have been called "dinosaurs" because, as it's being told, we want to repeat what this nation already did 40 years ago. If that were our mission objective, such a characterization might be valid. But who really is the dinosaur?

At a recent Senate hearing, Norm Augustine told anecdotal stories[2] in regard to lunar return, of how "our committee received many informal inputs, particularly from young people, questioning why we would have a space program whose centerpiece is something that was accomplished over a half-century earlier." NASA Administrator Charles Bolden states publicly[3] that trips to new (non-lunar) destinations is exciting, while there are already "six American flags on the Moon." The President himself, referring to our efforts to return to the Moon, remarks with disdain, "we've been there[4]," the implication being that only new destinations in space are sufficiently exciting for the American public.

The administration's new direction[5] calls on NASA's manned space program to: 1) stop what they are doing; 2) transition into technology study groups with a window of five years for sketching out the hardware and roadmap that NASA will follow for visiting a variety of new and ever-more-distant destinations; and 3) as soon as a rocket is built, commence with the objective – a series of intermittent (though

spectacular) space "firsts" (as they believe this formula is needed to recreate the emotional pull Apollo had on our nation) that will eventually lead to a human setting foot on something beyond LEO. The "new" direction requires that we launch everything we need on these voyages directly from the surface of the Earth. Afterwards, the small vehicle that returns the crew to Earth will be all that remains of the mission hardware. Then comes the next challenge: a more distant destination to keep a paying (but not going) public "engaged and excited." The "new" template is nothing more than a very old one – keep the Roman public amused with circuses and gladiatorial shows.

In contrast, those of us who support the Vision for Space Exploration (VSE)[6] want to return to the Moon to use its abundant resources to incrementally create a sustainable, permanent human presence in cislunar space. We want reusable, extensible, maintainable and affordable systems in space. We want to unlock and harvest the enormous wealth of the Solar System for the benefit of all humanity. We want to do what has never before been done – extend our civilization into the universe. In place of one-off stunt missions, we want to create something of value[7] – lasting and continuous access to space and its resources to expand our economy and create new wealth. Instead of repeating Apollo, the

VSE is an engraved invitation that will encourage participation from the paying (and probably going) public. We want to build a real space economy.

The concept that our space program should excite people is a long-held faith in many space policy circles. Several annoying facts remain though, suggesting that regardless of their desire for public excitement, its existence (or lack thereof) does not historically track with our national space program. President Kennedy did not assemble a focus group or enlist a write-in campaign to gauge and prompt support for his call for a lunar landing program. In fact, he himself wanted to find another venue for Soviet-American competition, one that would produce more tangible and practical benefits, such as desalination of seawater[8].

Ever since it ended, NASA has doggedly tried to re-create the Apollo program. But the facts do not bear out their collective rosy memories[9] of those days. Polls taken before and during the Apollo missions to the Moon, found, at best, a plurality of public opinion[9] in support of the lunar landing effort. Many polls found a majority against the effort. Media interest was intense for the first landing (Would anyone expect otherwise?) but tailed off afterwards. During the totality of NASA's existence, public support of the space program[1] has hovered around 50-50 favor/oppose, regardless of what the agency was doing at the time. My conclusion from these results is that, in broad terms, people don't really care that much about space; they do not oppose it, but they are not wildly enthusiastic about it either. Perhaps they can't picture a time when they will move beyond their current role as mere spectators.

There is a belief in space circles that public excitement is a critical and driving factor in selecting goals and objectives in space. Threads on various space forums[10] repeatedly argue for or against some path forward on the grounds that a certain program or effort will excite people. This belief is closely related to its corollary belief that excitement equals political support and hence, more funding for space efforts. There are two issues with this kind of thinking. First, regardless of the excitement factor, one cannot set goals and objectives that are technically impossible. For example, if the public decided tomorrow that only interstellar voyages were their hearts' desire, we would not set that as a goal because we don't know how to do it. More seriously, excitement does not necessarily correlate with value. We

all buy and pay for many things that are exciting, such as watching or participating in sporting events, but after they are over, they are over. They may have some long-term value in improving our own health or satisfying a need to be entertained, but eventually, we turn away from them and go back to attending to the necessities of our daily lives. As adults, we need to spend time and money on practical matters as we plan and prepare for our futures.

In other words, it's not excitement that we need from our space program, it's value for the money spent. Many in the space community (and even many inside the agency itself) parrot the falsehood that lunar return under the VSE was all about repeating Apollo. In contrast to the trite "been there, done that[11]" formulation of such misdirected thinking, the real purpose of return to the Moon under the VSE was to learn how to create sustainable human presence off the Earth, including learning how to harvest and use its material and energy resources. Such an objective has never been attempted. In fact, we're not even certain that it can be done – that's why it was given to an agency reputed to be our premier technical problem-solving agency – NASA.

So, who is the dinosaur here?

Topics: Lunar Exploration, Lunar Resources, Space Politics, Space and Society

Links and References

1. Gallup Polls, Majority of Americans Say Space Program Costs Justified, http://tinyurl.com/2areaoh
2. anecdotal stories, http://tinyurl.com/2bmdxpv
3. states publicly, http://www.denverpost.com/headlines/ci_14385263
4. we've been there, http://www.nasa.gov/news/media/trans/obama_ksc_trans.html
5. new direction, http://www.nasa.gov/news/budget/index.html
6. VSE, http://www.spaceref.com/news/viewpr.html?pid=13404
7. something of value, http://blogs.airspacemag.com/moon/2009/07/16/space-program-vs-space-commerce/
8. desalination of seawater, http://www.dow.com/liquidseps/prod/sw.htm
9. facts do not bear out, http://tinyurl.com/mhoyqf
10. threads on various space forums, http://nasawatch.com/archives/2010/05/media-bias-in-s.html
11. been there, done that, http://tinyurl.com/2a8tvdb

Comments

1. Typical political position, basing decisions on popularity for the goal of re-election and not on analysis with the goal of the best possible outcome. As a Gen-Xer, reusability and efficiency is high on the priority list and wasting material to break Earth's gravity is unpractical. Further more, how much fuel

can be saved from launching from the Moon, where gravity is minimal when compared to the Earth's, which then can be used toward deeper exploration? The government doesn't have the slightest clue on what is best for the American people, let alone proactive and sustainable space exploration.

Comment by Toby Barnett — May 21, 2010 @ 4:37 pm

3. *In contrast to the trite "been there, done that" formulation of such misdirected thinking, the real purpose of return to the Moon under the VSE was to learn how to create sustainable human presence off the Earth, including learning how to harvest and use its material and energy resources. Such an objective has never been attempted.*

I'm curious what your thoughts are on the lunar ISRU testbed planned for 2015 as part of the ETDD missions in FY2011: http://www.spaceref.com/news/viewsr.html?pid=34056

Also, what are your thoughts on how lunar ice harvesting could deal with the problems of extreme cold and lack of sunlight in the permanently dark craters?

Comment by Neil H. — May 21, 2010 @ 5:27 pm

4. Neil,

I'm curious what your thoughts are on the lunar ISRU testbed planned for 2015 as part of the ETDD missions in FY2011

That proposed mission has not been defined or funded, so as of right now, I have no opinion. However, I have thought in some detail about what kinds of robotic missions are needed, and in what order, to fully understand where and how we might conduct lunar resource prospecting, harvesting, processing and handling. See the appendix in this presentation for details:

http://www.spudislunarresources.com/Papers/The%20Vision%20and%20the%20Mission.pdf

Although this exploration strawman mission set was made almost 6 years ago, it is still valid today, after the LRO mission has completed our initial reconnaissance from lunar orbit.

Also, what are your thoughts on how lunar ice harvesting could deal with the problems of extreme cold and lack of sunlight in the permanently dark craters?

It's easier and less power costly to keep warm in space than it is to keep cool. Machinery will need thermal control to stay warm and operative in the cold traps, but this is a straight-forward issue to deal with; spacecraft need thermal control systems anyway. Likewise, we will use artificial illumination to conduct whatever surface operations that we need. However, even in the polar cold traps, there is some illumination derived from light diffusely scattered from the adjacent, illuminated areas.

Comment by Paul D. Spudis — May 21, 2010 @ 5:44 pm

5. When we did the Apollo program, that was truly 1st time exploration. We found that the Moon was not made of cheese, we would not sink into pools of Moon dust, we could hop and drive around the lunar surface, and that we knew how to return to earth safely. We know these things now, and although there are far more things that we still don't know, the same could be said for the Earth, which we have been exploring for a much longer period of time.

There is no question we will eventually return to the Moon. The question for NASA is what are they to do with their limited budget. If we gave them as much money as the Department of Defense, then yes, the Moon would be one of the many places they would go.

But NASA has a limited budget, so the President has decided, in a rip-off of STNG "To boldly go where no one has been before". We've been to the Moon, and have a good idea where to pick up where we left off. We can start doing that by using robot precursors, and then follow up with human exploration and exploitation. We already have the Google Lunar X Prize, one of many indications of non-government interest in going to the Moon, so the government does not need to lead the way back.

We have not learned how to live outside the confines of the Earth-Moon system, and that is where the President wants to focus NASA's limited resources. Just like with every step we take into uncharted areas, what we learn is passed down to the commercial space industries, and used to lower the costs to those new places. The Moon is waiting, as it has for millions of years, and a couple more years won't make a big difference.

Comment by Coastal Ron — May 21, 2010 @ 6:39 pm

6. Thank you for the response!

Comment by Neil H. — May 21, 2010 @ 7:35 pm

7. I don't think anyone thinks your are a dinosaur, Paul and I too, would like to see a return to the Moon. I think though that when the administration says: "Been there done that" what they are more interested in getting across is that Constellation is dead i.e. if they said they were still going to the Moon people would say: "why not use Constellation." I think the plan is to find ways to build cheaper rockets hence inviting commercial space to do Leo and find cheaper ways to do deep space by: 1)Building fuel depots in orbit 2) Treating Leo and the Space Station as the harbor for a True Spaceship that stays in orbit and doesn't have to be lifted out of Earth's gravity well every time you want to go some where.

How would it be if like Apollo, the even more expensive Constellation was just getting to do some interesting things on the Moon when the bean counters started hollering it's too expensive and funds were cut off? They say that those who don't learn from history are forced to repeat it and that's what was happening with Constellation.

Some people say space can't be made cheaper that it will always be expensive that the Space Shuttle was supposed to be reusable and cheaper but look how much it costs to run. Well the Space Shuttle itself was the only part that was reusable if replacing tiles all the time is reusable. The reuse of the solid boosters was an after thought but their refurbishment is so labor intensive that no money is saved. There was nothing that could be done about the tank. When they were deciding to build the Shuttle Nixon was given two choices an expensive one with a fly back first stage that was to be flown back to land on a runway for reuse and a cheaper one that he chose i.e. the present Shuttle. If at the time NASA had had the foresight to design an in line shuttle system with an extra second stage that could have been morphed into the more expensive version we would still be using it.

No we have to bring costs down first that is the only way we are going to get back to the Moon or Mars or Europa.

Comment by Gary Warburton — May 21, 2010 @ 8:20 pm

8. Thanks mostly to poor presidential leadership, NASA has spent the last 38 years trying to be socially relevant instead of pioneering the New Frontier. And even President Bush underfunded the Constellation program and then allowed it to disintegrate from a Moon base program into an Apollo redux program.

I conducted a poll last year on the rather liberal blog site, the Daily Kos:

http://www.dailykos.com/story/2009/8/27/773450/-Colonizing-the-Moon

And I asked the question, "Could lunar colonization be a social and economic benefit to humanity?"

Out of 92 responses, 59% said yes. 23% said no. And 16% said maybe. So in that little poll, 75% of people thought that lunar colonization might be socially and economically beneficial to humanity.

I'm not so sure if the politicians want us to colonize the solar system, but I do think the people are ready. NASA just needs to make the obvious case for a simple, potentially self-sustaining, base at the lunar poles that could help reduce the cost of space travel. If Bolden is for space depots, how can he be against cheaply resupplying those depots with oxygen and hydrogen from the lunar poles?

But there's no doubt in my mind that if a permanent lunar base is established, the space tourist and the entrepreneurs will follow. And the first nation to establish that base will reap most of the economic rewards.

Comment by Marcel F. Williams — May 21, 2010 @ 10:54 pm

9. I accept your proposition that space exploration can't be wholly justified on the basis of "excitement", and instead should be linked to value. But what, pray tell, is the value of a "return to the Moon to use its abundant resources to incrementally create a sustainable, permanent human presence in cislunar space"? I don't need a sustainable, permanent human presence on the Moon, nor does anyone else. It offers me no value.

I believe that human space exploration is ultimately justified by developing the ability to leave the Earth. As a matter of insurance for the species. The Moon may or may not play a significant role in that. But if that's why we need human exploration of the Moon, it isn't going to influence the Dow Jones average, or whatever one uses to assess the health of the space economy.

That the solar system can provide "riches" that could sustain a real business case are far from established.

Comment by Trevor Ruhl — May 21, 2010 @ 11:53 pm

10. [...] May 22, 2010 by The Space Geek Leave a Comment I have today, among other things, a very appropriate and well-stated explanation, I think, for my own pro-lunar-return opinions. I feel like I didn't really know why I felt the way I felt until I read this. I'm not sure yet if it's the be-all-end-all of my opinions visa-vis a lunar return, but I certainly found myself nodding along. Here's the link. [...]

Pingback by Now THIS is more like it! « The Space Geek — May 22, 2010 @ 2:32 am

11. Coastal Ron,

But NASA has a limited budget, so the President has decided, in a rip-off of STNG "To boldly go where no one has gone before".

In contrast to Augustine, the existing budget is adequate to implement the VSE. But such program changes were not seriously examined or considered. As far as "boldly going" is concerned, learning to live off the land on another world is certainly bold and creates capability for the future. One-off stunts to distant destinations that leave no legacy capability or infrastructure are not.

We have not learned how to live outside the confines of the Earth-Moon system, and that is where the President wants to focus NASA's limited resources.

Wrong. We are doing "missions" to other destinations — there is no "living there" involved.

The Moon is waiting, as it has for millions of years, and a couple more years won't make a big difference.

Congratulations on coming up with the lamest excuse yet for the "new path." Of course you realize that this facile formulation applies to all the other destinations (NEOs, martian moons) of Flexible Path, right? But while we're all waiting, NASA can still find ways to throw away $ 20 billion per year. Indefinitely.

Comment by Paul D. Spudis — May 22, 2010 @ 4:29 am

12. Gary,

I think the plan is to find ways to build cheaper rockets hence inviting commercial space

Yes, I hear this all the time from supporters of the new direction, but have yet to get a credible explanation from them of how this will come about under the new path.

How would it be if like Apollo, the even more expensive Constellation was just getting to do some interesting things on the Moon when the bean counters started hollering it's too expen-

sive and funds were cut off?

I'll repeat yet again that the spending assumption of the Vision was that NASA funding would grow only with inflation. It was up to the agency to craft an architecture that fit under that budget envelope. In contrast to what everyone seems to think, that IS possible and there are several ways to approach lunar return under such constraints. But it was much easier bureaucratically to simply abandon the VSE without giving any serious thought to how you might do it differently and come up with this Potemkin program in its place.

No we have to bring costs down first that is the only way we are going to get back to the Moon or Mars or Europa.

No sane person is against this — the debate is over how such cost savings are accomplished.

Comment by Paul D. Spudis — May 22, 2010 @ 4:37 am

13. Trevor,

But what, pray tell, is the value of a "return to the Moon to use its abundant resources to incrementally create a sustainable, permanent human presence in cislunar space"?

I've been making this case for the last 2 years on this blog. In brief, by creating a system that can routinely access the lunar surface (including refueling and propellant export), we also get the capability to routinely access all of cislunar space, where 95% of all our space satellite assets reside. This means that we can service, maintain, extend and build new orbital systems of greater capability and power. It also allows us to protect fragile space assets vital to our national security.

A detailed exposition of this argument can be found here: http://www.spaceref.com/news/viewnews.html?id=1376

Comment by Paul D. Spudis — May 22, 2010 @ 4:43 am

14. No, I'm sorry Paul. Creating that "system" does not give us the capability to routinely access all of cislunar space. In some respects, it may make it a little easier. But it certainly doesn't give us that capability. In fact, what such routine access largely depends on is driving costs down for access from the surface of the Earth to LEO and GEO, where your 95% live. Your system doesn't help much with that.

We're still learning what we want to do with regard to servicing and maintaining orbital systems. We don't know if what we want to do can be done entirely robotically, for which access to GEO or HEO is, in fact, routine. Of the 95% of our space satellite assets you refer to that are in cis-lunar space, the vast majority are defense and commercial surveillance and communication satellites. Neither the defense department nor commercial enterprises have shown a lot of enthusiasm for the capability you say lunar development would "enable". That may change, but the market isn't driving your idea. The "system" you're proposing is to do things more easily that we don't really want to do right now. That's a false economy.

You've said yourself that we simply do not know how difficult it is to develop this lunar system, or what practical problems might arise. I'd strongly endorse investments to try to understand those things, but that's a big step away from plans for creating a "system".

I know you've been making this case for the last two years in this blog. I'm just saying that I don't believe it.

Comment by Trevor Ruhl — May 22, 2010 @ 10:20 am

15. *I'm just saying that I don't believe it.*

Fine with me.

People can find fault with just about any long term plan or strategic direction. My contention is that attempting to create a cislunar space faring infrastructure using the resources of the Moon is at least worth trying, especially as we are going to be spending an enormous amount of money on space anyway that could be used to create such a system. Your belief that the "new direction" in the proposed budget will create such capabilities is touching, but completely contraindicated by the historical record of NASA — no money "invested" in technology development has ever resulted in a single piece of flight hardware.

Comment by Paul D. Spudis — May 22, 2010 @ 2:43 pm

16. Go, Paul, go!

Comment by Itokawa — May 22, 2010 @ 3:11 pm

17. If NASA and USA Govt. wants to focus a New Horizon. I think they are gently telling us... If we want to go to the moon, the privates/citizens/corporations would have to do it. The pioneers have paved the way, now its time for the settlers to come and look for the gold.

Comment by Chester — May 22, 2010 @ 3:50 pm

18. "...learning to live off the land on another world is certainly bold and creates capability for the future. One-off stunts to distant destinations that leave no legacy capability or infrastructure are not."

Apollo could be classified as a stunt under your definition, since we didn't learn to live off the land. The goal of Apollo was for "landing a man on the moon and returning him safely to the earth." The goal of going to an NEO or the moons of Mars would be the same, and along the way you create the knowledge and technology that let's others follow behind.

This gets back to what the role of government is in space. For the U.S. government, I see exploration as one of the engines of our commercial sector, inspiring the knowledge and technology needed to sustain the exploration mission. The government itself is not the creator of that knowledge and technology, but the financier. Once created, commercial companies can use that knowledge and technology to create new products and markets, and eventually create industries that sustain themselves without major government support. Commercial companies are the exploiters, and that's what is really going to drive the demand for returning to the Moon, not the need for a bunch of scientists to walk around chipping rocks (nothing against them doing that, but it's not my first priority).

For the Moon, we did our initial exploration and pathfinding back with Apollo. Though we have had the ability to go back to the Moon, we have not had the market demand to do so. It's kind of a chicken and egg situation, and I view having commercial access to LEO as the first of a number of steps that will push our commerce into space. Some people seem like they are ready to open up a McDonalds on the Moon, but how do you supply it, and who will be your customers? I see the next step as a hotel in LEO. After that, when people get bored with shooting up to the fringes of space, they'll want to orbit the Moon. Finally, with people getting closer and closer to the Moon, enough demand will allow some enterprising company to start a lunar motel. With people on the Moon, the need for indigenous supplies will be driven by the normal supply and demand market forces.

I agree with what Chester said – "The pioneers have paved the way, now its time for the settlers to come and look for the gold." History has shown us the way.

Comment by Coastal Ron — May 22, 2010 @ 6:15 pm

19. *Apollo could be classified as a stunt under your definition, since we didn't learn to live off the land. The goal of Apollo was for "landing a man on the moon and returning him safely to the earth.". The goal of going to an NEO or*

the moons of Mars would be the same, and along the way you create the knowledge and technology that let's others follow behind.

Apollo WAS a stunt. The difference between now and then was that Apollo had a driving political imperative (i.e., beat the Russians to the Moon), one supported by a significant majority in Congress and the Executive.

We are not in such a situation now. The current space program is a low level-of-effort operation (0.5% of the budget or less) that does not command the political momentum or support that Apollo did. So PR stunt missions are not appropriate activities for NASA.

My point is that we will spend a certain amount of money on government space and thus, what we buy should have some long-term value. Repeating Apollo – either on the Moon, on a NEO, or Phobos – doesn't. Building a lunar outpost to harvest resources and create a permanent space faring infrastructure does have it.

Commercial companies are the exploiters, and that's what is really going to drive the demand for returning to the Moon, not the need for a bunch of scientists to walk around chipping rocks (nothing against them doing that, but it's not my first priority).

Nor mine. And I have never written anything in this blog to suggest otherwise.

I view having commercial access to LEO as the first of a number of steps that will push our commerce into space.

We already have this. A thriving commercial launch and satellite business already exists. What you mean is that you want cheap commercial access to space. But cheap is a relative term – how low must orbital costs go before you declare them to be "cheap?" And what government program ever gave us anything cheap?

History has shown us the way

I agree with that. But you have to be able to read the right lesson for history to be a useful guide. The history of NASA shows that they do well with specific goals and destinations. And that they do poorly without them. The "new path" is spending without direction or purpose. We'll still spend $100 billion over the next decade – we just won't get anything from it.

Comment by Paul D. Spudis — May 22, 2010 @ 7:29 pm

20. Manned space flight will never be cheap as long as the demand for manned space flights is extremely low. NASA only had 5 manned spaceflights last year. That's really not enough demand to lower cost. But there's no logical reason why rocket engines should cost tens of millions of dollars each. But they do because the demand for such engines is extremely low.

Space tourism, IMO, is the key to dramatically increasing demand. Polls show that 7% of those wealthy enough to do so would fly into space if they had the chance. That's nearly 7000 people world wide. But there are billions of people around the world who would also like a chance to fly into space if there was some sort of space lotto system that would give them a chance to travel into space too. And most folks that I've polled prefer the Moon as their primary tourist destination.

I applaud the President for advocating giving the emerging manned spaceflight companies around $1.2 billion a year to develop their manned vehicles for accessing LEO. But, again, my concern is what the President is not doing with the other $18 to $19 billion a year.

Comment by Marcel F. Williams — May 22, 2010 @ 7:36 pm

21. Both you Paul, and Marcel, have some interesting topics that I add some comments to:

Paul said "Apollo WAS a stunt. The difference between now and then was that Apollo had a driving political imperative…". I'll agree that it was political reasons that lead us to the Moon, but the knowledge and technology that was created to do that persists, and now going back to the Moon is less of a challenge.

Living off the land is a goal, but it's not something that we can do today, nor in ten years. We can start the effort, but I think we have lots of research, technology and hardware that we will have to evolve before we can say we're living off the land. I see the Constellation program as an example of a "goal too far", in that everything took place either on Earth, or the surface of the Moon. Because of that, there was no infrastructure that was needed or used in between, and so the next "program" would have little to build upon. ISRU would not be needed at that point.

"We already have … A thriving commercial launch and satellite business already exists. What you mean is that you want cheap commercial access to space. But cheap is a relative term – how low must orbital costs go before you declare them to be "cheap?" And what government program ever gave us anything cheap?".

I should have been more explicit. I meant that we need commercial HUMAN access to space, which we don't have. The only way someone can buy access to space on the commercial market is through a former communist country (Russian w/Soyuz), which has to be the ultimate irony.

Comments from Marcel: "Manned space flight will never be cheap as long as the demand for manned space flights is extremely low. ". I think it's both supply and demand. If you look at the cost/lb of cargo from the Shuttle till Falcon 9 Heavy, you will see a halving of the price from generation to generation. Shuttle was about $17k/lb, Atlas/Delta Heavies are about $6k/lb, and Falcon 9 Heavy should be below $3k/lb. I don't know how much lower it can go, but actual $$/seat will decline too. "Space tourism, IMO, is the key to dramatically increasing demand."

I think space tourism will have it's place, but I think government sponsored/funded exploration will be the initial big money that will support the space industry, and cause it to develop the infrastructure that let's us move onto the Moon & beyond.

Back to Paul:

The decision to do ISRU on the Moon should come from economic forces, i.e. the cost of delivering our food, water and fuel is more expensive that creating and running ISRU. At this time, there is no need, even if Constellation were to continue. Once we create the ability for any company to routinely access LEO, then entrepreneurs will step forward to satisfy market needs, either through ISRU or cheaper transportation. This is how capitalism works.

I want to take a vacation on the Moon as much as the next person, but I won't have a chance to afford it until a robust and competitive infrastructure is established. When I said "History has shown us the way", I was not talking about NASA, but about the expansion of human civilization. The fastest expansions happened when there were strong economic incentives, like the gold rushes in Alaska and California. These were dependent on supply lines that could support the expansions, and entrepreneurs stepped forward to create the new products and technologies that the markets wanted. Expansion into space will be no different.

Comment by Coastal Ron — May 22, 2010 @ 11:21 pm

22. Low Earth Orbit: WE'VE BEEN THERE ALREADY!!! Why is going there, OVER & OVER & OVER AGAIN, the only thing that we can keep right on doing, with no complaints from anybody?? BILLIONS of federal budget dollars go to the ISS yearly!! Astronauts

have done NOTHING ELSE but LEO station stays for the past 40 years!! Only a mere 4 years were EVER devoted to trekking cislunar space—1968-1972. How the Anti-Moon space lobby can keep right on shooting down renewed Lunar exploration in favor of one-time-only circus stunts to asteroids and such, is gigantically stupid!! THE WEST WAS NOT WON, BY GOING TO A CERTAIN SPOT ONCE, AND NEVER EVER GOING BACK THERE AGAIN!! Pioneering a frontier ALWAYS has required a major return trip! When Captain James Cook's ocean-traversing crew eventually got back to England, the future solution was NOT to never send men back to Hawaii & Australia ever again;—NO—in order to DEVELOP previously surveyed lands, one's nation has to plan out return missions. The Moon needs to be further explored, and rediscovered by a new generation, whose overall plan will take the endeavor to FAR beyond merely reaching the destination again. (Hell, Project Apollo even went far beyond merely getting to the surface of the Moon! Remember the J-expeditions?! People really forget just how expanded & extended the Apollo expeditions were becoming, when the Nixon administration decided to pull the plug.) THIS TIME, WE GO TO THE MOON TO STAY!! Project Constellation, becomes the Operation Highjump of OUR time! Admiral Dufek, sir: we are ready to fly to the South Pole!

Comment by Chris Castro — May 23, 2010 @ 1:29 am

23. *Living off the land is a goal, but it's not something that we can do today, nor in ten years. We can start the effort, but I think we have lots of research, technology and hardware that we will have to evolve before we can say we're living off the land*

If you had ever read a single thing I have ever written about this, you would know that this is exactly my position (except that we can start doing it now, not in 10 years.) Learning how to live off the land in space is the goal of the VSE. And that goal was to "start that effort" — not by using viewgraph development as the new path proposes, but by actually doing ISRU on the Moon.

But you have to be on the Moon to use its resources.

When I said "History has shown us the way", I was not talking about NASA

I was. And my point is that the new direction takes us away from, not towards, what you claim to want.

And with that, I believe that our discussion is concluded.

Comment by Paul D. Spudis — May 23, 2010 @ 4:39 am

24. Right on, Mr. Spudis!! Project Constellation gets us STARTED with utilizing the Moon, via further scientific surveying, much like oil prospecting. Flexible Path totally AVOIDS any such needed resource assessment; even avoiding having to ever develop landing craft for strong gravity field planets. FP has us doing nothing but circus spectacle jaunts to asteroids, which will teach is nothing about dealing with Mars & Moon-sized bodies for manned & large-cargo landings, since you can't actually "land" on a negligible gravity field. Furthermore, FP has zero, zilch, nada goals for otherworld bases. FP delivers nothing but the cheap, cheap euphoric thrill of: "Look at us, we're the first to ever get here!" The sooner Flexible Path is defeated in Congress, and Project Constellation is re-instated: With adequate yearly funding; the sooner U.S. astronauts will accomplish majestic things again.

Comment by Chris Castro — May 23, 2010 @ 4:50 pm

25. @ Coastal Ron

Human beings have been melting rocks and ice on Earth for thousands of years. So I don't think we'll have too much difficulty melting rocks and ice on the Moon for air, water, and rocket fuel.

A single government just don't have enough manned space flights to drive cost down. Originally, in order to reduce cost per flight, the space shuttle was supposed to fly up to 60 times per year. But the most flights ever flown by the shuttle was 8.

Still, in theory, the shuttle could carry people into orbit a lot cheaper than any theoretical vehicle by Space X vehicle that could carry people for $20 million each.

The Shuttle cost $450 million per flight and can carry up to 11 people into orbit. That would be about $41 million per passenger. However, there were many proposals to use the cargo cabin to also carry passengers. Since you could land 20 tonnes of payload in the cargo bay, at least 40 passengers could have been carried in the cargo bay. So the shuttle could carry 51 passengers into orbit for less than $9 million per passenger. And if the shuttle was flying 60 flights per year, those passenger cost would have been a lot cheaper than that.

Comment by Marcel F. Williams — May 23, 2010 @ 11:21 pm

26. Hi Marcel,

Advocates of the new direction claim that lower launch costs are the *sine qua non* of true space commerce. Yet we already know why launch costs are high (the marching army of highly trained and paid people). We also know in principle how to lower them (either automate launch vehicle assembly and preparation or use the economies of scale provided by large payloads in one vehicle [heavy-lift]). This chicken-or-egg issue has been with us for the last 50 years and is really not a technical issue at all — it is a fiscal and managerial issue.

By abandoning lunar return, we lose the one chance we have to lay the groundwork for a totally different approach — using off-planet resources. Many in the space industry simply cannot envision this or are deathly afraid of it. Hence, we have an agency and aerospace industry fixated on access to LEO when the real frontier in knowledge and technology is beyond it.

Comment by Paul D. Spudis — May 24, 2010 @ 4:59 am

27. Dr. Spudis,

I just think its sad that NASA has suddenly become part of the rather silly commercial vs. government philosophical war in America. And the irony in this, of course, is the fact that the anti-manned spaceflight wing of the Democratic Party (you know, the folks who argue that we should be spending more money on the poor than in space) has suddenly seized this issue as their own.

Thanks to your research and the research of your colleagues, the lunar poles are starting to look like some extremely valuable real estate. And right now, we're probably the only nation in the world in a good position to really take advantage of the lunar resources there in order to lower the cost of space travel and for pure economic gain.

But if we once again pass up the opportunity of establishing a permanent presence on the Moon, I think future historians may well look at this missed opportunity as one of the significant reasons for the great decline of America and the American economy.

Comment by Marcel F. Williams — May 24, 2010 @ 1:48 pm

28. Marcel,

I just think its sad that NASA has suddenly become part of the rather silly commercial vs. government philosophical war in America.

I think this is all quite deliberate. Divide the human spaceflight community against itself and let them destroy each other. Combine this with NASA's known propensity to do nothing, at great expense, high bureaucratic overhead and maximum time, and you have the perfect formula to destroy human spaceflight for a generation.

But if we once again pass up the opportunity of establishing a permanent presence on the Moon, I think future historians may well look at this missed opportunity as one of the significant reasons for the great decline of America and the American economy.

When Rome fell, the existing generation could not accomplish the engineering feats that their fathers and grandfathers did. Public facilities first fell into disrepair, then ruin. The stones and columns making up aqueducts, buildings and roads became sources for building materials for hovels and shacks.

Comment by Paul D. Spudis — May 24, 2010 @ 4:00 pm

29. It seems that there are a number of readers to this blog that (I think) don't fully connect the message Paul is saying to their own concept of the Space picture. The message to me is that use of local resources (at the Moon or wherever off-planet) is the best way to lower cost for human activity in space (or on a planetary body). That is to say, NASA may tinker with technology, and may make some inroads in incrementally improving technology, and commercial ventures like SpaceX may moderately lower (my opinion is at best it may lower very moderately) launch costs, but the rocket equation is still the rocket equation. (Falcon 9 has not yet addressed the hidden cost of recycling hardware in the space economy. See launch prices for them rise after they do realize this.) Unless one looks at nuclear launchers, or sub-orbital tethered pickoffs, or space elevators, or mass drivers, the cost of launching a kg into orbit is still too high to stimulate more demand. Much as we space enthusiasts would have it otherwise, commercial ventures in space with humans cannot yet clear the initial cost hurtles that are HUGE and have a very long term ROI. Historical evidence: Study the rationale for Delta 4 and Atlas 5 as commercial cargo efforts. Were they both successful? (I would offer: NOT) Study the commercial human space efforts in the last 50 years since Apollo (not too many, correct?). Separating the emotion from the logic (which should be influenced by historical fact), commercial crew into space will not fare well, given current launch technology or even near-term extrapolation of technology. The high human launch cost may have a component due to NASA culture and bureaucracy, but it is not the dominant term in the cost equation. What Paul and others advocate is a shift in thinking that says, "instead of reducing the cost/kg launched, reduce the number of kgs launched

in the infrastructure needed to support human endeavors in space by using ISRU." Will this approach work? Don't know that, but it certainly won't work if we don't try and at least investigate what it would take, and how to do it (ISRU).

Comment by Tony L — May 24, 2010 @ 6:42 pm

30. *Marcel F. Williams said :*

"Human beings have been melting rocks and ice on Earth for thousands of years. So I don't think we'll have too much difficulty melting rocks and ice on the Moon for air, water, and rocket fuel." Except we've been doing it in 1G gravity with an atmosphere that already has copious amounts of oxygen. Let's stick you on the Moon and see how quickly you can start a fire and refine metals... ;-)

I don't know if you keep up with operations on the ISS, but we have problems just operating a water purification unit – if we can't make good water from urine, I don't see how you think it will be easy to make it from extremely cold slush on an airless body with 1/6 gravity.

Right off hand I can think of thermal conductivity, gravity separation and surface tension issues that we'll have to tinker with, and it may take a couple of iterations of equipment before we start generating small quantities of anything. Not to mention a whole lot of energy that this will take, either with solar panels or sunlight reflectors. All of this takes lots of supply runs, and the only way we can do that (as is my theme) is by having a robust commercial space industry operating in LEO. We can't have ISRU without a logistics system in place, and, paradoxically, ISRU is not as important if we do have one.

Don't get me wrong, I think ISRU will happen at some point, but there is no economic need for it to be a #1 or #2 priority when we do return. With the robotic precursor missions that NASA is planning, we'll start finding out what there is that we will want to mine, and that will help tell us what will be needed to exploit it.

Comment by Coastal Ron — May 24, 2010 @ 10:24 pm

31. @Coastal Ron

We've developed all kinds of ways to melt regolith that do not require oxygen. Microwaves have been used to melt rocks, plasma arc furnaces have been used to melt rocks, and concentrated sunlight can used to melt rocks. There are already NASA competitions amongst private institutions to see which machines can melt rocks and extract oxygen most efficiently.

Extracting water from the poles might be as easy as simply microwaving the surface with a magnetron and collecting the water vapor. Solar or nuclear energy could be used to supply power for such machines.

It would probably cost between $50,000 to $100,000 per kilogram to import water or oxygen from Earth to the lunar surface. So there's no doubt in my mind the a polar lunar base could manufacture oxygen and water substantially cheaper than that.

All machines malfunction– even your automobile. That's why you always build more than one machine so that you can continue to operate your facility while you fix the other one.

Comment by Marcel F. Williams — May 25, 2010 @ 12:08 pm

33. Thanks for a great article. It kind of helps me to understand why it was that, back in the days when I was active on Usenet, I would find myself and Mars advocates talking past each other when comparing things like lunar resource use or SPS vs. manned expeditions to Mars. All that High Frontier stuff I was talking about was boring, so it was never going to happen. The logic they were using was: exciting goal > excited public > greater public support > more funding. But I think this article very convincingly deflates that argument by demonstrating that not only is there no reason to expect it to be that way in the future, it's not even the way it was during Apollo.

Comment by Mike Combs — May 26, 2010 @ 8:50 am

34. Mike,

Thanks for your comments. This notion that public excitement drives the space program is very deeply ingrained; nearly all space advocates and most of the agency buys into it. The sooner we all recognize how things really are, the better off we'll all be.

Comment by Paul D. Spudis — May 26, 2010 @ 10:19 am

35. It would be lovely to go back to the beginning of the VSE and do it right. Unfortunately it was not embraced by NASA or Griffin. I hope that after a round of new technology development we can get back on that path.

I will believe NASA is serious about building a space economy when they hold the Interagency Space Solar Power Conference which was the most popular on the NASA Open Government Ideascale and the most popular idea for the Department of Energy and Office of Science and Technology Policy as well as the Government as a whole.

See the Top Ideas for the Government on OpenGov Tracker http://opengovtracker.com/

Open Government is another idea NASA doesn't care about since it took them 6 weeks to open comments on their open government plan.

Comment by Karen Cramer Shea — May 26, 2010 @ 11:23 am

36. It was nice to see Neil Armstrong today, once again, speaking in favor of returning to the Moon. His presence seemed to have a powerful and even emotional effect on the Federal legislators that were at the hearing.

Comment by Marcel F. Williams — May 26, 2010 @ 5:06 pm

37. The problem with the VSE is that putting some people on the Moon won't get us much closer to getting value out of space. It will cost a bundle, though. The estimate just out is over $200 billion up to the first flight.

There are things to do in space with a lot of potential value:

1. Tourism. The first sub-orbital flights may start in a bout a year. One vehicle is in flight test and another couple are in development. A lot of people would like to go to space (for example, me) and some are willing to pay a lot to go. The private orbital vehicles to be developed under the new space policy are the perfect next step.

2. Energy. Recent developments in solar power satellites may make them economically competitive with nuclear even with current launch vehicles. If SSP (space solar power) can get the price down this could be huge benefit economically, geopolitically and environmentally.

3. Zero-g manufacturing. We have a space station to do the research to develop 0g products. Now might be a good time to use it. Under the VSE out $100 billion space station was going to be dumped in the ocean in 2005.

Note that none of these need the moon, or even want it for a long time (SSP eventually might use lunar materials). Furthermore, if what you want is materials, the near earth asteroids have much more valuable materials (lots of water, carbon and metals) in more easily accessible form and many NEOs are closer than the moon when distance is measured in energy to get there.

Comment by AL Globus — May 26, 2010 @ 8:07 pm

39. *"People can find fault with just about any long term plan or strategic direction"… "The history of NASA shows that they do well with specific goals and destinations"*

So, make short term goals. Something that can be done in the very short term is to turn all the NASA centers into federally-funded research and development centers like JPL.

Next, extend the TRL scale past 9, with TRL-10 being the publication of specifications and enactment of procedures and standards and regulations regarding the new technology's use, and TRL-11 being NASA's administration of the use of the new technologies by commerce and industry (concentrating on the second A in NASA).

As for the short term goals, here are about two dozen enabling technologies, some of which could be finished in two or three Congressional terms and some of which will have to wait a bit, which would apply logistical and economic leverage. Any one of them makes space operations easier and cheaper. Even having just a working propellant depot alone will apply enormous leverage – think of it as taking the next logical step in the idea of staging.

NASA can apply leverage to its R&D dollars to these technology development programs, by expanding the Centennial Challenges program and possibly offering further contracts to Challenge winners. Some of this is already happening with the astronaut glove challenge (and apparently the lunar lander challenge although no official word is out yet).

In other words, the idea is to turn NASA into a catalyst for future space endeavors, rather than having to shoulder the lion's share of the work indefinitely.

Comment by Ed Minchau — May 27, 2010 @ 1:52 am

40. AL,

The problem with the VSE is that putting some people on the Moon won't get us much closer to getting value out of space. It will cost a bundle, though. The estimate just out is over $200 billion up to the first flight.

First, that's not all we would get from doing the VSE correctly — the purpose of lunar return is not "to put a few people on the Moon" but to open up the space frontier through the use of off-planet resources. Creating sustainable human presence requires that we learn these skills eventually, so why not now?

Second, we're going to spend this $200 billion anyway over the next 10 years! With the VSE, we'll make progress on sustainable presence. With the administration's plan, we'll get nothing.

Comment by Paul D. Spudis — May 27, 2010 @ 4:42 am

42. Ed Minchau,

In other words, the idea is to turn NASA into a catalyst for future space endeavors, rather than having to shoulder the lion's share of the work indefinitely.

I fully understand that this is the belief of New Space advocates who support the administration's new path. I simply do not agree with them that such will result from it.

In my opinion, technology development without goals or destinations results in the creation of nothing, not new missions or space faring capabilities. We will still spend ~$ 20 billion per year on NASA for the next decade or so, but have nothing to show for it except piles of paper studies and some really pretty artwork, showing human missions beyond LEO "sometime" in the future.

I base this opinion on the last 30 years of agency history.

Comment by Paul D. Spudis — May 27, 2010 @ 4:57 am

43. Paul,

The Shuttle led us into a quagmire which resulted in the failure of constellation. We have been stuck for decades with no real technological development. The new path of technological development first, is easily fixable, the goal of Mars is obviously not important to Obama, but the technologies his plan develops will work equally well with the Moon. Sure this will delay a return to the Moon by a few years (maybe since the constellation program was making little progress), but when we do get there we will have a much broader of technologies which will help us develop the Moon. None of these technologies would be available if we stayed on the Constellation path.

The key is to make sure we get hardware experiments not just paper. That the technologies are not Mars specific at the expense of the Moon. That we keep the development of heavy lift, long term life support, and in situ resource utilization in the mix. Once we have a new generation of launch vehicles which are cheaper and more reliable and a portfolio of new technologies going back to the Moon will be easier and cheaper.

The Moon is the key to developing space. Orbital fuel depots make little sense to do with terrestrial fuel. They make a lot of sense with lunar fuel. There are currently Helium-3 shortages on Earth and Helium-3 fusion could be an important long term power source. Space solar power is our best hope for long term clean energy and the Moon is the ideal place to manufacture solar power satellite parts in the long term.

Mars on the other hand will never be more than a subsistence colony. It is somewhat scientifically interesting but the will never lead to development. Its main appeal is in boyhood fantasies. We need a mission to Mars to get it out of their systems.

The Moon is will be the heart of space commercial activity in a hundred years. To do that we need new technology and to be more business like about space, Obama is taking us down that path. He is not perfect. He picked the wrong destination and his administration is ignoring space solar power. The destination is not important to Obama's plan. A few years of heading for Mars and it will start being obvious why we need to go back to the Moon and stay.

Comment by Karen Cramer Shea — May 27, 2010 @ 10:37 am

44. Karen,

To do that we need new technology and to be more business like about space, Obama is taking us down that path.

I do not agree. I think that we are headed down a very different path. The new plan is all about keeping the agency engaged in bureaucratic "busy work," industry happy with new government contracts, the scientists bought off with a few robotic missions, and the public misled by the idea that we're going to some (distant) somewhere in the (all-too-distant) future. In short, it is only the facade and simulacra of a space program, all smoke and mirrors and no substance.

Comment by Paul D. Spudis — May 27, 2010 @ 11:12 am

45. Paul,

This path will shed us of a lot of bureaucracy. This path will lay down a foundation for real space development.

Comment by Karen Cramer Shea — May 27, 2010 @ 11:26 am

46. Karen,

This path will shed us of a lot of bureaucracy. This path will lay down a foundation for real space development.

What is your basis for these beliefs?

Comment by Paul D. Spudis — May 27, 2010 @ 11:28 am

47. Paul,

How many NASA contractors are about to be laid off? Obama is getting rid of layers and layers of bureaucracy associated with launch by contracting launch out to the private sector. The private sector is trying many new launch technologies.

Obama is also investing in new heavy lift launch technology, Long term life support, in situ resource development, all these things are needed for lunar development.

There is a set amount of money the public is willing to spend on space. Technology development costs money. The shuttle architecture is expensive to use and has blocked us from investing in new technology development for decades. Taking a break from pushing outward for a few years will allow us to develop the technology to go outward for less cost. It is hard to be patient especially as all of who remember Apollo are getting old. But Obama is leading us down the quickest path to actual lunar development.

Why don't you believe that the Obama plan will lead to new technologies? Do you not trust NASA to do what Obama is calling for? Do you not trust the private sector?

Comment by Karen Cramer Shea — May 27, 2010 @ 12:02 pm

48. *Taking a break from pushing outward for a few years will allow us to develop the technology to go outward for less cost.*

That's a statement of faith, not fact. By "developing technology" in the absence of an identified specific need or envisioned mission, we end up developing many things that will never find application in actual spaceflight. Moreover, by eliminating real flight programs that exist now for the promise of flight programs that may never happen, we are dispersing personnel and destroying industrial capability that cannot be re-assembled later except at great cost and expense, if at all.

Why don't you believe that the Obama plan will lead to new technologies?

Because I've been to this movie before — Dan Goldin went down this path in the 1990's with numerous programs designed to develop "routine access to orbit" vehicles. Billions were spent, and a thousand viewgraphs bloomed, but no flight hardware ever resulted from it.

Comment by Paul D. Spudis — May 27, 2010 @ 12:10 pm

49. Paul,

There is a set of technologies which need to be developed to industrialize space regardless of destination. Obama's plan is to develop just that set of technologies. The standing army is the problem not the solution. Launch costs are high because we have a standing army to pay for. The only way to reduce launch costs is to figure out how to launch stuff into space without a standing army. Under Goldin the shuttle was still flying. Cheap access to space was not in the best interest of the standing army so they sabotaged the effort at every turn. Obama is getting rid of the standing army, putting all the incentives for NASA and the US aerospace industry to make COTS work.

Yes, it is an act of faith to trust that by letting go of the shuttle technology we will be able to develop a new generation of launch vehicles. But the cost and safety issues of the shuttle architecture will surely delay real space development as long as we use them.

We are definitely taking a leap of faith but it is a leap we have to take if we are going to grab the brass ring of lunar development. We will never reach the ring if we are weighed down by the requirements of the outdated standing army. I trust that Elon Musk can deliver lower cost access to space. I have faith in his abilities because of his track record. I have faith that Boeing and Lockheed Martin will figure out how to make money delivering what ever the President wants.

Comment by Karen Cramer Shea — May 27, 2010 @ 12:37 pm

50. If that's what you choose to believe, fine with me.

Comment by Paul D. Spudis — May 27, 2010 @ 12:41 pm

51. Moon Conn-young, Arirang News.

MAY 27, 2010 Japan Draws Plans to Build Research Center on the Moon

Japan, also known as the 'Land of the Rising Sun' is looking to have its hands on the moon. According to local media reports, the Japanese government unveiled its plan to send a robot to the moon in five years time and build an unmanned base there in the next ten years. The island nation is expected to invest 2-hundred billion yen, or 2.2 billion US dollars in the space mission.

The plan, to be carried out in two phases would send a mobile robot to the moon by 2015, which would send video images of the surface and analyze the moon's inner structure with a seismometer.

The following five years will be devoted to setting up a self-powering base, from which the robot would explore the surface within a 1-hundred-kilometer radius.

Samples of moon rocks will be sent back to Earth for further study.

Japan's recent move towards the new space mission is seen as an attempt to secure its position on the moon before China and India, who are also rapidly moving forward with their own lunar research over future usage of Earth's only natural satellite for peaceful purposes.

Comment by Marcel F. Williams — May 27, 2010 @ 4:17 pm

52. *"technology development without goals"*

Paul, what I'm saying is that technology development itself should be the goal, just as it is in the first A in NASA.

Comment by Ed Minchau — May 27, 2010 @ 4:17 pm

53. @AL Globus

1. The Moon could be one of the primary destinations for the emerging space tourism industry. In fact, in the polls that I've taken, the Moon is the preferred destination for space tourist.

2. Supplying reusable tugs with oxygen and hydrogen from the Moon would reduce the cost of transporting satellites– including solar power satellites– to geosynchronous orbit.

3.Capturing small asteroids and transporting them to the Lagrange points with light sails only requires the free energy of the sun. Still, the average asteroid probably only contains about 2% hydrogen. So you might have to import 1000 tonnes of asteroid material per year to manufacture 20 tonnes of hydrogen. But you could probably export 20 tonnes of hydrogen to a Lagrange point from a polar lunar base every month using a reusable Altair vehicle.

Oxygen derived from asteroids, however, would probably have a clear advantage over oxygen from lunar resources unless mass drivers would utilized on the lunar surface.

Comment by Marcel F. Williams — May 27, 2010 @ 5:16 pm

54. @Karen

Trying to develop a commercial manned spaceflight industry based on government contracts is a road to nowhere. There's simply not enough manned space missions commissioned by the Federal government to sustain more than one company.

Elon Musk needs to stop begging for tax payer dollars and start focusing on space tourism.

The Federal government should help develop a market for space tourism by starting a Space Lotto system so that average folks in the US and around the world can finally get a chance to travel into space aboard an American commercial launch vehicle. There's no doubt in my mind that there are billions of people around the world who would be willing to spend a few meager dollars every year for a chance to travel into space and especially to the Moon.

Comment by Marcel F. Williams — May 27, 2010 @ 5:53 pm

55. Ed,

what I'm saying is that technology development itself should be the goal, just as it is in the first A in NASA.

I understand you. My response is that we get relevant technology development when the agency has clear goals and a destination, but we get irrelevant development in the absence of such goals. I am also saying that declaring that you are "going to Mars" sometime in the distant future is not a goal; it's a motherhood statement.

Comment by Paul D. Spudis — May 27, 2010 @ 5:57 pm

56. Paul, I'd like to ask a favour.

Would you consider writing a piece on "The Commercial Advantages of Utilizing Lunar Regolith for the Commercial Planetary and Near-Earth Market"? Obviously, when I mean "Commercial Planetary" I mean the Earth's consumer market and obviously not an unpopulated planet anywhere else.

I realize that the cancellation of the VSE appears to be a rebuttal of common sense, but I believe that that's an easily potential misconception.

As always, education is the key, so my genuine question that I'd like you (kindly) to address for everyone's comprehensive benefit is, based on existing assets (Moon and ISS), within the framework of the above questioning article's title, how does one:

a) Fully understand the Lunar resource's commercial potential?

b) How does one exploit the presence of the ISS to develop this endeavor?

c) And combining both a) and b), what is the best way to process Lunar regolith on board the ISS for evaluation purposes?

I personally am a great believer in the 'Pragmatic Approach' and it's on this basis that I pose these issues to you Paul.

Comment by Marcus — May 27, 2010 @ 6:37 pm

57. Marc,

The Commercial Advantages of Utilizing Lunar Regolith for the Commercial Planetary and Near-Earth Market

I have written on several occasions about the commercial potential of the Moon. My personal belief is that for the near term, it lies almost completely with harvesting and processing water, specifically for export and use as propellant in cislunar space. Regolith per se does not have much economic value, unless you can find a cheap way to hurl it off the Moon for use as building and shielding material in cislunar space.

Water mining on the Moon creates a logistics train that can supply a cislunar transportation system. Once we have that, we can not only routinely access all of cislunar space (where all of our satellite assets reside) but we will also have a system that can take us to the planets.

Dennis Wingo, Gordon Woodcock and I elaborate on these ideas here:

http://www.spudislunarresources.com/Opinion_Editorial/status_quo.pdf

In addition, I outline the strategic economic value of the Moon here:

http://www.spaceref.com/news/viewnews.html?id=1376

Finally, I refer you to my general web site, where I have many pieces relevant to your topic of interest:

http://www.spudislunarresources.com/index.htm

Many thanks for your comment and reading the blog!

Comment by Paul D. Spudis — May 28, 2010 @ 4:16 am |Edit This

American Heroes

May 28, 2010

Neil walks on the Moon, July 1969

Memorial Day weekend is upon us, so thoughts of heroes and remembering them are foremost in my mind. As a kid growing up in the Sixties, I saw a lot of change in our country. There was upheaval and tension here at home and around the world but the U.S. space program was a shining light that inspired many of us. America was going to the Moon. My friends and I dreamed of going and did the next best thing by launching rockets in vacant fields, excited and inspired by the idea of going into space and to the Moon. Astronauts who flew into space, braving certain death in exploding rockets to fight the Soviets for control of the heavens, were our heroes.

The decade following was disappointing in many respects, but none more so than our apparent retreat from space. In my exuberant but ignorant youth, I did not realize that Apollo wasn't about space but rather about geopolitics right here on Earth. America had won the war of space supremacy. Unaware of the implications of what that reality meant to manned space exploration, we pressed on, eagerly preparing for a future that simply was

not to be. But our feelings for those men who braved the unknowns of space remained true. They blazed the trail, setting us on our course, and they hold a special place in our hearts and memories.

The Apollo astronauts were a varied lot and all of us had our favorites — and a few we didn't particularly care for. But we admired all of them. They did much more than simply play Russian roulette with rockets. All of them were technically trained people, keen sharp guys with thorough educations and long experience in handling, managing and using advanced technology. The Apollo astronauts were intimately involved with the design of their spacecraft; they were not "button-pushers" or "appliance jockeys," or "spam in a can." They knew the principles of how their systems worked and could adapt and improvise when things went wrong. They had the "Right Stuff."

Listening to Neil Armstrong, the first man on the Moon, testifying before the House Space Subcommittee[1] the other day brought back so many memories. Although I know many of the

Apollo astronauts personally, I have only met Neil Armstrong once, very briefly at a technical meeting. I was struck by his testimony during this House committee hearing. His words rang so familiar and true. I have said many of these things myself over the years and most recently argued for them in this blog.

Neil Armstrong is not only a famous, experimental test pilot, he has decades of experience in aerospace engineering and in the management of complex technical projects. He faced critical life-and-death decisions on both of his spaceflights. In 1966, his Gemini spacecraft malfunctioned[2], sending him and co-pilot Dave Scott tumbling end over end, out-of-control while they were near another space vehicle in Earth orbit. His piloting skills brought the vehicle under control, ending the mission early but saving his and Dave's lives. Three years later, as his Lunar Module continually rang out with program alarms[3] of unknown origin, he coolly guided his vehicle over a crater full of large boulders[4] before making a soft touchdown on the Moon for the first time in history – all with less than 20 seconds worth of fuel to spare! Because his training and experience gave him the ability to decisively, competently and quickly weigh his options, his corrective actions saved the mission.

His testimony before Congress reflects a grave concern over the "new direction" proposed for NASA[5]. He believes it is a mistake to abandon the Moon and the Vision for Space Exploration without a thorough review of all options and alternatives. He makes the case[1] that the Augustine committee[6], whose report allegedly is the basis of the new direction, was configured and given terms of reference in such a way as to assure that some options would be found untenable. Specifically, that the cost estimates provided by The Aerospace Corporation[7] and used by the Augustine committee to support the case for commercial transportation are unjustified and unsupported by serious analysis, a critical point that others have also made[8].

Armstrong is particularly mystified by the President's casual and unconsidered dismissal of lunar return on the grounds that "we've been there." And oh yes, he is also very aware that "Buzz has been there" – he was there with him. Armstrong made an analogy about this change in direction, using the example of courts of Europe in the early 1500s dismissing new trips to the Americas on the ludicrous grounds that, "We've been there."[9]

The Moon is a continent-sized landmass, where we have touched only six spots near the equator on the front side. Moreover, we have found that the polar areas of the Moon are even more interesting and useful than we had ever imagined or hoped for. The water found in the polar cold traps could enable the building of an extensible space transportation system, giving us access to all of our space assets in cislunar space, as well as taking us to the planets. We need to return to the Moon for fundamentally the same reasons Europe returned to the New World – for prosperity, knowledge and the expansion of our civilization – the very reasons so many other space faring countries are currently working toward building a base on the Moon.

Neil Armstrong had a distinguished career serving his country in space on his Gemini and Apollo missions. He seldom speaks out on public policy, so his emergence in this debate on our national space program is thus significant. By coming forward to warn us how this proposed new direction will eliminate our nation's space faring capabilities and leadership in space, he is once again striving to hold a steady course for the good of our country. This warning comes from someone with considerable experience in space engineering, as well as a former employee of an agency that he knows well, for both its strengths and its weaknesses. We should reflect on and consider his counsel carefully. He is a tried and true American hero.

Topics: Lunar Exploration, Space Politics, Space and Society

Links and References

1. Armstrong testimony, http://democrats.science.house.gov/Media/file/Commdocs/hearings/2010/Full/26may/Armstrong_Testimony.pdf
2. Gemini 8, http://en.wikipedia.org/wiki/Gemini_8
3. program alarms, http://history.nasa.gov/alsj/a11/a11.1201-pa.html
4. crater full of boulders, http://lroc.sese.asu.edu/news/index.php?/archives/101-Apollo-11-Second.html
5. new direction, http://www.nasa.gov/news/budget/index.html
6. Augustine committee, http://www.nasa.gov/offices/hsf/meetings/10_22_pressconference.html
7. The Aerospace Corporation, http://www.aero.org/
8. a point others have made, http://blogs.airspacemag.com/moon/2010/03/11/stuck-in-transit-%e2%80%93-unchaining-ourselves-from-the-rocket-equation/ See Chapter 51.
9. "we've been there" http://www.nasa.gov/news/media/trans/obama_ksc_trans.html

Comments

1. [...] This post was mentioned on Twitter by Robert, Holly Lamb. Holly Lamb said: American Heroes–a beautiful piece by

Air&Space mag #space http://bit.ly/adp8Y5 (via @spacefuture) [...]

Pingback by Tweets that mention American Heroes | The Once and Future Moon -- Topsy. com — May 28, 2010 @ 4:12 pm

2. Why do so many people equate ending the development of two particular rockets with abandonment of the Vision for Space Exploration? If anything, the Constellation program was progressing in the exact opposite direction of the VSE and the Aldridge commission and even the Space Act of 1958 (amended).

In fact, the Space Action section 203a, Functions of the Administration, parts (4)[seek and encourage, to the maximum extent possible, the fullest commercial use of space] and (5) [encourage and provide for Federal Government use of commercially provided space services and hardware, consistent with the requirements of the Federal Government] prohibit development of the Ares 1 given the existence of the Atlas and Delta.

The Vision for Space Exploration was not about one particular destination. It was Moon, Mars and Beyond. It isn't a sprint, it's a marathon. You prepare differently for a marathon than a sprint. And, you conduct a marathon race differently too, with depots along the way for athletes to replenish their supplies of water and nourishment.

I posted a list of technological stepping stones to space that NASA and industry could do in a reasonable amount of time and for a reasonable budget. Of all the things on that list, propellant depots are probably at the highest technology readiness level right now. And I know Buzz Aldrin had a company that spent years working on a flyback liquid fueled strap-on booster (his was a manned idea, I'm proposing using UAV technology) to replace solids.

We still do Moon, Mars, and Beyond. However, instead of just picking just one and sprinting, we set up the technological base and beginnings of infrastructure needed to do the Moon AND Mars AND Beyond.

In fact, just getting the propellant depots in orbit and getting flyback sidemounted stages alone would quickly facilitate the establishment of a moon base, much faster than developing Ares 1 and 5. It establishes a supply chain – real trailblazing.

And instead of developing two brand new rockets for a limited number of missions, NASA

engineers can spend the same 40 or so billion on things like ISRU and substantially enclosed life support systems and a lunar lander that refuels in lunar orbit and make each future mission easier and less expensive.

You can have it good, or fast, or cheap, pick two. We've spent plenty of time on the good and fast – but for the long run we'll have to pick good and cheap. Which means we have to stop sprinting, and spend the time (which is going to pass by anyway) preparing to do this properly for the long haul.

Comment by Ed Minchau — May 28, 2010 @ 5:27 pm

3. *Why do so many people equate ending the development of two particular rockets with abandonment of the Vision for Space Exploration?*

Because the Vision was abandoned by the new budget and then specifically repudiated by the President last month.

That was your rhetorical question. Here's mine:

Given the history of the last 20 years of this agency, why do so many in the space community think the "new direction" will result in anything useful whatsoever?

Comment by Paul D. Spudis — May 28, 2010 @ 5:54 pm

4. "Insanity – doing the same thing over and over expecting different results" – Albert Einstein.

If NASA does the next 20 years the way they've done the last 20 years, we won't get different results. Something has to give.

Comment by Ed Minchau — May 28, 2010 @ 5:59 pm

5. *If NASA does the next 20 years the way they've done the last 20 years, we won't get different results.*

The "new path" is exactly that — a decade of paper studies and viewgraph making, accompanied by promises of trips beyond LEO "sometime in the future."

Comment by Paul D. Spudis — May 28, 2010 @ 7:47 pm

6. Neil Armstrong was a magnificent hero, who is now lending his voice to the grand cause of America once more making great strides in space, via utilizing the Moon as the Antarctica of tomorrow. Project Constellation

merely STARTS OUT resembling Apollo, but will move far beyond merely reaching Luna by manned lander craft. Sortie missions will BE THE BEGINNING. Indeed, two flights which will resemble Apollos 9 & 10, will need to be mounted, in order to test out the dual spacecrafts, in both Earth orbit & Moon orbit. The Orion-Altair astronauts will accomplish tremendous things on the Moon's surface and in Lunar orbit! The expanded science that we could achieve, through these expeditions will be dramatic, plus the surveying of the Satellite's natural resources will make the effort so very worthwhile! Bases will be built, utilizing the Altair L-SAM in an unmanned cargo softlander variation. The Orion CEV will also be capable of long, unmanned phases of low lunar orbit flight, during multi-month long surface stays of the main crew; and will be useful for aerial-type of scientific observation of the Moon, both while attended with a crewman or not. In short: PROJECT CONSTELLATION GETS THE BALL ROLLING AGAIN, after 40 years of NASA being trapped in Low Earth Orbit. I support enthusiastically this effort of Returning To The Moon!

Comment by Chris Castro — May 28, 2010 @ 10:40 pm

7. Well, then, we need a different approach. The Aldridge commission recommended turning all NASA centers into FFRDCs like JPL, which would be a heck of a good start. Expanding the Centennial Challenges to provide leverage on R&D dollars would be another good start.

I think we are in agreement that PowerPoint engineering is no way to run a space program.

Comment by Ed Minchau — May 29, 2010 @ 12:01 am

8. One thing that has astonished me is the line by ObamaSpace supporters that goes like this: "Neil Armstrong? What does he know about—well—space exploration?" Thanks for providing the answer.

Comment by Mark R. Whittington — May 29, 2010 @ 12:05 am

9. My respect for Neil Armstrong grew when I watched him on television, many years ago, actually host an excellent PBS television miniseries produced by the BBC 'The Voyage of Charles Darwin'.

Mr. Armstrong is a great pioneer who apparently also appreciates the great adventurers and scientific pioneers who came before him.

Comment by Marcel F. Williams — May 29, 2010 @ 3:20 am

10. @ Ed Minchau

Actually, Obama is doing to Bush's program (Constellation) what Bush did to Clinton's program (X-33) and what Nixon did to the Kennedy-Johnson program (Apollo). So canceling programs is nothing new.

No new 'game changing' technologies are required to return to the Moon. And there were plenty of cheaper architectures than the Ares I/V for returning to the Moon. But Obama decided that returning to the Moon wasn't important and decided to choose none of them!

Comment by Marcel F. Williams — May 29, 2010 @ 3:31 am

11. No new 'game changing' technologies are required to return to the Moon if we want to do it rarely and as expensively as possible.

FIFY.

Comment by Ed Minchau — May 29, 2010 @ 4:31 pm

12. "Given the history of the last 20 years of this agency, why do so many in the space community think the "new direction" will result in anything useful whatsoever?" – Paul Spudis

"We see the Obama space policy as rescuing human spaceflight, allowing the private sector to take over low Earth orbit and allow NASA to go push the envelope, and do what NASA does best." – Michael Gold, Director of Washington DC Operations for Bigelow Aerospace (emphasis mine)

http://www.technologyreview.com/blog/guest/25255/

This is why, Paul. It has the potential to be a fundamental change in the way NASA operates. In fact, there is no way NASA can continue to operate the way it has over the last 20, 30, 40 years.

Marcel is right, canceling programs is nothing new. Multi-decade tens-of-billions-of-dollar programs that have to be funded through ten Congresses and sustained through at least three Presidential administrations will almost certainly follow the same fates as Space Station Freedom and NASP and X-33 and even Wernher von Braun's post-Apollo vision – in the end, PowerPoint hell.

However, a few programs totaling a few billion a year (taking the place of the Ares-1 and Ares-5 funding) that produce a few tangible stepping-stone results and expand the industry every Congressional term – that's a different NASA. And that's a different next 20 years.

Comment by Ed Minchau — May 29, 2010 @ 5:28 pm

13. Obama is the WORST President, with regard to manned spaceflight!!! If McCain would have gotten elected, none of this rhino-dung would be happening right now!! Project Constellation would have simply gone forward. Mr. Obama is thoroughly controlled by the Anti-Moon people & the Trekkies. There is NO hope of him backing off from this suicide march order he has given to NASA. It is all in Congress's corner now. May the wise statespersons there see it fit to rescue NASA's future, by restoring Constellation, and getting us started with prospecting the Moon's wealth of natural resources!

Comment by Chris Castro — May 30, 2010 @ 12:51 am

14. Ed,

allowing the private sector to take over low Earth orbit and allow NASA to go push the envelope, and do what NASA does best."….. This is why, Paul. It has the potential to be a fundamental change in the way NASA operates.

I know of no particular reason to give any credence to your "why" statement. People have been trying to reform NASA for years, laying out new management strategies to make the agency "leaner," more efficient, and more innovative. Didn't seem to take, did it? And in contrast to what you believe, the path of "technology development" HAS been tried before — NASA in the 1990's was focused on "technology development" in Earth to LEO transportation, with several programs designed to "revolutionize" the system and lower costs. We got nothing for it but a pile of paper studies.

I keep hearing the same thing from the advocates of the new path — it is indeed the triumph of hope over experience.

But even all this is secondary to my fundamental point. With the Vision, we had a real strategic direction, with goals, mileposts, a schedule, and a strategy to achieve them. All that has been discarded in the new path and not for a different set of goals and missions, but for nothing. You can harp on the mantra of technology development and lower costs for LEO access all day but in the final analysis, NASA will not change — they ARE a mission-oriented agency and that's what they do, for good and ill. The new path is a prescription for random activity, Brownian motion on a cosmic scale.

Comment by Paul D. Spudis — May 30, 2010 @ 6:31 am

15. In the 1960s "beating the Soviets" was the geopolitical imperative facing JFK and the US. Nonetheless, the US still signed the OST of 1967 to disclaim any intention of seeking control over lunar resources.

Today seeking American pre-eminence in controlling lunar resources is something that certainly seems to lack traction inside the Washington Beltway. Buzz Aldrin said as much last October in his Huffington Post piece.

Seeking to change such attitudes strikes me as being a more difficult task than putting a man back on the Moon. Even if Congress rejects FY2011 in toto, I cannot foresee Obama's NASA being willing to implement Marburger's plans for Luna.

Lunar advocates can bemoan and decry the politics of all this or face such political constraints, pivot, and design lunar architectures that embrace the geopolitical realities (and domestic politics) of the 21st century.

This is why I have been touting an EML-1 Gateway (depot and transfer station) deployed to support global lunar surface access without launch window constraints or concerns about lunar orbital inclinations. If operated by a "neutral" power (Singapore, for example) this facility could buy and sell materials and services to every current and potential spacefaring nation as well as to and from private companies.

Supplies would be sent from Earth to the Gateway by one of several "slow boat" trajectories, creating a logistics destination (depot) as desired by the NewSpace community in order to stimulate RLV development. In turn, logistics support would be provided to (sold to) anyone wanting to put boots on the Moon or establish facilities in one of the many craters that potentially contain ice.

Given the apparent distribution of ice in various discrete and easily distinguished craters various nations and various companies can be assigned primary responsibility for a particular crater allowing competition in the development of those resources even as the Gateway offers collaborative logistics support to all. The Gateway would also be available to purchase, stockpile and re-sell ISRU fuels.

Perhaps a 21st century "Great Game" could be stimulated with various nations wrangling for the prestige that would come from having the most productive lunar facility. Such commercial and scientific competition would occur under the aegis of a collaborative EML Gateway operating agreement.

And remember, Christopher Columbus wasn't Spanish.

Comment by Bill White — May 30, 2010 @ 8:31 am

16. PS — Yesterday at ISDC Robert Zubrin made the same point as this (and other points as well, for good or ill):

You can harp on the mantra of technology development and lower costs for LEO access all day but in the final analysis, NASA will not change — they ARE a mission-oriented agency and that's what they do, for good and ill. The new path is a prescription for random activity, Brownian motion on a cosmic scale.

I would add that NASA also is not a mining company and therefore expecting NASA to lead the way in developing the Moon's resources could also be expecting too much.

Comment by Bill White — May 30, 2010 @ 8:38 am

17. Bill,

I would add that NASA also is not a mining company and therefore expecting NASA to lead the way in developing the Moon's resources could also be expecting too much.

I have never "expected" that. My point is that as there are significant technical and operational issues with and questions about lunar ISRU, it is appropriate for the federal government to conduct research on the task. In other words, it is not NASA's job to industrialize the Moon — it is to determine if the Moon can be industrialized.

Comment by Paul D. Spudis — May 30, 2010 @ 1:17 pm

18. *And no game changing technologies are required if you don't want to go to the Moon at all.*

"Now, I understand that some believe that we should attempt a return to the surface of the Moon first, as previously planned. But I just have to say pretty bluntly here: We've been there before." President Barack Obama on April 15th, 2010.

Comment by Marcel F. Williams — May 30,

2010 @ 2:45 pm

19. At the height of the Apollo development era, NASA was spending more than $33 billion a year in today's dollars. NASA is currently spending less than $18 billion a year while operating a space shuttle program, an international space station program, and while funding another return to the Moon program.

So if anything, NASA is a model of government efficiency compared to practically all other government programs whose expenses have grown– enormously– since the 1960s.

Comment by Marcel F. Williams — May 30, 2010 @ 3:18 pm

20. Tangentially related… I like the Japanese approach:

http://www.universetoday.com/2010/05/27/japan-shoots-for-robotic-moon-base-by-2020/

The Japanese are quite interested in robotics. It seems that here in the U.S., however, we're still infatuated with Human Spaceflight and where it is going or not going.

The Japanese have always led the world in little gadgets. They like technology. That seems to be their culture. I hope to see great things from them on the Moon.

Comment by Itokawa — May 30, 2010 @ 10:48 pm

21. Marcel,

You forgot to extend out the budget horizon and see what was happening with Constellation:

No Shuttle No ISS A couple of flights on the Ares I/Orion to test them out, but nowhere to go No Moon landings until somewhere in the 2020's Nothing left in space after the program was done.

At least with the Shuttle we have built an orbiting space station that attracts and requires astronaut researchers, and continues to expand our knowledge of how to live and work in space. This is something the new budget builds upon and expands.

Constellation swept away all that came before it, and was only a temporary program.

p.s. Paul, I had a post from last night that was deleted – what's up?

Comment by Coastal Ron — May 31, 2010 @ 11:59 am

22. Neil Armstrong is absolutely a national hero. You know who else is? Buzz Aldrin. His

service to his country is no less than Armstrong's, before, during, and after his NASA career. He also has a doctorate in astronautics from MIT, is credited with developing the Aldrin cycler trajectory for trans-Martian cyclers, and has made many other contributions to science and space policy over the decades.

Despite the respective accomplishments and service of these two great men, their views on engineering and space policy are not necessarily correct. They are just as prone to error as anyone else. While I am sure you understand this, your choice to write a tribute to the one who wants to send NASA-made rockets to the moon, while neglecting to offer any sort of honorable mention of the Apollo-11 astronaut who disagrees with you(especially given the very public disagreement between the two on this issue), strikes me as rhetoric in extremely poor taste.

If you had left off after paying homage to Armstrong's career and character, without tying everything in to your ongoing rhetorical war on anything that doesn't involve going to the moon, it would have been a perfectly appropriate gesture, especially at this time of year. It was unfortunate that you felt the need to play the character card to further your own arguments. If you feel, as your post clearly implies, that Neil Armstrong's character and record make his opinions on NASA policy worthy of exceptional consideration, then Aldrin's own writings on the same subject are worth much the same, and certainly deserve more than the cursory dismissal that they have so far received on this blog.

Comment by Jared — May 31, 2010 @ 1:08 pm

23. Ron,

I had a post from last night that was deleted – what's up?

Don't know — didn't see it. Sometimes the spam filter of WordPress catches normal posts in addition to the spam, even from people who've had other posts approved.

But as long as this topic has come up now, I should mention that I approve all posts that appear here and sometimes I do not post those that are off-topic, repetitious or irrelevant. Sometimes we've reached the end of our discussion because the same points are being repeated to no productive end. I want discussion and debate, but I want it civil, reasonable and on-point. If this policy is objectionable, you are free to post at your own blog.

Comment by Paul D. Spudis — May 31, 2010 @ 6:29 pm

24. Jared,

your choice to write a tribute to the one who wants to send NASA-made rockets to the moon, while neglecting to offer any sort of honorable mention of the Apollo-11 astronaut who disagrees with you(especially given the very public disagreement between the two on this issue), strikes me as rhetoric in extremely poor taste.

In case you missed it, the title of this blog is "The Once and Future Moon." I post on topics that interest me from the perspective of lunar science, exploration and utilization. I am not obligated to provide your arbitrary idea of "balance" on this blog nor do I believe that all viewpoints are equally valid. On the issue of the proposed "new direction" for NASA, I believe that some are correct and some are wrong and I will not stop from discussing which is which on this blog. If you are offended by this "poor taste," you are free to seek enlightenment elsewhere.

Neil Armstrong is absolutely a national hero. You know who else is? Buzz Aldrin.

You praise your heroes and I'll praise mine.

Comment by Paul D. Spudis — May 31, 2010 @ 6:35 pm

25. Paul,

Rules are good, and yours are quite acceptable. Maybe a stray neutrino changed the storage bit. Luckily it was not my ultimate opus, so civilization has not lost much… ;-)

Comment by Coastal Ron — May 31, 2010 @ 8:02 pm

26. Save you looking Paul,

Why do so many people equate ending the development of two particular rockets with abandonment of the Vision for Space Exploration?

Because the Vision was abandoned by the new budget and then specifically repudiated by the President last month.

That was your rhetorical question. Here's mine:

Given the history of the last 20 years of this agency, why do so many in the space community think the "new direction" will result in anything useful whatsoever? —

I've read the words/comments posted by a

number of people on your blog over a number of articles and you're very fortunate to have such a erudite response. I read the information that you directed me to previously and I was unable to find what I was looking for, which surprised me.

Within this next decade, professionals in the field of Lunar Resources will need to engage constructively with their prospective domestic and international customers, giving the most astute advice money can buy.

My question, in light of the lead-in post, is will you be planning on engaging with these paying clients during this forthcoming decade, or have you other plans?

I am aware of an increase in 'non-lunar mineralogists' taking an interest in this field based purely on the issue of chemistry. And there's a lot of Chemists in this World Paul.

Kind regards,

Marc.

Comment by Marcus — May 31, 2010 @ 10:07 pm

27. @ Coastal Ron

You can't get me to defend the Ares I/V architecture since I was strongly opposed to it. There were many cheaper alternatives to the Ares I/V. I'm also not a big fan of space capsules.

Comment by Marcel F. Williams — June 1, 2010 @ 12:02 am

28. @Jared

The only problem with Buzz is that you never know what he's for or against from week to week.

Comment by Marcel F. Williams — June 1, 2010 @ 12:03 am

29. Marc,

I read the information that you directed me to previously and I was unable to find what I was looking for

I guess that depends on what you're looking for.

My question, in light of the lead-in post, is will you be planning on engaging with these paying clients during this forthcoming decade

I already am.

And there's a lot of Chemists in this World Paul.

Good to know.

Comment by Paul D. Spudis — June 1,

2010 @ 8:13 am

30. Paul:

Another great article!

I found a certain resonance in the comment at the House hearing last week that a thousand years from now, Armstrong will be the only person in the room who will be remembered.

Nelson

Comment by Nelson Bridwell — June 1, 2010 @ 2:39 pm

31. Paul, I read your Washington Post article with Dr. Zubrin, and your posts regarding the cancellation of Ares, and I can't make that square with your 2007 ISDC presentation. And I truly don't understand your objection to this change in direction.

I don't understand, for instance, how you figure that "(b)y adopting the new program, we will lose – probably irretrievably – this space-faring infrastructure and, most certainly, our highly trained, motivated and experienced work force."

I mean, how is the infrastructure such as launch pads lost if SpaceX and ULA and other companies (and the USAF and DoD) are actively using the pads and associated facilities? And if, for instance, the NASA centers are all turned into FFRDCs like JPL as the Aldridge commission recommended, how will NASA lose the best of their workforce?

And if asteroids are an eventual goal, then you're not talking about a sprint, you're talking about being able to assemble large movable structures in orbit, you're talking about propellant depots, staging areas at L1, ISRU on the moon for propellant production, a permanent lunar base – infrastructure necessary to extend the human reach into the solar system on an ever larger and larger scale.

I just don't understand why removing what you viewed as the major obstacle in 2007 (Ares) and actually enabling the "incremental, cumulative program" you called for then is viewed so differently today.

Comment by Ed Minchau — June 3, 2010 @ 2:15 am

32. Ed,

I truly don't understand your objection to this change in direction.

I have explained precisely why I am opposed to it in this blog, specifically the posts here:

http://blogs.airspacemag.com/moon/2010/02/03/vision-impaired/

http://blogs.airspacemag.com/moon/2010/02/13/confusing-

In brief, I object to the abandonment of the Vision for Space Exploration, a logical and well considered strategic path. I contend that it has been replaced with nothing; a technology development program for a decade followed by the "promise" of one-off missions to distant destinations is not a space exploration program. I do not believe the hype by certain sectors of the commercial space sector that they can deliver services cheaply (note well: I am NOT saying that they cannot do it at all). If they could, they could obtain private capital to develop their systems and wouldn't need NASA money. Finally, I reject the conclusion of the Augustine report that implementing the VSE is impossible under existing budgets; they were presented with alternatives that would do exactly that — and dismissed or ignored them in their report.

I mean, how is the infrastructure such as launch pads lost if SpaceX and ULA and other companies (and the USAF and DoD) are actively using the pads and associated facilities?

Because right now, their sole customer is government and as I argue above, NASA isn't going to be doing anything in space — but they will be spending huge amounts of money right here on Earth.

if, for instance, the NASA centers are all turned into FFRDCs like JPL as the Aldridge commission recommended, how will NASA lose the best of their workforce?

Because that won't happen. It was rejected when we proposed it in 2004 and I hear nobody advocating it now.

if asteroids are an eventual goal, then you're not talking about a sprint,

I'm not, but they are — both Augustine and Bolden talk about a "steady stream of space 'firsts'" — flags and footprints forever. Obama's statement ("we've been there") implies that ever new, unvisited destinations are the object of NASA missions. How can you build up capability and sustainable human presence with such an approach?

In sum, I do not believe the hyped up rhetoric surrounding this "new direction." I've worked with this agency for over 30 years. I see what they do well and what they do poorly. Give them a budget and no specific direction or mission, and you will get nothing in return. It abandons the VSE, a logical path for which a very hard-won political, bipartisan consensus was achieved for a worthless promise of a human Mars mission sometime beyond 25 years in the future. It is quite literally — meaningless.

This is NOT the "incremental, cumulative program" that I advocated in 2007 — and still believe in. This is a set-up to cancel human spaceflight in the not-too-distant future.

Comment by Paul D. Spudis — June 3, 2010 @ 4:42 am

33. Unfortunately, the President's plan is all smoke a mirrors with goals set so far into the future that they are almost guaranteed to be terminated by future administrations.

There's no logical reason why we can't return to the Moon before the end of the decade even with the current relatively meager $18 billion a year NASA budget– if we adopt one of the substantially cheaper architectures.

A permanent Moon base is a logical first step towards Mars and beyond and a perfect destination for the emerging space tourism industry which I believe will eventually dwarf NASA as far as manned spaceflight traffic in cislunar space once it gets going– largely thanks to the pioneering efforts of our Federal space program.

Those interested in the case for the Moon should check out the excellent GiaSelene Moon Colony website which features videos that include our good blog host, Paul Spudis!

http://www.gaiaselene.com/index.html

Comment by Marcel F. Williams — June 3, 2010 @ 6:38 pm

35. Marcel,

There's no logical reason why we can't return to the Moon before the end of the decade even with the current relatively meager $18 billion a year NASA budget– if we adopt one of the substantially cheaper architectures.

Absolutely correct. The Augustine committee heard about these alternatives and ignored them in their final report. In their unseemly haste to terminate Constellation, they threw the VSE overboard with it (the lunar return part of the VSE might also have been a prime target for termination by the committee as well.) The administration used the Augustine report to give high cover to their new budget.

Comment by Paul D. Spudis — June 4, 2010 @ 5:35 am

36. The irony is that the administration actually increases the 2011 budget by about $2 billion a year on average over the 2009 budget over the next 5 years. This would make it even easier to fund a return to the Moon ($20 billion extra dollars over the next 10 years).

This makes me suspect that the administration

is setting NASA up for deep budget cuts by Congress in the future. Since there's no goal to build anything or to go anywhere, Congress wouldn't feel any pressure to maintain a $19 or $20 billion a year NASA budget. And Congress could then use the budget deficit as a reason to cut the NASA budget.

Comment by Marcel F. Williams — June 4, 2010 @ 10:07 pm |Edit This

37. How many Clementine-class satellites could have been sent to the moon for the roughly 9 billion so far spent on Ares? Which programs would have to be cut to keep Ares going – Would you cut Cassini? Would you cut Spitzer? Would you cut the ISS (which was the plan a year ago, abandonment in 2015, thus giving Orion nowhere to go until a moon mission in the 2020s)? Would you have cut everything else NASA does just to keep Ares 1 and Ares V development going? Because that's what keeping Ares 1 and Ares V going meant in just a few years – no funding for anything but Ares.

Honestly, those who are equating the end of the development program of these two particular rockets with the end of American manned spaceflight are quite simply incorrect. Constellation would still require years and billions more dollars just to get to where SpaceX got the other day with Falcon 9. Constellation's continued existence and drain on NASA resources would have just ended everything else NASA did eventually. I sometimes think that Griffin's goal with Ares was for NASA to fade into obsolescence.

Those who equate the Flexible Path with the end of the Vision for Space Exploration are implying that Constellation was somehow consistent with the VSE (or the Space Act). If anything, the Flexible Path is much more closely aligned with VSE than Constellation ever was. Persisting with Ares with its yearly 1-year schedule slip and wasted billions with Atlas, Delta, and now Falcon (nearly) available simply makes no sense.

NASA was once an organization that took what was largely a military technology and changed and adapted not only technologies but organizational structure – the group that sent men to the moon for the first time could not operate as an army and a new organizational structure had to be created. It is not outside of NASA's power to adapt itself away from an Apollo mentality and organizational structure and into a sustained space technology development role and envelope-expansion role as it does with Aeronautics.

However, I have watched NASA over the years fight any sort of change, as if NASA had perfected itself with Apollo 11, as if Apollo was the best that humanity would ever do – that we could not hope to surpass it except by re-doing it "on steroids".

Comment by Ed Minchau — June 7, 2010 @ 12:45 am |Edit This

38. *Those who equate the Flexible Path with the end of the Vision for Space Exploration are implying that Constellation was somehow consistent with the VSE (or the Space Act).*

No, I am not implying anything of the sort. You are buying into the "either this (PoR) OR that (FP)" logical fallacy. I have consistently advocated that the lunar part of the VSE is capable of being implemented under the existing budget; we simply trade cost for schedule. This is discussed extensively in a previous post:

http://blogs.airspacemag.com/moon/2010/03/24/value-for-cost-the-determinate-path/

You continue to miss my principal point: the problem with the new direction is NOT the Ares rocket, or commercial transport to LEO or even Flexible Path (if properly implemented) — it is that we have eliminated outright a clear and logical strategic direction in space for (literally) nothing. The object of the VSE was to return to the Moon to use its resources to create new space faring capability. Thanks to the current fleet of robotic spacecraft that are mapping the Moon from orbit, we now know that the lunar poles have the resources we need to do this in abundance and accessibility. But lunar surface return and resource use has been specifically eliminated from the critical path in the "new direction."

It is those of YOU who favor the new path that are advocating continuation the old Apollo paradigm — the launch of everything from the bottom of the deep gravity well of Earth. I'm the one who wants to fundamentally change the spaceflight paradigm, by returning to the Moon to harvest its material and energy resources, create a reusable, extensible space transportation system in cislunar space, and extend that system to the planets.

I reject the idea that the new path IS a strategic direction; it is an excuse for NASA to blither and do paper studies for a decade, then face cancellation for non-progress.

But believe what you want to.

Comment by Paul D. Spudis — June 7, 2010 @ 4:43 am |Edit This

39. @ Ed Minchau

NASA is what the politicians want it to be. If Nixon or Obama and Congress tell NASA that they can't go to the Moon then NASA can't go to the Moon. Its that simple!

Of course when NASA finally did get a chance to return to the Moon, they quickly forgot about the Moon base part and turned it into an Apollo redux program probably because some folks at NASA really wanted to go to Mars. And this caused Griffin to pick the most expensive architecture possible.

And then George Bush decided to underfund the program, focusing his priorities on his adventure in Iraq. And now President Obama wants to shut everything down and start from scratch with meager goals set so far into the future that they are practically meaningless.

Comment by Marcel F. Williams — June 7, 2010 @ 2:42 pm |Edit This

40. Paul, I guess what is frustrating me the most, and what keeps me coming back to this argument, is that we are actually damn near in agreement. I agree that ISRU is absolutely critical, particularly at the lunar poles, and that the Hydrogen signature is a complete game-changer.

But under Constellation we weren't getting lunar ISRU tests started until Ares V, something that is four or five Presidential elections away and eight to ten Congressional elections away. And in the meantime, everything else productive – Hubble, ISS, Cassini, Dawn, all the robotic precursor missions, everything would have to be cut to get Ares finished at all.

You want a sustainable NASA? Give it something to do where a Congressman can vote for funding on a stepping-stone project and have results so quick that the smiling Congressman can be present at the big splashy press conference just before the end of the next Congressional election or the next one after that. Contra Marcel above, it was not George Bush who "underfunded" NASA, it is Congress that controls the money.

That means small projects – enabling technologies that allow us to use the existing LEO launch supply market to get started doing the necessary steps to go anywhere.

The VSE wasn't just about ISRU, it was "the Moon, Mars, and Beyond" – not "the Moon then Mars then Beyond". That means development of everything from propellant depots to substantially-enclosed life support systems to lunar lander stages to ISRU to power beaming to a thousand other things.

By themselves none is as sexy as an Apollo or Hubble, but each is a critical technology that needs risk retired, just as much of the risk has been retired for LEO launch. And cumulatively the capabilities produced add up to a much more robust space (not just launch) industry – and therefore in the long run much cheaper ways of doing everything from lunar ice ISRU to a mission to Phobos. And cumulatively each contributes to the ease of development of further technological advances.

This is getting too long. I feel a lengthy blog

post coming on. I guess to summarize my position I would have to say that Constellation really was a mission to nowhere and that the Flexible Path allows us a chance at a mission to everywhere.

Comment by Ed Minchau — June 7, 2010 @ 6:45 pm |Edit This

41. Ed,

But under Constellation we weren't getting lunar ISRU tests started until Ares V, something that is four or five Presidential elections away and eight to ten Congressional elections away

I don't know how many more different ways I can say this, but I am NOT arguing for the Program of Record — Ares I and V, a Winnebago Orion, the whole schtick. My point is that "Constellation" as you call it COULD be configured to work and work under the existing run-out budget. And in contrast to your formulation:

The VSE wasn't just about ISRU, it was "the Moon, Mars, and Beyond" – not "the Moon then Mars then Beyond".

On the contrary, the original VSE was about exactly that — the Moon, THEN Mars and beyond. NASA and Griffin ignored their executive direction in implementing the Vision and substituted a "Mars uber alles," Apollo-on-steroids rocket-building entitlement. I have covered this at length in multiple places, first in the ISDC 2007 presentation that you referenced earlier, and second in this presentation (note the date — we've known about this issue for some time):

http://www.spudislunarresources.com/Papers/The%20Vision%20and%20the%20Mission.pdf

Why am I covering this yet again? Because I contend that *what NASA did to their original marching orders for implementing the VSE, they will similarly do to the "new direction."* They will continue to employ large numbers of middle managers. They will spend a ton of money, upwards of $18 billion US per year, indefinitely. They will do "studies." They will create technology road maps. They will award a thousand small grants and contracts, with a wide geographic distribution, not one of which really allows any significant technical accomplishment, but the more contracts that proliferate, the more it looks like they are accomplishing something. They will deliberate and debate and conduct working groups and management retreats at length.

And you will get nothing. Zero. Nada.

Let me close with two points. First, Constellation was just a name that the agency gave to the system they were making to go beyond LEO. It is not engraved on tablets from Sinai. It can be configured in any way we need to accomplish its mission. The problem was that the "mission" was not understood by the agency (or ignored deliberately, in some cases). Second, NASA gets a lot of money. Yes, they could always find ways to spend more. But $18 billion per year is a significant amount. I simply do not believe the Augustine propaganda that lunar return is "unaffordable." The existing program may be unaffordable (I contend that the Augustine report didn't even really demonstrate that), but that does NOT mean that lunar return is unaffordable. I have tried to show a path that works in this blog over the last 18 months. There is a way forward; for the reasons I have discussed here, I don't believe the "new path" is that way.

Comment by Paul D. Spudis — June 8, 2010 @ 4:37 am |Edit This

42. "Because I contend that what NASA did to their original marching orders for implementing the VSE, they will similarly do to the "new direction." They will continue to employ large numbers of middle managers. They will spend a ton of money, upwards of $18 billion US per year, indefinitely. They will do "studies." They will create technology road maps."

This really is the crux of the matter. You're saying NASA will fight this (in my opinion entirely necessary and ultimately unavoidable) change of direction and dither.

I can definitely see that happening. I've seen NASA do it several times over my lifetime.

This time, though, it's different. If NASA dithers for yet another ten years the agency will be split up and parts shut down. I'm 41 years old. Nobody my age remembers the first moon landings firsthand. I was a little kid when Skylab crashed. In ten years well over half the baby boomers will be retired.

If NASA has nothing to show for its efforts in ten years there won't be a NASA at all. It will be cut to keep the pension checks flowing for a few more days.

Perhaps NASA can change itself. Perhaps the Aldridge report recommendation of turning all NASA centers into Federally-Funded Research and Development Centers like the Jet Propulsion Lab would eliminate those middle managers. Perhaps a massive changeover of personnel is required to bring the average age back down to 27 or so.

Something's gotta give.

Comment by Ed Minchau — June 8, 2010 @ 5:40 pm |Edit This

43. *If NASA has nothing to show for its efforts in ten years there won't be a NASA at all.*

In a sane world, yes, but we don't live there.

You are also assuming that the "new direction" is a sincere effort to reform the agency. I see no evidence of that. Instead, I see voodoo about magic propulsion systems and 39-day trips to Mars, routine and cheap commercial access to LEO and bold explorations to destinations for which no one can articulate a valid scientific or utilization rationale. In short, I see the usual.

And guess what? NASA **won't** go away. It will continue to spend money and do nothing.

Comment by Paul D. Spudis — June 8, 2010 @ 6:21 pm |Edit This

44. The Obama policy is really just a continuation of what NASA has been doing since the end of the Apollo era– except with far fewer resources since he's decided to outsource even NASA's ability to access orbit.

Nixon took away NASA's ability to fly to the Moon and now Obama is taking away NASA's ability to fly to orbit, crippling an organization that changed America and the world for the better and served as the ultimate symbol that America can do anything!

What the President is doing to NASA is sending a powerful signal to the rest of the world that America truly is a nation in decline! And that powerful signal is going to hurt this country around the world. And even the emerging private manned spaceflight companies in the US, which I strongly support, will suffer in the long run from the crippling of this great government organization.

Comment by Marcel F. Williams — June 8, 2010 @ 11:19 pm |Edit This

45. Paul writes:

"The object of the VSE was to return to the Moon to use its resources to create new space faring capability. Thanks to the current fleet of robotic spacecraft that are mapping the Moon from orbit, we now know that the lunar poles have the resources we need to do this in abundance and accessibility."

If NASA can't or won't pursue this goal, savvy Americans will need to find another way to get this done.

Comment by Bill White — June 9, 2010 @ 11:33 am |Edit This

46. @Bill White

Since President Obama terminated that goal, I guess you're right. But Russia, Japan, China, and maybe even India may beat us to dominating those resources.

Comment by Marcel F. Williams — June 9, 2010 @ 10:26 pm |Edit This

A Wetter Moon Impacts Understanding of Lunar Origin

June 19, 2010

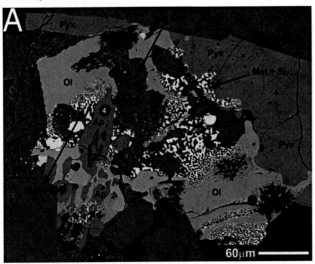

Microscopic view of a lunar sample, showing the various mineral phases present. Over 1300 ppm hydroxyl (OH) ater was found in the black circle marked "4."

Is there water on the Moon?

We know now that the answer to that question is a resounding Yes! As information continues to emerge from a wide range of studies, it's evident that we've just begun to understand the process of the creation, movement and history of water on the Moon and its prevalence.

A paper recently published in the Proceedings of the National Academy of Sciences[1] describes lunar samples containing the calcium-phosphate mineral apatite. Using a sensitive technique, they detected water (in the form of its ion hydroxyl[2], -OH) within the crystal structure of this mineral. Moreover, these hydroxyl-bearing apatite grains are found in several different rocks from a variety of geological settings. This indicates that the presence of water in the lunar interior is not some fluke, but a general property of the Moon. So the story of water on the Moon advances.

Why did scientists believe for so long that the Moon was bone dry? Largely because the samples returned from the Apollo missions contained no obvious hydrous phases, such as mica or amphibole, common water-bearing minerals in terrestrial igneous rocks. In addition, the chemical composition of lunar samples indicated that they formed under very reducing conditions, indicating very low partial pressures of oxygen. Typically, water oxidizes metals in magma on Earth, creat-

ing minerals that contain ferric iron (Fe^{3+}); the exclusive presence of ferrous (Fe^{2+}) iron in lunar samples indicates that no water was present during their formation. Finally, there is the extremely reducing nature[3] of the current surface environment of the Moon, in which solar wind protons (hydrogen ions) continuously impinge upon the surface and reduce metal oxides in the soil. This hydrogen reduction creates "free" metallic iron ($0Fe$) and hydroxyl ions, most of which are lost to space through a variety of mechanisms. But at least some of these ions migrate to cooler regions of the Moon, the poles.

We now have found traces[4] of water in volcanic glasses[5], mare basalts and an alkali highland rock. All these samples are very different types of material, formed from different parent materials at different places within the Moon at different times and under different conditions. The rocks possess apatite[6] grains that show evidence for the presence of water at the time of their formation. One reason that we are finding this water in lunar samples now is that the technology of laboratory instrumentation has vastly improved since lunar samples were first studied 40 years ago.

The ion microprobe used in the study by Mc-Cubbin and others[1] resolved a spot size on a mineral grain about 8 microns (about 100th of a single millimeter) in diameter. Additionally, the composition of this spot was resolved with extremely high precision, measuring the presence of water at about 3 parts per million or better. We now have at our disposal a variety of brand new lunar samples in the form of meteorites – rocks that were blasted off the Moon during impact events. More than 130 individual lunar meteorites[7] are now known; one of the samples in this new study is a piece of a lava flow (mare basalt), found as a meteorite from Northwestern Africa.

The results of the new discoveries indicate that water is (or at least was) present in the deep lunar interior. This water probably existed as gas as the pressures and temperatures within the Moon do not permit the existence of liquid water. The total amounts of water implied by this work are still very low; the bulk lunar water content is estimated to be between 0.064 to 21 parts per million, a low amount by almost any standard – except when compared to the previous estimate for the Moon, which was less than one part per **billion** of water. Thus, the new estimate suggests water contents inside the Moon that are several orders of magnitude higher[8] than previously thought.

So what does all this mean? For models of lunar origin, some mechanism preserved primordial water inside the Moon immediately after it formed. It had been thought that if the Moon originated during a collision of a planet-sized object with the proto-Earth[9] 4.5 billion years ago (the currently favored model), the high temperatures extant during such an event would "boil away" most, if not all, volatile substances. Indeed, the near absence of volatile substances in the Moon has long been cited as prima facie evidence for a high-energy environment of lunar formation, such as would be expected from a giant impact. It now appears that regardless of high temperatures prevalent during this time, some water was incorporated into the Moon. Does this make the giant impact model less likely? Perhaps. Clearly we do not fully understand the conditions created by such an event. Work continues on the problem of lunar origin with the handicap that a planetary-scale collision is something well beyond human experience or observation.

Some of this endogenous lunar water may have found its way into one of the permanently dark, extreme cold "traps" near the poles of the Moon. Thus, in addition to the water made on the Moon from solar wind reduction and deposited on its surface by the collision of water-bearing asteroids and comets, we must also consider the addition of water from the deep lunar interior. Considering that our estimate of the abundance of internal lunar water is still very low, it is likely that the vast bulk of the water found at the poles[10] is of external origin. Therefore, the possible finding of indigenous "Moon water" in the polar areas makes detailed study and examination of the poles even more attractive.

The Moon continually surprises us as she reluctantly (but always provocatively) reveals her secrets. In recent months, a wholly new and totally unexpected picture of the processes and history of our nearest planetary neighbor has emerged. We are in the midst of a renaissance of lunar science.

Topics: Lunar Exploration, lunar science

Links and References

1. paper recently published, http://www.pnas.org/content/early/2010/06/07/1006677107
2. ion hydroxyl, http://en.wikipedia.org/wiki/Hydroxyl
3. reducing nature, http://en.wikipedia.org/wiki/Chemical_reduction
4. found traces, http://blogs.airspacemag.com/moon/2010/05/02/the-four-flavors-of-lunar-water/
5. water in volcanic glasses, http://www.space.com/scienceastronomy/080709-moon-water.html
6. apatite, http://en.wikipedia.org/wiki/Apatite
7. 130 individual lunar meteorites, http://meteorites.wustl.edu/lunar/moon_meteorites.htm
8. water contents inside the Moon, http://www.csmonitor.com/Science/2010/0614/The-moon-may-hold-100-times-more-water-than-previously-thought
9. collision of a planet-sized object, http://en.wikipedia.org/wiki/Giant_impact_hypothesis
10. water found at the poles, http://blogs.airspacemag.com/moon/2010/05/02/the-four-flavors-of-lunar-water

Comments

4. Still more exciting reasons to return to the Moon!

Dr. Spudis, I hope you don't mind me posting this link to another interesting blog on the Moon from Centauri Dreams called 'Protecting the Lunar Farside'. You might want to make some comments yourself.

http://www.centauri-dreams.org/?p=13011

Comment by Marcel F. Williams — June 19, 2010 @ 8:32 pm

5. Dr. Spudis,

For many years the focus of Mars exploration has been to "follow the water." Is a similar strategy emerging for lunar exploration? How would you like these findings to affect planning for future missions? What missions are planned that will build on this growing knowledge base?

Sincerely, Jason

Comment by Jason — June 19, 2010 @ 9:58 pm

6. Jason,

You should keep in mind that this new work on lunar water is only marginally relevant to the issue of abundant polar water — most of the water at the poles is likely of external, non-lunar origin. However, I agree that in a sane world, we would be planning a strategy to systematically characterize and inventory the water, its various sources, and the processes involved in their disposal and fate. Sadly, we do not live in such a world.

Comment by Paul D. Spudis — June 20, 2010 @ 6:14 am

7. These results are interesting, nevertheless, even if they aren't directly relevant for the polar ice fields. If we accept the result at face value, it would tend to favor the old Darwinian fission hypothesis

for the origin of the Moon would it not? Presumably, there would be less energy released during a fission even as opposed to a massive collision, and so the Moon could condense faster thus preserving a higher concentration of volatiles.

Yes, the fission hypothesis would apparently require some sort of exotic mechanism to set it off (like georeactors); on the other hand, Velikovskian planets careening around the Solar System is rather exotic as well. The fission hypothesis better explains the oxygen isotope concentrations in that the concentrations are pretty much identical. The fossil angular momentum that we observe today is also difficult to square with the impact hypothesis, if I'm not mistaken.

One thing the hew data would seem to make more likely is outgassing as an explanation for at least some of the reliable observations of transient lunar phenomena, including the existence of active fumaroles on the Moon. Obviously, as a geologist involved in drilling for natural gas, I'm no believer in the abiotic origin of natural gas here on Earth–at least for commercial purposes. But there may be a grain of truth to the idea; if the lunar mantle has a higher concentration of volatiles than was formerly believed, perhaps its possible that it is still slowly migrating its way to the surface. Such fumaroles might eventually be good places to prospect for other volatiles besides water, such as methane.

Which leads to this that caught my eye: *"This [primordial] water probably existed as gas as the pressures and temperatures within the Moon do not permit the existence of liquid water."* Are you sure about this Paul? A quick glance at the water phase diagram why there can't be liquid water on the surface, but deeper within the Moon, surely there must be some combination of temperature and pressure where liquid water would be stable.

A simple BOTE calculation (assume an average surface temperature of 250 K, that temperature increases at 2.7 degrees per kilometer, and that pressure increases at the rate of about 4.9 million pascals per kilometer) suggests that liquid water should be stable from about a depth of about ~9 km down to a depth of roughly 130 km, at which point the temperature would exceed 600 K.

Comment by Warren Platts — June 28, 2010 @ 3:31 pm

8. Hi Warren,

the fission hypothesis would apparently require some sort of exotic mechanism to set it off (like georeactors); on the other hand, Velikovskian planets careening around the Solar System is rather exotic as well.

One difference in favor of giant impacts is that we have scars of such events preserved on the Moon (South Pole-Aitken basin) and on Mars (Utopia basin). So large impacts did occur early in the history of the Solar System. I am unaware of any independent evidence that planets can split into two pieces, ala the classic fission model. Doesn't mean that it can't happen — it just seems sort of ad hoc.

Are you sure about this Paul?

No. I haven't done a detailed study — my statement was based on analogy to typical igneous systems on Earth, where water is almost always in vapor phase.

Comment by Paul D. Spudis — June 28, 2010 @ 4:10 pm

9. In order for the fission hypothesis to work, the body would have to be rotating so fast that the centrifugal force would almost match the force of gravity (on the order of 2 hours per rotation). Then some sort of catastrophe would have to happen to cause the split. I was thinking of the recent paper by de Meijer and van Westrenen "An alternative hypothesis for the origin of the Moon" (http://arxiv.org/abs/1001.4243), where a runaway georeactor (sort of like the Oklo georeactor, only much bigger) provides the missing energy source.

The fission hypothesis would better explain the similarity in composition, since most collision models require that most of the Moon's mass come from the impactor. Also, the best simulations require that the proto-Earth be rotating in a retrograde fashion.

One thing's for sure, we're not going to settle the matter from our comfortable armchairs here on Earth. The authors do predict enhanced levels of certain helium and xenon isotopes deep within the Moon. On Earth, we usually recover these inert gases in natural (methane) wells. Which brings me back to my earlier post about the possibility of liquid water. Assuming a surface soil temperature of 250 K, and a 2.7 K increase per kilometer, then at 9 km total vertical depth, the temperature would be 274 K, a little above freezing. And since the gravity of the Moon is 1/6 of Earth's, the pressure at 9 km would correspond to a depth of about a mile here on Earth. At these depths, we routinely drill for natural gas, and fractured basalt actually makes for good underground storage of natural gas, and even forms a reservoir rock in the right conditions. Meteor impacts might cause further, intense, local "frakking". Geothermally heated water that was able to take advantage of the impact fractured system might significantly raise the zone of liquid water locally. At the logical extreme, there might even be gas seeps or even geysers as a result.

You see where I'm going with this. The fundamental premise of astrobiology is that where there's liquid water, there is life. Yes, I know it's a long shot, but it would be an irony of ironies, if, after gallivanting all over the solar system in search of extraterrestrial life, we were to find it right under our noses on the Moon.

Comment by Warren Platts — July 4, 2010 @ 8:11 pm

Malice, Mischief and Misconceptions

June 26, 2010

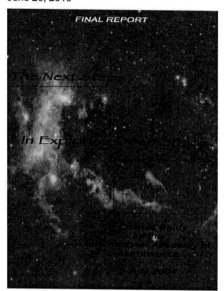

The "new direction" presaged six years ago?

The space community has fractured since the disastrous roll out of NASA's "new direction." Preceding the administration's budget announcement, endless delays and rampant speculation about administrators, rockets, and program design and direction kept people guessing. The current trench warfare is not a pretty sight, but it is not unexpected given the lack of a clear direction. Word has it that more detail will come out early next week, adding yet another layer to this growing space onion. The undirected, unfocused, unproductive spin cycle NASA (and the entire space community) has twirled around in for the last 18 months is instructive. It is real time, 20/20 insight on how the new direction will play out during the proposed five-year study hall being scheduled for NASA to find their "right stuff."

The latest attempt to explain NASA's new direction is an article published in Space. com[1] by Clara Moskowitz. She tries to "correct" some alleged "misunderstandings" about the Obama administration's new direction and budget for NASA[2]. Her article quotes several space luminaries, who opine that the new path is simply "not understood" by a few petulant detractors who stubbornly refuse to accept Flexible Path as advertised. Responding to the criticism that the new path was conceived in secret by a small cabal without detailed thought, Moskowitz quotes my friend Jim Oberg[1] as saying that the administration's space proposal is "extremely similar" to a report issued by the International Astronautical Academy (IAA)[3] and so (in effect) the new direction has been studied extensively by an "international astronautical group."

The IAA report Oberg referred to was an outgrowth of NASA's Decadal Planning Team[4] activities in the early 2000's and was being prepared for publication just as the 2004 Bush administration's Vision for Space Exploration (VSE)[5] was announced. There was no interaction between these two strategic plans. I re-examined the IAA report to see how closely it tallies with the Obama Administration's proposed new direction for NASA and more significantly, how they both differ from the VSE (a direction endorsed by Congress and both political parties in 2005 and 2008).

I find that the general outline of the IAA report corresponds to the proposed new direction quite closely. Both the IAA report and the administration's budget propose a "flexible path[6]" approach for human journeys beyond low Earth orbit. Both plans outline a variety of possible destinations, including the Sun-Earth Lagrangian points, near-Earth asteroids, Phobos and Deimos (the moons of Mars), and finally the surface of Mars. The IAA report does not address transport to and from LEO, the starting point of these missions, but acknowledges that commercially procured transport of people and cargo is highly desirable.

A glaring difference between the IAA report and the administration's budget proposal is that the IAA report specifically recognizes the Moon's surface as a valid objective[3] (as does the VSE). Many in the blogosphere continue to insist to those who know better that the new direction does include the Moon. By rejecting the Moon as a destination with such trite and unthinking casualness, the administration's proposal has left those who understand the national economic, scientific and security implications of lunar return reeling. One could be forgiven for concluding that there is some mischief in how these reports are being conflated.

Oberg is correct in that there is commonality between the activities described in this report and the purported activities promised by the "new direction." But the administration's proposal did not go "through years of analysis, modification, and critiques by a worldwide team." The IAA report retains the Moon as a destination and does not discard the national spaceflight system immediately in favor of a non-existent commercial transport system – it ignores the Earth to LEO segment completely. Neither the impact on our national aerospace industrial base and workforce, nor the bureaucratic effects of an unclear and indeterminate direction to NASA's productivity were considered by the IAA report – or are considered by the administration's new path, for that matter. The Obama administration has offered to assist the thousands of displaced aerospace workers affected by the new direction.

Let us examine the objectives of the IAA flexible path versus those of the VSE. The IAA report states that this plan is undertaken to "articulate a vision for the scientific exploration of space in the first half of the 21st Century" and that "scientific objectives are used to determine the destinations for human explorers" (IAA report, Executive Summary, page 3)[3]. Moreover, the report states (as does the new direction) that although many destinations are envisioned, "the ultimate goal [is the] establishment of a human presence on Mars for science and exploration." To this end, all technology development, infrastructure creation and scientific exploration are undertaken with the goal of humans on Mars as the ultimate end point.

In contrast, the VSE was undertaken to "to advance U.S. scientific, security, and economic interests through a robust space exploration program."[7] This included the "implementation of a sustained and affordable human and robotic program to explore the solar system and beyond, to extend human presence across the solar system, starting with a human return to the Moon by the year 2020 in preparation for human exploration of Mars and other destinations, to develop the innovative technologies, knowledge, and infrastructures both to explore and to support decisions about the destinations for human exploration, and promote international and commercial participation in exploration to further U.S. scientific, security, and economic interests." The Moon is the necessary starting point for what is (in essence) the doorway into our entire Solar System.

The purpose of lunar return under the VSE is not to collect rocks or relive past space glories. Simply put, because we can't take everything with us, humans must learn to use what we find in space to create new space faring capabilities, starting on the Moon. And our goals are not simply Mars, but everywhere – wherever human presence is needed or desired. Using the resources of the Moon (specifically, making consumables and propellant from lunar materials) enables routine access to all of space – not merely for science, but for economic and national security interests as well.

Oberg's statement that the new direction for NASA received detailed study and thought[1], follows from his evaluation: 1) of the amount of study put into the IAA report; and 2) that the IAA report and the new path are equivalent. I do not deny the former, but strongly question the latter. Unlike the IAA report, flexible path as articulated in the new budget proposal not only eliminates the lunar surface as a destination (one chosen by the VSE specifically for its ability to enable new and greater space faring capability) but it also has a much narrower rationale: scientific study of Mars as opposed to the Vision's objective of creating an extensible, reusable space faring infrastructure to conquer the budget-busting limitations imposed by our residence at the bottom of the gravity well of the Earth.

The administration's proposed program for

NASA indefinitely defers trips to destinations that have gravity wells. It is shortsighted and limited. The IAA report (and the new direction) focuses heavily on scientific aims while the VSE seeks to advance human exploration for "scientific, security and economic interests." Architects of NASA's new direction may not have understood or appreciated (or approved of) the objectives of the VSE, but those who read and understood the original reports and documents did. The VSE was about incrementally expanding the reach of people and machines by learning how to use the inexhaustible materials and energy of space[8], starting with the nearest, most accessible place beyond LEO that has what we need: the Moon. Breakthroughs and new understanding of the world and space around us rise to the fore and challenge us when we explore the unknown.

Civilizations thrive and advance when not in retreat. The administration's chaotic proposal for NASA retreats from human space exploration. Many in the space community have serious doubts and concerns about this new direction. Labeling these doubts and concerns as "misconceptions" does not make the new direction valid nor change the reality that we are in danger of losing our capability as a space faring nation.

Topics: Lunar Exploration, Lunar Resources, Space Politics, Space Transportation, Space and Society

Links and references

1. article in space.com, http://www.space.com/news/nasa-obama-space-plan-misunderstandings-100624.html
2. new direction, http://www.nasa.gov/news/budget/index.html
3. a report, http://www.lpi.usra.edu/lunar/strategies/AdvisoryGroupReports/iaa_report.pdf
4. Decadal Planning Team, http://www.thespacereview.com/article/521/1
5. Vision for Space Exploration, http://www.spaceref.com/news/viewpr.html?pid=13404
6. flexible path, http://blogs.airspacemag.com/moon/2009/12/16/arguing-about-human-space-exploration/
7. VSE goals, http://georgewbush-whitehouse.archives.gov/space/renewed_spirit.html
8. learning how to use, http://www.spaceref.com/news/viewsr.html?pid=19999

Comments

1. From the IAA report: "Four key destinations emerge as the most important targets for human explorers: the Sun-Earth Libration Point L2 (SEL2), the Moon, Near-Earth Objects (NEO's), and the planet Mars."

Unlike the Obama plan, the Moon is a "key destination" in the IAA plan.

Obama broke his promise to support lunar missions:

http://www.politifact.com/truth-o-meter/promises/promise/339/support-human-mission-to-moon-by-2020/

"With the release of his fiscal year 2011 budget, President Barack Obama dramatically broke one of his campaign promises."

"Obama had said during the campaign that he would "endorse the goal of sending human missions to the moon by 2020, as a precursor in an orderly progression to missions to more distant destinations, including Mars." But the proposed budget he presented to Congress would shift course significantly."

Comment by Pete — June 27, 2010 @ 4:02 am

4. So why not send a robotic ISRU demonstration first?

In the meantime build the orbital "gas stations" (1st @ LEO & 2nd @ EML-1) where, if the ISRU demonstration is successful, the LH2/LOX can flow from newly-built Lunar surface ISRU plants to EML-1 and onwards to LEO.

If Lunar ISRU does not pan out, then ship LH2/LOX from Earth to LEO and onwards to EML-1.

What is wrong with this?

Comment by Alan — June 27, 2010 @ 11:34 am

5. So why not send a robotic ISRU demonstration first? Where do I argue against that?

As for propellant depots, I think that they make sense if we can supply them with propellant made from space resources, in this case, propellant derived from lunar water. If we end up launching all the propellant from Earth, then nothing is fundamentally changed, except to eliminate the need for a heavy lift launch vehicle. But you still have to lift all your supplies from the bottom of the deepest gravity well in the inner Solar System. We know that will always be a costly task — the real leverage in space transportation comes from freeing ourselves from that necessity by using local resources. That's where the biggest payoff and the largest, most significant unknowns are. Thus, that is where I think we should concentrate our research efforts.

Comment by Paul D. Spudis — June 27, 2010 @ 12:02 pm

6. Paul D. Spudis said:

"As for propellant depots, I think that they make sense if we can supply them with propellant made from space resources, in this case, propellant derived from lunar water."

I think ISRU will happen at some point after we start colonizing the Moon, but I differ from your viewpoint in that I think it will take far more resources to set up and run ISRU than it would take to set up and supply fuel depots from Earth. At least for quite a while.

Maybe I'm missing something in your ISRU thinking, but how many metric tons of mining & processing equipment are you thinking it will take to create a water extraction/fuel refinery on the Moon that can process enough lunar material to exceed local supply requirements?

And you want to do this with people, who require a colony infrastructure to be set up, supplies and replacement parts delivered, and crew rotation shuttles. All of this has to be in place BEFORE you get your first liter of exportable fuel, and that will be needed to power your tanker fleet.

How many years is it going to take to do this, and how much money?

United Launch Alliance (Lockheed Martin/Boeing) proposed their Moon colony plan called "Affordable Exploration Architecture 2009", and in it they use fuel depots as an enabler to getting to & from the Moon until an alternative fuel source is found. Their plan also uses existing launchers, and only spends $15B on launchers to support one year of building up in-space assets, and the 2nd year of landing and supplying a permanent colony (with 120 day crew rotations).

This plan could is also based on many of the features in the "Flexible Plan", which is the basis for the proposed NASA budget.

I noticed that while blasting the proposed budget, you did not defend Constellation, so at least we agree that NASA has been on the wrong track with Constellation, and that the President had to make a change.

Comment by Coastal Ron — June 27, 2010 @ 1:48 pm

7. *How many metric tons of mining & processing equipment are you thinking it will take to create a water extraction/fuel refinery on the Moon that can process enough lunar material to exceed local supply requirements?*

Probably a couple of metric tonnes of equipment. You can start off small, with ~200 kg rovers and feedstock draglines. We need an oven to vaporize the ice and then radiators to cool it, tanks and piping to feed and store it. We can lay out an incremental approach that has distinct milestones: first water replace-

ment and radiation protection for the outpost (few mT), then re-fueling for local use (ascent; 10's-100's mT), then export to cislunar (>100's mT).

How many years is it going to take to do this, and how much money?

For a) I don't care — however long it takes. Time is the free variable. For b) whatever the space budget is over the next 20-30 years.

you did not defend Constellation, so at least we agree that NASA has been on the wrong track with Constellation

I never defend a single-point solution. My point is different — that "Constellation" (or whatever name NASA gives to its beyond LEO spaceflight system) can be fixed to meet cost envelopes and technical performance specs. The Augustine report is simply wrong when they say otherwise.

The problem is, the President didn't fix anything — his plan will result in the same amount of money spent over the same amount of time. The difference is that, with the Vision, we would have a functioning lunar outpost and ISRU system. With the new direction, we'll end up with nothing.

Comment by Paul D. Spudis — June 27, 2010 @ 2:09 pm

8. Paul D. Spudis said:

"With the new direction, we'll end up with nothing."

Being a little absolute there, aren't we?

Lunar robotic precursor missions are nothing? Commercial crew capability to LEO is nothing? Keeping the ISS for space research is nothing?

I know you're a "Moon or Bust" type of person, but you don't see anything in the proposed budget that would support your ISRU goal? Yikes, what kind of glasses are you wearing?

Comment by Coastal Ron — June 27, 2010 @ 2:57 pm

9. Paul D. Spudis said:

"How many years is it going to take to do this, and how much money? For a), I don't care — however long it takes. Time is the free variable. For b), whatever the space budget is over the next 20-30 years."

I guess this gets back to what your definition of NASA is. You apparently see NASA as a mining company, and I don't. Why should the U.S. Taxpayer be funding a mining company?

Why don't you want NASA out pushing the envelope of exploration, finding the ways for HSF to venture out past the Moon? You know, the hard stuff that private enterprises can't afford and don't have the resources to attempt.

The role of private enterprise is to follow into the newly opened wildernesses, and create viable businesses. Those businesses could include ISRU, if it makes economic sense, but you seem to have predetermined that ISRU is the number one product or service outside of LEO. That normally is a market decision, based on the prices of alternative supplies (like from Earth), and by forcing your economic solution on the marketplace, you run the risk of being wrong. And doing it with MY tax money.

I don't buy into your vision of government-funded mining on the Moon. I think ISRU has a future at some point, but through market needs, not government (or your) mandates.

Comment by Coastal Ron — June 27, 2010 @ 3:07 pm

10. Frankly I think referring to the '04 IAA report and implying it shows the Obama proposals is based on it, and all the study that went into it, is suspiciously disingenuous. I notice a lot of Obama plan supporters read between the lines to project a lot of depth and thought folks at NASA have never implied was ever there. If it was there, they would have thrown some of it out to congress – who appeared eager to grill them for dinner. ;)

Comment by Kelly Starks — June 27, 2010 @ 3:36 pm

11. > The problem is, the President didn't fix anything — his   > plan will result in the same amount of money spent over   > the same amount of time. The difference is that, with   > the Vision, we would have a functioning lunar outpost   > and ISRU system. With the new direction, we'll end up with nothing.

Frankly worse then nothing. It could force decimating layoffs across the aerospace industry.

On a positive, if unrelated, note; it looks like the military is serious about wanting to field a RLV replacement for the EELV's in about 15 years. one they want to drop per flight costs by a factor of 2 or 3, with much better on call flight ability. If they will drop costs that much with the mils flight rate, if any commercial market of any size starts up, or NASA really want to do a lunar base, this would make them much more doable in the 2020's?

Comment by Kelly Starks — June 27, 2010 @ 3:41 pm

12. *You apparently see NASA as a mining company, and I don't. Why should the U.S. Taxpayer be funding a mining company?*

You're very good at missing the point. NASA's role is **to determine if the Moon can be mined**, not to do the mining itself. The Vision directed the agency to return to the Moon to understand how difficult it is to use lunar resources to create new capability. NASA's job is to return to the Moon with people and machines, characterize what's there, how we might get to it and extract what we need, and demonstrate the feasibility of lunar resource extraction in an end-to-end systems test. Once developed, it is then possible to privatize it. How is that different from what you think the "new direction" is — government has shown Earth to LEO launch is possible and how to do it; now, it is time for the commercial sector to use this knowledge and construct their own capabilities.

Being a little absolute there, aren't we?

Just making a prediction based on 30 years observing how the agency works, what it does well and what it does poorly. Reject it, disbelieve it, accept it as you will.

Anyway, this is the place where I give my opinion. It's still a free country — put your opinions on your own blog.

Comment by Paul D. Spudis — June 27, 2010 @ 3:48 pm

13. Kelly,

Referring to the '04 IAA report and implying it shows the Obama proposals is based on it, and all the study that went into it, is suspiciously disingenuous.

Agreed, that's why I wrote this piece. But the supporters of the new direction will grasp at any straw to promote it and denigrate their opponents. Check out this quote from OSTP Head John Holdren:

In an analysis of the White House's proposed 2011 budget, the AAAS R&D Budget and Policy Program reported that Obama wants to end NASA's Constellation program, which had targeted a return to the moon by 2020. The budget plan would save $3.1 billion by ending the return-to-the-moon mission and retiring the space shuttle. But it would increase overall NASA R&D funding by $1.7 billion, or 18.3%. Obama would invest $6.1 billion over five years to transition regular near-Earth orbit missions to private industry; increase funding by $812 million, or 35.1% over three years, for the International Space Station; and invest

$559 million in "heavy-lift" and propulsion systems, including research into new engines, new propellants, and advanced combustion processes.

That shift has drawn criticism in Congress and from some sectors of the public—and recently from Neil Armstrong, who became the first man on the moon in 1969, and Eugene Cernan, who in 1972 was the last man on the moon.

In response to a question from the audience, Holdren strongly defended the administration's decision, both on scientific and budgetary grounds.

After a close look at Constellation, he said, the White House concluded that, "notwithstanding its rapidly escalating cost, to three and four times the original estimate, it was still not going to able to land U.S. astronauts on the surface of the moon until 2030 or later.

"To get them to the moon by 2025 would cost an extra $60 billion over the next 10 years," he said. "The president and his advisers, including me, made the decision that there are other destinations in deep space—destinations beyond low-earth orbit, that will allow us to do more science sooner, with more missions, more visits, more exciting discoveries—than going back to the moon 50 years later.

"It's not real surprising that the American heroes who were the first people to set foot on the moon might think the most exciting thing we could do now is to go back there. But not everybody agrees with them. In fact, the second person to set foot on the moon, Buzz Aldrin, is a strong supporter of the president's program. And a large array of other astronauts, including Sally Ride, the first American woman in space, and Mae Jemison, the first African American woman in space, and John Grunsfeld, the Hubble [space telescope] repairman, who had five missions and 60 hours of spacewalking...all are strong supporters of the president's program."

http://www.aaas.org/news/releases/2010/0521stpf_holdren.shtml

In other words, my "celebrity experts" are better than your "celebrity experts." Appeal to authority is a classic tactic in theological debate and its use here by Holdren demonstrates that we are entering that realm of such discourse.

Comment by Paul D. Spudis — June 27, 2010 @ 3:58 pm

14. So we are too dumb to "understand" Obama's PowerPoint? Give me a break! I am a laid off Space Shuttle Worker and I cannot believe that Obama wishes to cancel our Human Space Flight Program for a Power Point.

You can read more of my thoughts at http://www.rv-103.com/?cat=98

Very good article and well done!

Comment by Gregory N. Cecil a.k.a. "Rocketman" — June 27, 2010 @ 4:39 pm

15. Kudos to Paul, on another well-written article.

I spent all of my career on space projects, the majority of my career at JSC. Since college, I've been hoping I'd get to work on an exploration program. I finally started working on Lunar outpost preliminary design a few years ago. Now, as my retirement looms, all that has been yanked away, replaced with confusion and internal cannibalism (as people, labs, and branches jockey to find a source of funds for next year, and contractors scramble to have a job that will last past the layoffs).

Realistically, preliminary designs for human exploration beyond LEO won't happen again for about a decade. Historically, cancellation of Lunar programs by Nixon (1970+) delayed human efforts to live off-world for 40 years and counting. President Obama now has the dubious distinction of adding decades to that sidelining. Since the Moon is such an obvious stepping-stone, we can only hope that some other country will take the step that President Obama has foolishly denigrated with his pronouncement "been there, done that".

Although some of the creators and backers of the new plan have openly denigrated establishment of a Lunar outpost (John Holdren, Ed Crawley, Barack Obama, Buzz Aldrin), they can't be so naive as to think that their FP2N (Flexible Path to Nowhere) will last beyond the next President. It may not even last beyond the next Congress.

And even if a crewed mission to a NEO survives all the budget cycles, how many taxpayers will be inspired by such a one-shot stunt; inspired enough to pour further billions into a future one-shot stunt to reach Martian orbit but not land? If they actually wanted to make a meaningful contribution to HSF, it should have been by setting achievable goals with near-term deadlines, not goals whose first meaningful action comes a decade after they leave office.

And if they aren't naive, a cynic would say FP2N is a clever way to cause the US space program to self-destruct. The in-fighting in the space community is one example of that self-destruction. Another example: NASA's budget may actually shrink as the years go by and ISS nears the end of its life. That's because NASA may not be able to obligate the promised funds (so it won't get them the next year), or because future Congresses (and Executive Branches) may find it even easier to claim NASA isn't doing anything inspiring.

Unfortunately, the wheels of self-destruction are already in motion, and we are already losing our capabilities. Many of my colleagues have been reassigned, to protect them from layoffs. The most expert contractors are leaving or being moved away from NASA work.

Lucky people like me will retire, sadly taking our experiences away. The less fortunate will rebuild their lives, just as the thousands did after Apollo. Some opened grocery stores or motels, some pursued other professional lives; but few ever returned to aerospace, and only a few passed on their knowledge to the next generation.

It's hard to say what we could have done then, or what we might have done now. Instead, we are on a path to becoming less relevant to space exploration. Best wishes to the next great empire.

Comment by John — June 27, 2010 @ 5:45 pm

16. Rocketman,

I am a laid off Space Shuttle Worker and I cannot believe that Obama wishes to cancel our Human Space Flight Program for a Power Point….Very good article and well done!

Thank you. And let me return the compliment — the piece on your blog posted last April about the Presidential speech at KSC was outstanding. You cut through the fog with crystal clarity — much better reporting and analysis than anything else I read in the media on it.

Comment by Paul D. Spudis — June 27, 2010 @ 5:49 pm

17. Rocketman, as I'm sure you're aware, it was Bush/Griffin that laid you off, not Obama. Layoffs can be emotional, especially when you have worked at a place for so long like you. However Congress also decided to end the Shuttle, and did nothing two years ago when the program deadline to start shutting down suppliers was reached. However I see you're turning lemons into lemonade, and are going to start RV'ing around the country, so you're making the most of it.

The comment "Obama wishes to cancel

our Human Space Flight Program" is always a weird one for me too, since the Constellation program would have ended the ISS, and our permanent presence in space. When Ares V finally was ready to launch, then we would have been out of space for 10 years or more. Is that a good thing? No astronaut corp for 10 years? And during that time, plenty of other nations would be gaining valuable experience in space, with hardware that we paid for – that was the worst part.

The other thing I find funny is when people talk about "commercial companies", but conveniently leave out the two biggest – Boeing and Lockheed Martin. Why is that? Is it because then that invalidates the argument that "only NASA can launch people into space"? To think that only a government agency can launch people into space is also strange, as if only government personnel are smart enough – geez, this is 50 year old technology we're talking about.

Blinded by the light…

Comment by Coastal Ron — June 27, 2010 @ 7:21 pm

18. Paul D. Spudis said:

"NASA's role is to determine if the Moon can be mined, not to do the mining itself."

In real life, mining companies pay for their own exploration and resource mapping. It's one thing for NASA to devote some of their budget to determining the composition of the Moon and heavenly bodies, but you also stated " whatever the space budget is over the next 20-30 years." in regards to how much NASA should spend. That to me sounds like you want to do ISRU exploration as the exclusive program for NASA. Weird.

Also, considering that the current HIGH estimate for lunar water content is 5 parts per million, it's going to take a lot of mining (and energy) to extract anything useful from the Moon. Just like fusion power, water on the Moon is a distant goal, not a near term reality. We need our current budget for hardware, not vaporware (get it, water vapor… see, I have a sense of humor).

"Anyway, this is the place where I give my opinion."

And you allow comments, so you must want debate and discussion – unless Air & Space (who I subscribe to) requires comments… ;-)

Comment by Coastal Ron — June 27, 2010 @ 7:41 pm

19. *"scientific objectives are used to determine the destinations for human explorers"*

It is true that the plan presented in the IAA report is based around science alone, but the authors do explicitly state that it is not their opinion that science alone should be the guiding principle. It is a report written by a bunch of scientists addressing how the science part of an exploration program could be structured. They invite others to offer similar plans for commercial development of space.

Note that I don't actually think the Obama plan is very good, the IAA plan was much better, as was OASIS and as was Steidle's vision.

Comment by Martijn Meijering — June 27, 2010 @ 8:35 pm

20. "If we end up launching all the propellant from Earth, then nothing is fundamentally changed, except to eliminate the need for a heavy lift launch vehicle."

You don't believe in RLVs then? I totally agree that lunar ISRU can revolutionise space exploration and even if you had RLVs using lunar propellants might still be cheaper. Hard to tell what the economics will be so far in the future. But lunar ISRU does absolutely nothing for large scale space tourism, since that requires affordable transport to LEO. I think the major benefit of depots is that they could lead to development of RLVs in the early phases of an exploration program, long before ISRU was operational. Cheap lift helps everything, ISRU only helps exploration and more generally activity beyond LEO.

Comment by Martijn Meijering — June 27, 2010 @ 8:44 pm

21. Paul: Why do you and others keep saying that the Bush "Vision" was endorsed by both political parties? Apparently you do not know the difference between appropriation bills and authorization bills.

Modifying a quote from a famous Vice President, "Authorization bills aren't worth a bucket of warm spit." The rubber meets the road in appropriation bills. They tell you what the congress really supports. They never included the required funds for the "Vision" to be successful. Therefore, both political parties never supported the "Vision."

Authorization language is designed for the mail-outs to the voters back home. "See, I supported what you folks want for the big fancy Post Office. It was all those bad guys that wouldn't give you the money." The mail-out doesn't include: "I had to vote not to include the actual funds because that's what my leadership told me to do."

Cheers

Comment by Don Beattie — June 27, 2010 @ 9:51 pm

22. > In other words, my "celebrity experts" are better than   > your "celebrity experts." Appeal to authority is a classic   > tactic in theological debate and its use here by Holdren   > demonstrates that we are entering that realm of such discourse.

Yeah – and that's the high road, low road just goes to name calling.

Assumptions taken as gospel and vicious attacks on nonbelievers and heretics who ask questions. Folks reading tea leaves to deduce the true meaning behind Obama's words.

Yeah, this is a great way to plan a space program.

Comment by Kelly Starks — June 27, 2010 @ 11:08 pm

23. There is no single reason to return to the Moon, there are multiple reasons to do so. We need to return to the Moon to:

1. Find out if a 1/6 hypogravity environment is inherently harmful or harmless to long term human occupation. In just one or two years, we could find out if colonizing low gravity worlds is possible for the human body .

2. Find out exactly how much lunar regolith, or water, or even hydrogen is required to effectively mass shield humans from galactic radiation. This has huge implications for interplanetary travel.

3. Find out how efficiently we can extract oxygen from the lunar regolith. Supply oxygen for space depots at LEO, GEO, and the Lagrange points could dramatically lower the cost of space travel

4. Find out how efficiently we can extract water and hydrogen from the lunar regolith. Again, supplying hydrogen for space depots at LEO, GEO, and the Lagrange points could dramatically lower the cost of space travel

5. Determine how much carbon and nitrogen are stored in the polar regions. Carbon and nitrogen are essential for growing food.

6. Find out if catapults or cannons can be used to cheaply toss lunar material into cislunar space. This could further reduce the cost of exporting oxygen and shielding material from the Moon.

7. Find out how efficiently we can mine and manufacture aluminum on the Moon that could be used for the manufacturing concentrating mirrors for geosynchronous solar power satellites which could potentially be a multi-trillion dollar a year industry. Lunar aluminum could

also be used to manufacture interplanetary light sails that could completely open up the rest of the solar system manned interplanetary travel and the commercial exploitation of the asteroids.

8. Find out how efficiently we can grow food on the Moon and raise animals on the Moon. Astronauts, tourist, and colonist have to eat.

9. Find out how large we can build inflatable human habitats on Moon and well we can protect such structures from galactic radiation and micrometeorites. The larger such structures are the more tourist and possibly even permanent colonist are going to be attracted to the Moon.

10. Find out how comfortable, both physically and psychologically, can humans be living months or even years in artificial structures on the lunar surface.

11. Deploy solar powered robots, teleoperated from Earth could travel thousands of kilometers all over the Moon collecting samples and then they could return their samples to a Moon base for return to Earth.

The first nations to establish permanent bases on the Moon are going to dominate cis-lunar space and will also be the first to reap the stupendous economic benefits of utilizing the Moon's resources. The pioneering of the Moon by governments along with the commercialization and colonization of the Moon by private industry is going to be one of the most important and exciting events in the history of our species.

Comment by Marcel F. Williams — June 27, 2010 @ 11:18 pm

24. Thank you Paul. That is high praise to come from a professional space journalist. I am very humbled and flattered.

Over 10,000 people the world over have seen that post with only a few negative reactions. I know it has been seen and printed at the various NASA centers (and usually pulled down from the bulletin boards by management the next business day), and read by several congressmen/senator's offices.

I would strongly suggest to your readers that they go to http://www.Congress.org and find out who their representatives are. Contact them and let them know where you stand in this debate over the future of our Human Space Flight Program.

Be safe and well.

Comment by Gregory N. Cecil — June 28, 2010 @ 12:11 am

25. The critical thing is to get the Moon back into NASA's new direction. BUT...political face must be saved. On the one hand, I have heard Obama say "We've already been there". On the other hand, I have heard Bolden say in congressional testimony that the Moon is still part of the plan. There is still a sliver of opportunity to get the Moon back in the picture. Here's how.

Commercial space has had a very public and dramatic success in the form of SpaceX's Falcon 9. Maybe Obama would be willing to keep the Moon in the picture if it doesn't cost too much. Remember, the main reason that the Moon was taken out of the picture was because it would cost a lot of money to develop the Altair and the heavy lift rocket.

So, can we convince the administration to put the Moon back in the picture if it doesn't cost so much as to jeopardize the stated goals of the Flexible Path (i.e. asteroid, martian moon, & eventually Mars)?

Yes, here's how. Introduce a "new and bold" approach to enshrine commercial space within the NASA budget. Where commercial development of space makes sense, this is where NASA facilitates companies — namely cargo and crew to LEO AND cis-lunar space including the lunar surface.

There is practically no market for the glory seeking of footprint and flag expeditions to an asteroid, martian moon, or the Martian surface. But cis-lunar space has plenty of commercial potential including:

– orbital hotels, – space tug, – orbital refuel & maintenance, – lunar water ice fuel to LEO, – lunar metals to LEO / GEO, – circumlunar tourism, – lunar hotel, and – SPSs.

Do we want to leave it to the Chinese to "own" cis-lunar space? Do we want our children to be watching the American Mars mission buying LEO fuel from a big red Chinese fuel depot? Oh my!

The bold new direction would confirm manned commercial flights to the ISS and would set percent (20%) of NASA's budget would be dedicated to facilitating US commercial development of cis-lunar space (enabling technology for everything else). This would be done in a manner like COTS/CRS and prizes. Development of cis-lunar space would proceed at the pace that the 20% would facilitate. The glory seeking part of the new vision would not be jeopardized. And international partnerships would be more likely to support us if they could send one of their own to the surface of the Moon rather than be a junior partner of a martian mission. Obama likes international co-operation doesn't he?

The money to facilitate commercial development of cis-lunar space would come from certain programs being axed or delayed. In my mind that would include Ares I (beyond an escape capsule although maybe a Dragon would make even that unnecessary), Altair and Ares V.

Can human exploration of an asteroid and martian moon be done without a heavy lift vehicle? Yes. Can lunar development (e.g. tele-operated equipment) be delivered to the lunar surface on medium lift launchers (e.g. F9H)? Yes. Can humans be delivered to the lunar surface without a heavy lift vehicle? I believe the answer is yes with two F9Hs (EDS & command/lander/resource modules), but certainly if fuel for EDS, landing and ascending comes from lunar water ice to LEO. Indeed, by my calculations, an F9H could carry all of the original Apollo upper stack to LEO if it were unfueled. Can a martian landing be done without a HLV? Maybe not although that normally wouldn't come until after 2030 and, hopefully by then, the Moon would be providing the fuel, so maybe a HLV wouldn't be necessary. There is no commercial market for HLVs so NASA couldn't leverage company's financial input.

If these are good ideas, they need to be considered by the administration very quickly before they politically lock themselves into some specific policy.

Comment by JohnHunt — June 28, 2010 @ 2:10 am

26. Coastal Ron,

Also, considering that the current HIGH estimate for lunar water content is 5 parts per million, it's going to take a lot of mining (and energy) to extract anything useful from the Moon

From where do you get this number? It's wrong — way off. The polar deposits hit by the LCROSS impactor indicate between 5 and 10 weight percent water, about 4 orders of magnitude higher than the estimate you give. Moreover, the radar data indicate nearly pure water ice (that's on the order of 100%) inside some polar craters. I fully grant that these numbers need to be verified by ground measurement and that the physical nature of the polar deposits needs to be characterized, but there is abundant water there and in concentrations that make its collection and use feasible.

I know that terrestrial mining companies spend their own exploration money. But they know that the deposits they seek can be ex-tracted and utilized. For the Moon, we do not yet have that knowledge and obtaining it is an appropriate activity for NASA.

Comment by Paul D. Spudis — June 28, 2010 @ 5:01 am

27. Don Beattie,

Why do you and others keep saying that the Bush "Vision" was endorsed by both political parties?

Because it's true.

Apparently you do not know the difference between appropriation bills and authorization bills.

On the contrary, I do know the difference. Not only was the VSE endorsed as a policy by both parties in two different authorization bills in 2005 and 2008, but Congress fully funded the administration's request for development for Exploration in all five years of the agency budget after the Vision was announced.

They never included the required funds for the "Vision" to be successful.

That's your value judgment, not an objective statement of fact. The Vision directed NASA to return to the Moon under the existing budgetary envelope and then provided that amount of money to them. They were not instructed to develop an unaffordable architecture and then whine about not having the money to implement it.

Comment by Paul D. Spudis — June 28, 2010 @ 5:11 am

28. Rocketman,

That is high praise to come from a professional space journalist.

Thanks for the compliment, but I am not a journalist. My background is in geology but I have worked in the space program, studying the Moon, for almost all of my career.

Please keep calling them like you see them and I'll do the same.

Comment by Paul D. Spudis — June 28, 2010 @ 5:29 am 29. Coastal Ron,

"Rocketman, as I'm sure you're aware, it was Bush/Griffin that laid you off, not Obama. Layoffs can be emotional, especially when you have worked at a place for so long like you. "

I had to reply. I knew the shuttle program was ending and had 6 years to prepare. I went back to school and obtained my Masters, saved my money, invested well, etc. I was and am at peace with that.

I don't know why people try to muddy the waters when someone criticizes Obama's new plan and say its sour grapes over the Shuttle program ending. It is not. We are criticizing Obama's plan because of its lack of merit; He is ending our Human Space Flight Program and just dressing up the casket with flowery words he doesn't mean. This has nothing to do with the Shuttle and everything to do with America's leadership in HSF.

I've had 5 1/2 extraordinary years working on the Shuttle Program (22 years in the medical profession prior to that) and feel I have the education and experience to add my voice to the debate. It's not hard to see Obama's plan has no substance. Even the television show South Park agrees. http://www.youtube.com/watch?v=rjzggXBpRN4 (Well, you still have to keep a sense of humor don't you?)

Be safe and well.

Comment by Gregory N. Cecil — June 28, 2010 @ 10:30 am

30. Paul: Sorry, you still don't know the difference between authorization and appropriation bills. Appropriations are only good for one year. Sometimes congress permits carry-over if all funds not appropriated. A five year appropriation is a myth.

Waiting for a copy of your Apogee book. Cheers

Comment by Don Beattie — June 28, 2010 @ 12:13 pm

31. Don,

Appropriations are only good for one year. Sometimes congress permits carry-over if all funds not appropriated. A five year appropriation is a myth.

Now you're starting to embarrass yourself. I never said anything about a "five year appropriation." I said that the Congress funded NASA exploration activities in all the five remaining years of the Bush administration after the VSE was announced.

When I say that the VSE had strong bipartisan support, I backed that claim up by noting the authorization and appropriations actions of the Congress. So what is it about them that I do not understand (except for your erroneous claim it supports your viewpoint)?

Comment by Paul D. Spudis — June 28, 2010 @ 12:39 pm

32. I wonder how much money NASA could make from selling some lunar regolith on ebay? If not for furthering human exploration then perhaps for the exfoliating qualities of the moon we should go back. Skin cream with real moondust. Mmmmmm...

Comment by Phil Thomas — June 28, 2010 @ 2:54 pm

33. @ Coastal Ron

Why should the U.S. Taxpayer be funding a mining company?

The USGS has funding mineral exploration for nearly two centuries. NASA would be merely continuing a long tradition of government sponsored minerals exploration. Since you've read the ULA white paper, you know they estimate the cost of an aggressive lunar base at ~$7 billion USD per year. My own calculations suggest that such a base built near one of the "anomalous" craters discovered by Dr. Spudis and his colleagues could be fully self-sufficient for ascent and descent within 3-4 years of first landing. Removing the need for shipping 300 tons of propellant from Earth to L2 every year would save U.S. Taxpayer $3.5 billion USD per year at today's launch prices. If you truly respected U.S. Taxpayer, you would be onboard with the lunar ISRU mission.

Comment by Warren Platts — June 28, 2010 @ 4:50 pm

34. Phil, I know that your comment was made in jest...but allow me to take it seriously.

The market value of the stolen sample of ALH84001 was estimated at about $933,000 / kg or about a million bucks per kg or $1,000 per gram or just $125 per jewel-sized rock. Apollo missions brought back an average of 64 kg of lunar rock per mission. A commercial program focused just upon returning lunar rock for sell as collectibles could probably yield between $200 and $300 million per mission (until the market was saturated). Elon Musk offered to deliver 1,000 to 3,000 kg of cargo to the lunar surface for $80 million. If 1 out of 50 people in the developed world were willing to pay $125 for a moon rock "jewel", then moon rocks by themselves would pay for about 12 missions before the market became saturated. With the profit from these missions, many more lunar missions could be financed. It seems to me that moon rock collectibles could play an important role in developing lunar commerce.

Comment by JohnHunt — June 28, 2010 @ 5:18 pm

35. Consider this:

NASA spent a boat load of money to build an incredible orbital laboratory (the ISS) and thereby spawned a promised new industry to deliver "stuff" (COTS) and ultimately people

(COTS-D) to this fantastic destination.

This industry did not arise due to new discoveries in rocket technology (merely enabled by vast sums of money from the DotCom boom), and are completely unrelated to the $10M lure of the sub-orbital X-prize. It is directly related to there being a destination (the ISS) needing to be supplied. And other orbital destinations will benefit too.

If NASA were to spend the necessary treasure and effort to begin a permanent Lunar presence, several things would logically follow.

- We would learn answers to questions we can now barely ask. - This new destination would need resupply, and "commercial" will eventually step up to the plate. CLTS, CMTS, "Fly stuff/me to the moon?" - Support functions at such a base would also begin to be "outsourced" to "commercial"; cafeterias leading to restaurants, laundry service, recreation and perhaps guest suites for visitors (yes, lunar tourism – yes, a long time out). "Starbucks" could one day earn their name.

Sure it sounds like I'm smoking something without the required prescription; but there are McDonalds on overseas Army bases, the American West was tamed by the townsfolk not soldiers and cowboys, and the great beneficiaries of California's gold rush weren't the miners but the shop keepers who followed to sell them what they would need.

Or we can hope that "Lucy" will hold the football for us.

Comment by No-Ordinary-Joe — June 29, 2010 @ 2:45 am

37. From the 2030 NSP: All nations have the right to explore and use space for peaceful purposes, and for the benefit of all humanity, in accordance with international law.

Will mining the Moon would require "somebody" to be in charge ... the US? The UN? Who will grant permission and collect the taxes and fees (for all humanity) for a private company to mine?

Comment by LoboSolo — June 29, 2010 @ 1:44 pm

38. Lobo,

Will mining the Moon would require "somebody" to be in charge ... the US? The UN? Who will grant permission and collect the taxes and fees (for all humanity) for a private company to mine?

Excellent questions all. I have no answers for them, but you should pose them to the administration and your elected representatives.

Comment by Paul D. Spudis — June 29, 2010 @ 3:19 pm

39. @LoboSolo

Will mining the Moon would require "somebody" to be in charge ... the US? The UN? Who will grant permission and collect the taxes and fees (for all humanity) for a private company to mine?

A follower of John Locke might suggest that all humanity is entitled to harvest the resources of the Moon (which lying outside of any sovereign jurisdiction exist in what Locke calls the "State of Nature") without interference from any sovereign, provided enough is left behind for others to harvest, if they wish.

Thus, no taxes or fees should be due, except to the extent the government having jurisdiction over the miners chooses to impose such taxes.

It would seem the Isle of Man has recognized that there is opportunity, here. If the nation that "flags" the moon mining operation grants permission, what other permission is needed?

Comment by Bill White — June 30, 2010 @ 3:51 pm

40. An NEO mission will be NOTHING but a circus stunt! Hey!—instead of setting up building structures on the Moon, we'll catapult astronauts like a human cannonball to a gigantic shard of rock 2,3, or even 4 months travel time away! We'll cram them into a capsule tinier than an RV, and roll the dice with their life support system operating flawlessly that first time out! They might NOT EVEN be able to "land", but heck, it'll be 100% virgin territory! We'll just fly in tandem with the humongous pebble! (One big stomp, and the astronaut in question could be tossed into a fatal new solar orbit, but never mind THAT risk....) Look, people, why can't we just leave NEO's to robotic probes for now? These asteroidal stunt missions will NOT lead to bases NOR resource utilization. Manned asteroid missions will be just as big dead-ends as the ISS is right now.

Comment by Chris Castro — July 7, 2010 @ 1:55 am

Searching for the Moon's Mantle

July 7, 2010

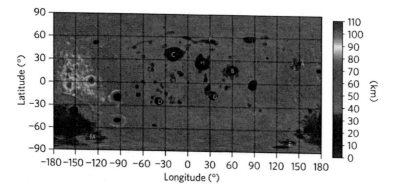

Red symbols on this map of lunar crustal thickness show where the Kaguya team have identified deposits of olivine

We've studied and examined the Apollo samples of the lunar maria (pronounced MAR-ee-uh) for thirty years but despite the thorough search of these collections, we have never found a sample of the deep mantle[1] from which these lavas were formed. How might such a deeply seated rock find its way to the surface? Large basin-creating impacts on the Moon might have dug through tens of kilometers of crust (down into its deepest layers) to excavate samples of the mantle. Another occurrence of mantle samples is as small chunks of rock included in lava. Fragments of rock may be ripped away during the ascent of the dense, liquid magma and become inclusions in the solidified lava flows; these "stranger" rocks are called *xenoliths*[2]. Despite an exhaustive search of the Apollo samples, no samples of the mantle have been found, either as a fragment of basin ejecta or as a xenolith in the mare basalts.

If no one has ever seen it, how do we know what the lunar mantle looks like? The properties and composition of planetary interiors are inferred by indirect evidence. Seismometers left on the surface by Apollo crews[3] measured the velocity of seismic waves inside the Moon, an indirect measure of the density of the deep interior. The density of the mantle is high enough so that common surface rocks cannot make up a significant portion of it; the rocks must contain large amounts of the minerals olivine and pyroxene. In addition, the mantle rocks were partially melted to make the mare

basalts that cover the surface in places. The chemical composition of these lavas show they were made by melting a rock rich in magnesium and iron. Finally, xenoliths of the Earth's mantle are sometimes found entrained in lavas – these pieces are made up of the olivine-pyroxene rock peridotite (after the mineral olivine (the gem form is peridot) that makes up most of it.) So the idea that the rocks of the mantle are olivine-rich is a well-grounded concept for which we have abundant independent evidence.

Now a new scientific paper[4] concludes that fragments of the mantle (the dense magnesium- and iron-rich portion of the Moon that lies below 70-100 km depth) are exposed on the surface[5], brought up from depth by the impact of giant asteroids 4 billion years ago. Such a finding would indeed be significant, as geologists always seek rocks from the deep interior to aid our understanding of the Moon's structure and composition. Data from the orbiting Japanese Kaguya mission[6] shows that olivine is present in the surface deposits of some lunar craters. But how do they go from this observation to the interpretation of lunar mantle? Basically, they mapped out the occurrences of these olivine deposits and found that many of them occur within the rims of large impact basins. Based on models produced from gravity mapping, the crust of the Moon is thought to be thin here and the mantle is close to the surface. Thus, these large impact basins could have excavated chunks of the mantle, throwing them out onto the surface of the Moon.

Why is the mineral olivine[7] important? Olivine is a silicate mineral rich in magnesium and iron; it forms one of the basic, silicate building blocks of the rocky planets. In magma (liquid rock), olivine crystallizes first and its composition is a key indicator of the composition of the magma. Current prevailing wisdom is that the early Moon was largely molten (the "magma ocean" phase); whether it was completely molten or merely liquid in its outer portion is uncertain, but in such a huge system of liquid rock, olivine is the earliest mineral that crystallizes. Being dense, crystallizing olivine would sink in the liquid magma, slowly accumulating deep in the Moon. As the entire Moon solidified, these "cumulate" layers of olivine and other iron-rich minerals would make up the mantle. Later, the mantle partially re-melted, creating liquids that erupted onto the surface as basalt lava and formed the dark lowland plains – the maria.

So how does the Kaguya interpretation hold up on examination? Estimates of crustal thickness are of the current Moon, after the basins formed. There is no particular reason to suppose that a given basin-forming impact occurred in terrain of thin crust – the crust is thin here because the basin formed. True enough, some of these impact features are very large – the Imbrium basin, a large crater on the western near side, is well over 1000 km in diameter, large enough to have punched through the thickest sections of the crust, one would think. Indeed, Imbrium is one of the sites that the Kaguya team propose as having excavated mantle. So they are finding these areas in the places where one might expect to find them.

Olivine is a very common mineral and abundant in the lunar crust. A curious fact is that olivine grains in lunar highland rocks tend to have high amounts of calcium, a minor element but a key diagnostic of the crystallization environment. In Earth rocks, olivine formed at depth has very low concentrations of calcium. My colleague, the late Graham Ryder concluded that the olivine crystals in dunite (a rock made up almost completely of olivine) from the Apollo 17 site – a sample proposed as a piece of the lunar mantle – likely came from the accumulation of crystals at a depth of only a few kilometers, far shallower than the tens of kilometers depth to the mantle.

Because the Kaguya spectral mapper is detecting only the presence of olivine, we cannot distinguish between pure olivine and olivine crystallized with plagioclase, what lunar scientists call troctolite. Troctolite is common in the Apollo highland samples, but is a relatively rare rock on Earth. It consists of (more or less) equal parts olivine and plagioclase, a calcium- and sodium-rich silicate mineral. Troctolites make up some of the most deeply derived rocks found in the Apollo collections, but all studied to date seem to be of crustal, not mantle, provenance. There is no objective evidence that the olivine seen by Kaguya is not derived from troctolites and/or dunites of crustal (not mantle) origin.

The long held desire of lunar scientists to sample deeper levels of the Moon is understandable, but we must proceed cautiously. Just as a sample return mission to the floor of the largest basin on the Moon is no guarantee that we will obtain the rocks needed[8] to answer questions about early cratering history, the new finding of abundant olivine on the Moon does not mean that pieces of the mantle are lying on the surface, awaiting collection by some future mission. The Kaguya findings are intriguing and very interesting, but not definitive evidence for the presence of mantle fragments on the lunar surface.

Topics: Lunar Exploration, Lunar Science

Links and references

1. mantle, http://en.wikipedia.org/wiki/Mantle_%28geology%29
2. xenoliths, http://en.wikipedia.org/wiki/Xenolith
3. seismometers, http://en.wikipedia.org/wiki/ALSEP
4. new scientific paper, http://www.nature.com/ngeo/journal/vaop/ncurrent/full/ngeo897.html
5. exposed on the surface, http://news.bbc.co.uk/2/hi/science_and_environment/10511064.stm
6. Kaguya mission, http://www.kaguya.jaxa.jp/index_e.htm
7. olivine, http://en.wikipedia.org/wiki/Olivine
8. no guarantee, http://blogs.airspacemag.com/moon/2010/01/11/robotic-sample-return-and-interpreting-lunar-history-the-importance-of-getting-it-right/

Comments

2. I have read (a long time ago) that some scientists expect that the central peaks of some large craters (those formed by impacts at a high angle?) may represent upthrust mantle material. Do you have a comment on that?

Comment by Peter Kokh — July 8, 2010 @ 10:53 am

3. Peter,

the central peaks of some large craters (those formed by impacts at a high angle?) may represent upthrust mantle material.

Central peaks form in craters between about 40 and 200 km in diameter; smaller ones do not have central peaks and larger features show multiple interior rings. We know from terrestrial impact craters approximately from what depths central peak material is derived; empirically, peak material comes from depths on the order of 10% to 20% of the observed rim diameter. Assuming such scaling roughly applies to the Moon, that means lunar central peaks come from depths as shallow as 4 km and as deep as 40 km. That range of depth is entirely crustal on the Moon; the average thickness of the crust is about 70 km.

That said, in areas where the crust has been thinned previously by a large basin impact, the mantle might be close enough to the surface such that a later-forming crater could contain mantle in its central peak. So it is possible, but would require special circumstances.

Comment by Paul D. Spudis — July 8, 2010 @ 12:09 pm

4. Would there be any primordial water left over from when it was trapped in the mantle when it condensed? Giant impacts might then cause such water to be released back into the Moon's hydrosphere.

Comment by Warren Platts — July 9, 2010 @ 8:35 am

5. Warren,

Would there be any primordial water left over from when it was trapped in the mantle when it condensed?

I guess it would depend on what you mean by "primordial." As far as we can tell, water is one of the very last things to condense from the nebula — that's why most of it is found in the outer Solar System, where the temperatures are coldest. Some of that water might be incorporated into the mantle, but the mechanism of how that might have happened is unclear. The Moon doesn't have mobile plates, so its water must have been incorporated into its mantle at the time that it was created. All of the current models of lunar origin have difficulties explaining this.

Comment by Paul D. Spudis — July 9, 2010 @ 8:58 am

NASA's New Mission and the Cult of Management

July 10, 2010

O Fortuna! Irony intentional?

During a recent interview on Al Jazeera television[1], NASA Administrator Charles Bolden outlined NASA's new priorities. His remarks became headlines[2] as the previously ignored story about the redirection of the space agency toward international diplomatic outreach and global climate change research finally reached the many who still hold NASA in high regard. Beyond the inane and vacuous policy comments[3], one statement by Administrator Bolden went virtually unnoticed: "We're not going to go anywhere beyond low Earth orbit as a single entity," he revealed, "The United States can't do it.[1]" If it's possible to shock Americans into paying attention to NASA and our national space program, those words might do it.

It's one thing to assert that the Unites States desires more international collaboration as a matter of policy for reasons of fostering alliances, developing new cultural ties, or even to promote world peace. It's an entirely different proposition to assert that the United States has lost the ability to reach for the stars, that America is incapable of exploring space alone. Have we become comfortable with the idea that it's politically incorrect to have pride in our nation's abilities – past, present and future?

The natural reaction of many Americans is to strongly protest such a notion by asserting that our technology is the greatest in the world

and that we can design and build anything we choose to. However, one can not escape the reality that the statement quoted above was made by the head of the agency charged with sending U.S. astronauts into space, a task that Americans have watched NASA carry out for almost 50 years. A more troubling thought is that maybe he is right.

I've worked with NASA for 25 years on experiments, research, and advanced planning and missions. I've been involved with studies to put a base on the Moon since 1984 (the year of the first conference on lunar bases[4] at the National Academy of Sciences in Washington DC). In all of these efforts, I believed that our future in space was at stake and that we were preparing for humanity's return to the Moon, followed by journeys to the planets.

As the years wore on, it began to dawn on some of us that some within and outside the agency really didn't want to go anywhere or do anything in space. Working for NASA means being involved in the process of working. How you structure a project, who gets assigned what work and how many meetings you conduct is the formula that elevates you faster than when (or even if) you deliver a product, a mission or a program. This culture becomes even more evident as one ascends the chain of command. The farther up the organizational ladder you go, the further removed from the productive segment of the agency you are, with product being replaced by an ever multiplying and bewildering variety of workshops, seminars and management training retreats. There exists in America today a thriving industry dedicated to convincing people who have no organizational or management skills that in fact, they are all excellent organizers and managers. Much of this sub-culture is accurately (and hilariously) portrayed in the comic strip Dilbert[5].

Management fads like Total Quality Management[6] regularly come and go. NASA has dabbled in Faster-Cheaper-Better[7], Spiral Development[8], Earned Value Management[9], and many others. Each new initiative is unveiled as The Answer, the magic beans solution that will re-establish the NASA of our forefathers – smart, productive, innovative, and competent. Management thrives by producing more middle managers. People joke about highway projects where one guy digs a hole while five others lean on their shovels, watching him work. That old chestnut always gets a laugh because we've all seen it at some point. You see it a lot within NASA. As administrative cost consumes more and more of the budget pie, the slice that represents funding for the productive sector of the agency keeps shrinking.

Bolden's comment is tragic, not in its misunderstanding but in its verity. America cannot go back to the Moon by itself not because we lack the wealth or the technical skills but because we have developed a culture of management bloat and process-orientation that proliferates bureaucracy. NASA is comfortably cushioned from the reality of their organizational maladies by the amazing legacy inherited from those who did great things in the past. They may believe they're immune to the Cult of Management, but as evidenced by Bolden's statements, the agency has changed. NASA has become a "feel-good" bureaucracy, stuck on the idea of doing one-off stunts in space for public approval and a guaranteed institutional lifetime, paid for by the public. This attitude both inside and outside of the agency has lead to the demise of our national spaceflight capabilities. Stagnant entities reflexively dwell on past glory; dynamic ones build upon them to create something both significant and permanent.

Occasionally, I'm asked why I stay in the space business. I do so because I believe in the mission of space exploration and in the importance of moving humanity into the Solar System. Every now and then, an opportunity

arises outside the boundaries of normal business to do something productive and create a lasting legacy – a time when the stars align and a leader refuses to follow the established rules or to unimaginatively subscribe to the conventional wisdom.

Currently, the "right stuff" manifests itself as a pattern of waiting for a propitious political opportunity and a few far-seeing people to seize the day and push through something that otherwise would be buried by "process." In my opinion, the Vision for Space Exploration[10] was such an opportunity – a path forward, achievable in stages, that would have created a legacy of space faring capability. The idea of actually doing something made the Vision a nonstarter to many within the agency – it challenged their worldview of process over product. They are content to merely manage an organization. The goal of leading America and the world into the Solar System has become a slogan in a strategic plan, an ever-fading banner over an office door.

Is any of this fixable? Perhaps. For now, the opportunity to strengthen the U.S. space program, an endeavor that has traditionally pushed technology development and stoked our economic engine, has been kicked aside. But by their very nature, new opportunities arise unbidden and unforeseen at irregular intervals. I wonder if we'll do better next time.

Americans need to wonder.

Topics: Lunar Exploration, Space Politics, Space and Society

Links and references

1. recent interview, http://www.washingtonexaminer.com/politics/NASA_s-new-mission_-Building-ties-to-Muslim-world-97817909.html
2. remarks became headlines, http://www.nasawatch.com/archives/2010/07/wide-range-of-r.html
3. inane and vacuous, http://newsbusters.org/blogs/jeff-poor/2010/07/05/krauthammer-rips-nasa-chief-declaration-improve-relations-muslim-world
4. first conference on lunar bases, http://ads.harvard.edu/books/lbsa/
5. Dilbert, http://www.dilbert.com/
6. Total Quality Management, http://en.wikipedia.org/wiki/Total_quality_management
7. Faster-cheaper-better, http://www.space.com/peopleinterviews/mccurdy_profile_000419.html
8. spiral development, http://en.wikipedia.org/wiki/Spiral_model
9. earned value management, http://en.wikipedia.org/wiki/Earned_value_management
10. Vision for Space Exploration, http://www.spaceref.com/news/viewpr.html?pid=13404

Comments

2. *"Stagnant entities reflexively dwell on past glory; dynamic ones build upon them to create*

something both significant and permanent."

We are totally on the same page here. Your assertions in some recent comments that the agency simply had to have a single particular goal and timeline and had to follow the Apollo model and that it just couldn't change to Flexible Path implied that NASA had gone stagnant.

Clearly the NASA of today is not the NASA of von Braun. Change can occur in one of two ways: gradually or in quantum leaps. The change that formed NASA in the first place was a quantum leap. The change since the mid to late 60s (I surmise bureaucratization started sometime shortly before Apollo 11) has been gradual and largely for the worse. Can NASA handle another quantum leap?

"How you structure a project, who gets assigned what work and how many meetings you conduct is the formula that elevates you faster than when (or even if) you deliver a product, a mission or a program."

This is a symptom of bureaucracy. It is also exactly backwards from the way things operate in any successful small business. How discouraging it must be for the truly bright lights to see the reward system operate this way at NASA.

"Management fads like Total Quality Management regularly come and go. NASA has dabbled in Faster-Cheaper-Better, Spiral Development, Earned Value Management, and many others."

To make matters worse, large companies tend to adopt these management fads too, because "that's how NASA does it".

"In my opinion, the Vision for Space Exploration was such an opportunity – a path forward, achievable in stages, that would have created a legacy of space faring capability."

I agree. I think that with Flexible Path, the achievable stages become more and smaller steps, leading to much more space faring capability and a much more robust space industry. The VSE goals are still there – but instead of groups of four heading to the moon ten congressional elections away, there are smaller, shorter-term goals that apply leverage to the whole effort – and instead we maybe get groups of thousands going there to stay.

"For now, the opportunity to strengthen the U.S. space program, an endeavor that has traditionally pushed technology development and stoked our economic engine, has been kicked aside."

I do not understand this view of the situation. What I see is quite the opposite: NASA finally going back to the edge of the envelope and the American economy stepping in behind to follow the path already well-marked by NASA, starting

with crew delivery to LEO.

Comment by Ed Minchau — July 11, 2010 @ 1:09 am

3. Mr. Bolden should be ashamed of himself for continuing to perpetuate the myth that NASA's relatively minuscule budget is unaffordable!

You could double NASA's manned spaceflight budget, an approximately $8.4 billion a year increase, and NASA's over all budget would still represent less than 1% of total Federal budget expenditures.

The scariest part of all this is the fact that the public believes it! A 2007 poll revealed that Americans in general thought that 24% of Federal budget expenditures were spent on NASA. 24%! That would be an $844 billion a year NASA budget!

There's a titanic amount of government waste in appallingly inefficient programs such as Medicare, Medicaid, and the totally unnecessary adventure in Iraq that's costing Americans hundreds of billions of dollars a year in waste.

NASA, on the other hand, actually creates wealth for the general economy.

To say that I'm extremely disappointed with Mr. Bolden– is an understatement!

http://blogs.discovermagazine.com/badastronomy/2007/11/21/nasas-budget-as-far-as-americans-think/

http://www.thespacereview.com/article/1000/1

Comment by Marcel F. Williams — July 11, 2010 @ 4:36 am

4. Ed,

I think that with Flexible Path, the achievable stages become more and smaller steps, leading to much more space faring capability and a much more robust space industry. The VSE goals are still there –

You and I simply disagree on this. I think that the Obama anti-Vision is fundamentally a non-serious proposal. It is all smoke-and-mirrors, designed to fracture the space community against itself (accomplished), pretend that we're going somewhere (a NEO mission in 15 years is not a space program, but a license for Powerpoint engineering ad nauseum), and spend the (increased) budget on widget-making and gadgets that may or may not be relevant to future spaceflight — all of this laying the groundwork for the termination of the human space program at some future date.

My point in this post is that by virtue of the prevalence of the Cult of Management at NASA, the agency is perfectly configured to spend and waste money, not to fly missions or develop technology (we get useful technology development by flying missions, not by research).

Comment by Paul D. Spudis — July 11, 2010 @ 5:37 am

5. You say " NASA has become a "feel-good" bureaucracy, stuck on the idea of doing one-off stunts in space for public approval and a guaranteed institutional lifetime, paid for by the public"

I dispute this, not that NASA is not an institution o one-off stunts, but that it has always been such. A read of the Apollo missions design history, where an initial idea (on paper only, not a serious proposal I believe) was to build a station in orbit, and go from there to the moon, reveals as much the one-off stunt thinking

The Saturn-V, humanity's most powerful machine, was a profligate waste of resources, throwing away most of itself to simply return one tiny capsule with 3 men, and leave nothing of productive use in a place we can use them. In and of itself, I believe Apollo was a great program. But it was a once-off, and did not lead to the things that we should have had – factories in orbit, hotels in orbit, and all the things that *finally* private industry is talking.

I reluctantly concede, much as I never thought to do so, that the current path is perhaps the correct one. Leave the space commercialisation to the private sector, where the passion still thrives

Comment by Mark Stacey — July 11, 2010 @ 5:47 am

6. Mark,

But it [Apollo] was a once-off,

I agree and have said as much in this blog on several occasions.

*and did not lead to the things that we should have had – factories in orbit, hotels in orbit, and all the things that *finally* private industry is talking. I reluctantly concede, much as I never thought to do so, that the current path is perhaps the correct one.*

Fine, except what you've just described is NOT the "current path" — instead of private investors putting up capital to develop a commercial spaceflight system, we're going to continue to spend NASA's $20 billion/year plus. The difference between the old path and new one is that with the new path, we won't get anything for the money.

Leave the space commercialisation to the private sector, where the passion still thrives

Perhaps. But you've just argued for not pursuing the "current path," which is to make NASA the agent for commercialization.

Comment by Paul D. Spudis — July 11, 2010

@ 8:24 am

7. Having worked for my share of "#1 Defense Contractors", I can tell you that program management is a challenge for any large program, NASA, DOD, whatever. The difference has typically been that DOD programs produce many of the eventual product, where NASA produces few, so the costs per unit end up looking extremely high for the eventual outcome (as opposed to only really high for DOD). This is the nature of large programs, and it's not unique to NASA.

The old saying "Give a man a fish and you feed him for a day. Teach a man to fish and you feed him for a lifetime" applies to NASA as it does for everything else in life, and why people don't want NASA to spread their knowledge to U.S. companies is shortsighted. Transferring the duties to U.S. companies of getting cargo and crew to LEO will allow NASA to focus on their goals outside of LEO, and that's really where NASA's talents and abilities shine.

The lesson of Constellation was that NASA has a hard time managing $100B programs that don't pay off for 20 years, so don't create all-in-programs that big again. Instead, break down the effort into smaller, easier to complete chunks that provide value when completed, and serve as building blocks for the next set of goals. The goals are easier to manage, the payoff is quicker, the capabilities create stepping stones that are valuable on their own, and large programs don't get a chance to overwhelm the entire agency. This is what the new NASA plan does, and part of the reason I fully back it.

Regarding the comment by Administrator Bolden that "We're not going to go anywhere beyond low Earth orbit as a single entity", I see that as a reflection of what has happened with Constellation, and the huge costs involved in attempting human exploration beyond LEO. It's not that we don't have the capability to do it, it's that we don't have the national will to spend that much money on it without a better payoff than "because we need to". The ISS, which has barely survived congressional funding challenges, is a good example of that.

Consider just the Ares I program, which was estimated to cost $40B. If Griffin had chosen to use the Delta IV Heavy instead, we would have saved at least $$35B, and would be planning the first launch of Orion this year on a Delta IV Heavy. That $35B would have paid for extending the ISS, and even the development of a smaller Shuttle replacement like Dream Chaser. Instead, decisions at the top (Griffin) lead to a program that was too big to manage for NASA, and too limited in it's end product to have universal support.

NASA can afford to do BEO missions, but it

cannot afford to do them by themselves. As soon as a task becomes repeatable enough (like sending crew to LEO), NASA needs to hand those tasks off to commercial companies (or willing partner countries). This reduces the management scope for NASA (and reduce their bureaucracy), introduces competition to lower prices, and allows NASA to stay focused on the things they do best. Commercial companies also bring marketing resources to bear, and they have a better ability than NASA to encourage new applications for products and services in space. It's a win-win for everyone.

Comment by Coastal Ron — July 11, 2010 @ 1:34 pm

8. Coastal Ron,

This is what the new NASA plan does,

I don't agree. The new plan purports to break down the task into steps as you describe, but the actual effect is to spread money around randomly, increase entropy and produce nothing.

NASA can afford to do BEO missions, but it cannot afford to do them by themselves.

That is entirely dependent upon what you think the mission is and how you chose to implement it. You still conflate Constellation Program of Record with the VSE. The former might not be affordable, but the latter always was and still is. But with the new anti-Vision, all that has been cast away for a promise of some accomplishment in the indefinite future.

Comment by Paul D. Spudis — July 11, 2010 @ 1:45 pm

9. Paul D. Spudis said: *You still conflate Constellation Program of Record with the VSE. The former might not be affordable, but the latter always was and still is.*

The Vision for Space Exploration is just that, a vision, and Constellation was Griffin's way to start achieving part of the VSE. Constellation, however, was not going to achieve enough for the money required – it was the wrong approach to achieving the VSE.

I agree with the overall goals of the VSE, but the vision talks about goals, and not the specifics of how to achieve them, nor the schedule of when they are to be done. NASA has a limited budget, and after this next congress gets done figuring out how to handle the growing deficit, it may be smaller still. Whatever we do has to build stepping stones that others can use to follow, otherwise we're wasting money re-accomplishing things we're already done. For instance, the ISS is a stepping stone, but Apollo was not. We need legacy, not memories.

You see the proposed budget as the "anti-

Vision", but I think that's FUD.

How is funding robotic precursor missions to the Moon, Mars and its moons, Lagrange points, and nearby asteroids, not supporting the VSE?

How is funding short turnaround programs (5 years or less) such as in-orbit propellant transfer and storage, inflatable modules, automated/autonomous rendezvous and docking, closed-loop life support systems, and other next-generation capabilities, not supporting the VSE?

How is keeping the ISS so we can continue to learn how to live and work in space, not supporting the VSE?

How is human-rating existing launch vehicles and developing new spacecraft that can ride on multiple launch vehicles, not supporting VSE?

How is modernizing the Kennedy Space Center to increase it's operational efficiency and reduce launch costs for all users, not supporting VSE?

If anything, what Obama/Bolden propose is to build the foundation for achieving the VSE, but not the details of how it will be ultimately achieved. This is smart, because the technology and knowledge we need to accomplish the VSE changes over time, and locking ourselves into a one technology ecosystem is not the cost-effective way to do anything in space. We need multiple capabilities for all of our transportation needs, and we need a commercial system in place that allows competition to gradually lower costs and expand the space marketplace.

With the proposed plan, we actually end up with a bunch of useable systems and technologies at the end of Obama's term, instead of an unfinished, outdated, over-budget grand space plan. Which is what Constellation was.

I think the VSE is a laudable goal, but it's not an endpoint, nor does it need to be done all in one chunk. The new NASA proposal works on laying a firm foundation to start the journey towards achieving the goal, and it will provide usable stepping stones for those follow.

Comment by Coastal Ron — July 11, 2010 @ 3:08 pm

10. Ron,

I think the VSE is a laudable goal, but it's not an endpoint, nor does it need to be done all in one chunk.

No one ever claimed it was an "end point" — lunar return was the beginning of the creation of a new space faring system, one built through the use of off-planet resources, that can routinely access cislunar space and the planets beyond. The anti-Vision discards all of this without any thoughtful alternative.

The new NASA proposal works on laying a firm foundation to start the journey towards achieving the goal, and it will provide usable stepping stones for those follow.

I know that you believe this. I don't. We disagree.

And with that, I believe that our discussion here is concluded.

Comment by Paul D. Spudis — July 11, 2010 @ 4:10 pm

11. *You and I simply disagree on this.*

Well, we could both be wrong, too. Maybe things don't work out as well as I think they could, maybe they don't end up as much of a CF as you think they could.

I guess I view space development as analogous to programming computers in some ways. For instance, I see things like commercial passenger service to low earth orbit, LEO propellant depots, CELSS, inflatable habitats, and so on as subroutines – breaking down a problem into component steps, solving them, and then re-using those components in other contexts.

Let's look down the road 20,30,40 years. What should routine access to the moon look like? To me, that means:

- fly to the nearest spaceport - transfer in the terminal to your launch vehicle, perhaps one of several commercial launches that day - launch to LEO - transfer to a LEO station, wait - transfer to a vehicle that only makes the LEO-L1-LEO trip, refueling at each end of the trip - travel to L1, transfer into station there, wait - transfer into a vehicle that only makes the L1-Luna-L1 trip, refueling at each end - land at de Gerlache station.

Can this be done with Flexible Path? Maybe. Could it be done with Constellation? Heck no. Do you have a better plan? Maybe, and I'd like to see your thoughts on how it should be done.

I've been posting my ideas on how it should be done for years, including some recent lengthy posts. From something I wrote three years ago:

The solution is to decouple the mission from the implementation. It matters that it gets done, not that NASA does it or that the agency does it in a specific carved-in-stone way. NASA can't do it all by itself anymore, so it shouldn't even try. No more of this business of NASA building their own brand new launch vehicles and their own brand new manned capsules and their own brand new moon landers and their own brand new moonbases and micromanaging every detail. It is a brittle way of doing things, and the slightest hiccup in the yearly budget process or the slightest failure along that critical path brings everything to a screeching halt.

Comment by Ed Minchau — July 11, 2010 @ 5:33 pm

12. The focus on management fads and process rather than product isn't unique to NASA. It's not a new problem at NASA, either. If anything, I think NASA's new approach helps address this problem.

NASA's new approach allows it to use commercial services rather than its most bureaucratic sections, such as NASA's government rocket divisions. It gives a lot of other opportunities for using more streamlined services from commercial space and academia. Meanwhile, it strengthens NASA's more productive segments, such as its robotic missions, by funding more robotic precursor missions, Earth observation missions, and general space technology developments and demonstrations.

Meanwhile, NASA's new plan gives plenty of opportunity for reform within its HSF areas by continuing the ISS, doing heavy lift engine development and related work, continuing Orion in a CRV role, and running various exploration technology demonstration missions. We will get to see which ones work and which ones don't work … and unlike megaprojects like Ares I/V, these projects will be small enough to shut down if they aren't working.

Comment by red — July 11, 2010 @ 6:29 pm

13. On an unrelated note, I'd be interested to see what ideas Paul might have for improving NASA's plans within the current overall framework. In other words, let's assume that NASA is flat-out going to fully support Earth observations and green Aeronautics based on Obama's general political inclinations. Also, the ISS will be kept, used, and improved. Thus, those budgets, and the baseline Science and Aeronautics budgets, are off-limits.

In addition, assume we're going to have a Flexible Path sort of plan that can include the Moon (but not first). We get $7.8B for exploration technology development and demonstrations, $3B for HLVs, $6B for commercial crew and cargo, $3B for robotic precursors, and so on through 2015. Partial use of Constellation hardware might free us from wasting some of the $2.5B on Constellation transition, and use of Shuttle-derived hardware might give us access to some of the $2B of KSC modernization funds.

In that context, how could things be improved? Are there specific robotic precursor missions we should fly, specific technologies we should develop on Earth or demonstrate in space, specific new missions we could give to the commercial crew, commercial cargo, AR&D space tug, or Orion-based CRV that aren't in NASA's current plans but that could be within the general approach that it seems we will be taking?

For example, one of the flagship technology demos is for SEP, and another is for aerocapture. Paul has spoken against aimless dabbling in technology development. In these cases, a certain amount of focus may be given because the demos will (possibly) fly to Mars and carry science instruments there. The SEP mission also has a certain amount of built-in focus because it's demonstrating military technology (from DARPA). Is this, in addition to NASA's hopes for SEP cargo transport for exploration, enough focus? If not, what would work, and if so, how can we do something similar for other NASA technology work?

Comment by red — July 11, 2010 @ 6:50 pm

14. Paul,

My problem with the VSE is the "E" part. I think that we need to press the pause button on Exploration and start doing some Development for a while.

When most people think about the term "exploration", it means going to new places where we've never been before for no other purpose than to discover or just to be able to say that we've done it (i.e. Glory). The concept of exploration is exciting to many people. It gives us a new knowledge of planetary bodies. It gives a nation great pride in accomplishing a difficult thing. It inspires our kids to go into math, science, and engineering careers.

But does exploration provide access to natural resources? Not unless it is specifically designed to do so. Does it make the next steps easier? Generally not. In fact, does exploration make development more difficult? Yes, because it diverts resources into non-productive ventures and away from sustainable steps. Does exploration make a plan vulnerable? Yes, because, fundamentally, exploration is probably not actually essential. It is mostly for pride or curiosity. But if there are budget overruns then the plan can be scrapped or switched to some other plan.

The exploration emphasis of the VSE was its Achilles' Heel. Obama could say that we shouldn't return to the Moon because, frankly we've been there before? True, we were there before but in an exploration mode not a development mode. That's the difference. But Constellation had huge landers because an Ares V could accommodate it. And Ares V was so large because it needed to also go to Mars. And we needed to go to Mars because...because we've never been there before. But since an HLV is considered to cost too much then we just scrap that and go to an asteroid because "we've never been there before". And we can always push off the date of going to an asteroid because...going now or going a bit later...what does it matter?

Constellation's lunar base wasn't really about real development but just gaining the skills necessary for the next step which was the exploration of Mars. So there was real talk about abandoning the Moon base while we spent money going to Mars. Neither Bush nor Obama really cared about the Moon because the Lunar exploration offered fewer exploration opportunities than Mars or an asteroid.

But from a development perspective the Moon becomes essential. It has remarkable natural resources. It is relatively nearby. Development can be done relatively cheaply because we don't need to safely transport and sustain people there in order to develop a base or mine. Work can be done there telerobotically. What's more (unlike the vice versa) development can dramatically support exploration through the provision of in-space resources.

BUT America's lust for Glory will not be denies. Apparently, we MUST be in the lead. We MUST have a manned program however much it costs and however much accidents might set back the program.

So, my solution is to grant the cheapest form of novel manned space exploration program. I see this as being a trip to an asteroid using EDSs which are launched without the need for a HLV. Then, the money saved by doing such a mission on the cheap should be directed to incentivize commercial companies to expand America's economic sphere through cis-lunar space including the lunar surface. We're talking fuel depots, space tugs, lunar landers, lunar water extraction, conversion to ascending fuel, and transport of lunar resources back to LEO. Lunar rock collectibles, large GEO communications satellites, supplying orbital hotels, even SPSs — all of these would be potential commercial spinoffs.

> instead of private investors putting up capital to develop a commercial spaceflight system, we're going to continue to spend NASA's $20 billion/year plus. The difference between the old path and new one is that with the new path, we won't get anything for the money.

I'm confused. My understanding is that SpaceX & company have put quite a bit of their own money into development of their own rockets and will end up owning them. Secondly, I'm not imaging commercial companies spending ALL of NASA's $20 billion/year plus. I'm thinking that 15-20% of NASA's budget incentivizing commercial companies to develop propellant depots, lunar landers, etc. "Won't get anything for the money"? Isn't NASA going to get reduced cost deliveries to the ISS because of its incentivizing of commercial companies?

Comment by JohnHunt — July 11, 2010 @ 7:27 pm

15. Paul, excellent post! I applaud your antidis-establishmentarianism. We ought not throw the baby out with the bath water.

If the world doesn't come to an end in 2012, the stars may line up once again, and a person, an organization, and the times will come together to make historic achievement possible once again.

Keep the faith brother!

Trackback: @ bautforum.com

Paul Spudis (of lunar water fame) takes on the (in)famous Bolden's interview for Al-Jazeera. He starts by noting that it has now been officially confirmed that NASA can't go beyond LEO...

Comment by Warren Platts — July 11, 2010 @ 9:19 pm

16. JohnHunt,

fundamentally, exploration is probably not actually essential. It is mostly for pride or curiosity.

Exploration includes much more than that, as I have discussed at length in a previous post:

http://blogs.airspacemag.com/moon/2010/01/25/have-we-forgotten-what-exploration-means/

Comment by Paul D. Spudis — July 12, 2010 @ 5:10 am

17. Ed,

Do you have a better plan? Maybe, and I'd like to see your thoughts on how it should be done.

I've outlined what we should be doing on this blog repeatedly for the last year and a half. This presentation, written over 4 years ago, includes a straw man architecture that outlines the basics:

http://www.spudislunarresources.com/Papers/The%20Vision%20and%20the%20Mission.pdf

But more importantly, it emphasizes that if you don't understand your mission, you're just wasting time and resources.

Comment by Paul D. Spudis — July 12, 2010 @ 5:15 am

18. red,

NASA's new plan gives plenty of opportunity for reform within its HSF areas by continuing the ISS, doing heavy lift engine development and related work, continuing Orion in a CRV role, and running various exploration technology demonstration missions. We will get to see which ones work and which ones don't ... and unlike megaprojects like Ares I/V, these projects will be small enough to shut down if they aren't working.

Congratulations. You've just outlined the essence of the anti-Vision. I have no doubt that all of these "efforts" are scheduled for eventual cancellation.

I'm a little confused on one point, however — if the administration gets what it wants, Constellation is gone. How then is it a "megaproject" that can't be shut down?

In that context, how could things be improved?

For a start, they could keep the Vision for Space Exploration, which was not about "Constellation" but a strategic direction and set of long-term objectives. If the new plan had such direction, it might be a serious proposal — it doesn't and it isn't.

Comment by Paul D. Spudis — July 12, 2010 @ 5:38 am

19. We have gone from one extreme to another. Both are flawed both will lead to over-all stagnation for our NASA. There is some middle ground one with some flex, a goal and some defined steps or milestones or "prizes" with some degree of accountability to at least focus the effort in a general all on board the same bus direction. What have now is anyone guess leading to total KAOS.

Comment by Doug — July 12, 2010 @ 7:39 am

20. Doug,

We have gone from one extreme to another. Both are flawed both will lead to over-all stagnation for our NASA....What have now is anyone guess leading to total KAOS

And that suits some people just fine.

If you lose CONTROL, why wouldn't you expect KAOS? :^)

Comment by Paul D. Spudis — July 12, 2010 @ 8:20 am

21. Paul D. Spudis said:

"I've outlined what we should be doing on this blog repeatedly for the last year and a half. This presentation, written over 4 years ago, includes a straw man architecture that outlines the basics:"

Excerpt from page 27 of your paper "The Vision and the Mission":

{Need a mission first; cannot judge whether a flight or widget is relevant to your aims if you don't have any.} Mission: Go to the Moon to learn to live and work productively in space.

Basic principles: Small, incremental building blocks Cumulative – each step builds on previous one Early accomplishment, early capabilities Robotic presence first, then people

Pretty much what you describe here is what I said the new NASA plan does. The difference is that you're talking about this in the context of it happening on the surface of the Moon, whereas the NASA plan is for many destinations.

And then you stated this in response to Ed Minchau:

But more importantly, it emphasizes that if you don't understand your mission, you're just wasting time and resources.

YOUR mission is to do stuff on the Moon. That has always been clear. NASA's proposed budget (their "mission") is to develop the basic technologies that NASA (or anyone) can use to go to the Moon, or anywhere else.

I noticed in your presentation that you only talked about what it would take when the "mission" was on the Moon. That reminded me of the cartoon of the guy doing a complicated math equation, and in the middle he writes "and then a miracle happens". Your Moon "mission" is the same, in that it talks about all the wonderful technology and capabilities that you see being landed on the Moon (and later people), but somehow the ability to get them there, affordably, is completely ignored.

You just "assume" low cost, multiple path transportation systems just "appear" and work at such an efficient level that the U.S. Taxpayer can afford to pour all that money into your Moon "mission". You're skipping a step.

Your Moon "mission" proposal is a fine strawman, and someday I look forward to something like that happening, but if we don't create a more efficient transportation system, you won't be alive by the time the U.S. Taxpayer can afford to do what you propose.

Comment by Coastal Ron — July 12, 2010 @ 12:00 pm

22. What you're describing here is Jerry Pournelle's Iron Law of Bureaucracy:

http://www.jerrypournelle.com/reports/jerryp/iron.html

A pity.

Comment by BD — July 12, 2010 @ 12:08 pm

23. *You just "assume" low cost, multiple path transportation systems just "appear" and work at such an efficient level that the U.S. Taxpayer can afford to pour all that money into your Moon "mission". You're skipping a step.*

The only thing I'm assuming is that NASA will continue to get the same budget, more or less, indefinitely. I base this assumption on the past 30 years of spending on space, which has hovered between 0.5 and 1% of the federal budget. I further assume that the objective stays fixed but that schedule is the free variable — if it costs too

much to do it by x, we do it by y. But we do it. This was the basic funding/schedule assumption of the original VSE.

In contrast, your assumption seems to me much more problematic — you assume that pouring money into a bureaucratic black hole at NASA in the absence of any real mission or destination will result in: 1) cheap, routine commercial access to space and 2) all the technology we need to go to the planets. I submit that the history of the agency (e.g., Delta Clipper, X-38, NSLI) is counter-indicative to your assumption.

Comment by Paul D. Spudis — July 12, 2010 @ 12:39 pm

24. Paul D. Spudis said:

"you assume that pouring money into a bureaucratic black hole at NASA in the absence of any real mission or destination will result in: 1) cheap, routine commercial access to space and 2) all the technology we need to go to the planets. I submit that the history of the agency (e.g., Delta Clipper, X-38, NSLI) is counter-indicative to your assumption."

Your second statement is an interesting one, in that NASA was funding specific technologies, and not specific services. The goals of those programs were to see if something could be done, not how inexpensively they could be done. This is an important distinction.

I do agree with your overall premise that large programs can turn into "bureaucratic black holes". Having worked at large defense contractors, and following the progress of many dead-end government programs, I know that the best of intentions can turn into money pits. Since good intentions don't seem to be the solution, I see transforming "programs" into "commercial services" as a good model.

If the military needs to move a corp of troops across the world, they rely upon contracted commercial transport. The same with cargo they need moved to non-hotzone areas. NASA has traditionally felt they need to create and own their own transportation systems, and this was so with Constellation. What I advocate is that they only need to own the new & unique vehicles, and that once they are perfected, they should be handed off to commercial market as soon as practicable.

NASA (Griffin) ignored two perfectly safe and reliable commercial launchers (Delta & Atlas), and decided to spend $30-40B just for a crew launcher. If they had gone with Delta IV Heavy instead, they would have saved $Billions, and would have freed up $Billions that could have been going towards your proposed Lunar Reconnaissance Orbiter or Lunar Outpost Landers.

Without a commercial & competitive transportation in place, we will never be able to afford a sustained presence beyond LEO. If NASA is not capable of managing large transportation programs, the only alternative is commercial. This should not be a big leap of confidence, since NASA & DOD already rely on Atlas & Delta for most of their satellite needs, and the commercial transportation model is how our country operates anyways.

NASA needs to get out of the transportation business, and the new budget starts that process. And with the money they save, you'll be able to finally start your lunar program... ;-)

Comment by Coastal Ron — July 12, 2010 @ 2:12 pm

25. Please do include posting other space exploration programs apart from NASA, Keep Your blog updated with INDIA's space exploration projects.. I appreciate the work done by you.

Check out my blog http://tarunvipparthi.wordpress.com/2009/09/28/mission-to-moon/

Comment by ramya — July 12, 2010 @ 4:02 pm

26. [...] if you just have to be angry at Administrator Bolden, read this. "The United States can't do it", eh? "Can't" and "Would [...]

Pingback by Conflict Resolution...of a Sort. « The Space Geek — July 12, 2010 @ 4:03 pm

27. Paul,

Thank you for referring me to your previous post re: exploration. I am relatively new to your blog and so had not read that before. ————
Let me suggest some practical definitions:

Explor(ation) – 1. to traverse or range over (a region, area, etc.) for the purpose of discovery (dictionary.com)

Prospecting – To further explore a resource to determine its value (my definition)

Develop(ment) – 1. to bring out the capabilities or possibilities of; bring to a more advanced or effective state (dictionary.com)

Sustainable Development – to develop resources and capabilities in a way where they will continue to be used even if the initial source of inputs is lost (my definition)

———————— What is needed is the incremental, cumulative build-up of space faring infrastructure that is both extensible and maintainable, a growing system whose aim is to transport us anywhere we want to go, for whatever reasons we can imagine, with whatever capabilities we may need.

To me, what you are talking about is Sustainable Space Development. If you were to ask most people (whether in NASA or out) what "space exploration" was, they would probably not come up with the above description. But if you were to ask them what "sustainable space development" meant, they would probably come up with something close to your paragraph.

This is not an esoteric issue. The big problem with both Constellation and the New Path is that their goal is not Sustainable Development but mostly Glory. That's clearly the case with the New Path but one could argue that Constellation would have built an architecture which could have enabled the develop of lunar resources on an industrial level. But they were discussing abandoning the base while money was being spent going to Mars.

BUT look at what is being accomplished with the COTS/CRS – SpaceX situation. When the NASA flights run out, they already have 9+Iridium flights manifested. And by 2014 there will probably be a lot more flights manifested. NASA puts in some initial inputs, technologies are developed, and companies start servicing a market which is sustainable.

All I am proposing is that the COTS/CRS approach be applied beyond LEO in order to develop a sustainable, commercial, cis-lunar economy. Saying that NASA needs to section out a part of its budget for commercial "development" incentives would be the way to achieve this. To achieve this we need to succeed in making the argument that "sustainable development" needs to be recognized as legitimate, discrete goal of the US space program.

Comment by JohnHunt — July 12, 2010 @ 8:13 pm

28. Mr. Spudis,

Another excellent article. Thank you. I really wish fans of Obama's space plan would stop with the precursor mission myths. Where do the precursor missions lead us? And more importantly, when does the end point happen? All I've seen of the Obama plan is talk about a mission to an asteroid 15 years from now. A mission to Mars... that's 40 or 50 years in the future. That's a plan? We're trading VSE for this? Without a specific timeframe to do anything, Obama's space plan is a waste of time and resources (other than commercial support for ISS resupply). Space anything is not on Obama's radar.

Bolden should be ashamed of himself. Obama and him are getting good at saying "we can't."

Comment by Jim R. — July 12, 2010 @ 10:04 pm

29. John,

To achieve this we need to succeed in making the argument that "sustainable development" needs to be recognized as legitimate, discrete goal of the US space program.

I'm with you on the statement of goals. My point in writing the other blog piece was that what you call "development" has been historically a part of the concept of "exploration." The idea that exploration is instead a series of PR stunts, akin to a modern gladiatorial contest, is relatively new and apparently subscribed to by many who know no history.

Comment by Paul D. Spudis — July 13, 2010 @ 4:59 am

30. President Nixon decided to decommission America's heavy lift rocket, the Saturn V, and the US has been stuck at LEO ever since. But now it looks like the Congress is going to mandate that a heavy lift vehicle and Crew Exploratory Vehicle be built– immediately.

Once NASA has a heavy lift vehicle again, then a new American space age will begin. And everything we do at LEO and beyond LEO will be a lot easier and cheaper.

Comment by Marcel F. Williams — July 13, 2010 @ 5:45 am

31. John Hunt,

I think you have grasped the dilemma NASA is facing. That is to say, I think the Administration and Bolden/Garver equate Exploration with "glory" and "science", but not the historical definition that Paul mentions. My read is that the intent of the VSE was to subscribe to the historical definition of "exploration", although in execution, the decision makers at NASA lost sight of that intent, and we were left with a suboptimal implementation of the VSE. While we may, can, and will argue about the proper phasing of commercial launch capability, that misses the major point on the VSE. We (NASA) like to build and play with rockets, but we haven't yet grasped the enabling significance of the importance of the intent of the VSE. The Moon is the best initial destination because it allows us the best chance to learn whether or not, and if so, how to use in-situ resources to enable cheaper long duration stays off-planet (in pursuit of the goal to learn how to live off-planet), and in so doing, create wealth for the country instead of expending it. Can this ISRU thing really be viable? We don't really know for sure until we try. However, if we don't try, then we certainly will not succeed.

That is why many of us are upset at the new "direction"; there is no underlying meaning or unifying rationale for human space flight. "Science" can be achieved very effectively and a lot less expensively with robotic missions. And how much is it worth for "the Glory of it"? If we aren't focused on the basic tenet to learn how to live off-planet, then why should be spend the large amount of

resources it takes to simply put a fresh set of footprints on Mars or an Asteroid, notwithstanding the fact that one can't "step on" and asteroid?

Comment by Tony L — July 13, 2010 @ 7:03 pm

32. I wrote about this very subject myself today, and it was a pleasure to read your take on it Paul.

All the best ;-P, bowlegged148.

Comment by Lauren — July 16, 2010 @ 12:40 pm

33. Major agree!! Excellent article!

The idea that a NASA administrator would say the US can't and shouldn't go beyond LEO by itself is nonsensical to insulting to the country as a whole. We certainly can afford to do it alone (frankly one lesson of ISS is that international programs cost are far higher the national programs – and no one saves money), and the idea we shouldn't do it alone – like it's presumptuous of us to step out ahead of others REALLY grates!

As to you're management points. Hey its a civil service organization. Managers are paid by the size of their staffs not their results – so they try hard to bloat their staffs. Is someone surprised? This is the post office in space!

NASA has shown itself to be behind the times in the fields, and actively working to move their technology back to a higher cost, fewer flights, Griffin model.

I desperately wish NASA was directed to do CATS, or commercials were contracted with a large scale flight contract that would justify a new RLV and drive down costs – but no ones talking about any of that..

Comment by Kelly Starks — July 16, 2010 @ 1:47 pm

34. One comment in particular really hit home with me on your blog this week:

Ed Minchau: This is a symptom of bureaucracy. It is also exactly backwards from the way things operate in any successful small business. How discouraging it must be for the truly bright lights to see the reward system operate this way at NASA.

You and others are seeing this from the outside. Some of us who have been in the midst of the disenchanting NASA bureaucracy have been facing this very personally from the inside for years.

Shuttle got off to a good and even handed start.

ISS has been a dismal place to work because of the proliferation of meaningless and unproductive meetings led by managers who never had any experience in the productive segments of the program since the management shake up in 1993. That is now 17 years.

But the worst was Constellation starting with people at the top who had never been a part of that productive segment; it started off on a bad foot and never recovered. It was openly visible to all concerned; program and project managers who had never managed anything in their careers. Design and development managers who had never designed or developed anything. And these same people, some called them NASA's "best and brightest", would tell those of us with experience, "what makes you think you can do the job".

Some of the people at the top of Constellation still today need to be kept from the leadership positions in whatever form the new program takes.

Comment by G-Man — July 16, 2010 @ 7:39 pm

35. Paul- You and I have a fundamental disagreement about what the Obama administration is doing. NASA has spun it's wheel for 20-25 years. They have funded a series of seemingly sound program ideas going back to Bush I, and yet we are still stuck in LEO. I think by getting money to smaller, leaner organizations that want to step beyond that self imposed exile, we have a opportunity to get past LEO and back to the moon and go to Near Earth Objects of interest. I think that NASA has become a rat hole that we need to stop flushing money into. I think the idea of funding American businesses to do our bidding is brilliant.

A few years ago, I had thought that the notion of foreign programs going back to the moon would be enough to overcome organizational apathy, but that has not been the case. I think we are about to see China, India, our private companies(and perhaps Russia) do what NASA has been unable to do- leave our gravity well, and journey outward.

I hope to finally see this in my lifetime (I'm almost 60), but it's likely too late for me to participate in.

Comment by Steverman — July 19, 2010 @ 10:26 am

36. Steverman,

I think that NASA has become a rat hole that we need to stop flushing money into.

You are aware, of course, that the President's proposal actually *increases* the NASA budget? I guess "flushing money into the rat hole" is fine, as long as you know which hole to pour it in.

I think the idea of funding American business-

es to do our bidding is brilliant.

Yeah, brilliant. We've been doing that for the last two hundred years or so — it's called "government contracting." The anti-Vision just trades one bunch of contractors for another.

Brilliant, all right.

Comment by Paul D. Spudis — July 19, 2010 @ 5:16 pm

37. The "United States can't do it" comment sure didn't go 'unnoticed' by me or my co-workers, that's for sure. And you nailed it on the head about the management fads — that goes for many contractor entities as well. I get called a "sucker" and "space junkie" when I try to find some molecule of optimism about the solutions put forth by NASA administrators or Congress… then I get treated to tirades about how we were better off when we competed with other nations instead of collaborated. Debatable, but a sad sign of human nature. "Fixable" appears to be relative, because even with budget increases, the money is spread more thinly across far more endeavours when compared to the glory days of moon landings. NASA's budget now encompasses space exploration, planetary science, oceans & climate research, astrophysics, heliophysics, green aviation, aeronautical engineering and education support. Too many cooks spoil the… tell me if you've heard this one.

Comment by Pillownaut — July 20, 2010 @ 10:35 pm

41. Paul wrote:

"Regarding the comment by Administrator Bolden that 'We're not going to go anywhere beyond low Earth orbit as a single entity', I see that as a reflection of what has happened…"

I rather interpreted this comment as a type of self-fulfilling prophecy. I suggest reading the comment again with the phrase "we're not going to go anywhere" taken as an active directive rather than as a passive observation. Of late, as my anecdotal observation goes, the Americans are being outvoted by the American'ts. In a way, Mr. Bolden's and Mr. Obama's comments seem to lean towards acknowledging this.

I agree with Paul's general approval of the VSE. In principle, I think that what I've been calling the BSE (Budget for Space Exploration), FY2011, the Flexible Path, whatever; can have a positive interpretation and outcome. However, VSE failed, largely for political reasons, and I think the new proposal suffers from the same propensity for probable political failure.

For example, FY2011 suggests canceling one HLV program, and substituting another HLV program, and doesn't address the management problem of large program accomplishment.

Haven't we Been There and Done That? In a way, FY2011 is a failure of capitalism; when the actual product is less important than corporate profit. In another way, it's a failure of socialism, when the process of keeping the worker bees busy assumes more importance than the products needed to fulfill the program. FY2011 doesn't really offer a mechanism for success, which would be a real game-changer.

Not only that, but the infighting among the HSF enthusiasts, partly encouraged by FY2011 it seems, has not convinced the American populace that HSF is a worthwhile endeavor, especially in light of the historically large other problems facing our country these days.

Our country, as a whole, does not really believe that HSF serves a valuable human purpose, and Congress seems to pick up on that.

Moving on, these are a couple of specific examples missing from FY2011:

As far as robotic precursor missions, what is the actual nature and disposition of that lunar water ice? And where are my AutoCAD r14 contour maps of the likely craters? And how about a simple demo ISRU water cracking mission?

The Moon is not the only object of attention either. My personal focus is on the near term goal of the Moon, because of the proximity argument. I have no fundamental problem with Paul's lunar proposal from several years back. It is a very useful talking point. But Mars bears serious attention too; some of the latest information was not available in 2004, which calls for a revision of Paul's ideas, but by no means calls for any dismissal. That time is virtually the only unknown variable in his presentation seems quite sound to me.

Dr. McKay is studying the possibility of perchlorate based life on Mars. There is also a methane cycle on Mars which does not conform to current understandings of abiotic processes; it should be studied carefully by an orbiting observatory. It seems that the question of life on Mars can be determined by robotic missions, carefully conceived and executed. Should life be discovered, it seems clear to me that the martian ecosystem should be studied, but also that there should be no human presence on the surface until that understanding should be more sound.

In my mind, the discovery of martian life might properly encourage the development of a lunar outpost and manufacturing abilities, in order to begin reaching out to the solar system in a more permanent fashion, with Mars as that next step. But it is by no means the only requirement for the development of the lunar outpost and ability.

Ultimately, I think that the HLV should be a passenger vehicle, and also carry high value, low mass cargo. Using it to deliver propellant and other unmanned lunar infrastructure would be a good way to demonstrate reliability and cost reduction. The martian mothership, whenever it is built, would be built on the surface of the Moon, and launched from there. If the people of our country do not wish this to happen, they will not ensure that Congress embark on a sustained effort such as this, which gets back to agreeing on how we should proceed.

Perhaps it really is the job of the entrepreneur to open up this new frontier. At the same time, I think that the recent Senate language shows that some sort of political compromise is emerging.

Comment by John Fornaro — July 22, 2010 @ 1:23 pm

42. Hello Paul,

I enjoyed your "Once and Future Moon" book. There are not many popular books on the moon in local libraries. I hope you update it one day.

You clearly dislike the Presidents' Fy2011 budget, but how do you feel about the Senate proposal? A HLV and a BEO Orion, but also adopts the asteroid goal. It also slashes the budgets affecting life support, ISRU and the lunar lander/rover with ISRU component planned for 2015. At least this would have given you an additional chapter.

Although the Presidents' budget had a weak vision with regard to the moon it did have a number of moon related budget items to be delivered within the President's term. My own hope was that several of these items could have been bundled into something more meaningful. If some more space technology prize money could have found its way to another commercial lunar lander prize then we would have all the components for a robotic lunar colony. Compared to VSE with a permanent manned lunar base, at least this scores 2 out of 3.

Right now I do not see any political force genuinely fighting for the moon. The senate seems to want an HLV, mission secondary. The house seems to want Constellation, reality secondary. (Constellation minus Altair of course.)

Comment by KelvinZero — July 24, 2010 @ 12:26 pm

The Moon, Asteroids, and Space Resources

July 23, 2010

The Moon: Useful and on the way

By abandoning the Moon, the administration's proposed space policy[1] has left the space community with a huge question mark over the important issue of learning how to harvest and use space resources. Clearly if we don't go to the Moon with people or machines, there is no way to use the abundant water, metals, and other lunar surface materials to create new capabilities in space. Supporters of the new path suggest instead that we can obtain all the materials we want from near-Earth asteroids[2], small, rock-like objects that co-orbit the Sun with the Earth. Indeed, some asteroid types appear to contain significant quantities of water, thus offering a possibly rich source of off-planet water.

Water is an extremely useful substance in space. By virtue of its varied utility, water enables extended human presence in space. Besides its obvious role as a sustaining substance for human life (both drinking and providing oxygen for breathing), water is also an excellent material to shield from cosmic radiation and a medium of energy storage, both by thermal storage and also through its use in rechargeable fuel cells, where hydrogen and oxygen are combined at night (producing water and electricity). Stored water is disassociated by solar generated electricity during the day and re-stored as hydrogen and oxygen. Most importantly, water can be converted into liquid hydrogen and liquid oxygen; in this form, it is the most powerful chemical rocket propellant known.

So what are the relative benefits and drawbacks of using asteroidal (not lunar) resources? The biggest advantage of asteroids is that they have extremely low surface gravity. As these objects are simply very large rocks, they don't have much mass and hence, virtually no surface gravity. A mission to an asteroid is more akin to a rendezvous in space than it is to a planetary landing. The advantage this confers is that vehicles can come and go to a given asteroid without the requirement to expend large amounts of propellant in a landing, with total changes in velocity measured in the few meters to tens of meters per second range. In contrast, a landing on the Moon requires a propulsive burn of over 2200 meters per second, both coming and going. This deep "gravity well" penalty is much smaller than launching from Earth (11,000 meters per second), but is still substantial compared with "dimple" dimensions of asteroid gravity wells.

If the propulsive energy of access were

the only (or even the main) consideration for resource exploitation, asteroids would win hands down. But there are some other issues to consider. Water is indeed present in the materials of Near-Earth asteroids[3], but in a chemically bound form. Water molecules fill sites in the crystal structures in rock-forming minerals, bound strongly to its encasing structure. These chemical bonds must be broken to extract the water and that takes energy. On the Moon, water occurs in bound form, but also in its native state as ice in the lunar polar regions. Ice-laden dirt can be scooped up and minimally heated to extract the water. In contrast, it takes 100 to 1000 times more energy to extract a kilogram of water from chemically bound asteroidal minerals than it does to scoop up the "free water" found in the lunar cold traps. The greater quantity of energy needed to extract water from an asteroid is annoying, but can be handled through the use of large solar arrays or even a nuclear reactor to generate copious amounts of electrical power. But both solutions bring significant mass penalties and a nuclear reactor significantly increases cost, both from the technical development it would require and from the hurdles raised by legal and environmental groups[4] it would have to overcome.

A more critical issue is the location of the two resource bodies. The proximity of the Moon is a major boon for its utilization. The Moon is both close and accessible. In terms of closeness, it takes 3 seconds for a radio signal traveling at the speed of light to go the Moon and back. This makes the remote, telepresence operation of lunar robots[5] from Earth feasible. Early steps in the location, surveying and harvesting of demonstration amounts of resources on the Moon can be done remotely with robots controlled from Earth. We do not have this luxury with asteroids.

Asteroids orbit the Sun (like the Earth does) and vary in distance from Earth by tens of millions of miles over the course of a year. At best, asteroids are several tens of light-seconds away and at times, tens of light-minutes. This long radio time-lag means that direct remote operation of robots on asteroids will be cumbersome, if not impossible. For well understood routine tasks, this may not be a serious issue, but space resource utilization is something we have yet to learn. It is unclear whether we will be able to harvest and process asteroid water using remote robots, but it is almost certainly possible to do so with robots on the Moon.

The other aspect of the Moon's proximity is accessibility, the ability to access a space destination routinely and often. As the Moon orbits the Earth, we can go to and come back from the Moon pretty much at will – launch windows are almost always open. In contrast, because even near-Earth asteroids follow their own paths around the Sun, launch windows are short and come at irregular (albeit predictable) intervals. Round trips to and from asteroids are even more difficult and after multiple weeks to months of travel, loiter times are either very short (on the order of a week or so) or very long (a year or more). This wildly variable duration of access may be handled on a robotic mission, but it precludes any significant human/robot interaction during the materials processing on an asteroid.

Finally, there is the issue of surface gravity. Much of the "dirty work" of resource processing involves separating some substance from another, or extracting something embedded. Having gravity usually makes this an almost trivial step, one that we don't think about very much – unless we don't have it. The Moon does indeed have a significant gravity well (about 1/6 that of the Earth) and although this works against us when we want to export product, it works in our favor when we need to process materials. The extremely weak surface gravity of an asteroid is almost microgravity and makes it very difficult to separate materials there without specialized equipment, again adding mass, power, complexity and cost to the processing chain.

In short, there are many considerations to take into account when planning an architecture based on resource exploitation. The seemingly damning case against going to the Moon to harvest material resources largely revolves around its relatively high surface gravity. It takes roughly two tons of water-equivalent liquid hydrogen-liquid oxygen propellant to lift one ton of water to the L1 point[6], where it can be used to supply and fuel a variety of spacecraft destined for many different places. That same ton of water lifted from the Earth would take over 19 tons of propellant to deliver it. The other side of that coin is that gravity is extremely useful – if not critical – for many materials processing techniques. Gravity can only be artificially created near an asteroid at some expense and mission complexity, whereas on the Moon, it's a feature that comes for free.

Learning how to access and use space resources is a critical skill for a space faring society – skills and knowledge that will reap rewards right here on Earth. The Moon offers us a school and a laboratory for acquiring this critical knowledge. By virtue of its proximity, accessibility and resource endowments, the Moon satisfies our early space ISRU[7] needs and allows us to create new capabilities to routinely access cislunar space[8], where all of our economic and national security space assets reside. The asteroids have much to offer for material resources and we will eventually journey to and use many of them. But we have business on the Moon first. Mining the unlimited wealth of the Solar System will become inevitable once we have learned the lessons of how to do this job on our nearest neighbor.

Topics: Lunar Exploration, Lunar Resources, Space Transportation, Space and Society

Links and references

1. proposed space policy, http://blogs.airspacemag.com/moon/2010/07/23/2010/02/03/vision-impaired/
2. obtain all the materials, http://en.wikipedia.org/wiki/Asteroid_mining
3. Near Earth asteroids, http://en.wikipedia.org/wiki/Near-Earth_asteroid#Near-Earth_asteroids
4. hurdles raised, http://www.spacedaily.com/news/nuclearspace-03b.html
5. telepresence robots, http://adsabs.harvard.edu/full/1992lbsa.conf..307S
6. L1 point, http://en.wikipedia.org/wiki/Lagrangian_points
7. ISRU, http://en.wikipedia.org/wiki/In-situ_resource_utilization
8. cislunar, http://en.wikipedia.org/wiki/Cislunar#Geospace

Comments

2. As far as manned missions are concerned, I think its obvious that the Moon should be our primary objective. Exploiting the Moon's oxygen and hydrogen resources should also be a priority. Obviously oxygen and water are essential for human survival beyond the Earth. But hydrogen and oxygen are also excellent rocket fuels that could dramatically lower the cost of traveling to and from the Moon and could even make transferring satellites from low earth orbit to geosynchronous orbit a lot cheaper.

Also, thanks to the Moon's low gravity well, a colony on the lunar surface could end up being the primary location for the manufacturing and launching of satellites destined for both low Earth and geosynchronous orbits before the end of the century,.

I'm also a strong advocate of using robots to grab small NEO asteroids 50 to 1000 tonnes in mass and bringing them back to stable Lagrange points like L4 or L5 for resource exploitation. The most efficient way to do that, IMO, is to simply deploy large light sails, one to two kilometers in diameter at L4 and L5. Such large sails could each probably transport a 50 to 100 tonne NEO asteroid to a Lagrange point every year. They shouldn't weigh more than 40 tonnes, and with the latest light weight materials, should probably weigh a lot less. A lunar colony is probably going to need carbon and nitrogen resources from the asteroids if those resources found at the lunar poles are insufficient.

I don't, however, see the logic of sending humans to a large NEO asteroid since it would require several months of travel and probably several hundred tonnes of mass shielding to protect the human brain and body from the deleterious effects of galactic radiation. Such a venture would be a lot more expensive and dangerous than setting up a lunar base. A manned asteroid journey also makes no sense as a precursor journey before we attempt to go to Mars since it would be equally and possibly even more dangerous than simply traveling to Mars orbit in the first place.

Comment by Marcel F. Williams — July 23, 2010 @ 5:57 pm

3. Of course the moon is a good place to test out ISRU. However, even there was already a habitat there, we still have a lot of work to make ISRU testing a reality. There is a lot that must be done on Earth before we can even begin doing ISRU on the moon or any asteroid.

For instance, any water in say de Gerlache crater will not only be mixed in with the regolith, it will also be super cold – down around 3 Kelvin. At that temperature, water is a mineral. It isn't slush, it's a rock. It may actually be more difficult to mine that than it would be to break chemical bonds in asteroidal material.

We don't know. There is so much we don't know – and need to know before trying ISRU anywhere – that there could be multiple Centennial Challenges to try to cover the trade space. And there is plenty that NASA can do in this regard before we ever go anywhere, too – mix some lunar soil simulant with some water, bring it down to 3 Kelvin, and test failure rates on diamond bits, for one.

At least, that would put the horse (research) in front of the cart (destinations).

Comment by Ed Minchau — July 24, 2010 @ 1:45 am

4. http://quantumg.blogspot.com/2010/07/dr-paul-spudis-responds-sorta.html

Comment by Trent Waddington — July 24, 2010 @ 5:17 am

5. It would be nice to see the public discussion focus on Lunar Robotic Precursor missions to extract and return reasonable amounts of regolith for developmental 'pro-

cess' research and analysis.

I think it would be somewhat prudent to utilize the Orbital National Laboratory (ISS) to evaluate micro-gravity processing techniques on celestial regolith.

Comment by Marcus — July 24, 2010 @ 5:32 am

6. Ed,

it will also be super cold – down around 3 Kelvin. At that temperature, water is a mineral. It isn't slush, it's a rock. It may actually be more difficult to mine that than it would be to break chemical bonds in asteroidal material.

Not likely. To get solid, dense ice, you need some type of freeze-thaw cycle for the ice to recrystallize. The polar cold traps have been as cold as they are now for at least 2 billion years and have never seen freeze-thaw. Moreover, the LCROSS results suggest a "fluffy," less-than-normal density of the cold trap regolith it encountered during impact, which released the observed water vapor and ice. Finally, the slow, incremental addition of cometary ice over extremely long times suggests that lunar polar ice deposits are aggregates of amorphous ice particles and dirt and may have a "fairy castle," high-porosity structure.

Radar cannot distinguish between solid, dense ice and fluffy, snow-like ice, so either is possible. However, having said all that, I fully agree with you that precursor robotic prospector missions are required to characterize the deposits and physical environment of the lunar poles. Even more robotic missions would be required to gather data for NEO asteroids, as our knowledge of them is much less than for the Moon.

Comment by Paul D. Spudis — July 24, 2010 @ 5:33 am

8. There are at least two additional challenges I see to processing NEOs for profit (in addition to those mentioned above):

1. Many NEOs spin, perhaps at high rates of rotation and along multiple axes. Actually mining a fast spinning NEO could be far more difficult than many expect.

On the other hand, if intact NEO fragments have survived impact with the Moon (as suggested by Dennis Wingo and others) we can mine NEOs by looking for those same fragments hopefully lying peaceably on or under the lunar surface. Another benefit of lunar gravity.

Note that I do acknowledge the "if" aspect of this.

2. Which NEOs should we fly to?

Arrive at a NEO that is NOT suitable for exploitation and you've wasted years on an expensive deep space prospecting mission.

Arrive at a suspected lunar cold trap crater that is not suitable for exploitation and within a matter of days your prospecting team (human or robotic) can be at the next crater. And this is only after lunar orbital assets have imaged the entire surface of the Moon, as LRO is doing now, in order to narrow the list of target craters worth visiting.

It seems to me that identifying potential mining sites on the Moon shall be far easier (at considerably less expense) than identifying which NEOs offer the best odds of containing materials worth harvesting.

Comment by Bill White — July 24, 2010 @ 7:29 am

9. Bill,

Many NEOs spin, perhaps at high rates of rotation and along multiple axes. Actually mining a fast spinning NEO could be far more difficult than many expect.

Thanks for adding this — I had meant to discuss this issue in the post, but forgot about it. You are absolutely correct; the spin rate issue is a problem that cannot be quantified until we go there.

I also agree with your second point and it is one that I had not considered. Multiple asteroid missions are required to assay which ones hold resource potential and as each one is unique, the results of one probe cannot be extrapolated to other bodies.

Comment by Paul D. Spudis — July 24, 2010 @ 9:25 am

10. The author said "By abandoning the Moon…"

This kind of statement is really misleading. The Administrations plan, which was based on the "Flexible Path" of the Augustine Commission, focused on creating the infrastructure to allow future administrations go anywhere, including the Moon. That their next exploration goal was an NEO reflected the desire to do something that has never been done before, which is leave the Earth-Moon system and rendezvous with an asteroid. You see this as abandoning the Moon, and I see this as picking a goal that has the highest amount of "New-ness" using NASA's limited resources.

One could make the same argument that you advocate less ambitious goals for NASA, in that NASA has already learned how to land and return from the Moon, and now they want to do something that has never been done before.

Resources abound in the Solar System, so I look at the question of where to extract them as really based on where do we need them? If someone funds an outpost on the Moon, then ISRU would definitely be a priority, but are no outposts planned or funded.

I was extremely displeased with the current Congressional bills that lower the funding for robotic precursor missions to the Moon, because even though I don't think it's time to return humans to the Moon, I do think we should be doing as much exploration and preparation as possible using robotic systems. Using what we have learned with our Mars exploration rovers, I see that we can build and deploy much larger and capable Moon rovers that could eventually do a vast amount of the exploration needed to start ISRU, and potentially even do the beginnings of ISRU without the need for humans.

Finally, just as you are a strong advocate for ISRU, you know I have been very vocal about lowering the cost to access space. In your articles about extracting water and minerals from the Moon, I always get the sense that you are glossing over the huge cost it will take to transport the continuous stream of equipment and supplies that will be needed to start even a pilot plant on the Moon. For anyone that has seen the TV show "Ice Road Truckers", instead of driving hundreds of miles over frozen land, imagine doing that over 238,000 miles of vacuum with no rest stops or rescue vehicles along the way – you have to have the road and the truck system before you can open a factory or mine, and that is what the Administration is focusing on first.

Comment by Coastal Ron — July 24, 2010 @ 12:01 pm

11. "If the propulsive energy of access were the only (or even the main) consideration for resource exploitation, asteroids would win hands down."

In my opinion this is rarely the case. It's true an asteroid has an extremely shallow gravity well. But to land on an asteroid you have to escape earth's gravity well and enter an elliptical transfer orbit about the sun. Then, upon reaching your destination, you must match velocities with the asteroid.

I would try to estimate delta V for different trajectories by setting up an iterative Lambert Space Triangle spreadsheet. Inputting earth and asteroid time and xyz coordinates as well as velocity vectors for earth departures as well as asteroid destinations, it seemed to me big delta V was the rule rather than exception.

Low delta V launch windows to asteroids seem to be very rare, so far as I can tell. And since most asteroid orbital periods aren't even multiples or simple fractions of earth's period, nice launch windows don't re-occur on a periodic basis. Should we land useful infrastructure on an asteroid, it could be quite some time before we could send additional infrastructure to the same asteroid.

Comment by Hop David — July 24, 2010 @ 1:02 pm

12. Coastal Ron,

so I look at the question of where to extract them as really based on where do we need them? If someone funds an outpost on the Moon, then ISRU would definitely be a priority, but are no outposts planned or funded.

You continue to miss the point. We need the resources – in this case water for propellant manufacture — in cislunar space. The true goal of the VSE is to develop a transportation system that can routinely access all of cislunar, including (but not restricted to) the lunar surface. If we do this, going to the planets becomes easy. Going to an asteroid or Mars for a one-off, flags-and-footprints stunt mission (as the anti-Vision calls for) does not serve this need.

You also seem to think that we need an oil refinery on the Moon to harvest significant quantities of water. Not so; in fact, we start small, with a couple of metric tonnes of equipment (rovers, haulers, solar arrays, ovens for ice melting), the total of which can produce a few tonnes of water per month. We build up capability gradually, first to support operations on the lunar surface, then to export to cislunar.

I've heard the mantra of "cheap access to LEO" for decades and we are no closer to it now than we were in the 1960's. The simple fact is that we need a different kind of game changer. ISRU might be that. But we'll never know if we don't try it.

Comment by Paul D. Spudis — July 24, 2010 @ 3:36 pm

13. IMHO, "cheap access to LEO" will require high launch rates from Earth. The high cost of LEO access also is a market condition not a technology problem, again IMHO.

If this is correct, then transporting "the continuous stream of equipment and supplies that will be needed to start even a pilot plant on the Moon" will require those same high launch rates that are needed to permit lower cost access to LEO to emerge.

Creating a genuine cis-lunar transportation infrastructure (LEO depots and EML depots working in synergy to move material to the lunar surface to do ISRU) will be a source of launch demand that will facilitate cheap access to space.

Comment by Bill White — July 24, 2010 @ 3:46 pm

14. *"Resources abound in the Solar System, so I look at the question of where to extract them as really based on where do we need them?"*

We need them in LEO. It would also be nice to have them in EML1 or 2 as these locations have an ~2.4 km/sec delta V advantage over LEO for deep space destinations like asteroids or Mars.

In terms of delta V, the moon is quite close to LEO as well as EML1 or 2. See this delta V map: http://clowder.net/hop/TMI/FuelDepot.jpg

Various Direct advocates point out it's easier to depart from LEO for a deep space destination than it is from the moon's surface. They then conclude lunar propellant is worthless. This is what I call the Tucson to Omaha by way of Austin argument:

http://clowder.net/hop/TMI/TucsonToOmaha.jpg

Austin may be out of the way, but Texas gas supplies stations in Albuquerque and Denver which are enroute. The same could be true of lunar propellant supplying LEO and EML1 depots.

Comment by Hop David — July 24, 2010 @ 5:52 pm

15. We need:

1. A rover that will find the consistency of the ice. Is it solid like a lake, which will require coal-miningesque machinery. Or is it light and fluffy bound up with the other regolith, requiring nothing more than a backhoe.(this can fall under a one-off mission)

2. The processing plant for fuel. The miner, shuttlecars, oven, storage. Hopefully enough to make 10+mT of fuel a month.

3. An OTV/lander which will fly off the lunar fuel. From then on, all we need to do is lift a package of bots to LEO to expand the works going on Luna. Fuel and launch costs should then drop. As the travel from LEO to Luna will be "paid".

Comment by Rhyshaelkan — July 24, 2010 @ 7:19 pm

16. Great post & great comments! I'm on the Moon-first side of this argument. A few quick questions...

1) There wasn't a great deal of water kicked up by the LCROSS mission. If that is representative of the density of water in permanently shadowed craters, will there be no advantage to going to the lunar poles than the lunar equator?

2) Can each of the components needed to establish propellant-quantities ISRU fit on a Falcon 9 Heavy? To develop industrial lunar ISRU, do we need to spend the billions on a new HLV or could that money be spent on prospectors, landers, extractors, ovens, ascenders, etc?

3) Do we really need the hydrogen which water ice provides? H2 is only 11% of LOX/H2 by mass. H2 could be brought up from Earth's surface. OR, is the amount of energy needed to extract O2 from lunar equatorial SiO2 so high (compared with mining from maybe low density water ice) that it is impractical?

4) Wouldn't just going with (soon to be) existing non-heavy launchers such as Falcon 9 Heavy sort of force us to have higher launch rates than if we build massive HLVs? So, wouldn't that eat into the cost savings argument of HLVs?

5) Surgeons can teleoperate on patients. Robonauts can have very dexterously manipulate an object. Is there any particular reason why humans are physically needed on the Moon during the establishment or ongoing operations of mining activities?

6) Finally, can even humans be delivered to the lunar surface using only non-heavy launchers? (e.g. using LEO-fuelled, dry Apollo lunar descent stage = 2034 kg)

Comment by JohnHunt — July 25, 2010 @ 10:49 am

17. JohnHunt,

A few brief answers to your questions, from my perspective.

1. LCROSS actually found between 5-10 wt.% water in its ejecta plume. That's 3 orders of magnitude greater than the highest non-polar concentrations of hydrogen. Moreover, we find km-scale craters that apparently have pure ice deposits within them. So there is no question that the "ore" bodes are at the lunar poles.

2. Yes, you could deliver all the mining equipment you need on an Atlas 5-class medium lift rocket.

3. Yes, because our ultimate goal is permanent sustainable presence in cislunar and we get the hydrogen when we get the water, so why not use it?

4. I am agnostic in regard to heavy lift. If we

have one, I would use it to put large payloads on the Moon. If we don't have one, we can use existing launchers.

5. We need humans to fix and maintain the robotic mining assets, just as we do with any large-scale industrial robotic infrastructure on the Earth.

6. Sure. We just assemble the vehicles we need in space and fuel them there.

Comment by Paul D. Spudis — July 25, 2010 @ 12:07 pm

18. > *4. I am agnostic in regard to heavy lift.*

Don't be. HLV and sustainable space development are competitors in a NASA budget of essentially fixed size. We should be arguing that the money being contemplated to be spent on a HLV would be much better spent incentivizing sustainable space development.

> *5. We need humans to fix and maintain the robotic mining assets.*

Why can['t] robonauts fix robotic mining assets?

Comment by JohnHunt — July 25, 2010 @ 12:20 pm

19. *HLV and sustainable space development are competitors in a NASA budget of essentially fixed size.*

It is not clear to me that it is NASA's role to capitalize private space companies. Let them demonstrate their ability to routinely reach LEO at low cost and I have no problem contracting for service. In the mean time, a "New Space" company is just another contractor.

Why can['t] robonauts fix robotic mining assets?

Who will fix the robotnauts?

Comment by Paul D. Spudis — July 25, 2010 @ 2:55 pm

20. > *Who will fix the robotnauts?*

Another robonaut!

Why not? Spare parts can be easily shipped to the lunar surface. Whenever there is a problem, one robonaut disassembles the defective part on the other robonaut, attaches the new part, and then screws it in. The defective part will probably be so small (e.g. a motor or chip) that it probably would be worthwhile just to replace the inventory of that piece on a latter mission rather than shipping it back to LEO for human repair.

> *It is not clear to me that it is NASA's role to capitalize private space companies.*

That's a legitimate perspective. But NASA's

already capitalizing private companies with SpaceX and Orbital. Now NASA's not paying for the entire development of Falcon 9 and Taurus II. But they're partly paying for the private development of technology which they will then be able to use for their own (NASA's) purposes at a fixed price. So far, it seems to be working and at prices which are considerably less than if NASA had done it all themselves. That's not just my opinion, it's also apparently the opinion of Iridium.

Now, I would only apply this approach where, at the end of the day, there is likely a commercial market which these companies with this new technology will be able to exploit on an economically sustainable basis. But once the market for lunar-derived fuel to LEO is established, I don't see the government needing to incentivize any further commercial space development. I think that it will proceed on its own.

I think that government money may well be necessary. The upfront investment requirement might be so high that too many private companies won't want to be the first to take the risk. NASA has its own need for these services. Why not allow it to partner with industry to develop this commercial capacity. I really don't care how cis-lunar space is developed so long as it is developed sooner rather than later. If government spending is helpful getting companies over the initial hump then I'm fine with that.

> *Let them demonstrate their ability to routinely reach LEO at low cost and I have no problem contracting for service.*

If we wait a couple or three years for them to demonstrate routine, low-cost of reaching LEO then, by then, we'll already be committed and spending billions of dollars on an HLV which is not economically self-sustaining. At that point there will be little budgetary room left to incentivize NewSpace companies to develop cis-lunar space. No, the time is now for the argument to be made that 10-15% of NASA's budget should go to incentivizing companies to develop the basic components necessary to exploit lunar resources and to master cis-lunar space on a pay-for-performance basis. I don't see how this can be done without holding off on an HLV for a few years. Look, once you get lunar-derived propellant to LEO then do you really need an HLV?

Comment by JohnHunt — July 25, 2010 @ 4:14 pm

21. *"Look, once you get lunar-derived propellant to LEO then do you really need an HLV?"*

Getting out of Earth's atmosphere and into orbit are the major hurdles. An HLV, if it is more efficient, is still useful to get things to LEO. From LEO on out you can use Lunar and asteroidal based propellants. Whether your space-tugs are solely chemical propellant based, or electro-chemical(ion, VASIMR, etc), an efficient HLV will move that much more goods, that much faster.

I am sure not getting younger, get 'er done! :P

Comment by Rhyshaelkan — July 25, 2010 @ 10:06 pm

22. Wow. Paul, you think the Obama is abandoning the Moon? The Moon is relatively easy to get to (yeah, it's a deep gravity well), and we all know, it's 3.5 days away, constantly. Let's see if some American corporations can get there. NASA, with all the resources of the US Government, can do the Asteroid missions. Let's give NASA something hard to do. After 35 years of LEO, let's see them do something that SpaceX can't be expected to do.

Me, I think they'll spin their wheels like they have been for the past 20 years. I expect SpaceX is up to the challenge. I actually doubt if NASA is up to the challenges that have been issued.

Comment by Steverman — July 25, 2010 @ 10:15 pm

23. *Wow. Paul, you think the Obama is abandoning the Moon?*

Yes.

NASA, with all the resources of the US Government, can do the Asteroid missions. Let's give NASA something hard to do. ....I think they'll spin their wheels like they have been for the past 20 years... I actually doubt if NASA is up to the challenges that have been issued.

Well, which do you believe? Does NASA need something "hard" to do, like flags-and-footprint missions to asteroids instead of the "easy task" of returning to the Moon and learning how to live on another world? Or are they incapable of doing anything at all? That's an argument for abolishing the agency, not for putting them in charge of developing a new commercial industry.

Comment by Paul D. Spudis — July 26, 2010 @ 4:43 am

24. *"Well, which do you believe? Does NASA need something "hard" to do, like flags-and-footprint missions to asteroids instead of the "easy task" of returning to the Moon and learning how to live on another world? Or are they incapable of doing anything at all? That's*

an argument for abolishing the agency, not for putting them in charge of developing a new commercial industry."

Exactly! If we're going to have a government space program then we should use it! Practically all of the studies show that NASA creates a lot more wealth for this country than it consumes. How many other government programs do that? There wouldn't even be a Space X or a Bigelow in this country if there were no NASA!

Comment by Marcel F. Williams — July 26, 2010 @ 2:25 pm

25. > *An HLV, if it is more efficient, is still useful to get things to LEO.*

A very good point Rhy. I've just been sort of burned by the Space Shuttle not living up to its promise of cheap access to space. I want to be rational not cynical. But when the Augustine Commission says that the first thing that they'd have to do with a fully developed Ares V would be to cancel it because operations would cost too much, I'm just not convinced that a NASA-developed HLV will be more efficient.

Comment by JohnHunt — July 26, 2010 @ 4:55 pm

26. Cheap Space Shuttle cost were based on the idea that there was going to be a demand for dozens of shuttle flights ever year. Of course such a high demand for space flights never showed up especially after other countries began to develop their own satellite launch vehicles.

Still, a three billion a year shuttle program represents less than 0.1% of annual Federal budget expenditures. In fact, the entire NASA budget represents less than 0.6% of total Federal budget expenditures. Yet for some reason, the public thinks we're spending a whopping 24% of the Federal budget on NASA?

Comment by Marcel F. Williams — July 27, 2010 @ 3:29 am

27. No one mentioned mass drivers. Gerry O'Neill and Henry Kolm envisioned that they could reduce the cost of putting lunar materials into space. As a member of Space Studies Institute, I helped in a small way to fund the development of the SSI mass drivers (and was standing in SSI's lab when one of them was demonstrated in the mid-80s... "don't blink or you'll miss it").

I haven't heard a peep about mass drivers for 20+ years, do the economics of combating the lunar gravity well weigh against them these days?

Comment by John — July 28, 2010 @ 10:12 pm

28. John,

do the economics of combating the lunar gravity well weigh against them these days?

Not to my knowledge. As far as I know, mass drivers are still on the table as a means for exporting large amounts of mass from the Moon.

However, we are largely debating the direction and activities of the space program over the next couple of decades. I would imagine that a mass driver on the Moon would be a feature of a more advanced stage of lunar surface presence. Right now, I am focusing more on getting us there in the first place, let alone devising a huge building project there.

Once we demonstrate that useful materials (in this case, water) can be extracted from the Moon, then we can worry about how to export it to cislunar space. I would think that rocket power would serve the early stages of such an industry, possibly transitioning to mass drivers once we have established a production chain and market for the product.

Comment by Paul D. Spudis — July 29, 2010 @ 8:54 am

29. Even the simplest of Moon bases could be the beginning of great things! I think many nations (Japan, China, Russia, India, etc.) are starting to realize this.

I predict that less than 20 years after the establishment of the first lunar bases, we'll have mass drivers launching thousands of tonnes of regolith and processed materials into lunar orbit annually. Lunar materials via mass drivers should be quite competitive with small asteroids being imported into the Lagrange points via light sails.

Comment by Marcel F. Williams — July 29, 2010 @ 12:35 pm

30. *"I haven't heard a peep about mass drivers for 20+ years, do the economics of combating the lunar gravity well weigh against them these days?"*

I still like the idea of mass drivers. I do not feel like doing the math at this time. However, we do know the km/s required to reach EML1. As the near side of the moon always points towards Earth and therefore always towards EML1. Imagine this.

The railgun crests the north or south pole pointing towards EML1. The railgun is long enough to generate sufficient velocity at 3 Gs acceleration to reach EML1. Now you can launch most everything you want, people or cargo.

It is all good stuff. However if you would try to present this to investors or agencies to cut costs. They would only see the costs of how much would it weight and how many launches, FROM EARTH, would it take to make it happen. And promptly say it is not feasible. Not taking into account lunar industry.

Same can be said of SPSs. Everyone says it is not feasible due to the costs of launching from Earth. However what if everything was built with lunar industry?

The easy retort is "lunar industry does not exist". Which it does not. I however, say "damn the torpedoes". Wanting SPSs is a win win situation.

To have SPSs we would:

Use the water present for fuel to cheapen spaceflight. Cheaper spaceflight will allow more launches of bots to start mining operations. Material is used to fabricate structures for humans to reside and work(there is also the possibility to sinter regolith itself into building material using microwaves emitters, whatever your poison is). Humans now arrive and take up residency without hardly lifting a finger, possibly only requiring the installing of airlocks. In workshop areas, I am sure there would be both pressurized and unpressurized ones, the old bots can be maintained and new bots can be fabricated using lunar materials. With the occasional Earth items that cannot be fabricated from lunar materials. New solar foundries using lunar aluminum as the concentrating mirror and regolith "kettles". Bigger mining bots are built, bigger foundries, more better faster. Seed lunar industry pulls itself up by the bootstraps.

Oversimplified I grant you. But not impossible and will cost only as much as you are willing to waste. I want privatization of space, as they look at bottom lines. It will still not be cheap, however a businessman willing to put the time and money in could eventually print his own money.

I feel SpaceX has made a huge first step. Rich and poor alike should join together to make an organization to bring mankind to space. The key is Luna. The gateway to mankind's eventual colonization of space.

Comment by Rhyshaelkan — August 2, 2010 @ 4:02 am

The New Space Race

February 9, 2010

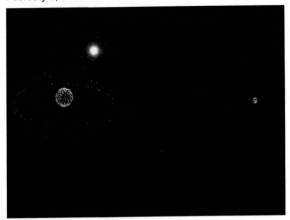

Cislunar space: Site of the New Space Race

Introduction

Recent media reports suggest that China is stepping up their program to send people to the Moon just as America appears to be standing down from it. This circumstance has re-awakened a long-standing debate about the geopolitical aspects of space travel and with it some questions. Are we in a race back to the Moon? Should we be? And if there is a "space race" today, what do we mean by the term? Is it a race of military dimensions or is such thinking just an artifact of the Cold War? What are the implications of a new space race?

Many in the space business purport to be unimpressed by the idea that China is going to the Moon and publicly invite them to waste money on such a stunt. "No big deal" seems to be the attitude – after all America did that over 30 years ago. NASA Administrator Charles Bolden recently professed to be unmoved by the possible future presence of a Chinese flag on the Moon, noting that there are already six American flags on the Moon.

Although it is not currently popular in this country to think about national interests and the competition of nations in space, others do not labor under this restriction. Our current human spaceflight effort, the International Space Station (ISS), has shown us both the benefits and drawbacks of cooperative projects. Soon, we will not have the ability to send crew to and from the ISS. But that's not a problem; the Russians have graciously agreed to transport us – at $50 million a pop. Look for that price to rise once the Shuttle is fully retired.

To understand whether there is a new space race or not, we must understand its history. Why would nations compete in space anyway? And if such competition occurs, how might it affect us? What should we have in space: Kumbaya or Starship Troopers? Or is the answer somewhere between the two?

Some History

People tend to think of Apollo and the race to the Moon when they hear the term "space race" but the race began with the October 1957 launch of a Russian satellite called Sputnik. The clear implication of this new Soviet satellite was that if they wanted to, they could lob a nuclear bomb at the United States. This situation led to near panic in America, with outraged demands that we technically catch up to the Soviets as quickly as possible and damn the cost.

The initial phases of the space race were not auspicious for America. In our publicized and televised launches, vehicles frequently blew up while the Soviets appeared to effortlessly achieve an endless series of headline-grabbing space "firsts." American officials working behind the scenes knew that we were not as far behind as it seemed but to reveal that knowledge was to disclose our national technical means of surveillance. So each new Soviet first was officially greeted with silence.

The Russians raised the stakes in the spring of 1961 with the launch of Yuri Gagarin, the first human in space. Although America followed a month later with Alan Shepard's ballistic hop, the new U.S. President, John F. Kennedy, wanted to issue a challenge, one carefully crafted to be beyond the existing capabilities of both the USA and the USSR, yet reachable by us (but not by them) over the course of a few years. A manned landing on the Moon was selected as the ideal target for such a race. Although no specific strategic goals on the Moon were identified, it was believed that the attainment of this difficult task would demonstrate the superiority of our open, pluralistic capitalist society in contrast to its closed, authoritarian, socialist opposite number.

The so-called "Moon race" of the 1960's was a Cold War exercise of soft power projection, meaning that no real military confrontation was part of it, but rather, it was a competition by non-lethal means to determine which country had the superior technology and by implication, the superior political and economic system. In short, it was largely a national propaganda struggle. Simultaneously, the two countries also engaged in a hard power struggle space race to develop ever-better systems to observe and monitor the military assets of the other. There was little public debate associated with this struggle, indeed, much of it was held in the deepest secrecy. But as the decade passed, military space systems became increasingly capable and extensive and largely replaced human intelligence assets for the estimation of our adversaries' strategic capabilities and intentions.

The United States went on to very publicly win the race to the Moon, giving rise to a flurry of rhetoric pronouncing everyone's peaceful intentions for outer space while the larger struggle continued to play out behind the scenes. NASA's replacement effort for the concluded Apollo program, the Space Shuttle project, promised to lower the costs of space travel by providing a reusable vehicle that would launch like a rocket and land like an airplane. Because of the need to fit under a tightly constrained budgetary envelope and for a variety of other technical reasons, the Shuttle did not live up to its promise as a low cost "truck" for space flight. However, the program resulted in a fleet of four operational spacecraft that flew over 120 missions over the course of its 30-year history.

Although widely cited in American space circles as a policy failure, the Shuttle had some interesting characteristics that led it to be considered a military threat by the USSR. One of the earliest missions of the Shuttle had its crew retrieve and repair an orbiting satellite (Solar Max). Later missions grappled balky satellites and returned them to Earth for refurbishment, repair and re-launch. This capability culminated with a series of Shuttle missions to the Hubble Space Telescope (HST), which conducted on-orbit servicing tasks ranging from literally fixing the worthless satellite (the first mission) to routine upgrading of sensors, replacement of solar arrays and main computers, and re-boosting the telescope to a higher orbit. The significance of these missions was that the HST is basically a strategic reconnaissance satellite:

it looks up at the heavens rather than down at nuclear missile sites from orbit. The Hubble repair missions documented the value of being able to access orbital assets with people and equipment.

Another relatively unnoticed series of Shuttle missions demonstrated the value of advanced sensors. As a large, stable platform in orbit (the orbiting mass of the Shuttle is almost 100 mT), the Shuttle could fly very heavy, high-power payloads that smaller robotic satellites could not. The Shuttle Imaging Radar (SIR) was a synthetic aperture radar that could obtain images of the Earth from space by sending out radar pulses as an illuminating beam. It could thus image through cloud cover, day or night, all over the Earth. In a stunning realization, it was found that it could also image subsurface features; in particular, the SIR-A mission mapped ancient riverbeds buried beneath the sands of the eastern Sahara from space. The strategic implications of this were immense; as most land-based nuclear missiles are buried in silos, they cannot be hidden from account because of sensors like imaging radar.

The construction of the International Space Station (ISS) became the next frontier for strategic space. One of the most complex spacecraft ever made, it was designed to be launched in small pieces by the Shuttle without an end-to-end systems test on the ground and assembled on-orbit. It worked perfectly the first time it was activated. The building of the ISS documented that not only could people assemble complex machines in space, they could also repair, maintain and upgrade them as well. As the ISS nears completion, much complaint continues about its cost and supposed lack of value, yet even if we get nothing further from it as a research facility, it has already taught us invaluable lessons about the building and maintenance of large spacecraft in orbit.

These new Shuttle capabilities had significant policy implications for the Soviets. To them, it seemed that the Shuttle was a great leap forward in military space technology, not the "policy failure" bemoaned by American analysts. With its capabilities for on-orbit satellite servicing and as a platform for advanced sensors, the Shuttle became a threat that had to be countered. The USSR responded with their own space shuttle (Buran), which looked superficially very similar to ours. The Challenger accident showed that Shuttle was a highly vulnerable system in many respects; even as the Soviets developed Buran, the American military decided to withdraw from our Shuttle program.

During the 1990's, we saw a revolution in tactical space – the use of and reliance on space assets on the modern battlefield. The Global Positioning System (GPS) has made the transition to the consumer market, but it was originally designed to allow troops to instantly know their exact positions. A global network of communications satellites carries both voice and data, and interfaces to the partly space-based Internet (another innovation originally built for military technical research). The entire world is connected and plugged in and spacebridges are now key components of that connection. Fifty years after the beginning of the Space Age, we are now, more than ever, dependent upon our satellite assets.

Space and the national interest

Most people don't realize how the many satellites in various orbits around the Earth affect their lives. We rely on satellites to provide us with instantaneous global communications that impact almost everything we do. We use GPS to find out both where we are and where we are going. Weather stations in orbit monitor the globe, alerting us to coming storms so that their destructive effects can be minimized. Remote sensors in space map the land and sea, permitting us to understand the distribution of various properties and how they change with time. Other satellites look outward to the Sun, which controls the Earth's climate and "space weather" (which influences radio propagation.) No aspect of our lives is untouched by the satellites orbiting the Earth. In a real sense, they are the "Skynet" of the Terminator movies – they are our eyes (reconnaissance), ears (communications) and brains (GPS and Internet) in Earth orbit. Fortunately, they are not yet self-aware. But the people who operate them are.

All satellites are vulnerable. Components constantly break down and must be replaced. New technology makes existing facilities obsolete, requiring replacement, at high cost. A satellite must fit within and on the largest launch vehicle we have; satellites thus have a practical size limit, which in turn limits their capabilities and lifetime. Once a satellite stops working, it is abandoned and a replacement must be designed, launched and put into its proper orbit.

Satellite aging is normal and expected but satellites can also be catastrophically lost or disabled, either accidentally or deliberately. Encounters between objects in space tend to be at very high velocities. The ever-increasing amounts of debris and junk in orbit (e.g., pieces of old rockets and satellites) can hit functioning satellites and destroy them. NORAD carefully tracks the bigger pieces of junk and some spacecraft (e.g., ISS) can be maneuvered out of the path of oncoming debris, but smaller pieces (e.g., the size of a bolt or screw) cannot be tracked and if they collide with a critical part, it can cripple a satellite.

It has long been recognized that satellites are extremely vulnerable to attack and anti-satellite warfare (ASAT) is another possible cause of failure. Both the US and the USSR experimented with ASAT warfare during the Cold War. Although it sounds exotic, ASAT merely takes advantage of the fragility of these spacecraft to render them inoperative. This can be done with remote effectors like lasers to "blind" optical sensors. The simplest ASAT weapon is kinetic, i.e., an impactor. By intercepting a satellite with a projectile at high relative velocity, the satellite is rapidly and easily rendered worthless.

Despite the fact that the destruction of satellites is relatively easy, it has seldom happened by accident and never as an act of war. Although most space assets are extremely vulnerable, they are left alone because they are not easy to get to. Some orbiting spacecraft occupy low Earth orbit (LEO) and are accessible to interceptors, but many valuable strategic assets are in the much higher orbits of middle Earth orbit (MEO) 3000 to 35000 km and geosynchronous Earth orbit (GEO) 35786 km. Such orbits are difficult to reach and require long transit times and complex orbital maneuvers which quickly reveal themselves and their purpose to ground-based tracking.

In 1998, a communications satellite was left in a useless transfer orbit after a booster failure. Engineers at Hughes (the makers of the satellite) devised a clever scheme to send the satellite to a GEO using a gravity assist from the Moon. This was the world's first "commercial" flight to the Moon and it saved the expensive satellite for its planned use. One aspect of this rescue is seldom mentioned but attracted the attention of military space watchers everywhere. This satellite approached GEO from an unobserved (and at least partly unobservable) direction. Most trips to GEO travel from LEO upwards; this one came down from the Moon, a direction not ordinarily monitored by tracking systems. This mission dramatically illustrated the importance of what is called "situational awareness" in space.

Our current model of operations in space is well established. Satellites must be self-contained and operated until dead, then completely replaced – a template of design, build, launch, operate, and abandon. With few exceptions, we are not able to access satellites to repair or upgrade them. Sometimes favorable conditions allow us to be clever and rescue an asset that had been written off, but the system is not designed for such operation. The current spaceflight paradigm is a use and throwaway culture. Yet thirty years of experience with the Shuttle program has shown us that such is not the case by necessity. What is missing is the ability to get people and servicing machines to the various satellites in all their myriad locations: LEO is easy, but MEO and GEO cannot be accessed with existing space systems. Yet from the experience of Shuttle and ISS, we know that if they could, a revolution in the way spaceflight is approached might be possible.

The Vision for Space Exploration and its implications

The Vision for Space Exploration (the Vision, or VSE,) announced by President Bush in January 2004, called for returning the Shuttle to flight after the Columbia accident, completion of the International Space Station, a human return to the Moon and eventually voyages to Mars and other destinations. This proposal was subsequently endorsed by two different Congresses (in 2005 and 2008) under the control of different parties; both authorizations passed with large bipartisan majorities. The preface to the founding VSE document states that the new policy is undertaken to serve national "security, economic and scientific interests."

Subsequent statements and writings elaborated on the purpose of the VSE. Despite concerted efforts to distort its meaning, the goal of lunar return was not to repeat Apollo but to create a long-term, sustained human presence in space by learning to use the material and energy resources of the Moon. The VSE was to be implemented under existing and anticipated budgetary constraints; the guidance given to NASA for this aspect of the mission was to stretch timetables if money became short. The idea was to create this new system with small, incremental, yet cumulative steps.

The intellectual underpinnings of the VSE began to be undermined by NASA almost immediately. The Exploration Systems Architecture Study (ESAS) made lunar return an Apollo redux, with the development of a large, 150-mT-payload heavy lift vehicle becoming the centerpiece and *sine qua non* of human spaceflight beyond LEO. An ambitious program to establish an early robotic presence to prospect for resources on the Moon was cancelled, along with the incremental approach outlined by the Vision. Thus, the Moon became a distant goal, with first arrival of humans occurring well after 2020, if then. NASA had chosen something familiar, an architecture very similar to Apollo with little effort made to develop reusable, refuelable spacecraft (although the Altair lander used LOX-hydrogen, so in principle, it could be modified for refueling).

In short, the purpose of returning to the Moon, i.e., to create a sustainable human presence based on the use of lunar resources, got lost in the ESAS shuffle. Lunar return became synonymous with "Apollo on Steroids" and heavy-lift rocket building while ESAS (Constellation) became synonymous with the VSE. Project Constellation, the agency project to develop the new Orion spacecraft and Ares I and Ares V launch vehicles, was a costly, throw-away space system that got us to the Moon with considerable capability, but with little or no thought given to planned surface objectives or activities. The idea of finding and learning to use the resources of the Moon became an experiment slated for the manifest of some future mission, not the primary driver or objective of lunar return. Lunar Reconnaissance Orbiter is currently mapping the Moon and sending us data on the extent and nature of lunar resources, but no lander missions are planned to follow up on its findings. The ingenuity of an incremental program was lost and we created no new capability in space.

The goal of the VSE is to create the capability to live ON the Moon and OFF its local resources with the goals of self-sufficiency and sustainability, including the production of propellant and refueling of cislunar transport vehicles. A system that is able to routinely go to and from the lunar surface is also able to access any other point in cislunar space. We can eventually export lunar propellant to fueling depots throughout cislunar space, where most of our space assets reside. In short, by going to the Moon, we create a new and qualitatively different capability for space access, a "transcontinental railroad" in space. Such a system would completely transform the paradigm of spaceflight. We would develop serviceable satellites, not ones designed to be abandoned after use. We could create extensible, upgradeable systems, not "use and discard." The ability to transport people and machines throughout cislunar space permits the construction of distributed instead of self-contained systems. Such space assets are more flexible, more capable and more easily defended than conventional ones.

The key to this new paradigm is to *learn* if it is possible to use lunar and space resources to create new capabilities and if so, how difficult it might be. Despite years of academic study, no one has demonstrated resource extraction on the Moon. There is nothing in the physics and chemistry of the materials of the Moon that suggests it is not possible, but we simply do not know how difficult it is or what practical problems might arise. This is why resource utilization is an appropriate goal for the federal space program. As a high-risk engineering research and development project, it is difficult for the private sector to raise the necessary capital to understand the magnitude of the problem. The VSE was conceived to let NASA answer these questions and begin the process of creating a permanent cislunar transportation infrastructure.

So where do we stand with the creation of such system? Is such a change in paradigm desirable? Are we still in a "space race" or is that an obsolete concept? The answers to some of these questions are not at all obvious. We must consider them fully, as this information is available to all space faring nations to adopt and adapt for their own uses.

A new space race

The race to the Moon of the 1960's was an exercise in "soft power" projection. We raced the Soviets to the Moon to demonstrate the superiority of our technology, not only to them, but also to the uncommitted and watching world. The landing of Apollo 11 in July 1969 was by any reckoning a huge win for United States and the success of Apollo gave us technical credibility for the Cold War endgame. Fifteen years after the moon landing, President Reagan advocated the development of a missile defense shield, the so-called Strategic Defense Initiative (SDI). Although disparaged by many in the West as unattainable, this program was taken very seriously by the Soviets. I believe that this was largely because the United States had already succeeded in accomplishing a very difficult technical task (the lunar landing) that the Soviet Union had not accomplished. Thus, the Soviets saw SDI as not only possible, but likely and its advent would render their entire nuclear strategic capability useless in an instant.

In this interpretation, the Apollo program achieved not only its literal objective of landing a man on the Moon (propaganda, soft power) but also its more abstract objective of intimidating our Soviet adversary (technical surprise, hard power). Thus, Apollo played a key role in the end of the Cold War, one far in excess of what many scholars believe. Similarly, our two follow-on programs of Shuttle and Station, although fraught with technical issues and deficiencies as tools of exploration, had significant success in pointing the way towards a new paradigm for space. That new path involves getting people and machines to satellite assets in space for construction, servicing, extension and repair. Through the experience of ISS construction, we now know it is possible to assemble very large systems in space from smaller pieces, and we know how to approach such a problem. Mastery of these skills suggests that the construction of new, large distributed systems for communications, surveillance, and other tasks is possible. These new space systems would be much more capable and enabling than existing ones.

Warfare in space is not as depicted in science-fiction movies, with flying saucers blasting lasers at speeding spaceships. The real threat from active space warfare is denial of assets and access. Communications satellites are silenced, reconnaissance satellites are blinded, and GPS constellations made inoperative. This completely disrupts command and control and forces reliance on terrestrially based systems. Force projection and coordination becomes more difficult, cumbersome and slower.

Recently, China tested an ASAT weapon in space, indicating that they fully understand the military benefits of hard space power. But they also have an interest in the Moon, probably for "soft power" projection ("Flags-and-Footprints") at some level. Sending astronauts beyond low Earth orbit is a statement of their technical equality with the United States, as among space faring nations, only we have done this in the past. So it is likely that the Chinese see a manned lunar mission as a propaganda coup. However, we cannot rule out the possibility that they also understand the Moon's strategic value, as described above. They tend to take a long view, spanning decades, not the short-term view that America favors. Thus, although their initial plans for human lunar missions do not feature resource utilization, they know the technical literature as well as we do and know that such use is possible and enabling. They are also aware of the value of the Moon as a "backdoor" to approach other levels of cislunar space, as the rescue of the Hughes communications satellite demonstrated.

The struggle for soft power projection in space has not ended. If space resource extraction and commerce is possible, a significant question emerges – What societal paradigm shall prevail in this new economy? Many New Space advocates assume that free markets and capitalism is the obvious organizing principle of space commerce, but others might not agree. For example, to China, a government-corporatist oligarchy, the benefits of a pluralistic, free market system are not obvious. Moreover, respect for contract law, a fundamental reason why Western capitalism is successful while its implementation in the developing world has had mixed results, does not exist in China. So what shall the organizing principle of society be in the new commerce of space resources: rule of law or authoritarian oligarchy? An American win in this new race for space does not guarantee that free markets will prevail, but an American loss could ensure that free markets would never emerge on this new frontier.

Why are we going to the Moon?

In one of his early speeches defending the Apollo program, President John F. Kennedy laid out the reasons that America had to go the Moon. Among the many ideas that he articulated, one stood out. He said, "whatever men shall undertake, *free* men must fully share." This was a classic expression of American exceptionalism, that idea that we must explore new frontiers not to establish an empire, but to ensure that our political and economic system prevails, a system that has created the most freedom and the largest amount of new wealth in the hands of the greatest number of people in the history of the world. This is a statement of both soft and hard power projection; by leading the world into space, we guarantee that space does not become the private domain of powers who view humanity as cogs in their ideological machine, rather than as individuals to be valued and protected.

The Vision was created to extend human reach beyond its current limit of low Earth orbit. It made the Moon the first destination because it has the material and energy resources needed to create a true space faring system. Recent data from the Moon show that it is even richer in resource potential than we had thought; both abundant water and near-permanent sunlight is available at selected areas near the poles. We go to the Moon to learn how to extract and use those resources to create a space transportation system that can routinely access all of cislunar space with both machines and people. Such a system is the logical next step in both space security and commerce. This goal for NASA makes the agency relevant to important national interests. A return to the Moon for resource utilization contributes to national security and economic interests as well as scientific ones.

There is indeed a new space race. It is just as important and vital to our country's future as the original one, if not as widely perceived and ap-

preciated. It consists of a struggle with both hard and soft power. The hard power aspect is to confront the ability of other nations to deny us access to our vital satellite assets of cislunar space. The soft power aspect is a question: how shall society be organized in space? Both issues are equally important and both are addressed by lunar return. Will space be a sanctuary for science and PR stunts or will it be a true frontier with scientists and pilots, but also miners, technicians, entrepreneurs and settlers? The decisions made now will decide the fate of space for generations. The choice is clear; we cannot afford to relinquish our foothold in space and abandon the Vision for Space Exploration.

(Published at SpaceRef.com: http://www.spaceref.com/news/viewnews.html?id=1376 *)*

About the Author

PAUL D. SPUDIS is a Senior Staff Scientist at the Lunar and Planetary Institute in Houston, Texas. He was formerly with the Branch of Astrogeology, U. S. Geological Survey in Flagstaff, Arizona and the Applied Physics Laboratory in Laurel MD. He is a geologist, educated at Arizona State University (B.S., 1976), Brown University (Sc.M., 1977), and Arizona State University (Ph.D., 1982.) Since 1982, he has specialized in research on the processes of impact and volcanism on the planets and studies of the requirements for sustainable human presence on the Moon. He has served on numerous panels and groups advisory to NASA, including the Committee for Planetary and Lunar Exploration (COMPLEX) of the National Academy of Sciences and the Synthesis Group, a special panel of the Executive Office of the President that in 1990-1991, analyzed a return to the Moon to establish a base and the first human mission to Mars. He was Deputy Leader of the Science Team for the Department of Defense *Clementine* mission to the Moon in 1994 and Principal Investigator of Mini-SAR, an imaging radar experiment on the Indian Chandrayaan-1 mission that orbited and mapped the Moon in 2008-2009. He was a member of the *President's Commission on the Implementation of U. S. Space Exploration Policy*, whose report was issued June, 2004 and in September 2004, was presented with the NASA Distinguished Public Service Medal for his work on that body. He is the recipient of the 2006 Von Karman Lectureship in Astronautics, awarded by the American Institute for Aeronautics and Astronautics. He is the author or co-author of over 100 scientific papers and five books, including *The Once and Future Moon*, a 1996 book for the general public as part of the Smithsonian Institution Library of the Solar System series, and (with Ben Bussey) *The Clementine Atlas of the Moon*, published in 2004 by Cambridge University Press.

His personal web site may be found at: http://www.spudislunarresources.com

The blog upon which this book is based can be found at: http://blogs.airspacemag.com/moon/